How to Interpret Family History & Ancestry DNA Test Results for Beginners

How to Interpret Family History & Ancestry DNA Test Results for Beginners

✦

The Geography and History of Your Relatives

Anne Hart, M.A.

ASJA Press
New York Lincoln Shanghai

How to Interpret Family History & Ancestry DNA Test Results for Beginners
The Geography and History of Your Relatives

All Rights Reserved © 2004 by Anne Hart

No part of this book may be reproduced or transmitted in any form or by any means, graphic, electronic, or mechanical, including photocopying, recording, taping, or by any information storage retrieval system, without the written permission of the publisher.

ASJA Press
an imprint of iUniverse, Inc.

For information address:
iUniverse, Inc.
2021 Pine Lake Road, Suite 100
Lincoln, NE 68512
www.iuniverse.com

ISBN: 0-595-31684-0

Printed in the United States of America

Contents

CHAPTER 1	How to Interpret Your DNA Test Results for Ancestry or Family History in Plain Language....	1
CHAPTER 2	Tracing the Female and Male Lineages:........	37
CHAPTER 3	Men and Women Have Different Genetic Histories	49
CHAPTER 4	How to Safely Tailor Your Foods, Medicines, and Cosmetics to Your Genes:	88
CHAPTER 5	Personalized Medicine from DNA Testing Companies	96
CHAPTER 6	Scientists and Physicians Comment on Pharmacogenetics	113
CHAPTER 7	Scientists and Physicians Comment on Pharmacogenetics: Question: Where would consumers (with no science background) begin to search and learn about pharmacogenetics?	129
CHAPTER 8	DNA Testing for Nutritional Genomics and Ancestry................................	142
CHAPTER 9	Menopause and Beyond Alternative Resources and Information Online	147
CHAPTER 10	Nutritional Genomics for the Consumer: How Are You Managing Your Gene Expression?	154
CHAPTER 11	Consumers Need to Be Involved in Quality Control	185

CHAPTER 12	Intelligent Nutrition or Smart Foods? Who Makes The Rules in Nutritional Genomics?	211
CHAPTER 13	What Products are Available Now for the Consumer?	237
CHAPTER 14	DNA Testing DNA for Ancestry	244
CHAPTER 15	Personalized Medicine from DNA Testing Companies	256
CHAPTER 16	Effects	285
CHAPTER 17	How Do Your Genes Respond?	298
CHAPTER 18	Nutritional Genomics for the Consumer	309
CHAPTER 19	Consumer Surveillance	320
CHAPTER 20	Intelligent Nutrition or Smart Foods? Who Makes The Rules in Nutritional Genomics?	345
CHAPTER 21	Your DNA Matches and Mine	371
CHAPTER 22	What's The Oldest HomoSapien mtDNA in Europe?	376
CHAPTER 23	From Which Families Do You Descend?	388
CHAPTER 24	Merging a Mosaic of Jewish Communities by DNA	398
CHAPTER 25	How Do You Use DNA Testing To Interpret Family History Records?	403
CHAPTER 26	Molecular Genealogy Revolution	417
CHAPTER 27	Personalizing Ethnic Family History Records with DNA Testing	438
CHAPTER 28	The Phenomics Revolution: My Positive Experiences with DNA Testing	449
CHAPTER 29	Finding Female Ancestors by Searching for Maiden Names	466

CHAPTER 30	The DNA Testing Companies of Interest to Family Historians	468
CHAPTER 31	What is DNA?	474
CHAPTER 32	Human Genome Project	478
CHAPTER 33	What We've Learned So Far	483
CHAPTER 34	After the Human Genome Project (HGP), the Next Steps	488
CHAPTER 35	Interviewing for Personal Histories: How to Interview Older Adults. Intergenerational Writing for Genealogy and Life Stories.	494
CHAPTER 36	Oral History	500
CHAPTER 37	Diaries Plus DNA Equal Time Capsules	518
CHAPTER 38	Mapping Your Personal Anthropology with Genetic Genealogy	525
CHAPTER 39	Managing a Genetic Genealogy Project: Participants with Poor Documentation	534
CHAPTER 40	Haplogroups and Markers	537
CHAPTER 41	Cancer Genetics Network (CGN)	544
CHAPTER 42	How to Open Your Own DNA Test Results or Molecular Genealogy Reporting Company	552
APPENDIX A		565
APPENDIX B		569
APPENDIX C		573
APPENDIX D		575
APPENDIX E		579
APPENDIX F		585
APPENDIX G		587

Appendix H 593
Index 641

1

How to Interpret Your DNA Test Results for Ancestry or Family History in Plain Language

How many DNA testing companies will show you how to interpret DNA test results for family history or direct you to instructional materials after you have had your DNA tested? Choose a company based on previous customer satisfaction, number of markers tested, and whether the company gives you choices of how many markers you want, various ethnic and geographic databases, and surname projects based on DNA-driven genealogy.

Before you select a company to test your DNA, find out how many genetic markers will be tested. For the maternal line, 400 base pairs of sequences are the minimum. For the paternal line (men only) 37 markers are great, but 25 markers also should be useful.

Some companies offer a 12-marker test for surname genealogy groups at a special price. When you order a home testing kit, you'll get mouthwash or a felt tip to rub inside your cheek and mail back. Find out how long the turnaround time is for waiting to receive your results. What is the reputation of the company?

Do they have a contract with a university lab or a private lab? Who does the testing and who is the chief geneticist at their laboratory? What research articles, if any, has that scientist written or what research studies on DNA have been performed by the person in charge of the DNA testing at the laboratory? Who owns the DNA business that contracts with the lab, and how involved in genealogy-related DNA projects and databases or services is the owner?

Will the company keep in touch with you and let you know by email each time you have a DNA match? What happens to your DNA after you test it? Is it destroyed? What projects are available for you to participate in using your DNA donation or that of relatives related to ancestry, genealogy, or family history?

How much will testing cost? What other projects can you donate your DNA to offering free testing for what uses?

After your DNA sample is sent to the return address, the DNA will be sent to a university or private laboratory to be tested. A report showing your sequences of the portion of DNA tested for ancestry only will come back in about six to eight weeks. The DNA-testing company will send you the report with your sequences. Now, it's up to you to find out what the sequences mean in terms of ancestry.

When you order a DNA test, you get a code number or kit number so your name remains private. Some companies let you sign a release form to allow others to contact you or you contact them by email each time you find a match with someone who shares your exact mtDNA (maternal) or Y-chromosome (paternal) genetic markers for ancestry. The DNA tested for ancestry shows only ancestry, not any risk or disease. Women have their mtDNA tested as they don't carry a Y-chromosome. Men may have their mtDNA (maternal line) or their Y-chromosome (paternal line) tested.

According to AncestryByDNA, "We've all originated from a common ancestor that lived some 200,000 years ago. The only way to know where you came from is by reading your genetic code." What might intrigue some is taking a racial percentages test to see what percentages of which 'races' live in your very ancient or recent past.

What Will Be On The DNA Report?

You'll find your sequences on the printout that you get back from your DNA testing company, but how do you interpret your sequences for ancestry? If you want more information on interpreting sequences than you find in this article, you can start with the free online message boards on DNA genealogy such as Genealogy-DNA Rootsweb.com at: GENEALOGY-DNA-D-request@rootsweb.com. You can watch my instructional videos on interpreting your DNA test results for family history and ancestry or on how to write salable life stories on my Web site at http://www.newswriting.net. Before you take a DNA test, enjoy these videos and look at my book or browse excerpts on *Creative Genealogy Projects: Writing Salable Life Stories*. Bring together DNA-driven genealogy reports with personal history for your time capsule.

Your biggest question could be, "What do you do with your DNA sequences in the field of genealogy?" You look at the ethnic databases online or find long lost relatives and email them. Then you put the DNA-for-ancestry report with the interpretations along with any genealogy information and keepsakes in your time capsule as part of social history.

You can send your DNA to a world wide database collecting the world's DNA. One such database is the Molecular Genealogy Project at the Sorenson Molecular Genealogy Foundation: The Web site is at: http://www.smgf.org/. It's a "nonprofit organization founded to build a publicly accessible genealogical database." You can contribute your DNA to their database free, but you need to have a known genealogy going back at least four generations.

Your Maternal Lineage—Mitochondrial DNA (mtDNA) and Ancestry

Since women only can be tested for mitochondrial DNA which shows only the female lineages that originated thousands of years, ago, find out how many base pairs of mtDNA will be tested. Usually the minimum is 400 base pairs of mtDNA.

If your mtDNA covers a wide area, it usually signifies that the DNA sequences are very ancient and had thousands of years to spread wide distances geographically. If your mtDNA sequences are found in a very narrow area, your mtDNA may have arisen relatively more recently.

It's the mitochondrial DNA (mtDNA) that is tested to find out your maternal line. The mtDNA is passed from mother to daughter starting with one female ancestor. That ancestor started your line of mtDNA sequences thousands of years ago. Since mtDNA mutates slowly over thousands of years, you are usually told in a report that your mtDNA sequences arose anywhere depending on the sequences from 10,000 years ago to 20,000 years ago.

How to Interpret DNA Test Results—Female

Your DNA test result will give you a letter of the alphabet called your 'haplogroup' or 'clan' as Oxford Ancestors calls it. If you're of European, Middle Eastern, (or from some parts of India) your deep maternal ancestry letters will be H, I, J, K, N, R, T, U, V, W, or X. Most European lineages of women have these letters. It only means your prehistoric female ancestors most likely came from Europe, Central Asia, or the Middle East.

If your letters are A, B, C, D or X, most likely you could be Native American or Asian. The letter 'L' is African, as in L1, L2, L3. The letter L3 is the same group that left Eastern and/or Southern Africa to populate the rest of the world thousands of years ago. And the letters M, A, B, C, D, E, F, G, O, P, Q and Z most likely are East Asian. P and Q mtDNA is found in Oceana, the Pacific Islands and Papua-New Guinea. N mtDNA types also are found in Australia, but N is found in Europe and the Middle East.

M1 is found in Africa and M in India and Asia as well as in the Americas. Some Asian lines are shared with the Americas, but with different mutations or sequence variations. mtDNA letters A and C are shared with East Asian and Native American, and Z and Y are found in Russia and Scandinavia at a low rate, and also in Asia. These haplogroups are very ancient. Most mtDNA sequences can be traced back 20,000 or more than 50,000 years into prehistory. And those have common ancestors that go back in time even further. The letter X is found in Europe and among Native Americans, in Southern Siberia, around the Caucasus, in the Middle East, and in Central Asia.

If your mtDNA covers a wide area, it usually signifies that the DNA sequences are very ancient and had thousands of years to spread wide distances geographically. If your mtDNA sequences are found in a very narrow area, your mtDNA may have arisen relatively more recently. Your point of origin geographically is the place where your mtDNA is most diverse, not necessarily where it is found most frequently.

Where Can You Match Your mtDNA to a Country in an Online Database?

For women and men interested in matching their mtDNA sequences of HVS-1 or HVS-2 (high and low resolution) there are databases online such as Macaulay's Tables database. These DNA databases online are matched with surname groups, lists, message boards, and other Web and online databases to help you match your sequences to a geographic location. I use **Macaulay's Tables** at: http://www.stats.gla.ac.uk/~vincent/founder2000/tableA.html.

Roots for Real, based in London at: <http://www.rootsforreal.com/english/eng-home.html> tests your low resolution mtDNA or Y chromosome and sends you a report and map showing the probable or possible geographic origin of your sequence by latitude and longitude, even naming the town that exists there today. The probable geographic center for the origin of my mtDNA sequences is located at 48.30N, 4.65E, Bar sur Aube, France with a deviation of 669.62 miles according to the map emailed to me by Roots for Real.

Which Company to Choose and Why?

I had my DNA tested at Family Tree DNA, Oxford Ancestors, and AncestryByDNA and geographical interpretations of the results done at Roots for Real. According to their Web site, Family Tree DNA coined the word anthrogenealogy "that combines the methods of the two sciences—anthropology and genealogy, "largely with individual or corporate sponsorship or carried out by avocational researchers."

Family Tree DNA gives a lot of choices. They sent me my sequences of both the high resolution and low resolution mtDNA called the HVS-1 and HVS-2. I was then able to look up on the Web "Macaulay's Tables," a database of sequences for HVS-1 and HVS-2 and find out in which countries people of today live who have my exact mtDNA sequences. The countries are England, Austria, Spain, and Bulgaria.

I chose to have my mtDNA interpreted by four companies so I could compare what they offered with what my goal was, to link genealogy to DNA and find out my matrilineal ancestry back to 21,000 years ago if that was possible as far as geographic location in longitude and latitude.

What I liked about each company was that they all offered different material. AncestryByDNA offered my genotype sequences on a CD and a racial percentages test. Oxford Ancestors offered a chart and a prehistory of the DNA that showed me how I link to the world's mtDNA clans. The company also showed me that 21,000 years ago my mtDNA lived in what was to become Spain and/or Southwest France.

The second company to test my mtDNA, *AncestryBy DNA* <http://www.ancestrybydna.com> in Sarasota, Florida, sent me a free book, titled, *The Great Human Diaspora*. It did help me understand how DNA is measured. For more information on the ancestry and migration of the male Y-chromosome, I found a newer book, *The Journey of Man,* by Spencer Wells, published in 2002.

Family Tree DNA tested my HVS-1 and HVS-2, my high and low resolution mtDNA. Roots for Real, London, sent me maps online that showed what latitude and longitude the probable origin of my exact mtDNA sequences appeared in the last 10,000 years and the town of probable origin that didn't exist in the distant past, the city of Bar Sur Aube, France.

What Will You Pay for a DNA Test for Ancestry?

In August 2001, Oxford Ancestors, London, became the first company to test my mtDNA for around $180. They noted my mtDNA sequences also showed up in England. They also sent me a chart showing where the mtDNA originated and how my mtDNA links with other mtDNA all over the world. I also received printed material on human migrations. I paid a little over $200 at Family Tree DNA. My husband paid $99 for a surname group-rate 12-marker Y-chromosome test at Family Tree DNA.

At most companies DNA tests can run from about $100 to over $300 for ancestry. Prices seem to be coming down and more markers are being tested for Y-chromosomes. DNA tests for nutrition or medical reasons are more, and a few companies even test the entire genome for a high price, more than $1,000.

How to Interpret DNA lineages—Males

Some DNA testing firms test 12, 25, or 37 markers of the male Y chromosome. The founder of your sequence might have originated around 10,000 years ago. To find the founder, DNA testing companies may work with university or private laboratories that test the Y-chromosome of males for ancestry. The laboratories test the Y chromosome of DNA for markers. Some of these markers are called short tandem repeats. These are, according to Spencer Wells in his book, *The Journey of Man*, "tandem repeats—short sections of the same sequence, repeated several times in a row in the DNA strand." Geneticists call them short tandem repeats.

Here's how you can do your own research independently of any laboratory that tested your own DNA if you're curious about Y-chromosome DNA tests. Males take Y-chromosome DNA tests to find out paternal ancestry lineages. Males also can have their mtDNA checked, but women don't have a Y chromosome. So women only can check their maternal lineages with mitochondrial (mtDNA) tests. According to Alastair Greenshields who runs Ybase at http://www.ybase.org, "Ybase is a free and open database which allows people to enter their Y-chromosome haplotype details independently of the laboratory they were tested at. Anyone can contribute to it and anyone can explore it."

The database can accept results for any of the 36 Y-chromosome markers currently in use along with other genealogical information. "Ybase is searchable for exact haplotype matches and/or near misses," says Greenshields. "Surnames, variant spellings and other relevant names can also be searched for, which is especially useful for the genealogist wishing to locate and contact others that share their surname and have had their DNA tested."

Most researchers that recognize their Y-chromosome cannot identify an actual individual and are happy to share their results online in an effort to find their DNA cousins and genetic roots. "Genealogical research coupled with DNA testing is already proving a very powerful method of substantiating ancestry," Greenshields explains. "And Ybase is sure to grow in line with this upward trend, benefiting the genealogical community as a whole."

Are male and female genetic lineages studied for different purposes? "Ybase is solely intended for Y-DNA and not mtDNA. A database of the latter would be so general and broad, given the nature of how mtDNA is passed on, as to be of little value to a researcher," Greenshields says.

Below are a couple of explanations on Y-DNA Alastair Greenshields wrote for two different people with entirely different backgrounds who needed the whole thing explained to them. "They explain essentially the same terms but have been 'dumbed down' to varying extents. Please feel entirely free to copy them verbatim or adjust as you deem necessary," says Greenshields.

To answer my question to Greenshields on how to interpret the results of DNA tests for Y-chromosome analysis for ancestry, he's explained it well in terms that most people can understand.

"Imagine a very long rope, some of which is lying across your desk. This is the DNA strand," Greenshields says. "It just happens to be the length of rope called 'Your Y-chromosome.'

"Now look at the bit that lies across your desk and grasp the rope with both hands, about half a meter apart. This is a 'marker' or 'locus' (Latin for 'place'). We'll call it DYS19.

"In between your hands, imagine that bit of rope is divided into 14 equally-spaced segments. If you look very closely at the segments, you can see that each one has a bit of writing on it, which reads TAGA.

"This is simply the DNA code for each repeat. Therefore the marker DYS19 = 14 repeats. Or if I ask you, 'What allele have you got for DYS19?' You can tell me '14'. (Allele effectively means the number of repeats.) For example, DYS19 has about nine possibilities (between 11 and 19).

"If you do the same at lots of different markers or loci (plural of locus), you'll get a whole series of numbers (DYS19=14, DYS388=15, DYS461=11 etc.). This is your 'haplotype'. It doesn't matter whether it is 20 or 200 numbers in length. This series of numbers is still called your haplotype.

"Now you are going to make some rope for your new son. You are pretty good at making rope and usually you can copy your own precisely, but this time you made it slightly too short. There are now 13 repeats. (Technically this is called a 'mutation' which can occur when an enzyme mis-types the DNA code). It still works perfectly well. So your son keeps it and is very happy. There, you have it—DNA in a nutshell." (Be aware that your repeats, like the stock-market, may go up as well as down, but for entirely different reasons.)

Genealogy and the Y Chromosome

"DNA for the use of genealogy usually requires an analysis of your Y-chromosome," Greenshields explains. "Only males have this particular chromosome and the DNA code held within it is passed down from father to son (virtually) unchanged. Provided there is an unbroken paternal line between two males, that is both share a g-g-g-g-g-grandfather, their Y-chromosome DNA will be the same."

When the DNA is analyzed, many small sections are looked at. Presently, the testing companies look for anywhere between 10 and 26 sections or 'markers.' "At any one of the markers, the code will repeat itself, for example, 15 times," Greenshields explains. "If the marker is called DYS19, we can give the result DYS19 = 15."

If you analyze several of these markers, you end up with a 'haplotype'. Thus you can compare haplotypes to see if you are related. "I say 'virtually' unchanged, as the DNA can change slightly over time due to 'mutations'—small errors formed when the DNA is copied," says Greenshields.

"When comparing the haplotypes from two people, this will show up as a 'mismatch'—where, for example, DYS 19 = 14."

These mutations are useful by themselves and occur at a fairly steady rate over time. "It also gives us the great variability that we observe over populations," says Greenshields.

"If the mutations did not occur, every male would have identical Y-chromosomes. Also, within DNA/genealogy studies, if many mismatches occur when comparing two male haplotypes, we can say that they are not related."

So you can see, DNA can be a very useful tool when comparing 'suspected' relatives. "But there is one caveat, however," Greenshields emphasizes. "You must share a surname, or have a very good reason to believe you are related. DNA alone will not identify your relatives from any other random person."

For example, you will probably share the same haplotype to at least someone in your home-town, but having a similar or same surname will raise the probabilities significantly. "DNA is a tool that should be overlaid on the existing genealogical records," notes Greenshield. "It can be an excellent way of deciding on further avenues of research, or indeed defining that there are several distinct lines within your family name."

For our family with the surname, Hart, we found that Family Tree DNA gave us the opportunity to start a surname group so that the male's Y-chromosome test of markers could be matched with the genealogy on the records. Here is the genealogy of records, and the DNA markers went into a database and could be put on our Web site. That way, all other males directly descended from the known founding father in the earliest records would have similar markers to the latest male descendants.

Hart Family Genealogy:

Homer Vincent Hart & Wife: Vera Dell Palmer—Family Roots.

We found the census for various dates more important than the old books, but it was the Stephen Hart Web site that gave us the best glimpse into the early years so that we could go directly to the primary sources, the old books to see who published what information in the last century. For your own genealogy source that you might combine with a DNA printout, look at the Web resources and old books, census records, city directories, and other records we used to give you an idea of what sources are available for your own searches. You can apply these searching techniques to your own genealogy research. Then match the records with the markers of people you make contact with through a surname DNA database or list that you can put on the Web using kit numbers instead of names for privacy. That way, you can keep a list of distant cousins or other descendants with the same Y-male chromosome markers and surname match. For females, the names would not match because maiden names are changed after marriage, but the mtDNA would show at least a distant ancestor who at one time in the last 10,000 or 20,000 years shared the same ancestor.

If the high and low resolution markers match, you are more closely related, but at what time in history—recent or prehistoric? That's a good question that can be cleared up when two males have the same surname and are a match in their Y chromosomes which shows they probably had the same male ancestor back in time. The last name and the Y chromosome markers would have to match.

Grandmpa Homer Vincent Hart is a descendant of Stephen Hart, b. 1605, and grandpa Homer V. Hart's wife, Vera Dell Palmer, is a descendant of Walter Palmer, b. 1585 through Vera's dad, Langford Wright Palmer. This couple starts our family's search back many generations. The genealogy research in the records took us back from grandfather and grandmother to 419 years ago for the Palmers. We started counting back from grandma Vera Dell Palmer, daughter of Langford Wright Palmer to see the paternal side of the paternal grandma in addition to looking at the maternal side of the maternal grandma. For the Palmers, Walter was born in 1585, and his father had the same first and last name. Walter's mother was Elizabeth Carter.

Next, we wanted to look at the male Y-chromosome side for the Harts that would have the same surname for all the males from the present back to 1605 when the records became readily available in books and records published in the USA. If we wanted to go back in time before 1605, we would have to turn to records in Essex, England, perhaps school, ships passenger lists, or church records and notary land grants or other business transaction records that could still be around from the 1500s and early 1600s.

For the paternal side of the family's grandfather, we looked back into genealogy records 400 years to 1605 for the Harts searching back from Homer Vincent Hart, our family's paternal grandfather, born in 1911 and moving back by each generation. We looked only at the direct fathers all the way back to Stephen Hart born in 1605. He arrived in Massachusetts in what is today Boston in 1632 according to records, on the ship called the Lyon. In the mid-1600s he moved to Farmington, CT.

Along the way, centuries ago, there was a marriage in the records that relates the descendants of Stephen Hart to the Spencers and the Lees, and eventually to the late Princess Diana. We used the old books written around 1875–1880 to trace the Harts of early Massachusetts and Farmington, Connecticut. Here are the techniques and records we used to track our family's grandparents back to their founding families in Massachusetts, Connecticut, and Michigan from 1632. Homer was born in 1875/6 in Michigan and married Vera Dell Palmer (b. 1881) in 1908 in Hudson, Lenawee County, Michigan.

We researched the male line of the surname 'Hart' of known relatives for the Hart family. For Vera Dell Palmer, who married Homer Vincent Hart in 1908, we searched the male Palmer line starting with her father, Langford Wright Palmer back to Walter Palmer. Langford Wright Palmer descends from Gershom, who is Walter Palmer's sixth child from his second wife, Rebecca Short. Here is the technique used. The Dunne and Allied Family's index at http://www.RootsWeb.com also shows the spouses and children of each of the Harts.

John Hart married Mary Moore who was born in 1654. They had Isaac Hart 27 Nov 1683 and **a son also named John Hart in born in 1684 in Farmington, CT.** They also had other children named Sara, Matthew, Samuel, Nathaniel, and Mary. The birthdates and Web links of information for each child are at the Dunne and Allied Family's site at: http://worldconnect.rootsweb.com/cgi-bin/igm.cgi?op=GET&db=dunne1&id=I08416. Those children of John Hart are: **Isaac Hart b: 27 NOV 1683**, John You can look at the Descendants of Stephen Hart Web site at: http://users.rcn.com/harts.ma.ultranet/family/harts/.

The biography at that Web site explains the highlights of Stephen Hart's life, his wife and children, the town, the environment, his home, and the events of his life as one of the early settlers. One of his children named John married Mary Moore. From John Hart and Mary Moore descend the various branches of the Deacon, Stephen Hart. If you search the records in Essex, England, you'll find the ancestors of Stephen Hart moving in time backwards from his birth in 1605, if you can find the records earlier than 1605 in the area of Essex.

At the Descendants of Stephen Hart Web site, you'll see the scanned old book presented with a new technology. The contents of the book "Stephen Hart and his Descendants," by Alfred Andrews are reproduced at the Web site. The pages online have been amended with corrections and additions that many have sent to Richard Hart at his Web site on the Descendants of Stephen Hart. The "Deacon Stephen Hart's Will" site, dated March 1682-3, is at: http://users.rcn.com/harts.ma.ultranet/family/harts/DeaStephenWill.shtml. It is thought that he arrived in Plymouth on the ship called the Lyon in 1632 coming from Essex, England. Essex is just east of London.

Stephen Hart, born about 1605, was the progenitor of many Harts now living in North America and other parts of the world. The book on Stephen Hart and his Descendants originally was published in 1875 to document all Harts then known to be descendants of Stephen Hart. According to a paper by John Corley,

"Emigration to New England on 'The Lyon' in 1632," prepared in 1984 by the Braintree and Bocking Heritage Centre, Braintree, Essex, England, "Emigrants on 'The Lyon' sailed in 1632 with the Rev. Thomas Hooker's 'Braintree Company' on board. It has been said that the ship carried only 350 passengers. However, many names are missing from the list in The Lyon."

This is partly due to the fact that several were members when only the head of the household was mentioned. Also others omitted went as servants. In 1635 two servants worked their passage for a John Brown. Another version written in *The Planters of the Commonwealth* by Charles Edward Banks, pp. 99–102 introduces a passenger list with a sailing date of June 22, 1632 going from London and arriving at Boston on September 16, 1632. The Banks passenger list mentions that William Peirce Lyon only brought 123 passengers—fifty children. More information on the Lyon and its passengers is at the "Passenger Lists for the Lyon" Web site at: http://www.whipple.org/docs/lyon.html

So how does this relate to my husband's paternal great grandfather, George Washington Hart, born in 1839 in Hudson, Michigan or George's son, my husband's grandfather, Homer Vincent Hart born around 1875 in Hudson, Michigan who married Vera Dell Palmer, a direct descendant of Walter Palmer who also arrived on the sister ship "The Four Sisters" and also again in 1632? I begin by tracing back from Hart, Homer Vincent (1911–1989)—male, b. 27 JAN 1911 in Hudson, MI to his father also named Homer Vincent Hart (1875/6–1920) also from Hudson, MI.

Tracing Back the Ancestors of Homer Vincent Hart: The Techniques of Researching.

The Hart Genealogy 1605–1938

See RootsWeb at www.RootsWeb.com, where you'll find a genealogy of the Harts descended from Stephen Hart and a biography at The Dunne & Allied Family's Web site at: http://worldconnect.rootsweb.com/cgi-bin/igm.cgi?op=GET&db=dunne1&id=I08416. Stephen Hart's son, John Hart was born in England and died in Farmington, CT on Dec. 15th 1666. He was the son of Deacon Stephen Hart and his wife Sarah, born in England. Stephen Hart was born around 1605. Stephen and Sarah Hart had three children, John, Sarah, and Stephen.

John Hart, born in England had a son also named John Hart born 2 APR 1655 in Farmington, CT. That John Hart died 11 NOV 1714 in Farmington, CT. That John Hart had a son, Isaac Hart.

HART b: 1684 in Farmington, CT, Sarah HART b: 11 DEC 1687, Matthew HART b: 7 DEC 1690, Samuel HART b: 18 SEP 1692, Nathaniel HART b: 14 APR 1695, and Mary HART.

Our family's line comes from their first-born son, Isaac Hart, born Nov. 27th 1683, son of the John Hart who was born on Dec. 15th 1666. **Stephen Hart is the grandfather.** Isaac Hart married Elizabeth Whaples on the 24th of November 1721 in Farmington, Hartford, CT. Elizabeth was born 15 Aug. 1697.

The source is the found at the Dunne and Allied Family's site on the RootsWeb.com Web site at: http://worldconnect.rootsweb.com/cgi-bin/igm.cgi?op=GET&db=dunne1&id=I08414

At this site you'll see the source as the Compendium of American Gen., Vol. VII, page 49. Look at this book for yourself in your research. Don't only stop at the Web site material as different Web sites could have different information on the same person. So be like a journalist and search the older, primary sources—those old books and records. The passage from the Compendium of American Gen. says: "Capt. John Hart (1655–1714), he alone of his father's family escaped being burned to death, Dec 15, 1666; Capt. Colonial troops in Queen Anne's War. 1703; dep. Gen. Ct. 4 yrs; m Mary Moore (d1738)." Yet there are other descendants from Thomas Hart, for example.

So you have to be careful to check the records as there are conflicting stories of what happened during the fire and who survived. You have to search for all the 17th century males' names in New England. Descendants are now living all over North America and in England and probably in other countries.

My sources included these books and records, also listed on the Dunne and Allied Family's Web site: Elaine Hart Kerskie of Victor, Ontario Co., NY, book 1940 Genealogys & Vital Records, FTM Genealogical Library, New England Families, Vol. III page 1549, FTM Genealogical Library,Compendium of American Gen., Vol. VII, page 49, FTM Genealogical Library, Families of Early Guilford, CT Vol. I, page 605, FTM Genealogical Library, Family History of Central NY, Vol. I, page 485–8. Genealogy record searching takes a lot of time, but is useful if you combine it with DNA testing of the Y-chromosome markers for the male line. It's easier than trying to test the descendants of all the wives as their names change each generation upon marriage.

So we soon found that Isaac Hart was the son to search for our family's paternal lineage. Isaac Hart's son is Job Hart, born 3 Jan. 1730/31. **Our family descends from Job Hart.** Isaac Hart's children are named: Ebenezer, Isaac, Elizabeth, Mercy C. and John, all born between 1722 and 1734. Job Hart married Eunice Beckley 20 Mar. 1755. Job Hart and Eunice Beckley had twelve children. Their first child was a boy named (alternate spellings) Jabez (Jabish) Jaluish Hart, born 1 Jan. 1757.

Our Hart family descends from **Jabez (Jabish) Jaluish Hart**. Job Hart's children born between 1758 and 1775 include: Jabez (Jabish) Jaluish, b. 1757, Canadua, b. 1758, Job, b. 1759, Harvey, b. 1760, Lucretia, b. 1762, Eunice, b. 1763, Joseph, b. 1765, Simeon, b. 1766, Reuben, b. 1768, Comfort, b. 1771, Hepzibah, and Betsey, b. 1775.

Jabez (Jabish) Jaluish Hart married Jemima Brace, b: 25 Oct. 1762, and Jabez (Jabish) Jaluish Hart had nine children. His first-born, a son named Harvey Hart was born Apr. 9 1784. **Our Hart family descends from Harvey Hart.** Jabez (Jabish) Jaluish Hart's nine children include in order of birth year: Harvey, b. Apr 9, 1784, Tryphena, b. Dec. 23, 1785, Theadoria, b. Dec. 15 1787, John, born Oct. 20, 1789, Demas, b. Nov. 25 1791, Cyrus, b. Nov25, 1794, George, b. July 4, 1797, Eunice, b. May 10, 1799, and Frederick b. Aug 6, 1802.

Harvey Hart married Polly Jackson on April 3, 1792. They had six children. Our Hart family descends from their second born, William Hart, b. Oct. 17, 1810. Other children include Martha M. Hart born Nov. 7, 1811 in New York, and the others including John, b. June 17 1812, Eunice, b. Sept. 10, 1819, Catherine, b. July 24th 1821, and Chauncy, born Sept. 3, 1826.

William Hart married Zillah Thompson on Dec. 16, 1826. Zillah Thompson was born on Apr. 29th 1810. William and Zillah had six children. **Our Hart family descends from George Washington Hart who was the fourth in birth order.** George Washington Hart was born in 1839. Other children include Harvey, born Sept. 11, 1831, Chauncy Benion Hart, born Dec. 1, 1834, Elizabeth Hart, born in 1836, Mary Cecelia, born April 3, 1841, and Jeremiah.

George Washington Hart married Addie Wydenbeck on June 4, 1874. Addie was born in 1849 and passed on in 1918. In the 1900 census the family was living in Hudson, Michigan. George's children include Homer, a son born in November of 1875/6, a daughter, Dora, born in Dec. 1877, Walter, a son born in May of 1883, and Arthur, a son born in Apr. 1886. **George Washington Hart's son, Homer Vincent Hart, is the man from whom our Hart family descends.**

Homer Vincent Hart was only four years old when he appeared on the **1880** census place taken at Hudson, Lenawee, Michigan. The source information is Family History Library Film number 1254591. NA Film Number T9-0591, page number 262A. His father and mother's birthplace is listed as Michigan. Homer Vincent Hart married Vera Dell Palmer in Hudson, Lenawee, Michigan on Oct. 15, 1908. Homer Vincent Hart and Vera Dell Palmer had five children: Margaret Dora, b. 1909, **Homer Vincent, b. 1911,** John George, b. 1913, Robert Langford, b. 1916, and Richard Kenneth, b. 1918. **From this Homer Vincent Hart, born in 1911, our family descends.** Note there is Homer Vincent Hart (senior) born around 1875/6 and his son, Homer Vincent Hart (junior) b. 1911-d.1989.

Homer Vincent Hart, born in 1911 in Hudson, Michigan is my husband's father. Homer Vincent Hart passed on in 1989. So going back generation by generation and finding the son that each male grand parent is descended from leads back to Stephen Hart. The source up to 1875 is Elaine Hart Kerskie of Victor, Ontario Co., NY, book 1940 Genealogys & Vital Records. Other sources could be the various government and military indexes online, especially for more recent dates.

As for Homer Vincent Hart, the **1910** census lists Homer's age as 34. He's in West Grants Pass Precinct, Josephine, Oregon with his wife Vera. This puts his birth date as about 1876. The 1900 census has a birth date for Homer as November 1875. (The 1900 census showed him then as age 24.) This Homer Hart born in MI in 1875 or 1876 has a sister named Dora. Homer Vincent Hart and Vera Dell had a child named Margaret Dora born in 1909.

According to the **1910** census "Family 23. Hart, Homer V. 34 M2 MI MI MI...Vera Wife, 29, M2, 1 child 1 living. MI MI OH.....Margaret. Daughter. 9/12 OR MI MI. Palmer, Langford W. Father-in Law, 60. MI NY NY...Harold Brother-in-law 22. MI MI OH.

In the census of **1910**, there's a listing "1910 Hudson, Lenawee, MI. Hart, George W. 71, NY NY NY. Langford Wright Palmer is living with his daughter and her husband, Homer Hart. Margaret is a toddler. Other people in the household are listed as Addie E. 61, MI NY NY...Arthur L. 24 MI NY MI (Fireman), and...Dora E. 32 MI NY MI (teacher, public school).

In the **1900** census at Lenawee, Hudson, Michigan, family 172: Hart, George is listed as being born in 1839 61 m25 NY NY NY. His wife is listed as Addie, No. 1849, age 51. m24...4 children...4 living MI NY NY. Children include: Homer, Son born Nov. 1875, 24, MI NY MI. Dora is listed as the daughter, born Dec. 1877, age 22, MI NY MI. Walter, a son was born May 1883, 17, MI NY MI. Arthur a son was born April 1886, age 14 MI NY MI.

So that verifies the line of Harts directly related to Homer Vincent Hart, born in 1911 in Hudson, Michigan, the father of my husband. Birthdates and names of his parents are on my husband's birth certificate. I can look up the 1880 census from http://www.familysearch.com where it lists Homer as a four-year-old with his dad, George. Now that I've found the links back to Stephen Hart and Walter Palmer, what will the DNA tests show? Will the Y-chromosome go in a line back to the founder? Can anyone actually be found who wants to take a DNA test to link genealogy with genetics?

In 1870 George is not in the census in Michigan. The year of the 1870 census there are no Harts in Lenawee County. He could be with his parents. Looking up George Hart, there were many born in NY around the same date. So the next step would be to see whether Homer Vincent Hart's 1911 birth certificate lists his parents birth dates. Since George Hart died after 1910, there could be an obituary in the newspaper that might mention the names of his children. If the death certificate is researched, it would probably have his parent's names, but if he's from New York, it might be difficult to find that line of research.

Another channel to research would be when George W. Hart applied for a Civil War pension. The place to look is the Civil War list to query about Michigan information—where to go to send away for data. Many applied for pensions, but on one pension application, records report that George W. Hart applied for a Civil War pension on April 26, 1889. The application number was #700929, cert: # 731396. Could that be him?

Another excellent Web site address might be http://awt.ancestry.com/cgi-bin/igm.cgi?op=GET&db=dunne1&id=I08612. Or I could try http://awt.ancestry.com/cgi-bin/igm.cgi?op=GET&db=dickdutton&id=I147864.

Regarding Homer Hart's father, George W. Hart (George Washington Hart), the first step would be to find out which unit would be in the Lenawee, Hudson, Michigan county. There also was a George W. Hart that filed for a pension. With so many George Harts in the Civil War, it had to be narrowed down to George Washington Hart of Hudson, Michigan who is the father of Homer Vincent Hart born around 1875. In the 1880 census Homer was four years old.

The **1880** US census notes that Homer Vincent Hart is the son of George Washington Hart, born Nov. 30th 1839. In another online index I later found out George died Feb. 25th 1920. Back on the **1880** US census, George's wife and children are listed along with the city in which they live (Hudson, MI).

The next step is to turn to the Dunne and Allied Family's index at www.RootsWeb.com. It shows George Washington Hart is the son of William Hart, born Oct. 17th 1810 who died Nov. 16th 1889. William Hart's father was Harvey Hart, born April 9 1784. Harvey Hart's dad was Jaluish Hart, born Jan 1, 1757. And Jaluish Hart's dad was Job Hart, born Jan.3, 1730/31. Job Hart's dad was Isaac Hart, born 27 Nov. 1683.

Sergt. Isaac Hart found in the Marriage Index: Connecticut, 1635–1860. Isaac Hart married Elizabeth Wheples on: Nov 24, 1721 in: Farmington, Hartford, CT. Information is found in the Family history library microfilm roll info: microfilm reference number: Roll number: 1315116 items 3 and 4. Isaac Hart's father was John Hart, born on April 2, 1655 in Farmington, CT. And John Hart's father was Stephen Hart born in 1605 in Essex, England. Stephen Hart was a church Deacon. A biography is presented at the Descendants of Stephen Hart Web site at: http://users.rcn.com/harts.ma.ultranet/family/harts/.

There is also a biography in the volume titled: New England Families, Vol. III page 1549. The page reads: "Deacon Stephen Hart, the immigrant, was father of John Sr., coming from Braintree, county Essex, England, to Cambridge, Massachusetts, in 1632, and to Hartford in 1636, finally locating in Farmington, where he died in 1682–83 aged seventy-seven." Also see **Family History of Central NY**, Vol. I, page 485–8 shows the family story of this line. Online, you can check out the name Stephen Hart and his descendants back to George Washington Hart from the Roots Web site http://worldconnect.rootsweb.com/cgi-bin/igm.cgi?op=GET&db=dunne1&id=I08612.

Also check out the RootsWeb.com Web site at http://worldconnect.rootsweb.com for other searches. My piece of the puzzle was to verify through the 1910 census whether the names matched with the descendants of Homer Vincent Hart who married Vera Dell Palmer in 1908. The last piece in the puzzle focused on verifying that this was the same Homer Vincent Hart who lived in Josephine, Oregon in 1910 with his wife, Vera and baby daughter, Margaret, who was a familiar relative to my husband when he was a child. Margaret was born around 1909. Various other relatives are mentioned in the census as living with the family.

My husband, the child of Homer Vincent Hart, born in 1911 is the son of Homer Vincent Hart born in 1875/6 as listed in the census of 1880, 1900 and 1910. A birth date of 1879 is listed in the Family History Library, Salt Lake City. That Family History Library records his marriage in 1908 in Michigan. So Homer Vincent Hart's dad is the George Washington Hart who applied in 1889 for a pension for service in the Civil War on April 26, 1889, the same year in which George Washington Hart's own dad, William Hart passed away on Nov. 16th 1889.

Homer Vincent Hart married Vera Dell Palmer in 1908 in Hudson, Michigan. My husband remembered their daughter, Margaret when he was young. By the time the **1910** census is taken, Homer Hart's family is living in Oregon with their daughter, Margaret and other members of the family mentioned in the census, including the father-in-law, Langford Wright Palmer. George W. Hart is living back in Michigan.

George Washington Hart is descended from Stephen Hart, born in 1605 who also is the early Cambridge, New England settler from 1632. George Washington Hart's son, Homer Vincent Hart, born in 1875/6 married Vera Dell Palmer in Hudson MI in 1908 according to records at the Family History Library, Salt Lake City, UT and census records.

The Palmers are easy to find and trace back to Walter Palmer by researching online the Walter Palmer Society. Vera is descended directly from the sixth child of Walter Palmer named Gershom, born in 1644. Now, the genealogy task is to trace back for the Hart family. Yes, this is an adventure. How about a DNA-driven genealogy time capsule? Thank you, Crystal from MyWebTree for guiding me by email to the 1880,1910, and 1900 census online which verified the recognizable names.

Are you related to Vera Dell Palmer and/or Homer Vincent Hart?

Vera Dell Palmer is related to Walter Palmer (1585–1661) born in Yetminster Parish, England. Walter's mother was Elizabeth Carter. Walter died in Stonington, CT in 1661. Vera Dell Palmer is descended from Walter Palmer through his son, Gershom Palmer 1644–1718, from his second wife, Rebecca Short, whom he married in 1633.

Genealogy links to all these Palmers may be found on the Walter Palmer Society Website. The Walter Palmer Society Web site is at http://www.walterpalmer.com/ or at: http://www.walterpalmer.com/WPS.wbg/wga87.html - 120605. According to the Walter Palmer Society's Web site, Walter Palmer, as a Separatist Puritan, in an effort to seek religious freedom, on April 5, 1629 sailed from Gravesend England on a boat called "Four Sisters"—one of six ships; the others being the Talbot, Lyons Whelp, George Bonaventure, Lyon, and The Mayflower.

Walter arrived in Salem, Massachusetts on June of 1629 and settled in Charlestown Massachusetts with his five children and Abraham Palmer, perhaps his brother. Vera Dell Palmer's father was Langford Wright Palmer, born January 20th 1851 in Dover township, MI. From Langford Wright Palmer, moving back in time by generations, each male Palmer is linked eventually to Walter Palmer, born in 1585, who had 12 children, five from his first wife and seven from his second wife, Rebecca Short. When seeking sources, start with the original books used as sources and birth certificates as well as searching similar resources online. Military records and the census also help locate people over time.

To continue the genealogy, here is a sample of the children of Homer Vincent Hart that links to Vera Dell Palmer from the Walter Palmer Society Web site. If you're related to a Palmer, check out the Walter Palmer Society Web site. It will lead you to your other relatives as well. It's incredible how many people are direct descendants of any one of the twelve children of Walter Palmer who lived in Massachusetts and Connecticut in the 17th century, and spouses with most birthdates, death dates, and marriage dates are included on the site. I highly recommend the site to search for anyone who might be related to, descended from, or married to a Palmer at any time in history on this side of the world.

Our family's original ancestor this side of the world was Walter Palmer whose mother was Elizabeth Carter, born in Yetminster Parish, England in the mid-1500s. Walter Palmer's genealogy Web links with our family's descendants are found at the Walter Palmer Society's Web site at: http://www.walterpalmer.com/WPS.wbg/wga29.html#I6787

Walter Palmer's son, Gershom Palmer, 1644–1718, born in Seacuncke, Plymouth Colony, Antient Rehoboth, was the sixth child of seven from Walter Palmer and his second wife, Rebecca Short. Gershom Palmer married Ann Denison, and had a son, George Palmer, the fifth of ten children born in 1678 who died in Stonington, Connecticut in 1728. George Palmer married Hannah Palmer in 1710 and had five children. The second child was Zebulon Palmer, born in 1714. He married Comfort Fairbanks in 1743 in Stonington, CT. Zebulon Palmer had a son also named Zebulon Palmer (junior) born in 1740. He was the third child of three.

Zebulon (junior) married Deborah York in 1743 at Stonington, CT. Zebulon Palmer and Deborah York had Jairus Palmer, the sixth child of seven born in 1758. Jairus Palmer married Sarah Spencer and had a son also named Jairus (junior) who was born the 18th of November, 1785 in Voluntown, CT, married Sarah Eells in 1808. Their son, John Celestine Palmer, was born in Ira, New York on June 18th 1824, the seventh of eight children.

John Celestine Palmer married Martha Ann Smith in 1849 in Ira, Cayuga Co., NY. The couple's first child, a son, born on January 20th 1851 in Dover township, Michigan, was named Langford Wright Palmer. John Celestine Palmer and Martha Ann Smith had four children. Langford Wright Palmer married Mary Permelia Higley in Medina, Michigan. They had three children, Percy Earl Palmer in 1876, Vera Dell Palmer in 1881, and Harold D. Palmer in 1888. Vera Dell Palmer was our family's grandma and link to the original Walter Palmer born in 1585.

Vera Dell Palmer was born January 17th 1881 in Hudson, Michigan. Vera Dell Palmer married Homer Vincent Hart on October 15th, 1908 in Hudson, Michigan. Homer Vincent Hart was born in Hudson, Lenawee, Michigan in 1875/6. The Family History Library in Salt Lake City lists the birth date as 1879 on one record that contains the date of marriage to Vera Dell Palmer in 1908 in Hudson, MI.

Each **1880, 1900, and 1910** census list his birth date as 1875/6 or list his age as 51 in 1900 or 61 in the 1910 census. **Vera Dell Palmer and Homer Vincent Hart had five children. The children's names are: Margaret Dora Hart, Homer Vincent Hart, John George Hart, Robert Langford Hart, and Richard Kenneth Hart, all born between 1909 and 1918. Their second child, born on January 27th 1911 in Hudson, Michigan also was named Homer Vincent Hart (junior).**

Homer Vincent Hart (junior) born in 1911, d. 1989, married Hazel Ridenour. Homer Vincent Hart (junior) and Hazel Ridenour had two children born in the 1930s. (Never put the names of living people online in a genealogy site as privacy must be preserved.) One of those two is my husband of many decades. That's the genealogy connection. At last I can visualize a long line of people lined up in a row representing each century and each branch joining a tree growing from a rock by the seashore.

Genealogy has revealed in a time capsule how my husband connects to both Stephen Hart and Walter Palmer, according to the surname list at the Walter Palmer Society Web site link at: http://www.walterpalmer.com/WPS. wbg/wgasurs.html. Check your own name as there are hundreds of names and links descended from Walter Palmer. Yes, we are all part of history anywhere in the world. This really is an adventure and a journey. Let's make time capsules that link all of us through the generations. That's what intergenerational life stories are all about.

The list starts with Stephen Hart and Walter Palmer arriving in sister ships, "The Lyon" and "The Four Sisters" in the new world which links the genealogy records back to Walter Palmer's mother, Elizabeth Carter in 1585 England. (I wonder whether he or his mom or dad of the same name ever experienced a Shakespeare play—live?)

Stephen Hart and Walter Palmer find themselves in Massachusetts around 1632 and then in Connecticut a few years later. Who would think that hundreds of years later that one of the male children descended from the male children of Stephen Hart would marry the one of the female children descended from one of the male children of Walter Palmer after so many generations?

Through Walter Palmer's son, Gershom Palmer, each son in each generation finally links to our family grandma, Vera Dell Palmer. The male lineage would continue through Langford Wright Palmer's brothers. The female link joins Vera Dell Palmer to Homer Vincent Hart. So there should be lots of relatives out there. Vera Dell Palmer's father was Langford Wright Palmer, born in 1851. Her mother was Mary Permelia Higley, born in 1853. Vera Dell Palmer was born Jan. 17th 1881 in Hudson, MI. She married Homer Vincent Hart Oct. 15, 1908 in Hudson, MI. The children of Vera and Homer born in Hudson, Lenawee, Michigan are Margaret Dora, born 1909, Homer Vincent, born 1911, John George, born 1913, Robert Langford, born 1916, and Richard Kenneth, born, 1918.

Genealogy projects can be an adventure in historical research and a journey in a time capsule. New projects could also include DNA-driven genealogy, old photos, and information about those relatives who served in various wars such as the Revolutionary War in the 18th century, the Civil War, or similar historical events and how they relate to records and/or photos or paintings of various ancestors. All these names and dates are part of keepsake albums that may be put into a time capsule and sent to future generations from any or all branches of this lineage. Visualize how many descendants must exist of all the children.

Each generation had anywhere from twelve children in the earlier generations to three to five children by the early to mid-19th century. Those children who survived began branches of the same tree. So there must be quite a number of descendants of Stephen Hart or just John Hart, his surviving son who had seven children. Our Hart family lineage comes from Isaac Hart. The other children have branches also, and these are more genealogy projects to explore. Like archaeology, DNA-driven genealogy may be another journey in time.

Primary Sources: Search Books Written Before 1941 and Before 1881.

Elaine Hart Kerskie of Victor, Ontario Co., NY, book 1940 Genealogys & Vital Records.
FTM Genealogical Library, New England Families, Vol. III page 1549.
FTM Genealogical Library, Compendium of American Gen., Vol. VII, page 49.
FTM Genealogical Library, Families of Early Guilford, CT Vol. I, page 605.
FTM Genealogical Library, Family History of Central NY, Vol. I, page 485–8.

Compendium of American Gen., Vol. VII, page 49 says: Capt. John Hart (1655–1714), he alone of his father's family escaped being burned to death, Dec 15, 1666; Capt. Colonial troops in Queen Anne's War, 1703; dep. Gen. Ct. 4 yrs; m Mary Moore (d1738). There's confusion as there are many John Harts descended from the earliest John Hart, and the sons of Capt. John Hart have many descendants, some also named John Hart among others. Also, there's no way to know whether the story about the fire or who set it and who survived as written on the Internet is clear. So you have to go to the primary sources to find out what happened or could have happened.

Sergt. Isaac Hart found in:
Marriage Index: Connecticut, 1635–1860.
Married: Nov 24, 1721 in: Farmington, Hartford, CT.
Family history library microfilm roll info: microfilm reference number
Roll number: 1315116 items 3 and 4.
The US 1880 Census is searchable at: http://www.familysearch.com.

April 26, 1889. The application number was #700929, cert: # 731396.

1910 Census, Hudson, Lenawee, Michigan.

West Grants Pass Precinct, Josephine, Oregon.

Walter Palmer Society Web link at: http://www.walterpalmer.com/WPS.wbg/wgasurs.html.

Descendants of Stephen Hart Web site is at: http://users.rcn.com/harts.ma.ultranet/family/harts

"Deacon Stephen Hart's Will" site at: http://users.rcn.com/harts.ma.ultranet/family/harts/DeaStephenWill.shtml.

"Passenger Lists for the Lyon" Web site at: http://www.whipple.org/docs/lyon.html.

The Dunne and Allied Family's index at www.RootsWeb.com

The Dunne & Allied Family's at: http://worldconnect.rootsweb.com/cgi-bin/igm.cgi?op=GET&db=dunne1&id=I0841 6.

National Genealogy Society

http://www.ngsgenealogy.org/

http://www.personalhistorians.org/

What Information Can You Expect from DNA Testing Companies?

Oxford Ancestors <http://www.oxfordancestors.com> sent me a history of each of the 'clans' or haplogroups of my matrilineal DNA and a chart showing how my first female ancient ancestor who started the H mtDNA haplogroup linked to all the other world clans going back to the first anatomically modern human in Africa.

Next I went to Family Tree DNA. Their Web site is at: <http://www.familytreedna.com>. Soon Family Tree as well as AncestryByDNA offered the racial percentages test as well as the high and low resolution mtDNA testing. As time passes, more companies offer to test more Y-chromosome markers. Only men carry the Y-chromosome.

What I like about Family Tree DNA are the choices. You can test males for a variety of number of Y-chromosome markers such as 12, 25, or more. They have surname projects and various ethnic databases. There's a Jewish database, for example, that is linked to various research projects. There are tests for Native American DNA

Where Can You Match Your mtDNA to a Country in an Online Database?

For women and men interested in matching their mtDNA sequences of HVS-1 or HVS-2 (high and low resolution) there are databases online such as Macaulay's Tables Database. These DNA databases online are matched with surname groups, lists, message boards, and other Web and online databases to help you match your sequences to a geographic location. I use **Macaulay's Tables** at: http://www.stats.gla.ac.uk/~vincent/founder2000/tableA.html

What you can expect from a DNA test is to use molecular genetics to observe your DNA data trail of a lineage. The test results allow you to use genealogy and DNA test results together to connect unknown family members by locations on non-Recombining Y or mtDNA markers. You inherit your Y chromosome or mtDNA from your father and mother. It's the same markers that show up in your great grand parents all the way back to the first ancestor of your sequences that appeared 10,000 or 20,000 years ago depending on your sequences.

Should Genealogists Learn Anthrogenealogy?

Anthrogenealogy is Family Tree DNA's word of choice "for the study of deep genealogical origins through means of genetics." From AncestryByDNA, the third company that tested my DNA, I received a package containing a map of the world with the theory of how humans evolved from Africa to the rest of the world. The map showed that traces of these ancient African markers or genes are found in all people today. The fourth company I contacted, Roots for Real, asked for the results of my mtDNA tests at the other DNA companies.

Roots for Real is a London-based DNA testing company that presently tests your HVS-1, low-resolution MtDNA sequences. The company asked for the results of my mtDNA test from any other DNA testing company. I sent them my sequences. What I like about Roots for Real is that they emailed me the possible geographic center of origin on the map and a list of my mtDNA matches by geographic coordinates. This consisted of a list of how many people tested are living in various countries. It's a guess made from computing the center of the geographic area, because there's a deviation or migration from that theoretical center of 669.62 miles, and that center turns up in France, even though people with my sequence now live within a radius of 669.62 miles of that center.

No people's names were mentioned on the list or database, only code numbers, the country, and how many tested were living in that country at the time of the study. The geographic center of possible origin on the map Roots for Real sent me is 48.30N 4.65E, Bar sur Aube, France with a deviation of 669.62 miles. For further information, their Web site is at: <http://www.rootsforreal.com>. You can write to Roots for Real at: PO Box 43708, London W14 8WG UK.

◆ ◆ ◆

Are you Curious about Your Ethnic Family Origins?

Every family has its own Adam and Eve—the original founders of a particular family line on either the male or female side. Find your own family's Adam and Eve. Trace your ancestral founders through DNA-driven genealogy. What you are looking for is evidence, patterns, and anything not random. Genealogists now can use molecular genealogy—comparing and matching people by matrilineal (maternal) DNA lineages—mtDNA or patrilineal (paternal) Y-chromosome ancestry and/or racial percentages tests of an ever increasing number of genetic markers. People interested in ancestry now look at genetic markers to trace the migrations of the human species. Here's how to trace your genealogy by DNA from your grandparents back 10,000 or more years. Where did they wander and camp?

Anyone can be interested in DNA for ancestry research, but of interest to Jews from Eastern Europe is to see how different populations from a mosaic of communities reached their current locations. From who are you descended? What markers will shed light on your deepest ancestry? You can study DNA for medical reasons or to discover the geographic travels and dwelling places of some of your ancestors.

What you're studying is non-randomness. You use DNA as a tool to study ancestry and the history of your ancestors as part of a larger population. You look for similar patterns.

Ashkenazim and Sephardim separated about 1,000–1,500 years ago. What happened genetically to each branch since that time? Scientists currently are researching the possible origins of Eastern European Jewish genetic markers and DNA such as mtDNA and Y-chromosomes. The questions beginners have when studying how to interpret DNA test results for family history and ancestry often is "How do you find the first Adam and Eve of your own family?" Every family has a founder that is a single individual that becomes a couple from whom all the male and females in that lineage descend. Who was your founder? At what time in history did your male or female founding ancestor start a family that resulting in you?

Your founding Adam or Eve would be the man and woman who carries your DNA sequences and has passed on those sequences that change very slowly and are passed down the line from either mother to daughter or father to son over a timeline of thousands of years.

You'll learn how to start interpreting your ethnic DNA test results for family history and ancestry. Who were the mothers, the female ancestors of the Ashkenazim and their communities in Eastern and Central Europe? From where did they originate or migrate?

How does this compare to the father's, the male's points of origin? What do the details in male and female DNA and genetic markers reveal about possible origins or migrations?

From who are you descended? Here's how you can search your own ethnic group and find out what the genes show when the written records end. Your genes can show possibly where mass conversions took place, where your founding ancestor originated geographically, and trace your family back to a single Adam or Eve that has your DNA sequences. That person's markers would be the same as your own. Somewhere in time you and that DNA match shared a common mother or father—perhaps 10,000 or 20,000 years ago, or perhaps only a few generations back. The DNA markers tell a lot and maybe not so much about your founding mothers and fathers, your personal Adam and Eve.

Scientists comment on a merging mosaic of communities that have since homogenized and are continuing to merge. Take the name Levien. It's an old Sephardic name listed from the book, *The Jews of Jamaica*, by Richard D. Barnett and Philip Wright. The book, *Jews of Jamaica* contains tombstone inscriptions and dates of death from 1663–1880. Only names that appeared as Sephardic are included in *Jews of Jamaica*. See the Web site at: http://www.sephardim.com/, a research tool for Sephardic Genealogy/Jewish Genealogy. There also are alternative spellings, including derivatives of Levy and the name, Lewin. There are also names such as Levit and Levita.

Also see: "Sangre Judia" ("Jewish Blood") by Pere Bonnin. A list of 3,500 Jewish names as created and defined by the Holy Office (la Santo Oficio) of Spain. The list comes from a census of Jewish communities of Spain by the Catholic Church. Names listed are those of Jewish origin.

Note that Levien, Levin, Levine and other spellings also are used by Ashkenazim and once were used and are still used by some Prussian non-Jews, the French using the spelling, LaVigne, as well as Spanish and Portuguese. Did the Ashkenazim take the name, Levine/Levin from the Sephardim or from the Prussian non-Jews? The name Levim also appears in the book, *History of the Jews in Aragon*, by Regne. Essentially a series of royal decrees by the House of Aragon, it contains Sephardic names recorded during the period 1213–1327. You're looking at 800-year old names that existed with the Sephardim and later—after the 18th century—also with the Ashkenazim in Eastern Europe. Levine existed in university records in Austria in 1598 and in Prussia in 1635 on Christian baptismal records, but is listed as a Sephardic name in the 13th century listings of Jewish names in Spain and Portugal. The pronunciation was more important than alternative spellings.

It's estimated that Cohanim and Levites each make up ~4% of the Jewish people (Bradman et al. 1999.) See the article, *"Multiple Origins of Ashkenazi Levites: Y Chromosome Evidence for Both Near Eastern and European Ancestries,"* Doron M. Behar, Mark G. Thomas, Karl Skorecki, Michael F. Hamer, et. al. American Journal of Human Genetics 73:768–779, 2003.

Also, see the article titled, *"Geneticists Report Finding Central Asian Link to Levites,"* September 27, 2003, by Nicholas Wade, New York Times. According to that article which refers to a report published in the "current issue" of the *American Journal of Jewish Genetics*, a study was prepared by population geneticists in Israel, the US, and England, based on a six-year study. The conclusion was that "52 percent of Levites of Ashkenazi origin have a particular genetic signature that originated in Central Asia, although it is also found less frequently in the Middle East," according to the NY Times article.

Who introduced that particular Y chromosome? One man or several? Did it come from the Khazars, a Turkic tribe, in the eighth century? Or from the West Slavs such as the Sorbs? Or other? The genetic signature is R1a1. It's very common in the area just north of Georgia, a geographic region inhabited by the medieval Khazars. Only there's another angle: That same Y chromosome genetic signature also is found in the Middle East long before the founding of the Jewish community. Only it's rare in the Middle East.

No one can exclude the possibility that a Jewish male brought the genetic signature or Y chromosome to the Ashkenazi population. Except the researchers "consider that a less likely explanation," according to Wade's article in the NY Times. The Chuvash of Chuvashia were studied because their language is closest to the medieval Khazar language. According to Dr. Michael Hammer of the University of Arizona, one of the US authors of the report, the people who are descendants of the Khazars have not yet been identified. So until the descendants are identified, the mystery still asks, "Who contributed the R1a1 Y chromosome genetic signature?"

Levites have mixed ancestry. Dr. Skorecki (of the Technion and Rambam Medical Center in Haifa) and Dr. Hammer (University of Arizona) studied Cohens and Levites in the past. Only now the study revealed that the dominant genetic signature among the Ashkenazic Levites was the R1a1 signature. It's different from the Cohanim signature among Ashkenazi and Sephardi. Ashkenazi and Sephardi Levites have different paternal ancestry. Ashkenazi and Sephardi Cohens or Cohanim genetic signatures of the Y chromosome are similar. The branches originated before they split. Sephardic Levites are different from Sephardic Cohens or Cohanim.

There's only one problem: Who's the ancestor Y-chromosome R1a1? He lived 2,000 years more recently than the founder of the Cohanim Y chromosome. That brings us to about 1,000–1,400 years ago—in early medieval times. So will the real R1a1 ancestor please stand up and be warmly greeted? Was he a Khazar? a Sorb? an Alan? a Caucasus Mountaineer? Who is that smiling man? No one knows as yet. So if 52% of Ashkenazic Levites carry the R1a1 marker, what about the other 48% of Levites? And what about the rest of the Ashkenazi population male or female?

What a journey to find the first ancestor who originated the branch or branches. What was his name, his language, his geographic space? The study among the Chuvash of Chuvashia, who speak a Turkic language, is still ongoing at this posting. Why Chuvashia in the Urals? Chuvash is a modern language that is closest to the medieval Khazar language. All we know so far is that the Ashkenazim have multiple origins with evidence for both Near Eastern and European ancestry, and more studies are ongoing.

There is a lot of yet unpublished research on Ashkenazi mtDNA. So scientists in some published journal articles are finding that in one sample approximately 9% of Ashkenazi mtDNA follows the CRS and 27% of Moroccan Jewish mtDNA follows the CRS. U3 mtDNA found in 17% of Iraqi Jews shows the mutation 343. Other research shows that 51.4% of Georgian Jewish mtDNA tested from one sample study shows the mtDNA mutation 355. The mutations are compared to the CRS. Jews from India in the same recent study showed mtDNA of these mutations: 166, 223, 311 in 41.3% of individuals sampled. What does research show about the geographic or demographic origin of some founding mothers in non-Ashkenazi and Ashkenazi communities?

You have to look at the sample size and the frequency. The articles in which the studies appeared are reviewed at my site along with articles and excerpts from my book at http://www.newswriting.net or http://booksreviewed.tripod.com. I hope this helps in research for those interested in tracing the origins of their own DNA or those of their friends and loved ones.

Research is ongoing on this subject, and I found an excellent audio tape by Dr. Vivien Moses on this subject from 1999 on genetic research ongoing on the origins of Ashkenazim testing Y-chromosomes and mtDNA from Ashkenazi males. It's at www.audiotapes.com, from the 1999 Jewish Genealogy Convention. There's another convention in July 2003.

Research is ongoing comparing Ashkenazi Levites to Sorbs/Wends/Lusatians of East Germany and W. Poland. Currently about 52% of Ashkenazic Levites were found to be close genetically to the Sorbs, the Westernmost of Slavic peoples living in East Germany, near the Czech border, and in Western Poland. Yet only about 4% of Jews worldwide are estimated to be Levites or Cohens based on family oral traditions. Now genetics has found a Cohen Modal Haplotype and a Levite Modal Haplotype, and Cohens differ from Levites. Additionally, Ashkenazic Levites differ from non-Ashkenazic Levites genetically, but Sephardic and Ashkenazic genetic "Cohen Modal Haplotypes" are similar worldwide. Historically, in the 9th century, Sorbs of Wendland in Germany were resisting being Germanized and their own language (Slavic) replaced with German words.

Sorbs (not Serbs) also were resisting having their own pre-Christian religion changed to Christianity and their W. Slavic language replaced by German in early medieval times, and may have contributed words to Yiddish. There's a book by Paul Wexler on the possible Sorb contribution to Yiddish language. You need to keep in mind that when a language changes, it may not have anything to do with the genes of any people speaking a new language coming from the outside.

Is it possible that some Sorbs might have chosen to join the Jewish communities as an alternative to changing their religion and language to German? Jews from the Roman Empire were in their lands, and Germans were invading from the West. This happened in the 9th century when Jews were moving into Poland in small numbers from Byzantium and from the West. Later, they'd move in from Lithuania in the 11th-15th centuries in major routes of migration. More research is needed to see what happened. If so, this might be why the genes of Sorbs and some Ashkenazi (not all again) Levites—and Levites only, not Cohens and not most non-Levite Ashkenazim—may have some Sorb Y-chromosomes?

The only way to know is to compare mtDNA and Y chromosomes of Sorbs of today and descendants of Yiddish speaking Ashkenazim from that area of the world. Anyway, check out the references I give to Web sites on my Web site at http://www.newswriting.net.

In the book ***Genetic Diversity Among Jews***, there is an excellent article by geneticist, Batsheva Bonne-Tamir, the author, in the chapter, "*Types of Mitochondria DNA Among Jews.*" The book was published in 1992. Mitochondria abbreviated as mtDNA is the part of the DNA that shows female lineages—female ancestry back to the original woman who first arose with the particular mtDNA sequence and passed it on from mother to daughter along the generations for thousands of years. Mitochondria mutates slowly.

Yours is very ancient and could go back ten or twenty thousand years to a coalescence point somewhere geographically where a date in time can be estimated. For example H mtDNA arose in Europe 20,000 years ago in SW France and NW Spain. Yet H mtDNA had an ancestor, HV that lived in the Caucasus before coming to Europe, and before that time, perhaps 50,000 years ago, lived in NW India, migrating to Central Asia and then to the Caucasus, and finally to Europe. After the ice age, H expanded from Western Europe to Northern Europe and east to the Urals. That's why H today is found from Iceland to Bashkortostan and at levels of highest diversity in India, where it is found at lower levels today than currently appears in Europe.

Y chromosomes in males follow their mtDNA "mates" so that haplotype "I" Y-chromosome in the past has followed a parallel migration route with HV mtDNA, which in turn split into H and V mtDNA. Another example is U mtDNA followed a parallel migration route with "J" Y-chromosome. "I" and "HV" traveled from India to Central Asia to the Caucasus to Europe. "J" Y-chromosome and U mtDNA followed a parallel route from the Arabian Gulf through the Levant and from what was once Sumeria (Iran, Southern Iraq) through both the Levant and Caucasus to Europe as well as through parts of Central Asia to Europe. There are other Y-chromosome haplotypes and other mtDNA haplogroups following a variety of routes into Europe from the Caucasus, Central Asia, India, and the Arabian Gulf region. This happened thousands of years ago, before the last ice age. The height of the last ice age (Glacial Maximum) occurred about 25,000 years ago. By 12,500 years ago, the last ice age ended, and by 10,000 years ago, in Anatolia, the Levant and Fertile Crescent—from what today is Lebanon, Israel, Syria, and Iraq, and Turkey, agriculture began.

In this chapter it explains the difficulty of making any assumptions about Ashkenazi mtDNA because what can be found is that the variability of mtDNA types is smaller among Jews than among other peoples around the world. However, it states that some of the types that are *exclusive* to one single Jewish community are "probably derived from non-Jewish ancestral lineages," but the extent of admixture "cannot be estimated, because differences between communities may also be due to *different lineage extinctions* within each community."

So we don't really know whether extinctions in one community has narrowed the mtDNA to only a few haplogroups or whether the few female founders are derived from local, non-Jewish sources, such as local village women taken as wives in early medieval times. See page 53 of ***Genetic Diversity Among Jews***, published in 1992 by Oxford University Press, NY.

What we have is this history of Ashkenazi in Europe:

Roman Emperor Claudius in the 1st century CE estimated the world Jewish population at 8 million. Scientists think it was 4 to 5 million. But only 2.5 million lived in Israel and Judea in the first century when Claudius was Emperor.

At that time, Jews lived all over Europe and the Mediterranean area, in Egypt, Turkey, Greece, Macedonia, and wherever else was under Rome because perhaps that was where civilization was focused in the first century, and Jews went where there were books, trade, and civilization.

After the Bar Kochba uprising in the second century, Jews moved north as they were repressed by Rome. The Diaspora started in Egypt. When Rome turned Christian in the 4th century, Jews kept moving further north and further east, and a small—maybe 150 families—medieval Jewish population in Armenia began to increase.

From the 4th to 8th centuries, a low point in population expansion and increase of Jews occurred, and Jewish communities flourished in Roman colonies of Germany and France and moved to the N. of Italy, particularly in what was to become Venice, Florence and also in the long-established province of Thessalonika in Greece.

The Arab conquest of the 7th century saw the Jewish population expanding from Baghdad west, especially the men as Persian Jewish Rhadanite merchants traveled to Central Europe. During the 7th century the Jewish population was around 1.5 million with a half million margin of error. Most Jews resided in countries under Islam.

The small Khazar conversion would not have changed the genetic composition as Khazars spoke Turkic dialects but were genetically more likely to be part of the Iranic nomads of the steppes before they settled at the mouth of the Volga/Caspian area. Their lands filled with Jews from Persia and the Islamic lands also as well as Jews from Central Europe moving east.

The writings of Benjamin of Tudela in Spain accounts for the 12th century with a continued population of 1.5 million Jews in E and W Europe and the Medit. area. In the 13th century Jewish life shifted to Prague and Poland and E. European countries. Until the 17th century the population was very small, but not below 3/4 of a million or above 1 1/2 million.

Local Jewish communities were small and geographically mobile. The Diaspora out of Spain in the 15th century went to W. Europe and Turkey and N. Africa also to Romania. In the 18th century, Jews migrated from Poland and the Ukraine to Romania.

Keep in mind that by quoting percentages found in any population, it is based on one or several studies with small samples of people in many cases. So you have to look at the sample size and the number of studies done. You also have to research how sample size represents the population as a whole. Between 18th and 19th century, the population expands. The outcome of this is very few female founders and very little mtDNA variability, such as in one sample, X, T, K found with K mtDNA being found in several studies at 27% to 32%. H mtDNA is found also among Ashkenazim in Eastern Europe.

In one study, nine percent of Ashkenazim with the H haplogroup mtDNA is the same as the Cambridge Reference Sequence (CRS). Also in one study with a small sample of people, 27 percent of the Sephardic Jews of Morocco have mtDNA that follows the CRS with H haplogroup. Among Iraqi Jews tested in one sample, 17 percent have U3 mtDNA.

No one knows whether the fewer female founders are really due to extinctions that narrows the females down to a few founding mothers, or whether only a few mtDNA haplogroups went to Europe in the first place, or whether only a few women from a small pool of mtDNA converted from local non-Jewish European villagers. The scientists perhaps can find out whether the mixtures occurred during the 4th century or the 8th or when? Also, Ashkenazi mtDNA is a bit closer to the host population than non-Ashkenazi mtDNA, according to one sampled study.

Did some Sorb women or West Slavs who lived in E. Germany and W. Poland make the non-Jewish "blonde" contribution to the Jewish communities in N. and Central Europe? Can genetic markers discern Slavic DNA from Middle Eastern? What some studies have shown is that when looking only at Y-chromosomes which only males have, the R1a1 marker shows up in Eastern Europeans in varying percentages in different countries. For example 60% of Hungarian men and 56% of Polish men. Those with the M17 (R1a*) rather than R1a1, marker show up in Belarus and Armenia. Origin would be in Central Asia, but before that, near India if you go back 51,000 years ago for both R1a1 and R1a* with the M17 sequence. By 33,000 years ago R1a1 had already come into the Ukraine and headed into Western Europe. What has been found is that the genes and markers for immune systems in Jewish men from E. Europe match with the similar immune system genes of men from the Middle East/Palestine, and the Levant.

Yet many Ashkenazic males have "I" Y-chromosome, present in Germany, Poland, Scandinavia and other European countries. Where did they get this from, a mass conversion in early medieval times, or from some other geographic location? Also, R1a1 so prevalent in Eastern Europe is also present in Syrians at 10%. So you cannot exclude from Middle Eastern ancestry the single male ancestor or very few males that contributed the R1a1 Y-chromosome marker to 52% of the Ashkenazic Levites. It could have come from Syria or from Hungary or Poland. Scientists say it isn't likely, but you can't exclude any possibility when referring to one or a few males who contributed the R1a1 marker 1,000 years ago into the Ashkenazic population. Did he come from Syria, Poland, Hungary, the Khazars, or somewhere else? Estimates are based on guessing that a genetic marker appearing in 56% or 60% of a population of Eastern Europeans has a better chance of entering the Ashkenazic population living in the host country than the same genetic marker appearing from Syria.

Only if the marker appeared 1,000 years ago, it could have come from anywhere, nearby or far. Every estimate is based on likelihood and probability. And so the research continues.

Only time and science can show what they find. So far, the best way to find out (as a science reporter) is to ask the experts and find out which studies hold up. Historians also need to take into account the 1,000 years Ashkenazi Jews spent living in Italy compared to the 500 years they spent living in Eastern Europe compared to the 4,000 years they spent in the Middle East, from perhaps a starting point in N. Iraq with a common ancestor 7,800 years ago with all other Middle Easterners.

Perhaps mtDNA that could have come from local communities or could have come from extinctions to narrow down the mtDNA variability to only a few female founders showing up on mtDNA tests today. The question is: are they local? And if not, from where did they come? And if so, where did they come from in what villages

Twenty seven percent of Moroccan Jewish mtDNA follows the CRS. That's more than Europe's 20% following the CRS. Combined H haplogroup mtDNA includes up to 60% of European mtDNA. Iraqi Jewish mtDNA stands at 17%

U3, the mutation at 343 being found all over the Middle East/Levant. About 9% of Ashkenazim follow the CRS in their mtDNA. New research is coming in every day. We may all be related and linked to these female founders.

◆ ◆ ◆

In One Study, Ashkenazi mtDNA Appears to Be Close to Moroccan Jewish mtDNA

Women frequently ask what's the origin of Ashkenazi mtDNA? It's close to Moroccan Jewish mtDNA in one study. In the book, Genetic Diversity Among Jews by Batsheva Bonné-Temir and Avinoam Adam, Oxford University Press, 1992, chapter 4 is titled "Types of Mitochondrial DNA among Jews." Thirty-four different mtDNA types were found in their sample studied. The sample consisted of Ashkenazic Jews from Central and Eastern Europe, and Jews from Turkey, Iraq, Yemen, Habban, Morocco, and Ethiopia. The mtDNA was studied for heterogeneity, that is, genetic diversity. Researchers calculated the degree of heterogeneity.

Researchers look for patterns in several enzymes. Each pattern is given a different number. So a series of five numbers represents the gene fragment patterns that show up after the DNA sample is "digested" with the five enzymes, and this is what makes up one's mtDNA type. Then the researchers look at how the mtDNA types are distributed. Scientists then draw up tables to look at the degree of diversity each of the Jewish communities reveals. Researchers look at the number of people, the number of each person's mtDNA type.

Then they looked at how frequently each type appears. So the purpose is to study heterogeneity, that is, diversity, and arrive at values of diversity for each Jewish community. The study of mtDNA is based on looking at the fragment patterns of the enzymes studied. In the chapter, five enzymes were studied based on their fragment patterns and presented in a table.

The researchers used five restriction enzymes, and 34 different mtDNA types were looked at from women from seven Jewish communities—Turkey, Ashkenazic (Central and East Europe), Iraq, Yemen, Habban, Morocco, and Ethiopia. Nine of the mtDNA types showed up in several of the Jewish communities—six in single communities and 19 were found in single individuals. The most common mtDNA type was found in 160 individuals out of 268 studied. The second most common type was found in four communities. These were 31 women from Ashkenazic, Iraqi, Yeminite, and Moroccan communities. The third most common type was found only among the 12 Ethiopians studied. Scientists looked for variability—diversity—heterogeneity.

The overall variability of mtDNA types of these women was smaller among Jews than among the Caucasian population as a whole, Asians, Australians, or Africans. Variability is highest in Africans, lower in Jews. But some populations have mtDNA heterogeneity values even lower than Jews, such as Native Americans (Amerindians) or Finnish women.

The study found that the largest number of different mtDNA types (14 different types) were in Ashkenazic Jews. Ethiopian Jews were different from all the other Jewish groups. So there might have been an African contribution to the Ethiopian Jewish mtDNA pool.

Ashkenazic mtDNA is close to Moroccan Jewish mtDNA. This was found in the study when another set of "six restriction enzymes was used." If you want to read further about the closeness of Moroccan Jewish mtDNA and Ashkenazic mtDNA, see the article "Mitochondrial DNA polymorphism in two communities of Jews," Tikochinski Y, Ritte U, Gross SR, Prager EM, Wilson, AC. (1991) American Journal of Human Genetics 48:129–136.

When and if you take a DNA test, you'll find that what's tested is your mtDNA or your Y-chromosome. It's affordable and easy to test that part of your non-recombining DNA for ancestry than the expensive route of trying to test your nuclear DNA that recombines. So only the non-recombining portions of your DNA are tested.

Sometimes this part of your DNA is referred to as junk DNA. Note that your mtDNA or Y-chromosome only shows about 2% or 3% of your ancestry. The rest of your genes recombine and are much more difficult and expensive to test at this time.

What your mtDNA or Y-chromosome doesn't show is the rich tapestry of all your ancestors' origins. It just gives you the origin of your deep female ancestor when the mutations you have first originated. You're testing the matrilineal line—the mtDNA if you're female. For a male, he can test the mtDNA—his matrilineal line and also his patrilineal line. A male can test the DNA of his Y-chromosome to look at male ancestry possible origins. Women do not carry or pass on Y-chromosomes. What can a name tell you?

If you have a name that has a Prussian origin such as Levine, note that it also could have a Sephardic origin as Levien or other alternative spellings. Also, there were Sephardic migrations after the inquisition and sometimes before into Germanic-speaking geographic areas. The names could go back and forth. There is no such thing as a specific Sephardic gene marker, an Ashkenazic gene marker, or

Jewish gene markers. So-called "Jewish" markers refer to markers found throughout the Middle East and Meditteranean areas among non-Jews as well.

Jews from Poland have one of the most diverse genetic pools, including Mediterranean Y-chromosome markers such as Eu9 as well as R1a1 Y-chromosome markers showing possibly a West Asian origin. Yet before R1a1 lived in West and/or Central Asia, the genetic marker also existed in NW India. The further back in time you go, the more geographic areas it is found in higher diversity. And the higher the diversity in any geographic area, the more chance the marker has of being the founder or place of origin, the root of the particular sequence.

These genes or DNA existed before there was such a thing as organized religion or countries with boundaries. You almost always have a common ancestor who was another religion or ethnic group at one time or another in the past due to Paleolithic and Neolithic migrations and expansions.

If you're interested in an explanation of the DNA process with an emphasis on Ashkenazi genetic origins research, I highly recommend an excellent audio tape on DNA by Dr. Vivien Moses of University College, London, UK Microbiology Department, who with Dr. Neil Bradman, the geneticist, has focused on research with Ashkenazi DNA to see how diverse the Eastern European Jewish population is. The title of the audio tape is: <u>DNA and Our Ancestors: Are We All Jacob's Children</u> by Prof. Vivan Moses. It was recorded 9/8/1999 at 19th Annual Conference On Jewish Genealogy. You can purchase the tape at the Web site: <u>http://www.audiotapes.com</u>.

***

2

Tracing the Female and Male Lineages:

With whom in the world do you share a common ancestry or history? Looking at DNA to study ancestry or to study medical issues is all about tracing patterns—from patterns in the genes to patterns in the migrations. Where do the women—the founding mothers—of the Eastern European Jewish communities—the Ashkenazim—originate? How do their origins compare with the origins of the founding fathers, the patrilineal lines? I asked genetic scientist, Dr. Mark Thomas, Department of Biology, University College London what his latest research findings were on Ashkenazi women as I had read his recent study.

*The article referred to in his letter is titled, *"Founding Mothers of Jewish Communities: Geographically Separated Jewish Groups Were Independently Founded by Very Few Female Ancestors,"* American Journal of Human Genetics. 70:1411–1420, 2002. The study was researched by Dr. Mark G. Thomas, Martin Richards, Michael E. Weale, Antonio Toroni, David B. Goldstein, et al. Here is his reply.

Dear Anne:

Our study did not conclude that they are not related to one another and not related to anyone in the Middle East. Neither did it conclude that they are wholly Slavic.

Our study presented evidence for strong independent female founding events in most Jewish communities. This evidence was not as strong for Ashkenazi Jews and we proposed that Ashkenazi Jews were made up of a mosaic of different, independently founded communities that has since homogenised.

Because of the evidence for strong independent female founding events, assigning an origin to those founding mothers becomes a very difficult, if not impossible task. There was some suggestion in two communities, Indian Jews and Ethiopian Jews, of a local rather than a Middle Eastern origin for the founding mothers.

I have enclosed a pdf version of our paper.

I hope this helps.

Best wishes

Mark
—
Dr Mark Thomas
Department of Biology
University College London
Web: www.ucl.ac.uk/tcga/

Information I Received from DNA Tests

One of the best genetic feedback materiasl I received was from Roots for Real, a DNA testing firm that lets you see the geographic area of your DNA matches for ancestry from their Roots for Real mtRadius database. The map they sent me shows where in the world I have matches for my mtDNA, my matrilineal lineage. The idea is to see where my estimated maternal geographic ancestral roots would be on a map.

They enclosed an article by Dr. Peter Foster titled, "Mitochondrial DNA, the Peopling of the World and Your Own Haplogroup's Story." This article gave me some background in depth of how the world was populated. More important to me, it pointed out where my own specific genetic group or haplotype originated. The article is an excellent introduction for beginners trying to interpret the results of their DNA tests for ancestral roots study. For men, mtDNA or Y-chromosome can be tested for male and female ancestry. The fascinating marriage of genetics with archaeology provided an open door to view where my ancestors might have been 10,000 years ago—namely, in what today is the town of Bar sur Aube, France or the geographic center of my origins at 48.30N, 4.65E to the nearest location.

How could that be? I don't know anyone who is related to me that came from that area. I had been reared in a balmy, southern Mediterranean family eating the typical Mediterranean fanfare from grape leaves to feta cheese pizza, from olive oil to parsley and grain salads, and mostly vegetarian foods. My listening in music focused on Eastern Med quarter tones and Greek bouzouki. So why does my maternal line point to an origin about 10,000 years ago mostly in Northern Europe? If I go back 20,000 years ago, it points to Spain and France, and more ancient than that, somewhere near India about 51,000 years ago as H haplogroup mtDNA with the most diverse sequences of H resulting in that area.

To find the founder, you look at the geographic area that has the most diversity in mtDNA sequences or Y chromosome sequences, and that turns out to be near India for H haplogroup and in the Arabian Gulf region for U mtDNA. After India, H traveled across Central Asia and into Western Europe coming from the East, near the Ukraine, and finally settling in the refugium in France and Spain to spend the last ice age.

Fascinating. So what I discovered from that point, the tracing geography began as I mapped the route Mitochondrial Eve, the mother of all humans on Earth today took when she left southeast Africa more than 130,000 years ago. We know where the original founders probably came from—the Adam and Eve of humankind—Africa, at different points in time, perhaps a few 160,000 years ago, a few more 120,000 years ago who perished in the Levant during an ice age when the Levant became a desert, and they couldn't return to East Africa.

Then a migration out of East Africa into what is today Yemen that finally survived, the L3 mtDNA branch, about 80,000 years ago that survived to reach what today is called East Asia, India, Australia, and New Guinea. Then a branch from South Asia, India, and Pakistan in today's geographic names turned westward and veered into Central Asia, the North Caucasus, and then the Ukraine in Eastern Europe, and then West into Western Europe, meeting another branch that came up through the Arabian Gulf area. When the Fertile Crescent opened about 50–43,000 years ago, that group also went up into the Levant and the Caucasus, and also into Europe.

So groups such as HV and U mtDNA along with their Y-chromosome carrying mates paralleled their journeys from Asia into Europe after having come out of Africa via the southerly route to Yemen, it is theorized by many scientists. For more information on this human journey out of Africa about 80,000 years ago and subsequent migrations research, see the book *The Real Eve (Modern Man's Journey Out of Africa),* by Stephen Oppenheimer. Carrol & Graf, NY 2003. It also has an excellent chapter by chapter notes appendix with numerous resourceful articles listed for you to read to get you started on this wonderful adventure in tracing ethnicity by DNA.

The study of ancestry through DNA research is an open door to DNA-driven genealogy for beginners. Where the written records go blank, the genetic markers tell a story of expansions and migrations in different areas of the world. You look for where the diversity is highest to find the root or original homeland. For Europeans that is India and the Arabian Gulf, depending upon your genetic markers—U mtDNA for females denotes the highest diversity in the Arabian Gulf region, and H and HV mtDNA denotes the highest diversity in or near NW India, around Kashmir, the Indus River, and the Punjab.

These roots or places of origin is where the people set out from before they reached Europe by way of the Caucasus, Central Asia, the Arabian Gulf, landing in the Ukraine and eventually, also in the Balkans and SW France and NW Spain just before or during the last ice age. People sought a corridor between the sheets of ice to follow the animal herds. They found it in Southern France and Northern Spain, in the Ukraine and Balkans. When the ice age ended, the peoples expanded once more to all corners of Europe. Those in Northern Asia also expanded, and those in Southern Asia traveled over land when the sea level become so low that land bridges from Java to Australia and New Guinea were created about 60,00–68,000 years ago. Another ice age remission opened up the Fertile Crescent and Levant 43,000 years ago and Central Asia to Europe about 50,000 years ago. Then the ice age began again about 25,000 years ago and ended about 12,500 years ago.

You now can trace the geographic path of travel mitochondrial Eve took as you follow her descendants leaving genetic prints or mutations in their mtDNA as they moved from campsite to campsite from one end of the world to the other. That's the awe about studying the peopling of the world. Because scientists now have genetic dating and DNA sequencing, you can trace your own ancestry from origin to present day migrations at least statistically.

Thanks to Dr. Frederick Sanger, Nobel Prize winner 1980, you can trace and understand your family tree by DNA mutations and genetic markers. It's a lot like looking at fingerprints people made as they touched the world to gain balance. It's also watching their footprints trot around the globe. In it, you'll find your own identity, or at least a few good possible places your ancestors might have been.

What you'll see are matches—people living today whose exact mtDNA sequences match your own, showing that somewhere back in the deep maternal ancestry you and they shared a single lineage connected from mother to daughter. The same can be said for looking at the male Y-chromosome about sharing a single lineage connected from father to son. Once you find your matches, you can see what geographic location they are at today and where the probable origin spot was geographically in the distant past.

After you find your matches, you are free to find out whether they want to be contacted or whether you want to start a group for people with the same matching mtDNA or Y-chromosome. Of course, privacy is usually asked for, but some people might want to meet their mtDNA or Y-chromosome matches. Somewhere in time, you might have shared an ancestor, maybe 10,000 years ago or maybe 250 years ago or in-between. It's like looking at living archaeology joining genetics as a new field of archaeogenetics.

What science finds today may all be changed tomorrow. The field is so new, and the technology evolves fast. Data has to be updated frequently. You learn starting with genealogy and oral history and moving onto genetics that each haplogroup split at an early stage into small groups. The pioneers colonized the world and expanded demographically and geographically. In some tropical areas of the world, original DNA types didn't die out. If the population was large enough, and food was plentiful, the people thrived and retained an overall similarity to the original African ancestors. In Europe and Northern Asia, people were confronted with cold climate. In Europe and the Middle East, add the experience of interfacing with the Neanderthals.

Conditions were fierce, almost arctic with tundra and penguins in the Mediterranean. By 20,000 years ago only small groups could pioneer into Europe and Asia. Most took refuge in Southern Europe, in Spain and the Balkans. These areas became the homelands of haplogroup H and V. At the height of the Ice Age of 20,000 years ago, Native Americans were closed off from Asia by glaciers, leaving only a small group in Alaska until 11,400 years ago. Genetic variation continued to evolve in different areas of the world due to climatic changes. People followed the herds or the places where food and water were available as part of their migrations.

Take my haplogroup of mtDNA, H. It developed in a single female by 40,000 years ago who lived in the Middle East. That single woman had an ancestor whose haplogroup was R, and R had a female ancestor who was N, and that N haplogroup had a female ancestor whose ancestor came out of Africa as L3 mtDNA haplogroup. By the time the single female ancestor mutated to H, it was 25,000 years ago. She first came to Europe in the middle of the last Ice Age called the Glacial Maximum. Her descendants landed in Spain and France by 20,000 years ago. By 15,000 years ago, the descendants of H mtDNA haplogroup colonized northern and Western Europe. So H is the most frequent Western European sequence currently and back then. Sixty percent of Western Europeans have mtDNA haplogroup H, with or without some mutations, and 38 percent of all Europeans, including the Eastern Europeans, have H without some mutations.

This is fascinating, but I needed to narrow it down to a place, and fortunately, Roots for Real sent me an actual geographic center with the latitude. They even mentioned the nearest location—Bar sur Aube, France. Now, that city is not scientifically assigned to my ancestor, but that geographic area is close enough for me as the possible area of origin. Today, my high resolution HVS-2 of my mtDNA shows up in France, but not my low resolution HVS-1. They're parts of my mtDNA that I can request testing for when I have my DNA tested by a testing firm that emphasizes DNA-driven genealogy.

I can see on the map by the coordinates that there was a migration north and south and slightly to the east, but mostly a straight line going from Spain and Portugal to Norway and Iceland and Scotland/England and from France to Austria/Germany/Poland/Bulgaria. Another migration fans out east to the Ukraine and the Urals. Now if I could only figure out whether that ancestor is from Scotland/Norway, Spain/Portgual/ Central Europe or Bulgaria? So I look at the geographic center for the most probable place of origin—Bar sur Aube, France. On another database, my "origin" point focuses on North Central or North East Europe.

Which database do I choose to find my original female lineage founder, my personal family Eve? With no Y chromosome, as I'm female, I can't check the origin of my paternal side without testing the nuclear DNA, which is difficult to do and expensive. In the future, the whole genome could be tested at a price. At that time, a person would be able to look at male and female markers in a woman's DNA. Currently, only a man can look at his mtDNA and Y chromosome markers. There are racial percentage tests than can test 800 markers and discern a European from someone from South Asia, but they are in the university research stage currently.

So what you have to deal with at the present are the DNA-driven genealogy companies that send your DNA to a laboratory, often in a university research lab studying DNA for ancestry, and add your name to a database, if desired, to find out who else matches with you and where do they originate? Their names are kept private, of course. What you see online in the database is the country of origin and the DNA sequence for the mtDNA or Y chromosome markers. Currently you also can have a racial percentage test that looks and male and female ancestral markers. In a few years much more markers will be tested with better accuracy. Keep looking for the 800 or more DNA marker tests to become available in the years to come.

Which population has the most genetic variation of all the world's peoples? It's the Africans with ancestors living south of the Sahara. What's the pattern of variation in your ancestry? Variation decreases the further you travel from Africa—that is with people dwelling farther in distance from Africa. Long ago there was a bottleneck in Africa with only a few people surviving outside of Africa to migrate to the rest of the world over a long span of time.

So the more you have historic bottlenecks in an area which reduces the population to only a few, the less variation you'll see in descendants. Different researchers such as geneticists and oral historians, archaeologists and anthropologists study very different areas of the human genome. They study different markers for a variety of reasons, focusing on different areas or loci. When different people study different markers for different reasons, all this information gets published in different journals, databases, and other sources of information.

How do the archaeologists know what the geneticists are already researched which markers for what purpose? Yet science must compare populations to learn more about the history of how genetic markers traveled with migrations of people thousands of years ago and at the present time.

With all the findings scattered in a variety of publications that are not read by everyone, how do researchers learn from one another? The forensic scientists publish in forensic journals. The anthropologists have their own journals.

The medical geneticists may be publishing and reading different publications than the oral historians, and the archaeologists may not be cross-referencing all those articles in one huge database. Click on the Web site of the Human Genome Diversity Project. It's great to see the coordination of multiple markers. You need to look at the whole world's population in a database online or in a library to find where you fit in.

You need a database or data set. You can search populations at the data base called ALFRED. It's at the Kidd Lab Web site at: http://alfred.med.yale.edu/alfred/index.asp. For example, if you're European or Middle Eastern, you can do a general search on the Internet. Or search ALFRED at the Kidd Lab Web site for populations such as the following:

Abazians	Ingush	Cambodians, Khmer
Adygei	Irish	Chiangmai
Andalusian	Italians	Hakka
Athens	Jews, Ashkenazi	Han
Basque	Jews, Mixed	Indonesia
Bretons	Jews, Sephardic	Japanese
Bulgarian	Jews, Yemenite	Koreans
CEPH, related families	Kabardinian	Manado
Calabria	La Alpujarra	Moluccas
Catalans	Mari, Highland	Nusa Tengarras
Cherkess	Mari, Meadow	Sibo
Croatian, Krk pooled	Mordavians	So
Croatian, Southern	Omisalj	Sulawesi
Cypriot, Greek	Poljica	Taiwanese
Cypriot, Turkish	Portugal, Northern	Thai
Cyprus	Punat	Thai, Northeastern
Danes	Roma	Toraja
Darghinian	Russians	Uyghur
Dobrinj	Samaritans	Vietnamese
Druze	Sicilian, Agrigento	
Dubasnica	Sicilian, Caltanissetta	Agharia
English	Sicilian, Catania	Andhra Pradesh
Epirus	Sicilian, Enna	Ankara Turkey
European Americans	Sicilian, Messina	Armenian
European Canadians, Edmonton	Sicilian, Palermo	Asian Americans
	Sicilian, Ragusa	Assam
European Canadians, Vancouver	Sicilian, Siracusa	Azerbaijani
	Sicilian, Trapani	Bagdi
Europeans, Mixed	Spaniards	Bahraini
Europeans, Northern	Swiss	Baiga
Europeans, Southern	Vrbnik	Bihar
Finns		Brahmin
French	You can search other Asian and Central Asian populations such as:	Chamar
French Acadians		Chenchu
Garfagnana		Chuvash
Germans, Mixed	Ami	Delhi
Germans, Southern	Atayal	Dhimers
Greeks	Batak	Gaud

Georgians
Gond
Gujarat
Himachal Pradesh
Ho
Indian, Central Tribal
Indian, Eastern Castes
Indian, Eastern Nontribal
Indian, Eastern Overall
Indian, Eastern Tribal
Indian, Mixed
Indian, Northern Castes
Indian, Northern overall
Indian, Nothern Nontribal
Indian, Overall
Indian, Overall Castes
Indian, Overall Nontribal
Indian, Overall Tribal
Indian, Southern Castes
Indian, Southern Nontribal
Indian, Southern Tribal
Indian, Southern overall
Indian, Western Castes
Indian, Western Nontribal
Indian, Western Overall
Indian, Western Tribal
Irula
Jordanian
Kachari
Kapu
Karnataka
Kashmiri
Kazakh
Kerala
Khanty
Khatris
Khonda Dora

Komi-Zyrian
Kotas
Koyo
Kshatriya
Lambadi
Lodha
Madhya Pradesh
Madiga
Maharashtra
Mahishya
Mala
Manipur
Mizo
Moor, Sri Lanka
Munda
Pakistani
Parsi
Pashtun
Punjab
Rajput
Relli
Santal
Sikkim
Sinhalese
Tajik
Tamil
Tamil Nadu
Tanti
Tharu
Thoti
Tipperah
Toto
Turkish
United Arab Emirates
Vysya
West Bengal
Yadava

Yakut

Or search African populations such as:

!Kung San
Afars and Issas
African Americans
African Pygmies
Afro-Caribbeans
Algerians
Arabs (Mauritania)
Bambara
Bamileke
Bantu speakers
Bantu speakers, Democratic Republic of the Congo
Bateke
Benin
Beti
Biaka
Cameroon
Cape Coloured
Chadians
Chagga
Congo
Democratic Republic of the Congo
Djerna
Egyptians
Ewe
Ewondo
Fang
Fon
Fulani
Hausa
Herero

Hutu
Ibo
Jews, Ethiopian
Kikuyu
Kwengo
Malagsy
Malinke
Mbuti
Merina (Madagascar)
Moroccans, Central
Moroccans, Northern
Moroccans, South-Central Arabs
Moroccans, Southeastern
Moroccans, Western
Mossi
Mozambique
Nama
Ngbaka
Nguni
Nigerians
Saharawi
San
San, Central Kalahari
San, Vasekele
Senegal
Shi
Sokoto
Somali
Songhai
Sotho, North
Sotho-Tswana
Tanzanian
Toucouleur
Tsonga
Tunisian
Tutsi
Uganda, Mixed
Woloff, Senegal
Xhosa
Yoruba

Zimbabwe, Mixed

Or search populations such as: Siberia with the
Buryat
Chukchi
Kungurtug
Siberian Eskimo

Or search Oceania/Australia/etc among the:

Australian
Australian Aborigines
Australian, Caucasian
Filipino
Javanese
Malaysians
Melanesian Highlanders
Melanesian Islanders
Melanesian Lowlanders
Melanesian, Nasioi
Micronesians
New Guineans
Orang Asli
Papua New Guinean, Coastal
Papua New Guinean, Highlanders
Polynesians
Samoans

Search North America with the: Alaskan Natives
Cheyenne
Choctaw
Cree
Dogrib
Greenland Inuit
Hispanic
Hispanic Americans

Maya, Yucatan
Mexican Mestisos
Mvskoke
Na-Dene
Navajo
Ojibwa
Pima, Arizona
Pima, Mexico
Pueblo
Salishans, Coastal
Sioux
Southwest Amerindians
Zuni

Or search South America with the:

Abra Pampa
Amerindians Pooled
Arara
Arsario
Awa-Guaja
Aymara
Ayoreo
Bari
Birongo
Cajueiro
Cameta
Cayapa
Chimila
Chorote
Curiepe
Embera
Gaviao
Guahibo
Guarani
Guihiba
Huilliche
Ica
Ingano
Kaingang

Karitiana	Quechua, Peru	Wayuu
Katuena	San Salvador de Jujuy	Wichi
Kayapo	Sotillo	Xavante
Kogui	Surui	Xikrin of Bacaja
Koreguaje	Tehuelche	Yanomami
La Plata	Ticuna	Yuco
Lengua	Tilacara	Zenu
Mapuche	Trombetas	Zoe
Mocovi	Urubu-Kaapor	Zoro
Panaquire	Wai-Wai	
Paredao	Wayampi	Or search the "Unplaced."
Pehuenche	Wayana-Apalai	

Where will your ancesters be found among which living populations today? You'll be surprised if you think you're ancestors are from one place and the matrilineal or patrilineal lines turn up in another area, showing deep ancestral origins before your ancestors migrated to where they were for the past few hundred years. If you're wondering whether your DNA found in Bulgaria means you are related to the Chuvash, it might be possible. The Chuvash language is the closest to the medieval Khazar language, and the Chuvash are descendants of the medieval Bulgars living in the Middle Volga region whose homeland was in the North Caucaus. See the publication:—Shnirelman VA. *"Who Gets the Past? Competition for Ancestors among Non-Russian Intellectuals in Russia"*. Washington: The Woodrow Wilson Center Press (1996).

You never know, until your genetic markers give you a possible clue, but only a possibility, because your DNA may show up not only in Bulgaria but in northern France. Which is the real you? Will your real ancestor stand up? (The whole world rises). I began looking for mtDNA matches from anywhere, especially Spain, Portugal, Greece, Crete, Bulgaria, Black Sea area, Italy, and anywhere in the Mediterranean. Most of my mtDNA shows up in Scandinavia, NE Europe, North Central Europe and Germany. So let's look at what the various DNA tests found.

Research the back-migrations from Northern Europe to the Mediterranean and Middle East in the various articles listed in the various journals of genetics. You can read these publications in most any university library open to the public for the price of a library card through membership. Or use your main city public library and various university library interloan services. Another way is to subscribe to the online genetics journals on the Internet. You can read the abstracts free to find the subject you want.

About 15% of Europeans migrated back to the Middle East in prehistoric times. The Estonian unpublished database found my matches in Bashkortostan where the Bashkir population is perhaps often 65% Mongolian and the rest European. My mtDNA is also popular in Iceland. Take your pick. Iceland, Spain, or Baskhortostan? Our mtDNA H got around. Guess whose mtDNA most resembles the CRS in the entire world? No not the 20% you find in Europe.

The mtDNA haplogroup H with mutations is found at the rate of more than 45% in Europe, including the CRS)...but the CRS exactly is found 20% of the time all over Europe. Who has the highest match with CRS exactly and no mutations outside the CRS? Moroccan Sephardic Jewish people....27% CRS in one sample appearing in a research article.[*1] In the same research article, Ashkenazim mtDNA matches the CRS in only 9.0% of individuals sampled.

Interestingly....my 356 mutation of H is found in central Italy and Armenia. With the other two mutations, 189, 362, is found everywhere else from Central Europe to Turkey....Iceland to Germany...and mostly.....ta, da...in Scotland and in the Orkney Islands at the highest rate. Anyone for a Scottish reel?

3

Men and Women Have Different Genetic Histories

Since women usually followed the husband to his community, men and women have different genetic histories, but also parallel one another in major prehistoric migration routes. If you look at non-Ashkenazic mtDNA, you'll find that studies by Dr. David Goldstein, Dr. Mark Thomas and Dr. Neil Bradman of University College in London et al. show that the women in nine Jewish communities studied show different genetic histories than the men. Their communities had a small number of founding mothers with little exchange after the community was founded with the host country. Their genetic signatures are not always related to one another or to those of the non-Jewish host populations of Middle Eastern countries.

That's not always the case with Ashkenazic women's mtDNA who. If Jewish men had kept marrying local women over the centuries, the MtDNA diversity would increase among Jewish women. You have a bottleneck effect with the mtDNA of Ashkenazi women. When you look at the mtDNA Ashkenazic men inherited from their mothers you're seeing their mother's mtDNA. Yet 27% to 32% of Ashkenazic mtDNA is of the haplogroup K, and 9% of it is the same as the Cambridge Reference Sequence (CRS), which is haplogroup H, the most common haplogroup in Europe regardless of anyone's religion.

So where and when did the Ashkenazic mothers originate? Why do Ashkenazic women resemble their host populations slightly more than non-Ashkenazic women? How many of the non-Ashkenazic women were local women from the host country compared to Ashkenazic women? Scientists say that the early Ashkenazic mothers came from a mosaic of Jewish communities that homogenized and merged. Tradition says they fled oppressed lands to other lands that invited them. What do the genes say as scientists look at the details? Did traveling Jewish traders begin the early communities? From where did they seek their brides?

In the study by Dr. David Goldstein, Dr. Mark Thomas, you see in a table that 9.3% of Iranian Jewish women in that sample have mtDNA markers with mutations at 183C, 189, and 249. This shows up in 1.5 of Yemenite Jews but not in any Ashkenazim or in any non-Jews in that sample. In what direction does this type of information guide you? Perhaps you first look at the size of the sample.

For years the origin of Ashkenazim or those practicing the Ashkenazic rite and living in northern, north central, and north eastern Europe had been a subject of many explanations. Some say male Radhanites, merchants left Baghdad and Persia to peddle wares, mint coins, or trade and travel, coming to Poland in the eight and ninth centuries. Others say Jewish communities under oppression in the eleventh century moved East, formed communities, and intermarried with Jews arriving from Italy, France, the Rhineland, Spain, and the Byzantine Empire. Molecular remains (proteins and lipids) on stone tools and pots can tell us something about what our ancestors were hunting and eating.

The DNA tells us what groups of people were intermarrying with our family groups, and history may tell us why they came to join our settlements. Jews whose ancestors came from Poland are very diverse genetically. People came from many countries to Poland in medieval times, and in the 18th century migrated from Poland to other lands such as Rumania, where they found Byzantine Jews from Constantinople living there, some refugees from the Spanish Inquistion of the 15th century. There also was a Jewish migration from Lithuania into the area around Bialystock, Poland around 1495. See the Web site on the history of Polish Jews at: http://members.core.com/~mikerose/history.html.

History tells us that in the third and fourth centuries, Jews fleeing oppression in the Middle East settled in Southern Italy from where they moved to Northern Italy and then into the Rhineland and then further east into what is the eastern part of Germany and then into Poland.

Still others claim that Jews living in the eastern parts of Germany intermarried with Sorbs, Wends, and Lusatians, all members of the westernmost branch of Slavs living in eastern Germany and Western Poland, developed the Yiddish language combining German words that replaced most of the Slavic ones, and finally intermarried with other Jewish communities moving into the same general area. The point is Ashkenazic DNA is a mosaic of the Jewish communities it has shared for hundreds of years, but where is its point of origin—West Asia, Europe, or the Middle East—or all three?

Studies of the male Y chromosome such as the one by DM Behar and K Skorecki, Technion, Haifa Israel researched 18 binary and 10 STR polymorphic genetic sites from eight Ashkenazi Jewish communities and compared them to non-Jewish populations from Europe in order to see what relative contributions of common ancestry, founder effects, admixture and genetic drift shaped the men's patterns and variations on the non-recombining (NRY) region of the Y-chromosome. The analysis stronger supports a "striking homogeneity among all Ashkenazi communities with extensive inter-population male migration. The studies revealed "very low rates of admisxture with non-Jewish European populations, and a founder effect for at least three out of the six majr non-recombining (NRY) haplogroups."

I recommend reading the article, "Jewish and Middle Eastern non-Jewish Populations Share a Common Pool of Y-Chromosome Biallelic Haplotypes," Procedures of the National Academy of Sciences, USA, Vol. 97, Issue 12, 6769–6774, June 6, 2000. Then compare this article to those researching mtDNA and y-chromosomes for later years and see what new information comes in. Keep updated on the latest research. New research seems to tie in findings with practical applications from molecular genetics to phonemics to looking at the entire genome to how this information may be used by oral historians and genealogists, by physicians, linguists, and by researchers in many fields. See the recommended bibliography at the back of this book and in other genetics or genealogy publications you find in libraries.

What about the women? What do their genes show as to origin? The key is extensive inter-population male migration. Ashkenazim spent two thousand years moving from the Middle East to Eastern Europe. Along the way, the Jewish communities that lived in Eastern Europe until recent times had varying degrees of admixture and gene flow with their neighboring populations in which countries they dwelled.

Mostly, the admixture was between mobile Jewish communities. As the Jewish population left the Levant after 70 CE, they set up villages at intermediate locations along the way from the Mediterranean to North Eastern Europe. After traveling through the Levant, Anatolia, and Greece, they sailed to Italy. At that time (70 CE) Rome had a large Jewish population and some Romans and Greeks converted to Judaism, especially those living among Jewish communities or living in Roman colonies and villas in Palestine during the time of Roman rule.

Greeks in Alexandria may have converted to Judaism if they had close contact with the large Jewish community there in classical times. As the centuries past and Rome fell, the population of Rome decreased from a large, highly populated city with a large Jewish component to merely 25,000 people during the early dark ages after the fall of Rome when its Empire moved east to Byzantium. Some Jews moved to Anatolia to join the bustling cities of Byzantine Jews there. Some Jews kept moving north in Italy, although the Southern Italian peninsula provided a first destination for communities of Jews.

By the fourth century CE, Jews who were to become the Ashkenazi population moved north, settling in Northern Italy and living there until around the 10th century, they formed communities farther north in the Rhineland. By the 10th century, Jews set up communities in the Rhineland (Germany) around the Rhine valley. This became the cultural center of the northern European Jews, the Ashkenazim as they kept moving north and east, expanding after the 11th century. As they settled in Eastern Germany and Poland, this shifting community gave rise to the largest Jewish Diaspora community to live in Europe before World War II.

Scientists have studied the heterozygote advantage versus the theory of genetic drift. Healthcare professionals have researched the recessive disease genes in the Jewish communities. Paternal lineages have been studied. What about the maternal lines? Where did they come from? Science looks at mutation rates and most of all, patterns. Research looks at genetic markers. Along come family historians to look at oral history or family traditions of who's a Cohen and who's a Levite and other information. Who contributed what genes to shape patterns?

Researchers study "polymorphic sites" to look at genetic contributions. The goal is to find common ancestry or study variability—diversity or lack of it and to look for founder effects, admixture, or genetic drift. So we know there is homogeneity among all Ashkenazi communities. We know the males traveled within the various Jewish populations, and we know the rates of admixture is low with non-Jewish populations, but still a little is there, or is it drift?

What about the women? Did the males take their women with them from the Middle East or marry local women in Italy, France, Germany, Poland, or Russia? Or did they travel the length of the Jewish communities from Spain and N. Africa to the northern most reaches of Germany, Poland, or Russia to find a "nice Jewish girl"? Did they marry local Non-Jewish women, mesmerized by the tall blondes and redheads of the German forests or marry "Viking" or Finno-Ugric-speaking women of medieval Kiev or of the Ural mountains? Who were the brides of the earliest Jewish communities in Northern Europe?

It has been said that American Jews constitute about 27 to 30 percent of any group whose distinguishing characteristic is intelligence. Scientists reveal that the mean Jewish IQ is about one standard deviation above that of other Europeans. Does this show anything genetically? You can play with unrelated statistics—about 27 percent to 30 percent of Ashkenazim have mtDNA—matrilineal lineages with the K haplogroup. But does this mean anything other than random statistics?

What you have to do is look at reality checks from what the genes actually say. You can do tissue typing, look at white blood cells—leukocytes and look at ethnic origins. It's done by anthropologists trying to find out what ethnic groups from which geographic locations match—for tissue typing reasons to help people get the correct transplant or marrow match. In the case of mtDNA looked at across three thousand years, there's a lot to research. Too often when the word "Ashkenazi" is mentioned, the question is asked whether they have absorbed the Khazars who converted to Judaism in early medieval times and then disappeared shortly after their country was conquered by the Kievan princes. Did they flee to Poland, the Ukraine, and Romania? Or did they disperse and merge with the mosaic of Jewish communities across Europe?

What happened to the medieval Jewish community of Armenia? Who else contributed to the Ashkenazi population? Who converted to Judaism in the regions near the medieval Jewish settlements? The story begins as Jews are pushed out of Western Europe and flee to the East as oppression and the bubonic plague decimates Europe. Few migration routes of Jews lead from East (Khazaria) to the West. Most routes run West to East and from Lithuania to Poland or from the Crimea to the northern Russia, from Byzantium to the Ukraine.

Immigration to Poland/Lithuania before 1600 was largely coming from Germany and Central Europe. Migrations from Hungary went east towards Russia. Polish princes invited Jewish and non-Jewish settlers from Germany right after the Mongol invasion decimated the Poles followed by the bubonic plague of the mid-1300s. The situation in Western Europe was so bad for Jews in the 1340s and 1350s, the years of the bubonic plague and Mongol invasions in the East that Poland known in Hebrew as Polin, (Poylin in Yiddish) land of rest, looked like an open invitation to great farmland and forests. The Ashkenazim moved east.

The Khazars didn't as one huge community of 500,000 people suddenly move West into Poland in the 10th to 12th centuries at the rate that the German, French, Central European, and Italian Jews moved East to escape oppression in Germany and/or N. France and the bubonic plague. Geneticists are not finding traces of Khazari genetics unless they are testing for medieval Khazari genes. Yet numerous Khazars did convert. Some moved to Constantinople and lived near leper colonies in special neighborhoods set aside for Jews—set next to a leper colony. In studies of early medieval Byzantine writings, there are accounts of Khazar Jews living next to a leper colony in a suburb of Constantinople.

What shows up in some Ashkenazic males—possibly—are genes found in some Slavic peoples, such as the Eu 19 chromosome that appears in 56 percent of Polish men and 60 percent of Hungarians. But Eu 19 shows up in 10 percent of Syrians. The marker comes from the Gravettian wave of Paleolithic hunters who came to Eastern Europe from Western Asia and Central Asia 20,000–40,000 years ago. Western European males have Eu 18 in larger amounts. Eu 19 is a marker for Eastern European males. However, what also shows up in some Ashkenazic males also is the CMH, the Cohen Modal Haplotype. Some men carry these markers and some don't in the Ashkenazic population whose ancestors came from Eastern Europe. You have to look at where they were before they arrived in Eastern Europe.

What Eastern Europe was to Jews in medieval times was a refugium from oppression they found in Western Europe from Spain to England and especially during the crusades, the plagues, and before that, the fall of Rome that had a huge Jewish population. Southern Italy was the first route out of the Middle East into Europe, then northern Italy and onto the Rhineland valleys and France. When life in Western Europe became unbearable for Jews, they looked for lands that invited them. The princes of Poland invited them.

Poland has the most diverse Jewish genetic patterns in Europe. You find Sephardic Jews coming into Poland in small numbers, families with Spanish names such as Rangel and Angel. Rappaportes from Portugal and N. Italy. You have people coming from Lithuania and from Germany. Jews in Bialystock are diverse. Some still have ancestral features very similar to Assyrians or other peoples from N. Iraq. Only a look at the DNA will show to whom they are related the closest. You have to look at the genes before you accept at face value reports online that claim that Polish and Russian Jews are descended from Khazars.

Keep in mind that Ashkenazim are a mosaic of many Jewish communities migrating and merging, coming together to form a pattern. You have people from Persia and Spain in Poland intermarrying with people from Prague or Germany, France or the Ukraine, Romania or the Byzantine Empire. Marriage usually took place within Jewish communities, and travel was wide, especially by males from one Jewish community to another, especially for trade. Polish Jews traveled to Odessa in the Ukraine to find spouses, and Jews from Romania married Jews from N. Poland. So diversity exists, even as the mtDNA haplogroups go through a bottleneck.

Today you're told to marry outside your own diaspora to keep genes mixing and the diversity growing. What about languages spoken by Ashkenazim? Yiddish was spoken in Eastern Europe, Ladino after the 15th century and/or Greek before, in the Byzantine Empire. Romaniots in Greece still speak Greek. These Jews have been in Greece since antiquity, later joined by Jewish refugees from Spain. Polish Jews migrated to Rumania in the 18th century and found Jews living there who came to Rumania from Constantinople in the 15th century. In medieval times, Jewish communities were founded quickly. Did the women come with the men and families? Or did the young men set out alone and find a local bride?

What will tell the story is genetics. You keep looking at the women's mtDNA and comparing it to Y-chromosomes of men until you see patterns that shape the migrations, that tell a kind of oral history written in the genes. So even though most women's maiden names were not recorded (not many had last names until the 18th century), their genes identify them. They existed and have a core identity that lives on in their descendants. They were faithfully 'frum.' By 1600 nearly 20,000 to 30,000 Jews lived in 60 European communities of Eastern Europe.

Sometimes looking at the history of a language can give a clue to how the genes landed in a particular container. That answers the long pondered question of how Jewish genes landed in the most northern and eastern parts of Europe and on other continents. Yiddish has had an ancestor language it came from before it was Yiddish. So has Hebrew. You have Slavic, German and Hebrew words in Yiddish. You have a few ancient Egyptian words in Hebrew, such as Moses (from the water) in Egyptian.

What words does Yiddish have that are Khazari other than "davenin" for praying and perhaps yarmulke for kippot or skull cap? Linguists suggest some Khazar names such as alp for brave, but did the name Alpert or Halpern come from the Khazari Alp for brave or from the Germanic Heilbrun? Did Kaplan, the common Ashkenazi name, come from Kaplan, a Khazari word for tiger or from another language? The people speaking the language most closely related to the Turkic dialect of the Khazars today are Christians living in Chuvashia.

You be the judge and do your research at the linguistic and Khazari history sites on the Internet. Then check out secondary and primary sources for historical research. Were the Khazari Jewish names originally Slavic? Is it important whether one or two people converted to Judaism from a Khazar, Turkic ancestry or from a Sorb/Wend/Lusatian/Slavic Polish-German ancestry? Or from a Middle Eastern ancestry? After all, 20 to 26 percent of the population of non-Jewish Europe, came to introduce agriculture in Neolithic times from the Levant/Anatolia/Syria about 8,500 years ago and carried the J mtNA haplogroup. When they arrived, they met the Paleolithic European hunters. So 20–26 percent of Europeans have the similar Middle Eastern genes to what the arriving Jews are supposed to have.

It's true that Khazars living where the Volga meets the Caspian had royal families who converted to Judaism. However, after the destruction of Khazaria, the people living under the Khazars's rule were Slavs and Armenians, and for a time, Georgians. If these Slavic people also converted, they would tend to marry into the communities of Byzantine Jews living in their lands as well as in later centuries with the Jews moving East from Germany and Central Europe.

Whatever the DNA would show would be a mosaic, a rich tapestry of mixtures of Jews of Iraq, Persia, medieval Armenia, Germany, France, Italy, Greece, Turkey, and the rest of Eastern and Central Europe. Yet even the Sorbs of Eastern Germany and Poland have had an origin in Western Asia as well as Europe. So what can the DNA tell other than that after the communities of Jews were formed from the mosaic, they remained relatively isolated and intermarried within their group.

A bottleneck of mtDNA, of the female line resulted with few women, few founding mothers in the Ashkenazic communities. Today, you have the Himyar communities of Yemen, a Jewish group marrying with Polish Jews, Moroccan Jews marrying Lithuanian, and so on, especially in Israel. In the past, dozens of studies already have shown the common genetic origin of Jewish communities by studies of molecular markers, and studies also have shown the admixture between Jewish communities and their neighbors (Mourant et al 1978; Livshits et al. 1991. Bonné-Temir et al. 1992).

You can read these scientific articles and decide for yourself about the high level of paternally inherited Y chromosomes that reveal Ashkenazic and Sephardic Jews all have high frequencies of haplogroup J as pertaining to the Y chromosome, the male or patrilineal ancestry, and that both Ashkenazic and Sephardic males are similar in their Y chromosome haplotype J with Lebanese non-Jews. The populations of Central Europe don't have this pattern.

The Cohen modal haplotype (CMH) is a microsatellite haplotype within haplotype J in the Y chromosome that links Jews—both Ashkenazic and Sephardic to a common ancestor with other non-Jewish men from the Middle East. The CMH is a type of genetic signature of the ancient Jewish priests, perhaps of the ancient Hebrew people. Yet not all Jews have the CMH. For the ones that have the CMH, it doesn't matter whether they lived in Northern or Eastern Europe or in Spain or in the Middle East. On the other hand, in the early medieval past, conversions were common. The point is the communities mixed, intermarried, and took their religion seriously regardless of their common language or difference in language. They all had Hebrew in common as their written and sacred language.

What the studies didn't do is to decide which geographical origins brought forth the founding female mothers—the founding mtDNAs within the various Jewish groups. What studies did find out that the mtDNA of Ashkenazic women went through a bottleneck. And with the non-Ashkenazic women, the founder effects were so severe that they show a great difference from non-Jewish populations. This means there were very few female founders—at least eight—in the non-Ashkenazic population. However, in the Ashkenazic populations there were more female founders, still less than in the non-Jewish populations.

The non-CRS modal haplotypes in Jewish populations are rare in non-Jewish populations when you look at the women, the founding mothers. You can't assume that there were few female founders in the Ashkenazic communities, although there definitely was an mtDNA bottleneck in Ashkenazic shtetls or villages. When you look at the genes, the DNA of Jewish women of Eastern Europe and Northern Europe who say they belong to ancestors who dwelled in Ashkenazic communities, spoke Yiddish, and practiced the Ashkenazic rite, you find that the Ashkenazic population has a modal haplotype with a frequency similar to the non-Jewish area in which it dwelled—that is the host population. The study found 9.0% compared to 6.9%.

This revealed that there wasn't a strong founder event for the Ashkenazic women. So if there wasn't a strong founder event for Ashkenazi Jewish mothers, from where did they come? The answer is that all these women formed a mosaic, a pattern quilt, a salad bowl, a melting pot of a few independent shtetls or villages or if not communities, than founding events only on the maternal side. Does this mean that the Radhanites from Persia and Baghdad arrived in North Central or North East Europe and married local women?

To find out, you'd have to test the mtDNA of Ashkenazic women and compare them to Jewish women in the Middle Eastern communities. Most of the women tested were not related to one another. Some were not related to sequences found today in the Middle East and some were. So what you find is that mosaic, that salad bowl of isolated communities. With Europeans, if you want to see a bottleneck, you normally look at the extinctions that occurred during the Last Glacial Maximum ice age.

What many studies are finding about Jewish mtDNA is that 27% to 32% of Ashkenazim show mtDNA of K haplogroup, whereas only 2.6% of Ashkenazim have mtDNA U3. Compare that to 17% of non-Ashkenazi mtDNA in Iraqi Jews who have U3 mtDNA. What this shows is that the two Jewish groups have different founding mothers in those cases. Ashkenazic Jews also have mtDNA H and other types in different percentages. You can compare that with Russians who have mtDNA T haplogroup in 22% of the cases. One way to find out the origin of people today is to compare people's mtDNA for women and Y chromosomes for men. These are the ancestry markers, at least some of them.

You can also do racial percentages tests, but they won't reveal the origin as far as geography in ancient or medieval times. People move around and intermarry as well as create communities where they can remain isolated. For example, 51.4% of the Georgian Jewish women had a transition at mtDNA site 355, a particular mutation, when compared to the CRS, that standard Cambridge Reference Sequence against which mtDNA is compared to see mutations and differences from mtDNA that makes up about 46% of Europe and 6% of the Middle East. (I use Near East and Middle East interchangeably here). West Asia is not the Middle East. It's the Urals and N.W. Asia, including Turkey.

The Middle East is the Levant, Fertile Crescent, and Iran. North of the Middle East you have Western Asia, the steppes, the Urals. Beyond that is Central Asia. You'll find H mtDNA spread as a pan-European mtDNA haplogroup from Iceland to Bashkortostan. If you want to find the origins of Ashkenazi women, you can also go back further in time and read, "Tracing European Founder Lineages in the Near Eastern mtDNA Pool," Martin Richards, et al, American Journal of Human Genetics, 67: 1251–1276, 2000. If you're looking for the mother of all Europeans and Middle Eastern people, this article's for you.

What you'll find is that we ultimately all descend from the same single woman, and that lines are traced from the Near East to Europe and from Central Asia to Europe, but to get to Central Asia, you had to come either from the Middle East or back-migrate from Pakistan to Central Asia and across the steppes to Eastern Europe. Or you migrated from Southwestern Europe (Spain or S.W France) straight to North East Europe. The two peoples there met and intermarried, had children, and again, we come to the mosaic communities and melting pots.

Look further back at the genetic split occurring when people migrated from Africa to the Middle East and to Europe. The study found that Georgian Jewish women had very few founders. A late Paleolithic explosion took people from southwestern to northeastern Europe. See the article, "mtDNA Analysis Reaveals a Major Late Paleolithic Population Expansion from Southwestern to Northeastern Eruope," American Journal of Human Genetics 62:1137–1152, 1998.

You're trying to find out the history and geography of your Ashkenazic Jewish DNA, especially the mtDNA or matrilineal side—the mother's side. The mtDNA is how your ancestry is passed from one generation of daughters to the next. If you are male, you'll want to also look at the origins and migrations, the diversity in your Y chromosome. Women don't inherit a male Y chromosome, so for women, you'll be looking at your mtDNA, your matrilineal side and any other markers you are able to have tested. Perhaps you want to contemplate your most recent common ancestor or most ancient, long ago before nations had borders and places had religions with which we are familiar today.

What do the scientists tell us about the founders or origins of the Ashkenazic Jewish communities—especially about the sometimes nameless women in ancient or early medieval times and right up to the present? Research studies on the founding mothers of the Eastern European Jewish communities look at the mtDNA and report that the genetic patterns called the "modal haplotype" in Ashkenazic Jews has a frequency similar to that of its host population (9.0% vs.6.9%), according to page 1418 of the article, "Founding Mothers of Jewish Communities: Geographically Separated Jewish Groups Were Independently Founded by Very Few Female Ancestors," American Journal of Human Genetics. 70:1411–1420, 2002. The study was researched by Dr. Mark G. Thomas, Martin Richards, Michael E. Weale, Antonio Toroni, David B. Goldstein, et al—the rest of the team. Read the entire article for a scientific explanation.

For those with little science background, the point is for the non-Ashkenazi women the study recommended above found that there is a high frequency of particular mtDNA haplotypes in the Jewish populations, at least according to those people sampled. What that means is that a high number of Ashkenazic Jewish women have the same haplotype. They belong to the same mtDNA haplogroup in large numbers. In particular women of certain Jewish groups such as the Georgians, Moroccan Jews, and Bene Israel of India have high modal frequencies. This means that there were very few founding mothers.

When you compare male Y-chromosome and female mtDNA patterns, you see a striking contrast between the maternal and paternal genetic heritage of Jewish people. If you look at the Y chromosome, the study didn't find a consistent pattern of lower diversity in Jewish communities compared to non-Jewish host populations—the countries in which they lived for hundreds of years. This study contrasts with older studies showing many males having a Y chromosome relationship in 70 to 80 percent of the samples close to Lebanese and Syrian non-Jews and in other cases, a Y chromosome showing Eu 19, representative of 56 percent of Polish men and 60 percent of Hungarian men. What does this mean? Ten percent of Syrian men also have the Eu19 marker in their DNA.

With Jewish women, the pattern in the mtDNA is different than in the Jewish men. When you look at all Jewish communities, you see a significant lower mtDNA diversity than in any population with which the Jewish community is paired with other women's mtDNA markers. The finding shows that mistakes in associating any non Jewish population with any Jewish population would not influence the results of the latest study.

So what does this have to do with northern and Eastern European Jewish women, particularly those from northeastern Europe? When you look at female Israeli military recruits and see the Moroccan Jewish females with a cholesterol level of 128, and then compare them to the same age Eastern or Central European Jewish women with a cholesterol level of 162, eating similar foods, you wonder what is it in the mtDNA of the North African Jewish women that is keeping their cholesterol to such a low level and what health benefits will it bestow in later life? Were these women prepared for a tropical climate whereas the North East European Jewish women had bodies set up to withstand the famine and rigors of a cold Ice Age climate? Both live now in Israel, but their ancestors came from different areas. Are these women related genetically at some time in the past three thousand years or not?

Among the nine Jewish groups studied by Thomas's team, there are eight different mtDNA types that are modal with a very high frequency. What this means is that their mtDNA, their matrilineal ancestry by genetics was compared to the standard Cambridge Reference Sequence (CRS). A part of the DNA was looked at called the HVS-1 CRS, occurring which occurs at 16% in Europe and 6% in the Near East (Richards et al. 2000). What the study found was that all of the seven European and Near Eastern non-Jewish populations have the CRS as their modal haplotype. Compare that to only two of the nine Jewish populations studied have the CRS as their modal haplotype. What this means in non-science language is that of the nine groups of Jewish women, eight of those groups have different mtDNA types "that are modal" with a very high frequency.

If you're now wondering what all this talk about Cohen Modal Haplotypes and Atlantic Modal Haplotypes in a man's Y chromosome (the chromosome that determines his gender) has to do with one's right to be a "real Jew" it's about as philosophical as asking who has the right to be a Christian, Moslem, Hindu or Buddhist. A "real Jew" is someone who takes Judaism seriously and practices it. The orthodox rabbis say a real Jew is a Jew who either has a mother who practices Judaism or a grandmother who practices Judaism or who converts to Judaism.

It doesn't matter whether your hair is blonde or black or whether you look like you just stepped out of Babylon wearing the mask of King Sargon, whether you have a honey complexion, melanin deposits, wavy dark hair, Mediterranean features, or whether you look like a freckled, silver-eyed, russet-tressed Sorb milk maid from medieval East Germany or Poland. Sorbs, the most Western of Slavs, have an ancient origin perhaps in West Asia or Europe. Slavs lived in Central Europe before the 6th century CE.

When they migrated south and east, there were Iranic-speaking people living there in ancient times. Don't let anyone call you intimidating names such as "a self-styled Jew." Some people like to label others with names to make them feel that they don't have a right to practice their religion or have a country of their own. If the tiny two percent of Y chromosome genes that make up your enriched tapestry of inheritance happen to be close to the Sorbs of Germany and Poland or anybody else, it shouldn't matter when you practice your religion or seek your core identity. These names get labeled on Ashkenazi Jews. Don't let anyone take away your core identity. No one in the world would call someone a self-styled Christian or any other religion.

How often do you find non-Jews calling the Ethiopian or Indian Jews self-styled? Not much. And these two groups have genes that differ from most Jewish groups outside their geographic areas. The fact is that the genes of most Jews cluster together when the Y chromosomes of the males are studied for ancestry, and most of these sequences also occur among non-Jewish males in the Middle East. However, the matrilineal lineages—the mtDNA of the Ashkenazic females often shows they are not related to one another and often not related to anyone in the Middle East.

Some of these women were descended from local women in communities founded independently who then merged together forming a mosaic. You can't say that they are all Slavs because they are not related to one another. If these women are not related to anyone in the Middle East, then to what geographic area are they related? It depends on their individual sequences. Until you look at the sequences of individuals, you won't know whether that person is related to sequences found in the Middle East.

The answer is a community that merged together from different origins has become a mosaic of diversity, and out of that diversity has come one people. Not as diverse as the United States, but nobody calls an American a self-styled American because of a diverse genetic ancestry. Still, looking at the women's lineages, you find a bottleneck in the past and today figures among Ashkenazi female lineages of 27 percent to 32 percent having K mtDNA. Georgian Jewish female lineages so very few female founders. The whole idea is to do detailed genetic analysis to find out how useful population-based gene mapping is in tracing the founding mothers of various Jewish communities.

The plethora of hate groups direct their diatribes to Jews who look the most like northern Europeans with labels like "Khazar warrior." Genetics so far has not compared the genes of "Khazar warriors" to the genes of Ashkenazim. What would they find? Two percent traces? Or would they more likely find a rich tapestry of diversity? Or certain groups within Ashkenazim such as Ashkenazic Levite males whose Y chromosomes are found in some cases to be close to modern Sorbs of East Germany and Western Poland?

Real Jews come in all colors and hair types. The fear factor is played upon when some group gets on the Internet and announces you don't have the right to be a Jew because you come from Northern Europe instead of Iraq, that your European blood marks you as a self-styled Jew. Don't listen to that hogwash. There are no such things as self-styled Jews any more than there are self-styled Christians or any other religion. It's like saying you don't look Catholic because you aren't dressed like a Roman and speaking Latin so you must be part of the "barbaricum."

The argument comes from those who say that if you're Ashkenazic and have one drop of blood from a northern European, you have no right to call yourself a "real Jew" (and come to pray or live in Israel.) Your answer would be, show me that Palestinians or other Middle Eastern persons don't have one drop of Greek or Roman blood, as European as you can get, and therefore they become "self-styled Arabs" whose real language is Aramaic, reflecting the Greek, Roman, and Arab genes in their admixture. The point is we are all mixed mosaics of time, and the mixture often occurred long before there were any organized boundaries or religions. Back migration from Europe to the Middle East in prehistoric times is about 15% for all populations in that area. You are what you take seriously.

In the book, Genetic Diversity Among Jews by Batsheva Bonné-Temir and Avinoam Adam, Oxford University Press, 1992, chapter 4 is titled "Types of Mitochondrial DNA among Jews." Thirty-four different mtDNA types was found in their sample studied. The sample consisted of Ashkenazic Jews from Central and Eastern Europe, and Jews from Turkey, Iraq, Yemen, Habban, Morocco, and Ethiopia. The mtDNA was studied for heterogeneity, that is, genetic diversity. Researchers calculated the degree of heterogeneity.

Researchers look for patterns in several enzymes. Each pattern is given a different number. So a series of five numbers represents the gene fragment patterns that shows up after the DNA sample is "digested" with the five enzymes, and this is what makes up one's mtDNA type. Then the researchers look at how the mtDNA types are distributed. Scientists then draw up tables to look at the degree of diversity each of the Jewish communities reveals. Researchers look at the number of people, the number of each person's mtDNA type. Then they look at how frequently each type appears. So the purpose is to study heterogeneity—that is diversity—and arrive at values of diversity for each Jewish community. The study of mtDNA is based on looking at the fragment patterns of the enzymes studied. In the chapter, five enzymes were studied based on their fragment patterns and presented in a table.

The researchers used five restriction enzymes, and 34 different mtDNA types were looked at from women from seven Jewish communities—Turkey, Ashkenazic (Central and East Europe), Iraq, Yemen, Habban, Morocco, and Ethiopia. Nine of the mtDNA types showed up in several of the Jewish communities—six in single communities and 19 were found in single individuals. The most common mtDNA type was found in 160 individuals out of 268 studied. The second most common type was found in four communities. These were 31 women from Ashkenazic, Iraqi, Yeminite, and Moroccan communities. The third cost common type was found only among the 12 Ethiopians studied.

Scientists looked for variability—diversity—heterogeneity. The overall variability of mtDNA types of these women was smaller among Jews than among the Caucasian population as a whole, Asians, Australians, or Africans. Variability is highest in Africans, lower in Jews. But some populations have mtDNA heterogeneityh values even lower than Jews, such as Native Americans (Amerindians) or Finnish women.

The study found that the largest number of different mtDNA types (14 different types) were in Ashkenazic Jews. Ethipian Jews were different from all the other Jewish groups. So there might have been an African contribution to the Ethiopian Jewish mtDNA pool. Ashkenazic mtDNA is close to Moroccan Jewish mtDNA. This was found in the study went another set of "six restriction enzymes was used." If you want to read further about the closeness of Moroccan Jewish mtDNA and Ashkenazic mtDNA, see the article "Mitochondrial DNA polymorphism in two communities of Jews," Tikochinski Y, Ritte U, Gross SR, Prager EM, Wilson, AC. (1991) American Journal of Human Genetics 48:129–136.

It's only by reading these articles in the various human genetics journals that you can follow the studies for a decade and see how new inroads are opening to finding out more about the founding mothers, the mtDNA of Ashkenazi women or other groups in which you are interested so you can perhaps get a handle on your own origins from any of these communities.

The question is how many mtDNA types in Ashkenazi or other groups of Jewish women are derived from or exclusive to one single Jewish community and how many are possibly derived from non-Jewish ancestries? What geneticists aren't able to estimate to the degree that most people want is the extent of admixture by looking at the markers. When the total human genome becomes affordable, individuals can see their own total genetic profiles and decide for themselves how rich their genetic heritage is from so many migrations, expansions and mixing or isolation.

Right now the extent of admixture can't be determined because scientists as well as science journalists like me cannot assume the lineage is or isn't the way it is due to extinctions of mtDNA in each community or due to admixture. All we know is that there are differences between communities. We don't know whether the differences are due to extinctions of lineages. So we can't estimate the extent of admixture...yet.

We do know that mtDNA in Ashkenazic women went through a bottleneck. We know that there is a striking high frequency of particular mtDNA haplotypes in Jewish populations, such as K mtDNA haplogroup showing up in 27%-32% of Ashkenazim. But when we look at K, it did come in Paleolithic times from the Middle East to Europe. mtDNA K is a clade of mtDNA U. That is, it is a branch off of U mtDNA and rather like its sister group.

Keep in mind that 17% of Iraqi Jewish mtDNA is U3. U mtDNA haplogroup has many branches such as U6 in northern Africa, U2 in Europe and in India, U7 in the Middle East, U5 in Europe, U4 in Europe and Armenia/Caucasus/Europe, and other branches off of U elsewhere from Iran (22% U mtDNA) and Central Asia to all over Western and Eastern Europe. H is the most frequent mtDNA haplogroup in Europe—about 46% and about 25% in the Middle East.

You have to look at the sequences and then at the tables to see where the sequences are present, more in Europe or more in the Middle East. The origin point is not where you have the highest number, but where the mtDNA sequences are the most diverse. Where you have the highest number tells you where most of the people migrated to and are living in today. Where the mtDNA sequences are the most diverse shows you where the mtDNA haplogroup originated.

Haplogroup K mtDNA springs up 17,000 years ago in Northern Italy where Venice is located today. Then it migrates to the Alps. Jewish populations have high modal frequencies of mtDNA, particularly Moroccan Jews, the Bene Israel of India, and Georgian Jews. In "Founding Mothers of Jewish Communiites: Geographically Separated Jewish Groups Were Independently Founded by Very

Few Female Ancestors," Mark G. Thomas et al, American Journal of Human Genetics, 70:1411–1420, 2002, the findings were that most Jewish communities were founded by relatively few women. So we know that the founding process was independent for different geographic areas. We know that genetic input from the host country population was limited on the female side. You go back far enough and find out there was a major late Paleolithic population expansion from southwestern to northeastern Europe. Read the article, "Genetic Affinities of Jewish Populations," Livshits G, Sokal RR, Kobyliansky E (1991) American Journal of Human Genetics, 49:131–146.

The mtDNA is inherited maternally and is gender-specific. Mothers pass it on to their daughters without much change for thousands of years. It's a way of looking back at ancestry, but it mutates over time, but slowly over thousands of years. Each mtDNA haplogroup originates from a single female founder who lived thousands of years ago, sometimes as long ago as 50,000 or 20,000 years. When scientists look at the Y chromosome of males in Jewish communities, the Y chromosome shows diversity similar to that of neighboring populations, and shows no evidence of founder effects according to the article, "Founding Mothers of Jewish Communities: Geographically Separated Jewish Groups Were Independently Founded by Very Few Female Ancestors," Mark G. Thomas, et al, American Journal of Human Genetics 70:1411–1420, 2002.

The mtDNA is a female gender-specific X-chromosome microsatellite. The article also showed that the Georgian Jews have a female-specific founder that appears to have resulted in "elevated levels of linkage disequilibrium." What this all means is that when you compare Y chromosome and mtDNA diversity patterns you see a contrast between the male and female genetic heritage of the particular Jewish populations studied. Only with the Ashkenazic populations studied, the mtDNA diversity is more like the host country population. Also compare this with the studies of Ychromosomes done by comparing diffent Jewish communities and comparing Jewish and non-Jewish Middle Eastern populations (Hammer et al 2000; Nebel et al 2000).

That means that males who have the Cohen Modal Haplotype will share a common ancestor with other males who have had ancestors in the Middle East in prehistoric times, possibly a common ancestor about 7,800 years ago. It also means the males descend from a common ancestor who lived about 3,000 years ago in the Middle East who could have been Aaron, the brother of Moses, and that the 3,000 year-old CMH existed in ancient Israel and is the signature of the ancient Hebrews. The CMH also is found in some Arabs. So the CMH had an ancestor still further back in time that both Jews, Arabs, and other Middle Eastern men shared 7,800 years ago. Genes have coalescence points, a time of origin which are thousands of years old.

What geneticists are not finding yet is an uncut umbilical cord going from ancient Israel to modern Jewish communities anywhere in the world. (de Lange 1984, p. 15; Encyclopaedia Judaica 1972). The mobility of early Jewish communities is remarkable (Beinart 1992). You have conversions common in pre-Christian Rome and its empire. You have the Khazars and the Himyars adopting Judaism in ancient or medieval times. Jewishness is defined my maternal descent (Encyclopaedia Judaica 1972.) So what happens when a Jewish man from the Middle East travels on trade business to Northern Europe and marries a local pagan woman? He converts her. By Jewish custom, when someone converts, the person is obliged to forget past the former religion, which usually was pagan in ancient and early medieval times.

From then on, the children are reared as Jews. Only one question: When a Jewish woman marries a Sorb man, if that's the case (as in Sorb Y chromosomes being close to some (not all) Ashkenazi male Levite Y chromosome genes), does he convert, and for emphasis, become not only a Jew, but a Levite, a servant, reader, or temple musician, of the Cohanim in the medieval synagogue? Historically, Levites were the tribe who lived in Egypt before the Exodus and who assisted the Cohens in the temple in ancient times. They came from the same family or tribe as the Cohens.

So, if the Khazars had their own Levites and Cohens, did the medieval Sorbs—at least the ones who may have converted? And did they merge later when the Jews were expelled from Germany into Poland in 1349? The main waves of migration into cities with large Jewish populations show that Jews migrated from Germany in 1012 CE to Poland near Bialystock and Grodno. By 1495 Lithuanian Jews also migrated to Bialystock Poland. But the founding families of Lithuania by tradition are said to have come from Babylonia in early medieval times. You have the plague in 1347 followed by a 1348 migration of Jews gong from Eastern Germany to Lemberg and Temopol, and Hungarian Jews from 1349–1360 migrating to Temopol also. Earlier you have 1096–1192 migrations of Jews from North Eastern Germany going to Grodno, Poland, to Kalisz and Lodz in 1248, to central Poland and Krakow in 1159. Seems every time a plague broke out, the Jews were thrown out of some German town and went into Poland. You have Jews going from Austria to Temopol in 1421. Most of the movements are from Germany to lands East and North.

Then you have the Lithuanian migration of 1445 due Southeast to the Crimea. And you have a 1016 migration of Jews from the eastern cost of the

Black Sea near the Crimea in the Feodosiya area going to Kharkov. Interestingly both Byzantine Jews from this area as well as the Judaic Khazars who lived there might have gone to Khakov going from southeast to the north in Russia. At the same time the Jews of Kiev, a mixture of Byzantine, W. European, possibly Khazars and other Jewish communities living in Kiev moved into the Crimea in 1350. All this movement of communities focused around 1350, just after the plague receded.

You have to look not only at DNA but at rabbinic family histories, court Jews of medieval times and court Jews in the 16th century in the Germanic countries. In the Ashkenazic world, you have rabbinical dynasties as well as in the Sephardic or Mizrahi environment. Sometimes court Jews were members of rabbinical dynasties. According to an article that I highly recommend which you can find on the Web at: http://www.jewishgen.org/Rabbinic/journal/ashkenazic.htm titled, "Ashkenazic Rabbinic Families," by Dr. Neil Rosenstein, One Ashkenazi court Jew and member of a rabbinical dynasty was R. Samuel Oppenheimer. He also served as a military contractor and banker to Leopold I of Austria.

An Ashkenazic aristocratic dynasty held sway from medieval times to the end of the 16th century, especially in Austria and Germany. Court Jews married the children of other court Jews. According to Dr. Rosenstein's article, "Any Ashkenazic Jew living today has at least one in fifty chance of tracing back to a rabbinical dynastic lineage." Poland also had its Jewish scholarly aristocrats, especially in Bialystock. A long line of authors, musicians, physicians, and artists came from the Levin and Levine families of Bialystock.

Back in 16th century Austria, R. Samuel Oppenheimer's generation used the term "Judenkaiser," according to Dr. Rosenstein's article. Also see the Encylopedia of Jewish History. Aristocratic Jews also lived in Switzerland and the Alpine areas. In Germany, you can research the biographies and history of Leffmann Behrens, Court Jew of Hanover. Another cousin to research was Behrend Lehmann, Court Jew of Halberstadt. You can research various archives and Jewish encyclopedias such as the Encyclopedia Judaica, for biographical information. A nephew of Oppenheimer was Samson Wertheimer, said to be the wealthiest Jew of his 16th century era and Court Jew of Vienna. You might also look elsewhere at the genealogy of Chaim Muni Tucker (Hermann Tucker) born in 1880 whose mother was Belle Weitzmann, born in 1861. The genetics of that family carried a string of genetic or inherited "talent" to play music by ear at first hearing. What about the genealogy of famous cantors or rabbis?

Where are the descendants today? What if their descendant's DNA matches yours? After three hundred years, you could have a yichus, a pedigree. Would it matter or be important to you in current times considering the impact of World War Two on Jews in German-speaking nations? Descendants carried the title "Edler von Wertheimstein." Another family traditionist and oral historian was Belle Weitzmann, born in 1861 in Poland.

According to Dr. Rosenstein's article and various Encyclopedias of Jewish history and biography, if you track the female side by mtDNA, Leffmann Behrens' wife was the sister of Haim Hameln. His wife was a famous author of the book, Memoirs, written by Glueckel of Hameln. Her sister married the Court Jew Mordecai Gumpel of Cleves. He also was purveyor of Brandenburg and his son was Elias Cleve (Gomperz) who founded a banking house. That banking house became the largest in Prussia. So the children of court Jews married other children of court Jews. Is there a cultural component that becomes genetic when looking at the history of banking houses in Europe?

Ask yourself, when did the practice of court Jews marrying other court Jews stop, at what time in history and in what land? Why? Now you have a map of the cultural and biological components of not only Jewish history, genealogy, and biography, but the routes taken by some genetic markers. Records in churches also parallel non-Jewish genealogy. Where the written records stop, the genetic-based DNA-driven genealogy picks up.

Other court Jews had last names such as Margolis, Jaffe and Schlesinger, one was military purveyor to the Austrian Court in Vienna. Then came the 1670 expulsion. Margolis-Jaffe-Schlesinger was the great grandson of R. Mordecai Jaffe, author of the book called the Levush. You look to their descendants in the 18th century and come up with R. Akiva Eger (1761–1837). He was famous for hisYeshiva. His son-in-law was Moses Sofer. Was something else in the DNA? Interestingly, he was the ancestor of dozens of scholars, prominent scientists, and creative writers.

You get a lot of writers related to these people. If you look at the DNA together with the history and genealogy you get the 17th century Issachar Berish (Behrend Levi) Court Jew of Brandenburg, son of R. Levi Joshua, a writer. There is the famous author of Pnei Aryey. He was the son of R. Jacob Joshua, author of Pnei Yehoshua. His son R. Samuel Berenstein in the 18th and early 19th century preached sermons in Dutch in Amsterdam's Ashkenazic community. So it wasn't only the Sephardim in Amsterdam that had a history of family names prior to the 18th century.

You can read a lot more about these prominent individuals in Dr. Neil Rosenstein's article, "Ashkenazic Rabbinical Families," in the Rabbinic Genealogy Special Interest Group, Online Journal at the Web site: http://www.jewishgen.org/

Rabbinic/journal/ashkenazic.htm. Of special interest are the genealogical chart and the section of the article on the influence of book printing. Surnames and names of individuals are in the article in the online journal at the Web site. I highly recommend this outstanding article and site. The footnotes and bibliography introduce you to the biographies of several key persons in Ashkenazic history. Most people aren't familiar with the Jewish aristocracy that arose in the Ashkenazic communities of Europe from medieval times and beyond.

In the article's section on book printing, you realize the profound influence it had since the 15th century on scholarship in the Jewish communities. In other resources on Jewish genealogy and history you see patterns linking book printing or scroll copying, family history, and genetics or DNA. That bridge is oral history, diaries, and family history journals. When families talk about "the curse" from generation to generation, they may be referring to an inheritable disease or genetic event. For example, one Jewish family wrote of the "curse" on their family passed from father to son which turned out to be diabetes and glaucoma over generations. Tracing Jewish DNA is like investigating a mystery with many clues that may be found in records of journals and books from the time book printing began to influence the world of Jewish scholarship, writing, and study. The rabbinic world was where writing was taking place in the past centuries. Families tended to form particular cliques or groups within some Jewish communities to ensure certain positions for their heirs.

So tracing the DNA of your most recent common ancestor 250 years ago could point to something fascinating in your background possibly linked to involvement of Jews with book printing in Europe or elsewhere. Before book printing there was scroll copying with ink and quiver. You have to go back further to look at DNA and Jewish migrations, back to the 14th century. The years 1347–1350 meant huge migrations from West to Eastern lands for Jews. It also coincided with the bubonic plague. Jews were expelled because the people didn't know the plague came from fleas piggybacking on rats. Many blamed Jews for poisoning the wells. So large migrations or expansions of Jews coming out of Germany, Hungary, Lithuania, and Central Europe moved east into Poland and parts of Russia.

About one third of the Jewish population was said to have succumbed to the bubonic plague. There were also massacres, problems with the crusaders, and most communities were on the move. Jews from the West went East and those from the east went northeast. Nobody was going West from Russia or Poland back into Germany between the 11th century and the 15th, at least by the looks of the arrows on the migration map of Jewish waves of migration, and these were only the main waves of migration to cities with large Jewish communities.

So we know that the mtDNA of Ashkenazic women is more like the host country population as far as diversity. However, we can't assume it's because of admixture because the bottleneck in the past that occurred from possibly extinctions could have created the situation of a lesser number of mtDNA types in the founding mothers of the Ashkenazi European Jewish communities. What we can assume right now until other evidence comes forth is that the communities were made up of a mosaic of different communities of Jews from different geographic areas that eventually merged somewhat.

We can't assume that just because Sephardic and Ashkenazic Jews spoke different languages and had different prayer rituals that they didn't marry. They did. Sephardic Jews settled in Germany, Holland, and other European countries. Some married only with other Sephardim and a few married with Ashkenazim. After the end of the Spanish Inquisition, many Sephardic Jews migrated to Livorno, Italy, and from there, sailed to Aleppo, Syria where two groups emerged, the Signoreem from Spain and Italy and the Mustarabim, living in Syria for thousands of years. Ashkenazim moved to Turkey, Bulgaria, Serbia, Romania as well as Polish Jews moved in the 18th century to Romania where both Ashkenazim and Sephardim lived in the same country. The extent to which they communicated with one another varies from country to country.

Even in medieval times, Sephardic travelers from Spain visited Russia, and the Balkans were under the Ottoman Turks for hundreds of years, then under the Russians. Jews moved around. Radhanites peddled their wares in the Middle East and in Poland in medieval times.

You have groups such as the Bene Israel of India whose mtDNA is similar to Hindu people of India. Yet the men they married could have come from the persecutions of Epiphanus in 175 BCE. The Bene Israel have a local tradition that claims they are descendants of refugees from Antoichus Epiphanus. Over two thousand years, they might have taken local wives from that area of India where they live.

You'd have to test the Y chromosomes to see whether the males are linked to ancient Israelis or at least link to other Jewish communities. The Beta Israel (Ethiopian Jews) show mtDNA related to other non-Jewish Ethiopians. Again, you'd have to test their husbands or fathers to see how they relate in the scheme of genes. The point is anyone can become Jewish at any point in time. Jews come in all colors and mtDNA haplogroups or Ychromosomes. What you're looking for is heterogeneity when you look at Jewish genes. You're looking for diversity, variability in the various communities. Ethopian Jewish tradition says they are descended from Jewish noblemen who came with Menelik on his journey from Israel/Judea in the early years of the first millennium BCE.

So far few if any have compared ancient Israelite bones to any particular group of Jews living today, but Iraqi Jews are said to be closest to ancient Israelites in both mtDNA and Y chromosomes. There are Jews in Persia who are a branch of Iraqi Jews. Then you have Persian Jews who joined the community of Bukharan Jews in the third to seventh century of the Common Era. Jews of Central Asia, Persian Jews, and Iraqi Jews all have slightly different mtDNA sequences. Georgian Jews have mtDNA types that show they may have come from very few local women. For example, a Jewish community thrived in Tbilisi in the 4th century CE. What's missing from articles in English in various genetic journals are studies that compare the mtDNA or Y chromosomes in the bones or teeth from Khazari burial sites in Russia or Daghestan with Ashkeanzi Jews, with modern non-Jews today in various communities nearby what was once Khazaria, and with Jews from non-Ashkenazi communities.

Since so much literature abounds trying to related millions of Polish and Russian Jews to descendants of Khazar warriors, a Turkic tribe from Central Asia whose homeland actually was in what today is Daghestan—and also those lands where the Volga meets the Caspian. Cossacks presently occupy the lands that were in medieval times inhabited by Khazars (Khazari, Kosarin). Operas have been written about Khazar princes in love with Russian maidens. There is a Viking-like aura about the Khazar warrior whose nobles, perhaps about 4000 convert to Judaism in early medieval times. The Khazar empire ruled Slavic peoples who may or may have not converted to Judaism. Today, the language of Chuvashia is most closely related to the medieval Khazari language. Chuvashia is a country of mainly Christian peoples (Eastern Orthodox) called the Chuvash, a Turkic-derived Christian population.

Since Ashkenazim speak Yiddish, looking to the roots of Yiddish we find Sorbian (not Serbian) also known as Lusatian and called Wendish by the Germans (their word for Slavs) at the root of 13th century-derived Yiddish. The Sorbs lived in Prussia and in the areas in Eastern Germany near Poland and in Western Poland. In Yiddish, German words quickly replace the Slavic words spoken by Sorbs. They are the most Western and most little known Slavic peoples living in a country called Lusatia. According to the Columbia Electronic Encyclopedia, at http://www.factmonster.com/ce6/world/A0830646.html Sandy and forested or hilly and agricultural Lusatia is located in a region of East Germany and Southwest Poland. Jews lived in Germany and Poland, but did they live among the Sorbs?

The question for geneticists and historians is whether the Ashkenazim lived in medieval times in small communities in the vicinity of Lusatia, land of the Western Slavic Sorbs? The area in question extends north from the Lusatian Mts., at the Czech border, and west from the Oder River. Did Jews live in the sandy northern or hilly, agricultural southern Lusatia?

The particular area in question is where the Lusatian Neisse separates E Germany and SW Poland. The Sorbs were farmers and foresters, mostly raising livestock. Did the Jews come there to work the lignite mines, textile mills, and glass-making factories in medieval times? The towns in question would be Bautzen, Cottbus, Görlitz, Zagan, and Zittau—the primary Sorb towns. The Lusatians are descended from the Slavic Wends. Wendish is still spoken in the Spree Forest. Traditional Wendish traditional dress and customs are preserved. The word Wendish is what the Germans called the medieval Sorbs.

The Ashkenazi communities left Southern Italy in the 4th century and by the eight century migrated to France and the Rhineland. Did they take Sorb/Wend brides? Did the Sorb men convert to Judaism? Was it the Sorb men, not the Khazars who mixed more into the Ashkenazic population? Or is all this legend? By the 10th century, German and some Polish Jewish communities were becoming known for their rabbinical scholarship. We know Jewish communities existed in the Rhineland and France by names such as Speyer/Shapiro, Luria for the Loire valley, and other Ashkenazic rabbinic families in the areas. When the Jews may have first met the Sorbs, the Sorbs were not yet Christianized by the Germans. Did the Jews taken pagan brides and convert them? Or did the pagan men convert to Judaism before the Germans colonized the areas of these Western Slavs and converted the remaining pagan inhabitants to Christianity?

The Sorb/Lusatian/Wend region was colonized by the Germans beginning in the 10th century. German rules constituted the area into the margraviates of Upper and Lower Lusatia. Both margraviates changed hands frequently among Saxony, Bohemia, and Brandenburg. In 1346 several towns of the region formed the Lusatian League and preserved considerable independence. By 1349, Jews migrated in large numbers eastward to found the shtetls and villages of their communities in Poland.

Back in Lusatia, Under the Treaty of Prague (1635) all of Lusatia passed to Saxony. The Congress of Vienna awarded (1815) Lower Lusatia and a large part of Upper Lusatia to Prussia. After World War II the Lusatian Wends (or Sorbs, as they are also called) sought unsuccessfully to obtain national recognition. Since

Sorbs spoke a language somehow related to Yiddish, are the two peoples linked genetically? See the article, "The Origins Of Ashkenazic Levites: Many Ashkenazic Levites Probably Have A Paternal Descent From East Europeans Or West Asians" Bradman, N1, Rosengarten, D and Skorecki, K21, The Centre for Genetic Anthropology, Departments of Biology and Anthropology, University College London, London, UK. And 2 Bruce Rappaport Faculty of Medicine and Research Institute, Technion, Haifa 31096, Israel and Rambam Medical Center, Haifa 31096, Israel. (See the Bradman Index at http://dna6.com/abstracts/bradman.htm which is online as part of an index to a list of abstracts and posters at http://dna6.com/abstracts/index.htm also see at: http://dna6.com/abstracts/filon.htm the article, "An Unexpectedly High Frequency Of (-Thalassemia Carriers In Ashkenazi Jews," Filon, D, Jackson, N, Oron-Karni, V, Oppenheim, A and Rund, D. The Hebrew University and Hadassah Hospital, Jerusalem, Israel.

In the article, frequencies of high resolution Y chromosome haplotypes characterised by 11 biallelic polymorphisms and six microsatellites were analysed within the context of their genealogical relationships to investigate the paternal origins of the Ashkenazic Levites. When comparing Sephardic and Ashkenazic Levite datasets, the results of the study suggests that, "unlike the Jewish priesthood (Cohanim), there is no evidence for a shared patrilineal origin for this caste across Jewish communities."

So, are Ashkenazi Levites from Eastern Europe really sons of Sorbs? "In the research, Levite haplotype distributions were compared with distributions in Israelite Jews and candidate source populations (north Germans and two groups of Slavonic language speakers). The Ashkenazic Levites were most similar to the Sorbians, the most westerly Slavonic speaking group," the article reports. Linguists studying the grammar of Sorbian and of Yiddish, the language of the Ashkenazim propose that on grammatical grounds, for Sorbian to have contributed to Yiddish—the language of Ashkenazic Jews. According to the article, "Comparisons of the Ashkenazic Levite dataset with the other groups studied suggest that Y chromosome haplotypes, present at high frequency in Ashkenazic Levites, are most likely to have an east European or west Asian origin and not to have originated in the Middle East."

So just when we think, well, the Levites of Eastern Europe have Y chromosomes that show male (paternal) ancestry down through the millenniums, coming from the Slavic Sorbs of East Germany and Western Poland, when did this occur? How far back in early medieval times did they convert? Was it before the

Germans came to the Sorb lands to convert the pagan Slavs to Christianity? Then where geographically did the Jewish mothers originate and in what century?

In non-Ashkenazic communities, there were few female founders of Jewish communities. We know there was a bottleneck in mtDNA in the Ashkenazic communities. Yet, there is somewhat more diversity in the Ashkenazic communities in mtDNA than in the non-Ashkenazic. Only not too much diversity with 27% to 32% of Ashkenazic mtDNA being K mtDNA haplogroup and 9.0% being H haplogroup and following the CRS.

Could the fewer mtDNA haplogroups be due to extinctions? For non-Ashkenazic communities, there were at least eight founders. How many female founders were there in Ashkenazic communities? These are questions only research will answer as more unpublished information and research becomes available.

Many genetic research studies find that they negate a hypothesis rather than prove it. The article reads, "Out of 151 samples only one was found to carry the (anti3.7 allele, negating the hypothesis; b. founder effect and genetic drift; c. selective advantage against P. vivax malaria; d. non-malarial selection to a linked factor. As there are a number of developmentally important genes linked to the (-globin cluster, the last explanation is an attractive possibility."

Basically, geneticists look for attractive possibilities when they study genes. The scientific method usually starts with a hypothesis, and the research tries to find out whether the hypothesis is negated or whether an explanation is an attractive possibility. Can genes prove anything? Only that there are results in print in the scientific journals waiting until the next researcher brings in new information. What can we know about genes?

If you read the book, Genetic Diversity Among Jews, you'll find chapters on the major demographic trends of world Jewry, on types of mtDNA among Jews, but the book has a subtitle: "Diseases and Markers at the DNA Level." So I highly recommend this book on any books that come after it on the subject of discussing the genetic details of inheritable diseases such as the chapter, "Tay-Sachs Disease Mutations Among Moroccan Jews." It's not just an Ashkenazi mutation. Different mutations for Tay-Sachs show up in Moroccan Jews and still other mutations in the French Canadians. For example, more than one mutation underlies TSD in Moroccan Jews.

You have a population of approximately 1.5 million Ashkenazim compared to about 500,000 Moroccan Jews living in Israel. Intermarriage between Ashkenazim and Moroccan Jews is common in Israel. The studies on the book were done on people living in Israel. In Ashkenazim three mutations account for nearly al the genes that cause TSD. There's also a milder variant that strikes adults, a type of gagliosidosis. Moroccan Jews have another TSD mutation. The Moroccan Jews came from a variety of cities—Casabalanca, Marrakech, Rabat, and Res. Geneticists would have to trace their prior origins back and to relate their mutation with the geography and history of the people in their ancestral towns. It's difficult to do because the carriers come from large cities and populations that have intermarried for centuries.

So there are many other chapters such as "Genetic Complexities of Inflammataory Bowel Disease and Its Distribution among the Jewish People." Each of the chapters are by different authors. So the book is a wealth of information about not only the history and geography of Jewish genes, but mostly about the distribution of disease among Jewish populations from all over the world, but with the populations of this mosaic now living in Israel and studied there.

You find conclusions such as the less common Tay Sachs mutation could not have been kept in the Ashkenazi population by solely drift. So you may have to infer some advantage for carriers of the Tay-Sachs mutation. What was this advantage? The field of molecular genetics is studying this. If you consider a career in this area, you would be thinking about admixture rates, population size, and a variety of selective mechanisms. The gene you pinpoint, the hetertozygote advantage is part of ongoing science in looking at the whole genome. So as geneticists study these details, genealogists look at oral history and genealogy in a new way that combines DNA testing with looking at how diseases or mutations for disease are inherited. On the other hand, DNA testing can help you match your DNA to your most recent common ancestor.

If you have your DNA tested and find a match, perhaps this person shared a common ancestor with you, either male or female about 250 years ago, more or less. That's why databases used in genealogy and for DNA analysis find a common meeting point between those trained in science and those trained in family history.

So what is the conclusion here? We are a mosaic of different communities that have eventually merged. Let's take a look at the Middle Eastern Jews. Since Iraqi Jews are supposed to be the closest you can get genetically to the ancient Israelites, they have been living in Iraq since the destruction of the first temple in 586 BCE. Since late antiquity, you have Jews living in Morocco. Not all of them came there from Spain in the 15th century. Jews from Yemen dwelled in the ancient Jewish kingdom of Himyar which is in Yemen. There were several Jewish kingdoms in ancient and medieval times—countries ruled by a Jewish king or queen. Examples include Himyar in Yemen, a Jewish Queen in northern Iraq, and Khazaria, where the Volga meets the Caspian. There were Jewish kingdoms in the Arabian Peninsula.

See the Jewish Kingdoms of Arabia (from 4th to 7th centuries) Web site at http://www.eretzyisroel.org/~jkatz/arabia.html. The second paragraph in the article "The Jewish Kingdoms of Arabia 390–626 CE" at the Web site reads: "According to Moslem tradition, conversion to Judaism started under Abu Karib Asad (ruled 390–420), who became a Jew himself and propagated his new faith among his subjects. Arabic sources expressly state that Judaism became widely spread among Bedouin tribes of Southern Arabia and that Jewish converts also found with the Hamdan, a North Yemenite tribe. This time, many of the upper strata of society embraced the Jewish faith. The position of Judism in Yemen reached its zenith under DHu NuwAs."

Interestingly, the Habban (city in Yemen) Jewish community seems to have come from one single maternal origin. That means a single female actually appears in the study to be the founder of the Habban Jewish community. You can view photos of the town of Habban in Yemen at the Web site at http://www.alovelyworld.com/webyemen/htmgb/habban.htm. When the Jews of Habban left Yemen, they carried with them their special skills in working silver which had made the town famous for its silver crafts. Most of the Habbanite Jews settled in Israel.

Referring back to the book by Batsheva Bonné-Temir, et al. 1992 book, Genetic Diversity Among Jews, chapter four titled, "Types of Mitochondrial DNA Among Jews," what the mtDNA study found focused on diversity. While there was no significant difference between the heterogeneity—the variability—in the Moroccan Jewish community and those from the Central and East European (Ashkenazi), Iraqi, and Yemenite Jewish communities, (that is the diversity patterns were similar), there were significant differences between the Ashkenazim and the Iraqi Jews. And the differences between the Ashkenazim and the Yemenite Jews were statistically significant also. Read the chapter on "Nonisulin Dependent (Type II) Diabetes Mellitus in Jews."

I've read in popular magazines that in medieval times in Central and North Eastern Europe, Jews got diabetes and mental problems, and Christians got tuberculosis. You didn't find too many Jews with TB. Was it the custom of washing the hands three times before eating? Tuberculosis bacteria was often found in the soil in medieval times, and not too many Jews were farmers. Most were banished to small apartments in Jewish ghettos where they did close work such as copying books, minting coins, money changing, or sewing and diamond cutting.

Think about type II diabetes—possibly caused by lack of exercise and eating too many carbohydrates? Or mental illness—caused by oppression and living in tiny ghetto apartments instead of on spacious farms? When the plague broke out in 1347, Jews in the Rhineland were blamed and had to leave Germany. By 1349 they were streaming into Poland to join the small earlier communities founded by Jewish Rhadanite merchants from Persia and Byzantine Jews who came from the East at a time when Aramaic was still spoken by Jews in Baghdad. This was at a time when the Yiddish language seemed to have developed out of Sorbian, but the Sorbians had no written language, and the Jews had Hebrew. Did the Sorbs, fleeing Germanization join up with the Jews? Only the genes will tell. Now you have DNA tests of Y chromosomes on Ashkenazic Levite males showing some of them have Y-chromosomes close to what the modern Sorbs have. What does this all mean? A richer tapestry, a mosaic quilt of communities merging?

How do the Ashkenazim relate to the Iraqi Jewish community? Prior studies related Ashkenazic males to Iraqi Jewish males. Looking at the mtDNA of the women of Ashkenazic and Iraqi Jewish communities, you find a lot of U3 mtDNA, about 17 percent in the Iraqi Jewish mtDNA—the matrilineal lines. And you find about 27 percent to 32 percent of K mtDNA in the Ashkenazi Jewish female—matrilienal lineages.

In the study, "Types of Mitochondrial DNA Among Jews," Additionally, there were significant differences between the Iraqi Jewish community and the Yemenite Jewish community. You need to keep in mind you're looking at a sample that tested 75 Ashkenazim, 8 Turkish Jews, 26 Iraqi Jews, 65 Yemenite Jews, 17 Habbanite Jews, 22 Moroccan Jews, and 55 Ethiopian Jews. That's 268 people in the sample. The mtDNA is what was studied for diversity—heterogeneity—variability in that small sample. So you have to research how significant a small sample of 268 is before you reach a conclusion that involves a whole population.

Interestingly, past studies of Y chromosomes found only in men, showed much more similarity between the various Jewish groups. Look at the article "High-Resolution Y Chromosome Haplotypes of Israeli and Palestinian Arabs Reveal Geographic Substructure and Substantial Overlap with Haplotypes of Jews," by Almut Nebel, Ariella Oppenheim, Mark G. Thomas, et al, appearing in the publication Human Genetics (2000) 107:630–641. Y-chromosome variation in the Israeli and Palestinian Arabs was compared to the same in Ashkenazi and Sephardic Jews. Then all of these Y chromosomes were compared to the Y-chromosomes of North Welsh males.

The findings: The Y chromosome distribution in the Arabs and Jews was similar—but not identical. Then the study went further to look at the haplotype level that is determined by certain markers. Haplotypes of Y chromosomes are determined by what scientists call "binary" and "microsatellite" markers. These genetic markers showed a pattern in finer detail. So when the scientists looked deeper to reveal more detailed markers, what they found was that there is a common pool for a huge portion of the Jewish and Arab Y chromosomes found only in males. This common pool suggests a recent common ancestry. By recent we are talking about several thousand years.

What the researchers saw was that the two "modal haplotypes" in the Israeli and Palestinian Arabs were closely related to the most frequent haplotype of Jews which is called the Cohen Modal Haplotype or CMH. The only problem is that not all Jews have the Cohen Modal Haplotype. And you don't have to be named Cohen to have one. Neither do you have to be a Cohen, a direct descendant paternally of the ancient Hebrew priests of the temple to have the CMH genetic marker. Yet the CMH is said to be the genetic marker of the ancient Hebrew people as it originated around 3,000 years ago at a time when Moses's brother, Aaron become the first Cohen or high priest of Israel. The marker has a common ancestor in the Middle East that relates Jews to Arabs and other middle eastern peoples dating back to a paternal ancestor or group of ancestors who lived around 7,800 years ago.

In the article, however, the study found that the Israeli and Palestinian Arab clade that includes the two Arab modal haplotypes that makes up 32 percent of Arab chromosomes isn't found much among Jews. In fact it's found at a low frequency among Jews. So there is divergence or admixture from other communities in which the Jews lived, but how much is the question? Nevertheless, other studies show that Arabs and Jews share about 18 percent of there gene markers but not too much of the distinctive Arab clade shows up in Jews.

On the other hand, the Cohen Model Haplotype and another modal haplotype in the Israeli and Palestinian Arabs were closely related to the Cohen Modal Haplotype in the Jews. Will scientists be able to go further and show that a sizable number of Palestinian Arabs once were Jews in ancient times? We already know some Jews once were Middle Easterners before Judaism existed. Again, we see how closely people were related in ancient times and how diversity began to spread, even slightly, when people migrated to other communities.

In the article, it reads, "Based on two Y chromosome RFLP markers it was suggested that Ashkenazi and Sephardic Jews are more closely related to Arabs from Lebanon than to Czechosovakinas (Santachiara-Benerecette et al. 1993.)"

Also, the article reveals that "a recent survey of 18 binary Y-specific polymorphisms showed that Y chromosome haplotypes of Middle Eastern non-Jewish populations are almost indistinguishable from those of Jews (Hammer et al. 2000)."

The study also found a significant difference between Sephardic Jews and Arabs due to a higher frequency of haplogroup 2 in the Y chromosome of Sephardic Jews. Yet there was no significant differences found between Arabs and Ashkenazi Jews nor between the communities of Sephardic and Ashkenazic Jews. Yet Arabs, Ashkenazi Jews, and Sephardic Jews all were distinguished from the Welsh samples.

The major portion of Y chromosomes of Arabs and Jews belonged to haplogroup 1. Most of the Welsh Y chromosomes belonged to haplogroup 2. Yet the frequency of haplogroup 2 chromosomes in Arabs was lower than that in Sephardic Jews and slightly, but not significantly lower than that in Ashkenazim. However, Ashkenazim were closer to the Welsh than were Ashkenazim to Arabs. Sephardim were closer to Arabs than to Welsh.

So what does this say for Ashkenazi men? Does it say they have northern European admixture or genetic drift? You can't make this assumption because two Ashkenazi individuals carried the most common Welsh haplotype, and that finding can't be based on assumptions because if two men carry the most common Welsh haplotype, it's also the most common haplotype for most other European countries such as England and Norway or Frysland, all in northern Europe.

In the study three Arab individuals from a variety of areas carried haplotype 27 which is the Cohen Modal Haplotype, the signature of the ancient Hebrew people and also found in some Arabs. None of the Welsh in the sample carried the CMH. Neither of two modal haplotypes found in Arabs showed up in the Jewish sample.

When the Romans conquered Palestine and built their villas and cities there, you get gene inflow from the Romans and some Greeks. Interestingly, in ancient times numbers of Palestinians and Jews became Christian by the fifth century of the Common Era. (Bachi 1974). Then in the first millennium of the Common Era, immigration of various Arab tribes into Palestine increased with a huge wave coming in during the seventh century CE. Some of the Christianized Jews in Palestine along with the Christian Arabs converted to Islam at that time. (Saban 1971; McGraw Donner 1981).

Then you get more people moving into Palestine in later times such as Turks and Crusaders. You get recent migrations and gene flow from Europe. So Palestinian Arabs vary genetically as well as the Jews. What Palestinians share with Jews are linguistic backgrounds as Hebrew and Arabic share common origins, and geographic origins at their roots.

Researchers are studying their genetic relationships. Okay, there are genetic affinities between Arabs and Jews. Look at articles studying mtDNA (Bonné-Temir et al 1986; Ritte et al. 1993) Bishara et al 1997. There are affinities and there are differences. The point is Ashkenazi and Sephardic Jews are more closely related to Arabs from Lebanon than to Czechs and Y chromosome hapltoytpes of Middle Eastern non-Jews populations are almost indistinguishable from those of Jews (Hammer et al. 2000). But what about the founding mothers of the Jewish communities? What about Ashkenazic women? Are they Middle Eastern? Slavic? Germanic? Or? The answer is they form a mosaic of different communities formed from a bottleneck in ancient and medieval times.

What is known is that numerous people converted to Judaism at various times in ancient and medieval periods and in different countries varying from Southern Italy and Greece to Northern, Central and Eastern Europe. On the other hand, some people converted from Judaism to Christianity in ancient and medieval times to escape oppression and now exist in the Europe population with those same ancestral Y chromosomes or mtDNA.

Now the question remains, did more people convert to Judaism or convert to other religions in historic times? Whatever the answer is, it will show up when you compare gene markers of Jews and non-Jews in any European population or in the Middle East, considering back-migration from Europe to the Middle East taking place from prehistoric times to the present. During and right after the end of the Ice Age people in several waves back-migrated to the Middle East from Europe, which shows about 15 percent of the Middle Eastern peoples consist of individuals who migrated there from Paleolithic Europe.

Mutations are any heritable change in a DNA sequence. Polymorphisms are differences in DNA sequences among individuals that may underlie differences in ancestry or differences in health. Genetic variations occurring in more than one percent of a population would be considered useful polymorphisms for genetic linkage analysis. To study mutations or genetic variations in DNA sequences, scientists put your DNA through a polymerase chain reaction (PCR) test. It's a method for amplifying a DNA base sequence. The researchers usually use a hate-stable polymerase and atwo 20-base primers, one complementary to the (+) strand at one end of the sequence to be amplified and one complementary to the (-) strand at the opposite end.

So you are copying DNA. You're synthesizing DNA strands. These newsly synthesized DNA strands will serve as additional templates for the same primer sequences. You go through successive rounds of primer annealing and do things like strand elongation and dissociation. What's produced are rapid and highly specific amplification of your desired sequence of DNA. Then you can use your PCR that is your polymerase chain reaction test to detect the existence of the defined sequence in your DNA sample. Okay, now you understand what happens when DNA is tested to find certain sequences.

Polymerase DNA or RNA also refers to an enzyme that catalyzes or brings together the synthesis of nucleic acids on pre-existing nucleic acid templates. You assemble DNA from deoxyribonnucleotides and you assemble RNA from ribonucleotides. So that's how it all comes together to see the human genome or part of it.

Numerous Jewish men, including Ashkenazim are genetically indistinguishable from Arab Palestinians, and some share the Cohen Modal Haplotype (CMH). This is a subset of an insular group such as Jews, Finns, Sardinians, and Basques. The DNA tests shows up the genetic pedigrees going back around 3,000 years to the Exodus from Egypt. Europeans in general are mixed and intermarried for thousands of years so that's it much more difficult to tell a Swede from a Russian or any one country from another in Europe. On the other hand, a study of the DNA of the Ashkenazi Male Levites revealed that some of then had Y chromosome DNA markers showing descent at least paternally, from East Europeans or West Asians.

So are Levite Ashkenazi males the only Jewish males in Eastern Europe to possibly be descended from fathers who were Sorbs? What about non-Levite males? We know a percentage of the Cohens really do inherit the Cohen Modal Haplotype (CMH) that links them to the ancient Hebrews and also before there was organized religion to ancient Middle Easterns and to some of the Arabs. Let's take a look at what happened in the land of the Sorbs when the Jews might have lived there and picked up Yiddish from the Sorbs at a time when the Germans were trying to get Sorbs to replace their Western Slavic language with German words, words that might have shown up in Yiddish when the Jews moved to more Eastern towns in Poland.

One of Charlemagne's sons defeated the Wends and burned Bautzen in 806. Wends/Sorbs were pagan, and the Germans were moving in their territory to Christianize them. Also the area geographically had fertile farmland, forests, sand, hills and….potential brides. In the West, the Vikings were on the move also looking for trade and new lands for farming. They sailed their long boats down the rivers to Kiev in the East, eventually to Iceland in the West. What migrations did the Jewish communities at the time begin? You have Jewish scholars and traders from Spain visiting Russia, Syrians in the land of the Norsemen as ambassadors, and Ashkenazic scholar rabbis in France and Germany speaking old French, old German, and writing in Hebrew (or Arabic in Spain and the Middle East, Judeo-Persian in the Caucasus, Persia and Central Asia).

That's the same time that Khazaria reached its golden age and the nobles converted to Judaism much further East where the Volga meets the Caspian in Khazaria. They were trading with Kiev in the Ukraine by then and had a vast empire stretching from the Black Sea across the Caucasus to the Caspian. They inhabited parts of Georgia and Armenia, trading with the medieval Jewish population in Armenia. At the other end of Eastern Europe, here you have the defeated Sorbs. Charlemagne was Christian. The defeated Sorbs/Wends might have converted to Judaism in the 9th century just as the Khazars converted to Judaism at least in part (their nobles and some of the Slavs ruled by Khazars) around that time also. Jews from Byzantium and Persia, Anatolia, Armenia, and Baghdad moved into Khazaria.

Then as the Ashkenazic communities in the Rhineland were blooming, by the year 1100 the Wends had been subjugated. Did some of them convert to Judaism because the German Christians subjected them in the 12th century? German nobles dominated the Wendish peasants and relegated the urban Wends to homes outside the walls or to restricted sections of the city. They could become active in society only through German institutions and the German language. The guilds were German. Mercantile activity was conducted in the German manner. Under pressure, especially in the part of Lusatia under Prussian control, many Wends adopted German names and relinquished their Slavic traditions. Only a funny thing happened.

A lot of these German names taken by Sorbs/Wends (Wend is a German word for Slav) sound Yiddish....such as Lavine/Levine. Does Levine come from the word for Levite? (Levi, Lion, Loew, Loeb, Lewin, Lewinsky, Levin, Levinson, Levenstein)? You find the name "Levine" showing up in Austria in a university in 1598. It's listed in genealogy records as being a Christian name with a Prussian origin in 1635. Look up Catholic baptismal records for the name "Levine" in Prussia in the 17th century. It abounds. Why did so many Jews of Bialystock, Poland and other areas in the Pale of Settlement of Eastern Europe or in Germany choose the name Levine? Were they Levites? Why did they pick this Christian, Prussian name? Or is it merely a way of "Germanizing" Levi, according to rules in the 18th century when Jews were ordered to purchase or assume German or Slavic names in their respective countries? What happened to their secret Hebrew names? Did they remain only in the synagogue?

The Christianization of the Wends began prior to the German conquest, but it was vigorously promoted by the Germans. They also followed the Germans in the Reformation. The majority of Sorbs/Wends converted to Lutheranism in 1530 after the Council of Augsburg.

In medieval times, the pagan Sorbs/Wends did not yet have a written language. If they contributed to Yiddish as the linguists propose, the Jews who began to speak Yiddish, whether new Jews or ancient ones, used the Hebrew alphabet. In fact it wasn't until the time of Martin Luther in the 16th century that Luther emphasized using the vernacular.

The reformation and Luther encouraged the Wends to devise a written language, and in 1574 Luther's Small Catechism became the first work to be published in it. Perhaps those Wends/Sorbs/Lusatians who became Jews at a time in the medieval past before they became Christian and before they had a written language, were so happy to assume the Hebrew alphabet and to speak Yiddish, related to the medieval Sorb language in many ways.

What they have in common also is that the Germans required the Sorbs to replace what once were their Slavic words with German words—the Germanic of medieval times. Also, Yiddish has many medieval Germanic words, some Slavic ones, and Hebrew. It's almost as if the Slavic words in Yiddish had been replaced over time with German words.

That's what happened also to the Sorb language over time as Germanization of this Western Slavic community replaced some of the words. Linguists suggest that the Sorb language had some influence on Yiddish that's attributed to 13th century German. Yet the Slavic words in Yiddish shows an eastern influence. This now becomes a melting pot, a salad bowl of language so to speak. You get some words from Hebrew, some from Slavic, some from German, and one or two from ancient Egyptian and perhaps from Turkic, perhaps Khazari word for prayer similar in sound to "davenin"…for praying. yarmulke….Or perhaps yarmulke for the Hebrew "kippot" comes from the Ukrainian feast holiday called Yarmaka?

Without DNA evidence of Khazars compared to ancient Israelites and to modern Ashkenazim compared to Slavs compared to Germanic peoples to see who carries whose genes in this mosaic of communities and migrations, no one can make a statement saying a whole community derives from another community because the details of history are carried in the DNA—the mtDNA and the Y-chromosomes, the tissue typing tests, and the tests of racial percentage. So you assume a whole group of people are descendants of another whole group of people—unless you can show it by intensive DNA analysis of the entire genome.

What you can say is that either the Y chromosomes of the majority of Ashkenazi males cluster with Sephardic and Middle Eastern Jewish males or they don't. And you have read the results of ongoing DNA research. And with the females, all you can do is see how the mtDNA fits into a pattern to shape migrations at least until the entire genome is known. You can trace female ancestry of Ashkenazic women from Eastern Europe until you arrive at the single female ancestor of each woman who lived 20,000 or 40,000 years ago in similar Ice Age refugiums in the Pyranees and/or in the Ukraine.

The clan mothers you find gave rise to almost all other Europeans. And if you reach out to that mother's ancestor, she will be found in Western Asia or in the Middle East or Africa before she reached Europe, except for V mtDNA haplogroup who possibly arose 13,000 years ago in the Pyranees in Europe when her haplogroup split off from H mtDNA. After the Ice Age mtDNA V migrated north to Scandinavia and also is found among the Saami peoples, and a branch went south to N. Africa, found currently among the Berbers.

So you have back migration from Europe to N. Africa. Let's consider the male Levite Ashkenazim of Eastern Europe. Could they be descended partly from Sorbs? Is it this West Slavic people living in East Germany and Western Poland who contributed to the Ashkenazic gene pool and Yiddish language? Would the admixture in Ashkenazim, if any, come perhaps from Sorbs rather than Turkic Khazars? Would Yiddish have more Sorb words than Khazari Turkic words? Would some Ashkenazim even resemble Sorbs more than other Europeans? Or would they resemble more their more ancient Eastern Mediterranean ancestors?

So far no scientist has compared ancient Israelites to modern Ashkenazim and found an unbroken chain of ancestry going back until they found the Cohen Modal Haplotype that does go back——in the males to a common ancestor in the Middle East thousands of years ago, perhaps 1,500–3,000 years ago—about the time of the exodus. And that ancestor had a common ancestor in the Middle East about 7,800 years ago from whom most people in the Middle East are descended.

Could Ashkenazic Levite males be descendants of the Sorbs/Wends/Lusatians? If the genetic analysis is right and they are close genetically to the Sorbs of today, it would show how enriched the tapestry is of the mosaic of various Jewish communities of the world all acting as a catalyst, coming together, intermarrying, and forming a quilt. Let's look at the Sorbian language.

There are two versions of Sorbian, also called Sorbic, Wendish or Lusatian, corresponding to the divisions of the Lusatian region. Both versions belong to the

Western Slavic group. The southern area called Upper Lusatia speaks a dialect nearer to Czech (Luther's Catechism was translated into Upper Sorbian); the northern area, or Lower Lusatia, a dialect nearer to Polish.

Which version is closest to Yiddish? Traditionally the Wends call themselves Srbi (Sorbee) in their own language, but the Germans call them Wenden, a term widely used both by others and by, many of the Slavic Lusatians themselves, including those who migrated to foreign lands in the 19th century.

When the Jews left the Rhineland in medieval times to move a little to the East, say in East Germany or Western Poland, Wend was the German name for all West Slavs. The German word, "Wend" eventually symbolized the feared Germanization of the Wends that began in the 9th century with the Carolingians and continued through the Weimar Republic and the Nazi period. So if you were a Srbi a Sorb (Wend) in the 9th century, and a Jewish village from the Rhineland sprung up near you, and you were resisting Germanization, forced Christianity on your Pagan beliefs, wouldn't you at least consider becoming Jewish? You'd automatically get the chance to write your language with newly learned Hebrew letters.

The alternative would be the German alphabet, but you've just been defeated by the Germans and told to become Christian, to replace your Slavic language with German words. What would you do? Comprise, and start speaking what would develop later into Yiddish? Does this possibility sound more plausible than 500,000 Khazars turning Jewish and rushing into Poland after the destruction of Khazaria in 965 CE? Make up your own mind. Chances are a little bit of each might have happened, with traces of this enriched mosaic of genes joining the towns of the Jews fleeing oppression from the Middle East or Spain.

Lusatia was ruled at times by non-German princes. Since the Peace of Prague in 1635, it has remained under German control. In the 19th century when Jews were seeking surnames, many of these surnames came from Prussia such as Lewin, Lewinsky, Levine, Levene, Levin, and Levenstein. Any book on genealogy will give Prussia as the place of origin for many names now used by Ashkenazim. Some genealogy books will give surnames as originated in both Prussia and Austria for these same names.

Before German unification in 1871 Lusatia, land of the Sorbs/Wends were contained in parts of the kingdoms of Prussia and Saxony. Lower Lusatia was under Prussian administration and Upper Lusatia under that of Saxony.

After World War II, the original land of the Sorbs was in part of the German Democratic Republic (East Germany). The Sorb area was divided administratively between the districts of Dresden and Cottbus. Wendish ethnic awareness has been encouraged under the German Democratic Republic. The Slavs of Lusatia, in their own language probably would prefer to be called Sorbs. So the term Sorb has been adopted for the Slavs of Lusatia.

Sorbs may want to reflect their Slavic heritage as West Slavs of Central European ancestry. They are not the Serbs of southern Europe who are another Slavic people living in Serbia. And Sorbs have never lived in Serbia, which is near Macedonia. Sorbs live in Eastern Germany and Western Poland. In current times, nearly 60,000 people in Lusatia call themselves Sorbs. For more information, contact the Sorbian cultural center (the Domowina). Sorbs have a Sorbian-language newspaper, radio station, theater, and folklore tradition.

So what if you are an Ashkenazic Levite whose Y chromosome reveals your ancestry is close to the modern Sorbs? Do you invite a Sorb to your home to share family photos? Do you find out genetically whether the rest of your genes match by markers? Or do you take a DNA test to find out your most recent common ancestor or any medieval ancestor linking you to that Sorb you've invited for dinner exchange of family photos? How important is your DNA test result in your core identity and in your life? There are a lot of travels to ponder.

That's why the historic origin of the Ashkenazic communities is debatable. You have to let the genes tell what they can. You also have to look at linkage disequilibrium (LD). Kruglyak (1999) reported that populations don't keep their LD over long genetic distances unless there have been sharp founder effects called bottlenecks.

Posters sometimes go up at genetics conferences illustrating the bottleneck in Ashkenazic mtDNA—female lineages. And the bottlenecks usually are in the recent past. (Reich et al. (2001).

You look at the last Ice Age (LGM) or you go further back to the out-of Africa migration into the Middle East or West Asia to see bottlenecks in huge populations. The bottlenecks can be specific to male or female lineages. You see them in the mtDNA in matrilineal lineages or the Y-chromosome lineages in males.

If the bottle neck or founder effect is harsh enough, the extinctions severe enough, you see a change in the genes (Wright et al 1999). One example of a great reduction in mtDNA diversity is seen in Georgian Jewish females. When you get severe mtDNA reduction in a population, you also get long-range linkage disequilibrium. Only Georgian Jewish females showing the great reduction in mtDNA are not Ashkenazic females from Central and Eastern Europe, especially North Eastern Europe. Also, South Eastern non-Jews are genetically close to North Eastern non-Jews.

So you have to look at Ashkenazic women and men in their own community and see how close or not close their genes are to the host populations. You already know the males cluster with other Jews and with males in the Middle East in 70 percent to 80 percent of individuals studied. What about the origin of Jewish mothers from Ashkenazic communities? Are they related to anyone in the Middle East? Are they related to one another?

Or are they formed from independent local women in independent communities that eventually merged in medieval times? You be the judge after reading the latest genetic findings. If you want to dig further, read the latest articles. Many findings on Ashkenazic mtDNA are unpublished at the time this book is being written, so there's no way to read the articles of the very latest findings. One thing is certain:

The mosaic of communities forms a rich tapestry of having merged together. By the time of publication, the articles will probably be available to the public in the various journals of human genetics. Read them. I've learned a lot from reading them, and I highly recommend the articles and the books. For now, until any more information comes in, the mothers of the Ashkenazic communities present a mosaic of communities that have merged. As Dr. Mark Thomas wrote, "Ashkenazi Jews were made up of a mosaic of different, independently founded communities that has since homogenised."

Also, I highly recommend the audio tapes of the conferences on Jewish genealogy. Visit audioiotapes.com on the Web at http://www.audiotapes.com/conf.asp?ProductCon=77 to see the 111 available audio tapes for the 19th Annual Conference on Jewish Genealogy recorded in 1999.

Check out the Y-chromosome haplogroups at:
http://freepages.genealogy.rootsweb.com/~dgarvey/DNA/markers.htm
Check out your mtDNA at:
http://www.stats.ox.ac.uk/~macaulay/founder2000/tableA.html
http://shelob.bioanth.cam.ac.uk/mtDNA/toc.html
Also see the geographical index at:
http://shelob.bioanth.cam.ac.uk/mtDNA/geogind.html

***

4

How to Safely Tailor Your Foods, Medicines, and Cosmetics to Your Genes:

Scientists and physicians speak out in the following consumer's guidelines to genetic testing kits from ancestry to nourishment. This chapter includes resource material for the general consumer and/or beginner in DNA-driven genealogy to genetic testing kits from ancestry to nourishment.

Here's a letter from Daniel F. Hayes, MD, Clinical Director, Breast Oncology Program, University of Michigan Comprehensive Cancer Center. He is one of several physicians I asked to comment on the topic of 'pharmacogenetics'—tailoring your medicines to your genetic signature. Other physicians and scientists I've asked to speak out and explain about nutrigenomics for the consumer.

The point is consumers need to become more involved in both nutrigenomics and pharmacogentics. They need to become as involved as they are in learning to interpret their DNA tests for family history and ancestry or to make time capsules. Check out how your genes respond to skin care products or supplements and food interactions. Where do you start? As Dr. Hayes notes, pharmacogenetics is in its earliest stages. What's the consumer's role right now? Let's look at the doctor's letter on how consumers may be genotyped and analyzed for patterns in the DNA.

——-Original Message——
From: Daniel Hayes
To: Anne Hart
Sent: Sunday, August 24, 2003 1:07 PM
Subject: Pharmacogenetics.

Pharmacogenomics is in its infancy. However, I believe that in the future every potential patient may be "genotyped" by examining his or her DNA, which can easily be collected from either a simple blood specimen or a swab of the inside of your mouth. Developing technology will permit analysis of the patterns of the genes that control response to and metabolism of different drugs and medicines. These patterns differ from one person to another, according to what one inherited from one's parents, much like hair and eye color, etc.

Although it will take an enormous amount of careful and sophisticated clinical research, I believe doctors will then be able to prescribe either different drugs, or the same drugs at different doses and schedules, based on these inherited patterns. Currently we do so, but crudely, based mostly on a patient's size (kids take smaller doses than adults), or obvious organ dysfunction (people with kidney or liver problems might need less or more of an agent than those with normal function).

We and others have begun preliminary studies addressing these issues, and we hope that they will give us important insights not only into where we are now, but where our investigations should proceed in the future.

dfh

Daniel F. Hayes, M.D.
Clinical Director, Breast Oncology Program
University of Michigan Comprehensive Cancer Center
6312 Cancer Center
1500 E. Medical Center Drive
Ann Arbor MI 48109–0942

Here's how to tailor your food, medicines, and skin-care products to your DNA. This is your consumer's guide to genetic testing kits. Your DNA, including your ancient ancestry and ethnicity has a lot to do with how your body responds to food, medicine, illness, exercise, and lifestyle, but just how much? And how do you know which DNA kits and gene testing are reliable and recognized?

Learning about DNA to understand and improve your health is now interactive and available to the average consumer, not limited to students and teachers, but to anyone else. In the last few years genealogy buffs, parents, and anyone interested in DNA without a science background took an interest in DNA tests rests that reveal deep maternal and paternal ancestry.

Currently consumers with little or no science background are interested in learning about drug metabolism—pharmacogenetics. How your body metabolizes medicine is as important as how your body metabolizes food. Nutrigenomics is about how your genes respond to food and how to tailor what you eat to your DNA. Consumer DNA interest ranges from forensics and anthropology to nutrition, caregiving, family scrapbooking and healthcare knowledge. Nurses are becoming more interested in DNA.

The DNA consumer revolution began when media broadcasts revealed to the public that fast computers had revealed the human gene code. Once more TV opened doors. Suddenly, a gap between science and consumers had to be bridged by available interactive education.

A proliferation of products relating to DNA emerged. The internet shows DNA summer day camps for students and teachers. DNA testing companies and books emerged geared to the average consumer. Genealogists tried to interpret DNA for ancestry. People left other non-science-related businesses to open up DNA testing companies for ancestry research, contracting out to university research laboratories to do the DNA testing. Again, opportunities opened doors to the public.

Nutrigenomics product marketers sought those who wanted a diet tailored to their genetic signature. Pharmacogenetics reports customized medicines in order to prevent adverse drug reactions.

Pharmacogenomics studies the entire genome in relation to chemicals and drugs, whereas pharmacogenetics researches specific genes and markers to look for adverse drug reactions for individual clients or patients. Finally, DNA testing products emerged offering to tailor skin care products such as creams and cosmetics to your individual genetic signature.

If you've had an interest in learning about how to interpret your DNA test results for ancestry, you now can see the links to understanding how to tailor your food, lifestyle, exercise, medicines, supplements, and skin care products—in fact numerous environmental chemicals—to your genetic expression. It's not only about food anymore or ancestry alone, or medicine.

DNA testing also is about kits sent to you directly or to your physician. It's about tailoring to your DNA skin products, cosmetics and anything you put into or on your body that gets absorbed. It's about what chemicals are in your water and home-grown vegetables.

No science background? Don't worry. There's a DNA summer camp near you, or an educational experience in learning about DNA now available to the average consumer. Educators, scientists, and multimedia producers have teamed up to teach you the wonders of DNA, your genes and your lifestyle.

What's left? Physicians and genetic research scientists need to talk more to each other because most family doctors don't have time to read the proliferation of publications reporting new advances in genetics or other areas of science that directly affect consumers. It looks like it's the consumer's job to bring people together through the media and through consumer's watchdog organizations, professional associations, and support groups. Key words: action and public education about DNA through multimedia and consumer involvement.

I highly recommend the DNA Interactive Web site at: http://www.dnalc.org/. Consumers need to know more about how to interpret DNA test results for whatever purpose they seek—tailoring diets or drugs, skin care products, or seeking out ancient family history or ancestral lineages. The science is new enough to have many more applications on the horizon that consumers can digest in the future. What can the average consumer with no science background learn about applications of DNA testing currently? Start with the publications and the interactive DNA learning sites.

You'll soon become familiar with the DNA terminology. The goal is to bridge the gap between science and the consumer, let alone the gap between science and medicine. So you'll have to invite your busy family doctor to join you in creating consumer groups. It'll work fine. Doctors and scientists usually are found conversing together at parties.

The triangle now includes the consumer. If you have children, bring them into the fold. There are now wonderful DNA summer camps. So include your children's teachers. Learning about DNA as a consumer might make you wish you had majored in genetics. Link your beginning self-taught DNA studies to your special field of interest such as your healthcare or your ancestry.

For history buffs, there's always molecular genealogy for family history. History buffs can follow population genetics. Anthropology enthusiasts can read about archaeogenetics. Bring archaeology and DNA together.

Nutrigenomics links DNA research to nutrition for the diet-conscious. Pharmacogenetics helps you to tailor specific medicines to your genes. DNA gets into all walks of life and work. Start with population genetics and physical anthropology if you wish. I like the popular book titled: *The Real Eve*, by Stephen Oppenheimer. Carroll & Graf Publishers, NY 2003. Or start with nutrition and genetics. First find out what field you like best—nutrition, pharmacy, genealogy, anthropology, or heathcare—and read a book for beginners on DNA related to understanding your particular area of interest, such as nutrition or family history. It's about discoveries.

- A blizzard of discoveries are published monthly in recognized scientific journals found in local medical school libraries open to the public. Only a few consumers ever look at them, and still fewer physicians. Doctors are busy with so many patients and paperwork or bureaucracy. Consumers may not know information is accessible to them. And few can keep up with the proliferation of material in science publications.

- For information, resources, the research network, and references on pharmacogenetics (education) see the Pharmacogenetics and Pharmacogenomics Knowledge Base Web site at: https://preview.pharmgkb.org/resources/education.jsp. As far as education, the Web site features links and articles on the following subjects: What is Pharmacogenetics? Asthma Case Study, CYP2D6 Case Study, The National Institutes of General Medical Sciences (NIGMS), Medicines For You, Minority Pharmacogenomics, The Importance of Genetic Variation in Drug Development, Publications, and News Clippings.

- Click on the Dolan DNA Learning Center at: http://www.dnalc.org/. The Dolan DNA Learning Center at Cold Spring Harbor is entirely devoted to public genetics education. The gene almanac is an online resource that provides timely information about genes in education. The Dolan DNA Learning Center is the world's first science museum and educational facility promoting DNA literacy.

- Dolan's Saturday DNA program is designed to offer children, teens and adults the opportunity to perform hands-on DNA experiments and learn about the latest developments in the biological sciences.

- If you're interested in student "DNA Camps" see the Web site at: http://www.dnalc.org/programs/workshops.html. Student summer day camps have fun with DNA and enzymes and study DNA science or genetic biology. Students and high-school teachers can participate. There are a lot of ways to become involved in learning more on these topics. I wish there were DNA day camps for senior citizens newly retired with time free at last, or for parents, children, and teachers to participate and learn together.

The student summer day camp workshops feature such wonderful learning experiences as the genomic biology and PCR workshop. This new workshop is based on lab and computer technology developed at the DNALC in the past year. The workshop focuses on the use of the polymerase chain reaction (PCR) to analyze the genetic complement (genome) of humans and plants. DNA educational centers bridge gaps between scientists and communicators.

Many physicians have not yet been trained in nutritional genomics or in pharmacogenetics—how to correctly interpret a DNA test for consumers. Who advises consumers how to tailor their food or drugs to their genes in ways that consumers can immediately use? Who instructs them how to make time capsules out of their ancestry from DNA reports?

For the time being, it's the bioscience journalist who acts as a communicator, the liaison, the publicist, the broadcaster, the bearer of news, the turner of complex terms into plain language, the reporter and the publisher, showing consumers, physicians, and scientists how to bridge the gap between science and healthcare.

Who acts as the middle person between genetics and medicine or between nutrition and healthcare? Who bridges the gap between the dietician and the geneticist? For now, it's the media, the science writer writing for the mass media or an audience of general readers—consumers like you and I. Who should join the science media team? Consumer watchdog groups, concerned physicians, genetics researchers—scientists, DNA testing companies, and bioscience publishers.

Instead of worrying about physician apathy or consumers turning to alternative medicine, let's turn to scientific knowledge available to all consumers if you know where to look and what is accessible. If you want to take charge over how your body responds to food or medicine or lifestyle changes, you need to take control by finding out how your genes respond to what you put into your body from what you eat to the drugs you take to the chemicals in your environment.

Has a local or national newspaper reported on rocket fuel or other specific chemicals in your water supply that went into your home-grown vegetables that made your thyroid go wild, find out what research is being done on the situation.

If your body responds to various foods in certain ways or various drugs in other ways, you can take control by learning what new advances are available to you. Check out what is credible and what works for you. Minorities may have genes that respond differently to various dosages of medicines. Check out the Web site for research on minority pharmacogenetics at: <http://www.sph.uth.tmc.edu:8057/gdr/default.htm>. The Minority Pharmacogenetics website is devoted to issues of minorities, populations and pharmacogenomics.

Have you ever met a doctor who keeps up with all the latest advances? Who has time? Consumers do. Look at the breakthroughs in the journals published monthly. Why worry that medical discoveries aren't being delivered fast enough to patients or clients if you can see what's happening in the journals?

Consumers need to form watchdog groups and to link up with the media. You need to become involved in accessing scientific pioneers. As a journalist, my job is to convey information to the public to bridge the communication gap. As a consumer, your job is to take care of your body. Nobody should walk in medical or nutritional ignorance when the information you might need "is out there." You don't need any science degree or license to read public information. Let's all bridge the communication gap between consumers, scientists, and physicians.

Consumers want applied knowledge. Here's where to start. On the Web, look at: http://www.nigms.nih.gov/funding/medforyou.html. It's the page for The National Institutes of General Medical Sciences (NIGMS). It's the home page for the National Institute of General Medical Sciences, a component of the National Institutes of Health, the principle biomedical research agency of the United States Government. "NIGMS supports basic biomedical research that is not targeted to specific diseases, but that increases understanding of life processes and lays the foundation for advances in disease diagnosis, treatment, and prevention," according to the Pharmacogenetics and Pharmactogenomics Knowledge Base at: http://www.nigms.nih.gov/funding/medforyou.html.

If your doctor hasn't the time to take advantage of current treatment findings, you as a consumer can read what your physician may not get to read for a while. Picture one future fantastic scenario regarding DNA testing where spouses are chosen based on DNA reports. If this sounds too bizarre, visualize choosing foods based on DNA reports. That sounds credible. The point made here is that you could help to bridge the gap between research and conventional medicine. You and I have our work cut out for us, speaking as a consumer and as a journalist who loves reading scientific articles.

The scene at the "speed dating" table opens with a young woman waving her latest DNA test result under the nose of her blind date. "I'll show you mine, if you'll show me yours," she says, smiling. The fellow opens his DNA test result folder or database. They exchange DNA test result printouts and interpretations. "Hmm…regarding ancestry," he says, "Looks like we're both members of the same mtDNA matrilineal clan."

That could be haplogroup H, L, M, or with whatever worldwide mtDNA haplogroup letter or matrilineal 'clan' they match. "Now let's look at your specific genetic markers and genes tested for adverse reactions to this list of medicines or foods or…." She gazes up at him smiling. "I like your lack of risk for diseases," he says. "And I like the way your gene mutations put you at risk for all these inheritable diseases," she answers.

"Do you also have any inheritable real estate and cash? I'd like to be beneficiary." He looks at her askance and holds up a little green bottle. "With this gene equalizer, my risk is greatly lowered." They exchange DNA printouts once more and move to different tables to begin another Dating by DNA interview.

Is this Science fiction version of "blind dating by DNA?" Maybe, but consumers want practical results from DNA testing that they can apply to real-life situations. What kind of personalized medicine can consumers find today regarding genetic testing? How many ways can you use DNA tests? Tailoring your food, customizing your medicines by avoiding adverse drug reactions, looking at your ancient ancestry, family history, or genealogy by DNA through "molecular genetics" and perhaps, choosing your spouse by DNA?

Genetic screening of couples hoping to marry or couples seeking answers to fertility questions have been screened for inheritable diseases in several countries. How many ways can you market DNA tests? For what purposes can you use DNA tests? Who will review the products? Do consumers need a guide to genetic testing kits sold directly to the public or sold only to physicians and healthcare professionals? Both are needed.

***

5

Personalized Medicine from DNA Testing Companies

❖

Consumer's Guide to Genetic Testing for Adverse Drug Reactions, Health Screening, and Tailoring Food, Supplements, and Cosmetics to Your DNA: Pharmacogenetics/Nutrigenomics

How do consumers or consumer's watchdog groups regulate doctors' services? Are DNA test kits sold to consumers really doctor's services or are they medical devices? Consumers want their healthcare and nutrition tailored to their individual genetic signatures, and they want it to be affordable and available to everyone. The FDA regulates medical devices, not physician's services. Where does a DNA kit sent directly to consumers fit into a definition? Scientists and physicians study haplotypes when they map complex-disease genes. Genetic researchers look at the abundance of single-nucleotide polymorphisms (SNPs). They use terms such as "single-locus analyses."

Consumers want those terms translated into plain language by bioscience communicators, since most scientists and physicians are too busy to continuously write about genes for the public. That's why public education programs about DNA exist, including the summer day camps for students and teachers. We need the training programs and camps also for parents—for entire families from great grandma and pa to elementary school children and teens.

When it comes to studying how your genes respond to food or drugs, scientists avoid analyses techniques with limited power and instead focus on economical alternatives to molecular-haplotyping methods. So you, as a consumer, presumably with little or no science background, are in a position to learn about your own genes and about DNA in general.

You can start with some of the DNA public education learning programs open to anyone. Since the gene hunters began to explore the entire human genome, they opened a door to the public for knowledge accessible to all. At first consumers were interested in learning how to interpret their DNA test for ancestry and family history. Next, consumers wanted to know how to tailor their drugs and food to their genes—their genetic expression. Finally, consumers found they could customize skin care products to their DNA reports as well.

Consumers ask important questions. What's waiting for you in a DNA kit that will be valuable to your health? How will you understand and apply the scientific reports to your daily lifestyle? How do you find out whether the various DNA testing companies are credible and recognized by which group? Who are the watchdogs here? Who are the experts? What do you do with the information you receive after a DNA test? Should you test for deep ancestry, adverse response to drugs, or tailoring your food to your DNA? How do you apply the information in a practical way?

Should the consumer only deal with companies that sell their DNA test kits to physicians and similar healthcare professions such as dieticians and genetics counselors? Or should the average consumer buy a DNA testing kit directly from a company that also markets to consumers, bypassing physicians? What do you think as a consumer? Which stance are you taking? For resources and articles, see the Web site at: http://syndromexmenu.tripod.com.

What happens when a DNA testing company sells its DNA tests or kits to companies that turn around and sell personalized nutritional or cosmetics products so that consumers can buy skin products or nutritional supplements customized to their genetic signatures? Is this good for consumers? What do you think? I like DNA testing. The interpretation of the tests is tricky. Results can have a lot more than one interpretation, even for physicians and scientists. So how does the consumer learn to screen the screeners?

Should a company tell consumers which specific genes are tested? Some companies do and others don't. How do you know whether the dose of vitamins you take is beneficial or harmful? Do you need to cut back or add more? What can DNA testing do for you, the consumer, presumably starting with no science background? As a bioscience journalist, I'm not taking sides here, but I really like the variety of DNA tests.

Some types of genetic test kits are sold directly to the public, and others are marketed only to physicians and similar healthcare professionals. From the consumer's point of view, there is always the question of whether the tests are needed and if they are, how accurate and valuable are the tests, and who is trained and experienced enough to interpret the answers? If you're going to take a DNA test, it could be for ancestry, for nutrition planning, or to see whether any prescribed or over-the-counter drug, chemical, supplement, nutraceutical, food, or herb you take will have adverse effects on your body. In the past you could test for allergies to some substances and foods, some chemicals in the environment, but now there's genetic testing.

Specific genes are tested for risk or reaction. The type of consumer most likely to order a genetic test for drug response or a "pharmacogenetic test" would be an individual with medical conditions taking several drugs and concerned how his or her body handled the mix of medicines, food, and lifestyle. Each person has his or her own genetic signature. You rinse your mouth with a type of mouthwash and send the contents in a small tube to a laboratory, hospital, office, or company to be analyzed. Or you swab the inside of your cheek with a felt tip and send the swab to a hospital or genetic testing company so your DNA—specific genetic markers—can be analyzed.

In a few weeks, you get a report. The genetic testing company looks at specific genes and markers. If you're concerned about drugs, you want to know how slowly your body is breaking down any of the drugs you take or might take in the future. You don't want to gulp down a prescription or over-the-counter drug and have it build up in your body to the point you land in the emergency room of a hospital. The whole point of a genetic test for adverse drug reactions is to alert your physician to adjust the dosage of your medicine or change the type of medicine.

On one side, the consumer sees the physician and scientist diagnosing rare diseases by looking at particular genes. On the other, the consumer wants information on how genes work. Firms that sell gene tests directly to consumers need to fill an education gap. Certain tests can be sold directly to consumers, such as DNA testing for ancestry, genealogy, family, oral history, DNA matches, and surname projects where people with the same or similar last names are matched by DNA to see whether there's a relationship in the past.

What DNA testing companies can do for the consumer is to widely teach the consumer about population genetics—the peopling of the world. How about for disease? Genetic disease specialists have been in charge of this new science. Lately, companies that test DNA now market directly to consumers as well as to their physicians. Here's a chance to educate consumers as well as physicians in everything they need to know about their genes. Physicians want to know how to interpret DNA tests for disease markers. Consumers want this information directly. So there needs to be a level for education for both physician and consumer.

If a DNA testing firm markets directly to the consumer, the test should be interpreted for the consumer and the physician. Consumers are paying to learn what medications to avoid. In the past allergists told people what foods to avoid. What consumers want is to know what foods, vitamins, minerals, supplements, skin creams, herbs, medicines, and other nutraceuticals will work best with their individual genetic signatures. Who will teach them? If the physician has only a few minutes to consult with each patient, who else has the time? Either the DNA testing companies must educate the consumer, or the consumer as an autodidact, must teach himself from reliable sources.

Today anyone with a genealogy hobby an open a DNA testing company and contract out to a laboratory in a university to test DNA for clients. When it comes to test DNA for diseases or drug reactions, it's different than sending a printout regarding the client's ancient general ancestry such as the results of a DNA test for mtDNA (matrilineal ancestry) or Y-chromosome (patrilineal ancestry) in the deep past. Those tests are general and would be of interest to people matching ancestors or learning about population genetics an ancient migrations.

When it comes to tailoring food or drugs to your genotype, how far can DNA testing companies bypass the genetic disease specialists and market directly to the consumer? The answer is as far as it takes to teach the consumer about his or her own genetic expression. When you look for a DNA testing company, ask yourself whether the company diagnoses diseases? Does the company help you choose your medicines, food, or cosmetics according to your genetic signature? Or does the company work with your physician and you. Does the company pass you over entirely and only send the results to your personal physician?

Look at the marketing efforts of the company. Investigate gene research for yourself by the many articles available online and in the medical libraries and popular magazines that go to physicians. If you buy a test, is it affordable? Do you need the particular DNA test?

If one of your relatives has a specific condition you might have inherited that will show symptoms when you reach a certain age or presently, you need to know what you can do to delay or prevent the situation. You need to know that the results of DNA testing is for a particular purpose, such as testing adverse reactions of your body to certain prescription drugs. Find out the value of the tests from reliable sources. If you start with your personal physician, make sure he or she can correctly interpret a genetics test. Some physicians can't because they were never trained to do so.

Consult the various professional associations for referrals to genetic disease specialists rather than rely on a generalist when it comes to interpreting your test while you learn how to interpret your tests yourself. The purpose is to make sure your physician's interpretation and your interpretation agree. If you have any doubt, you need more information and a second opinion.

What you want to avoid is confusion. Are the tests available today reliable? Find out from several opinions of specialists. To reach these specialists, the professional associations for the various genetic disease specialists would be of help as would the publications. You can talk to the research labs at universities specializing in genetic testing. The institutes are online and have their own Web sites.

Talk to federal health officials about what's on the consumer market in your country and in other countries. Some people send their DNA swabs to companies and/or labs in other countries. Currently genetic tests are sold without a special 'watch' organization that reviews them. So look at any publications put out by the U.S. Food and Drug Administration (FDA). The FDA wants consumers to become more involved in their own food safety and security monitoring. Keep a folder on DNA testing company marketing claims and check them out.

Companies can tell you which mix of vitamins works best or what food to eat based on genetic testing and/or health questionnaires. For example, check out GeneLink Inc., New Jersey. Their Web site is at: http://www.bankdna.com/breakthrough_profiling.html.

According to Multex, an investments site at: http://biz.yahoo.com/p/g/gnlk.ob.html, "GeneLink Inc. is a bioscience company that offers the safe collection and preservation of a family's DNA material for later use by the family to identify and potentially prevent inherited diseases. GeneLink has created a new methodology for single nucleotide polymorphisms (SNPs)-based on genetic profiling. The Company plans to license these proprietary assessments to companies that manufacture or market to the nutraceutical, personal care, skin care and weight-loss industries. GeneLink's operations can be divided into two segments, the biosciences business, which includes three SNP-based, proprietary genetic indicator tests, and DNA Banking, which involves the use of its proprietary DNA Collection Kit." Contact GeneLink at: 100 S. Thurlow Street, Margate, NJ 08402. See the Web site at: http://biz.yahoo.com/p/g/gnlk.ob.html.

GeneLink also offers gene testing for skin care products. According to a March 6, 2003 media release posted at GeneLink's "In the News" Web page at: http://www.bankdna.com/news_articles/03_06_03.html, GeneLink, Inc. (OTCBB:GNLK) "entered into a collaborative agreement with DNAPrint Genomics, Inc. (OTCBB:DNAP) whereby the companies will combine certain scientific and intellectual property resources to develop and market "next generation" genetic profile tests to the $100 Billion plus personal care and cosmetics industry. Further details were not disclosed.

"GeneLink invented the first genetically-designed patentable DNA test for customized skin-care products, and DNAPrint brings its ultra-high throughput genotype capability and ADMIXMAP platform to the partnership. The companies anticipate screening millions of candidate markers to broaden proprietary product offerings.

"Tests are designed to assess genetic risks for certain skin and nutritional deficiencies and provide a basis for recommending formulations that have been specifically designed to compensate for these deficiencies."

I like this idea. I'm allergic to hair tint and aloe vera, and it would b wonderful to find skin care products that didn't turn my skin red and make it itch. Think of all the allergy warnings and patch tests on hair products. If everyone walked into a beauty salon with a list of products that can be safely put on the scalp or other skin without an adverse response, that would be my definition of comfort. Also, check out GeneLink's test which tells clients which mix of vitamins is appropriate for an individual's genes without getting an adverse reaction from the vitamin combination. I know certain supplements over-stimulate my thyroid. I'd love to have these kinds of tests.

How about you? Still, you need to do your homework on the usefulness of the tests or the claims. Perhaps you need to deal with a company in another country. Sciona Ltd. in England gives nutritional advice based on your DNA and a questionnaire related to your health. **The company sells tests only through doctors and dieticians.** How do you know which defect is associated with any specific gene when the gene is defective? With so many interpretations, consumers can become confused from test results. Sometimes scientists in various companies don't know which results are accurate predictions. That's why consumer watchdog groups are necessary.

The only problem is that the average consumer usually can't afford to hire genetics counselors unless they have a specific genetic disease. If you have a reduced ability to process something, it's not a disease, but unless you take action, your body could suffer in other ways resulting from the reduced ability to process an essential nutrient from your diet that your body needs to work with other processes. It's like a chain reaction.

That's why DNA testing is helpful. At the same time, you can't assume some people need more of one vitamin just because their body processes a nutrient at a reduced rate. The science is still very new. A lot of data is still in the works. You don't know whether certain genetic profiles need special diets. Consumers want to see rare clinical data. Consumers most of all want to know the value of what the DNA tests predict. And no one wants to take a test and then worry for a decade.

What consumers want are not only tests of prescription drugs and food as to whether adverse reactions happen with individuals based on their genetic testing, but personalized anti-aging formulas, cosmetics, creams, and other products customized to the individual's genetic markers are needed.

These tests must be reliable, consumers say. Feedback is helpful. Customer reviews of the companies are needed. Who will publish customer reviews of the products and the DNA testing firms? These consumer reviews could also include input from scientists and physicians. Physicians should take the DNA testing themselves after they are properly trained to interpret the tests before looking at their patient's results.

Nutritionists and dieticians show concern that medical schools only give very brief and shallow courses in nutrition to graduating physicians. What about a course in interpreting genetics tests? Will this be left to naturopaths and homeopaths? Or can medical schools include courses in DNA test interpretation to physicians other than genetic disease specialists? Healthcare consumers should ask for DNA testing from their HMOs before taking prescribed medicines. Will insurance companies pay for testing?

Look at the various DNA testing companies. Also look at health screening companies. One would be HealthcheckUSA, San Antonio, Texas for example, sells a mail-order test for iron buildup in the blood known as hereditary hemochromatosis. Results go to the consumer. If you need a test for cystic fibrosis or a blood clotting disorder known as factor V Leiden, check out this company. Their Web site is at: http://www.healthcheckusa.com/media.html, or you can write to them at: **HealthcheckUSA,** 8700 Crownhill Rd., Suite 110, San Antonio, TX 78209.

Practicing physicians started HealthCheck USA in 1987 to provide health awareness screening to customers throughout the USA. The tests provided to their customers are the same as those ordered by physicians across the country. Many physicians refer their patients to Healthcheck USA. Since 1987, Health-

Check USA has provided over 500,000 health awareness screening tests to satisfied customers nationwide.

The company is associated with the country's major fully accredited medical reference laboratories, to ensure quality, accurate test results. HealthcheckUSA has partnered with <u>Virtual Medical Group, Inc.</u>, to offer physician interpretation of results. According to HealthcheckUSA's Web site, "You can have your results reviewed and interpreted by a <u>Board Certified Physician</u>. By selecting the "Physician Interpretation of Results" option, you will be mailed instructions with the hard copy of your results. These instructions will give you a toll free number along with your username and password to access your interpretation 72 hours after your blood draw. This process is completely confidential, private, and secure."

Here the consumer has the best of both worlds. What tests are provided? Check out the list of tests provided at the Web site: http://www.healthcheckusa.com/testsweoffer.php.

When any company does health screening, find out whether it is DNA testing, blood testing, comprehensive tests, or whether you are sent test kits such as for colon cancer screening, prothrombin (factor KII) DNA test, cystic fibrosis (DNA test), Factor V R2 (DNA test) or other. Are the results sent back to you directly as a consumer? Or do you purchase additional services such as a physician's interpretation of the results?

What consumers want are genetic tests for metabolism and health. In the past, genetic testing was used by doctors when the patients were concerned about specific diseases diagnosed by genetic testing. Today, the general consumer of healthcare and healthcare alternative medicine wants genetic testing for general healthcare in the absence of a specific disease. Consumers want to know how to handle the risks they might have inherited to delay or prevent what happened to their grandparents. Also parents of children and couples worried about fertility issues also want these types of tests. Genetic testing shouldn't be used to screen people out of insurance, employment, or anything else.

Health screening should be focused on matching people to the best possible foods, nutraceuticals, and medicines when and if needed. It should be an inclusive not an exclusive process. Like personnel departments that focus on screening out applicants, genetic testing should not be used to screen people out or to exclude based on genetic risks. Instead, it should be used to draw people into learning about taking responsibility for their own health habits such as choosing the right foods. It's all about choice, like the science of nutrition.

In the past, genetic tests were used on people who suffered from rare genetic disease. Consumers are worried about claims by companies that sell DNA tests to the public. Presently, consumers are asking the FDA to review tests before they are sold to the public or to physicians. The FDA reviews drugs and other medical appliances, why not DNA tests or other health screening tests sold to consumers? Should the government be reviewing these tests? What do you think? Or should the tests be monitored by the companies offering them? Or should the approach be to the tests similar to vitamins and minerals found in health food stores? What side will you take? Do you have enough information to even begin to take sides? Who runs and regulates the DNA testing companies that may or may not be headed by physicians and geneticists?

Testing DNA for ancestry would not have the same impact on someone's health as testing DNA for a specific disease marker. Still, the people doing the testing should have the same qualifications. What about the people doing the interpreting? What stance do you take? Then there's the question of privacy which is essential.

If you're testing for DNA ancestry, you might want to meet your DNA match to correspond with online. If you're testing for a disease marker, you want privacy. You may want to join a support group of families with similar DNA markers for the specific risk or disease you've inherited. You want meetings held in a hospital or some other private, medical setting. Personal issues come up.

Research the various associations of physicians, such as the American Academy of Family Physicians and other similar groups. You can find opinions there. You can bring in lawyers who want to make sure laws are put into effect to keep employers and insurers from using your results to terminate your job or kick you off of health or life insurance plans if your genes put you at risk.

Set up groups made up of advisories of experts that include consumer groups. What consumers with no science background can do is to work with the US Department of Health and Human Services to monitor various panels of experts. Some people want regulations to change. Others don't. There are hundreds of groups out there who advise the US Dept. of Health on what policies they 'should' adopt.

Consumers ask for reviews. Experts ask for reviews. Consumers don't want their vitamins taken away from health food stores. In the midst of it all are the wonderful DNA tests. Talk to pharmacists as well. If you're having adverse reactions to drugs, you eagerly want a DNA test you can take at home as a consumer. What I like about Genelex Corporation's drug metabolism test is that the test can tell you whether your body is taking too long to break down the drugs you're consuming. Consumers need that kind of calmness that comes from knowing personal physicians aren't prescribing too much of what you may or may not need. Maybe you're worrying about your pharmacist giving you the wrong medicine or dosage. Whatever you're concern, a home DNA test is a tool for consumers.

Monitoring your own health is your responsibility, not merely your doctor's. You can get DNA tests sent to your home or scans from clinics. Prescription drugs are available on the Internet. How your body will metabolize what you put into it is important. Are you becoming your own doctor? Is the average consumer becoming more knowledgeable? Or is it more accurate to say most people haven't an inkling of what DNA is?

Pharmcogenetics and nutrigenomics offer tools to consumers. The movement is towards taking charge of your own healthcare. Are doctors worried consumers are taking money away from them? Do tests create more problems than they solve? What happens when scans or other types of tests show non-existent problems?

Research what comes out of the National Institutes of Health. Consumers want to know what the results of their tests mean and how they can apply the information in practical ways. Who interprets the tests? Who is reviewing genetic tests sent directly to consumers and physicians? Right now, it's the consumer's 'job' to ask these questions. If you don't know what vitamins to take, perhaps you need a DNA test to tell you what vitamins work best with your individual genes.

DNA tests seek out mutations in the genes. If a mutation is only remotely associated with a risk, do you ignore it or customize your food and vitamins to the mutation that is only remotely connected to the risk? What action do you take as a consumer? You talk to a physician who knows about genes and your particular risk or you go to a genetics counselor. If you have certain forms of a gene that results in cancer of a certain area of your body by a certain age, you take action right away. Some mutations are associated with disease and some are not, such as certain ancestry markers.

Some genes account for only a tiny percentage of certain cancers. Not everyone needs to be tested. If you have a family reason to be tested, then get tested. If one of your siblings or parents has an inheritable disease, get tested. You can check out Myriad Genetics based in Utah regarding their breast and ovarian cancer tests, available only through doctors. Their Web site is at: http://www.myriad.com/. Or contact them at: Myriad Genetic Laboratories, Inc. | 320 Wakara Way, SLC UT 84108-1214

The Controversy over Whether Genetic Tests should be Sold to the Public

Professionals working in molecular genetics may be divided among those who think genetic tests should be sold to the general consumer and those who think they should not. The professionals who think genetic tests should not be sold to the public may underestimate the intelligence of the public's drive to learn as much as possible about their own genes—the DNA, ancestry, health response to medicine and chemicals in their environment, their response to prescription and over-the-counter drugs, and their response to food, nutraceuticals and other supplements.

Those who are against selling genetic tests to the public fear that genetic testing is way too complex for the average consumer to understand. The average consumer consists of the person who says "it's way over my head," or "my eyes are glazing over" when you mention anything scientific, and the individual who wants to learn as much as possible about his or her own DNA, health, ancestry, and genetic response to food. Most consumers only need to be told that more than a hundred mutations can cause a specific disorder or disease. Average consumers can understand that. Those who underestimate the intelligence and desire to learn of an individual about his or her own health or the health of children and parents usually are the ones who are first to speak out in censorship of a consumer's right to learn all that is possible to learn at the moment about his own genes.

When a consumer is told that negative results don't always signal health, he or she is intelligent to understand that plain English or any language statement. It's a learning process. Some genetics companies are in direct marketing. Some test DNA only for ancestry. Other genetics companies test for genetic responses to drugs or for various genetic health risks that can be helped by changes in diet. The aim of these companies is cooperation. That means cooperation between patient, physician, genetics counselor, dietician or nutritionist and anyone else involved in the healthcare team working with the patient.

In fact in genetics testing, patients are really clients seeking information, feedback, and recommendations for changes in lifestyle and diet. In drug response genetic testing, the client wants to know how his or her body metabolizes the prescription or over-the-counter medicines used. The whole idea of genetic testing is to bring the healthcare team closer to the healthcare consumer. Instead of a one-size-fits all attitude with drugs, foods, or nutraceuticals, or the usual five to fifteen minute consultation with a doctor, the consumer can have the chance to learn everything publicly available about how his or her genes respond to food, drugs, chemicals in the environment, or lifestyle changes.

If a genetics testing company markets cancer tests, it should have the responsibility to let the patient know that the tests are meant for specific people with specific familial cancers and not for the entire consumer population. When a genetics company takes a family history, it opens doors to the individual not only to explore a familial inherited disease or risk, but the entire history and ancestry of the person's family and the DNA of the family members.

Individuals need to learn that there could be more than one gene involved. So teaching consumers about their DNA and genes is important. Classes could be springing up to train people what to expect before they undergo testing. It's a matter of learning and teaching. This is one more way to make use of retired professionals and other scientists and physicians or genetics counselors working in molecular genetics—helping consumers learn what they can expect from testing.

Consumers need to learn about how important it is to sequence certain genes from many family members. Consumers need to get in touch with affected family members and form a support group for DNA testing, perhaps creating a time capsule for future generations with genetic information resulting from testing.

Most consumers go to their family doctors first. Yet how many family physicians are interested in, let alone trained in molecular genetics? Not so many. Most physicians may not interpret a genetic test correctly because they haven't been trained to do so. In may be that the consumer is the first person to request training be set up to show physicians how to interpret genetic tests, especially for such diseases as cancer. Consumers have more power as a group. So consumers need to become involved, work together as an organization, and make sure not only the physicians are trained, but also themselves as consumers of genetic testing. Perhaps the physician and consumer can learn together in groups set up for both, where at a point, physician and consumer meet together. Will this present the physician differently in the eyes of the patient? Not really. The science is new enough so that consumer and patient could learn together to unravel the mysteries of genetics not unlike checking the clues in a mystery novel.

Consumers need to read more medical journals such as the New England Journal of Medicine. Spend some time in the library of your local medical school reading about how physicians interpret genetic testing. Look at surveys. As of now, not too many general consumers from the public at large even know what DNA is. So you have to educate yourself, perhaps as a new hobby. First find out what DNA is. Look it up in the dictionary.

Then start reading magazine articles about genetic testing. Look on the Web at the testing companies. Then you can graduate to reading articles in medical journals. The gap between the physician and the consumer needs to be narrowed at least when it comes to your own DNA and genetic testing for the things consumers put into their body such as food and prescribed or over-the-counter medicines, supplements such as herbs, and other nutraceuticals.

Everyone talks about the consumer's consent to testing. The problem is that the consumer needs to be given information. If you're going to eat 'smarter' you have to become informed first. If you have an inheritable risk or disease, you need to understand everything you can find that is publicly accessible about the risk, the disease, or the genes involved. Material is online, but is it credible? There are always the articles in medical journals, but can you understand the terminology? If not, look up the terms in glossaries and dictionaries. Make a list of terms and learn how to make the journal articles understandable in plain language.

Think about risk for a moment. Your risk could change. How are you going to receive the information from genetic testing? What are you going to do about it as far as changing your lifestyle, diet, or medicines? If you are concerned how your family might react to genetic testing, ask them whether they would change their food choices or other lifestyle changes based on the information. You can always keep the information private, but it could save the lives of family members to get tested and to change what can be changed, such as food choices or exercise routines.

Most consumers worry more about discrimination in employment and health insurance due to companies finding out their genetic risks. Privacy must be kept, and information only given to the consumer of the testing, not his employer or health insurance company. It's nobody's business but your own as far as your genetic information. So, measures to keep out strangers from using your genes against you must be in order. That's why numbers instead of names work better with testing.

The big issue is accuracy. If so many physicians cannot correctly interpret a genetic test and advise their patients, more training is needed for the physicians and the patients. It's a case of whose watching the watchers. Who is going to review the genetic testing itself as to accuracy? What if the tests are not accurate? Right now nobody is reviewing the testing companies other than themselves. Nobody is reporting directly to the consumer after checking the accuracy of the genetic tests. And nobody is reporting directly to the physician with that same information.

If you're going to get tested, at least learn as much as most professional genetic counselors know about interpreting your tests. It can be done without going back to school for a graduate degree in molecular genetics. The information is available to the public from libraries, the Internet, databases, and medical school libraries, journals, and professional articles. Many are on the Web in PDF files. Subscribe to the journals or read them in libraries and learn how to interpret your own tests. Then form a consumer's group to watch the watchers. Learn how to check the accuracy of tests done by the testing companies.

You have to become involved in your own healthcare and nutrition. Take charge and take responsibility. Consumers can't be passive. You won't know how your body responds to food or medicine or lifestyle changes until symptoms appear. Start by looking at the health of your family members and realizing what is inheritable, and what you can do to help yourself if you've inherited what they have or will get.

The first step for consumers is to start learning about DNA and genetic testing just as they have learned about genealogy and family history. Classes online or in adult education can be set up as well as support groups and organizations of consumers. The second step is to form a group to review the accuracy of information that comes with your genetic tests. Invite physicians and other healthcare professionals to join with consumers and become a team. Don't let separation between those with scientific knowledge and those without become a barrier to reviewing accuracy of information.

When several physicians or groups of geneticists band together to review the accuracy of information provided with genetic testing, what might happen is that the group then becomes made up only of professionals. When consumers group together, they can include professionals, but would never ban someone from the group because of professional training in the field they are watching.

So consumers need to create a group to review the accuracy of information that comes with genetic tests. And that same group needs to make sure education exists for consumers. You need to set up classes to train consumers in how to understand the results of their genetic testing and how to review the accuracy of the information given along with the genetic tests.

If trained physicians can't correctly interpret a genetic test for their patients in certain cases, it's time for consumers to take charge of their own healthcare education. Okay, you can't do surgery on yourself, but you can learn to read the information provided with genetic tests. And having read the information, you can learn also to question the accuracy of that information. Consumers can read not only the medical journal articles usually found in medical school and university libraries open to the public, publications such as the **New England Journal of Medicine**, but also some of the more popular articles in magazines that go to physicians such as **Physician's Weekly**.

All this autodidact education can be achieved with consumers involvement and support groups if people are interested enough to involve a wide range of members or participants. Genetic training should be available to everyone equally and widely available through the internet, through senior centers, classrooms focused on adult education, through hospitals and HMO programs, and through the genetic testing companies.

In the magazine, *Physician's Weekly*, October 7, 2002, Vol. XIX, No. 38, the article in Point/Counterpoint, titled, *"Should genetic tests be sold directly to the public?"* featured Howard Coleman, CEO, Genelex Corp., Redmond, Washington and Kimberly A. Quaid, PhD, Professor of Clinical, Medical, and Molecular Genetics, Clinical Psychiatry, and Clinical Medicine, Indiana University School of Medicine. Coleman's response was yes, and Quaid's response was no. See the article, copyrighted 2002 by *Physician's Weekly*, reprinted below with permission.

Should genetic tests be sold directly to the public?

Howard Coleman, CEO, Genelex Corp., Redmond, Washington.

YES. Every person has the right to know his or her genetic information, and the right not to share it. People have legitimate privacy concerns about their genetic information being loose in the medical records system, fearing for their jobs or insurance if it gets into the wrong hands. Physicians are also concerned, and don't want to risk compromising a patient's privacy. This contributes to their reluctance to order genetic tests.

There are compelling reasons for people to obtain reliable genetic information, whether they go through their doctors or not. According to a 1998 JAMA article, more than 100,000 deaths occur annually due to adverse drug reactions, along with an additional 2.2 million serious events that require hospital stays. These are not medical mishaps, as they occur within the labeled use of drugs.

When physicians begin to learn the genotype of their patients they will begin to solve this problem. The practice of their clinical art will be improved because in many instances the genotype of their patient trumps the many other characteristics they do know.

Despite the fact that science has known for the past half century that people react differently to drugs, doctors have been unable to put genetic testing into practice for the benefit of their patients. Doctors lack sufficient training in pharmacogenetics and drug metabolism. For example, most doctors don't know that genotyping for Coumadin metabolism is available, and that it will help patients avoid adverse reactions with this drug.

Adverse drug reactions kill many people in the U.S., but the threat goes largely unrecognized. Making gene tests available directly to patients not only provides a valuable service, but helps push the medical community into the 21st century era of personalized and evidence-based medicine.

Kimberly A. Quaid, PhD, Professor of Clinical, Medical, and Molecular Genetics, Clinical Psychiatry, and Clinical Medicine Indiana University School of Medicine

NO. No. Genetic testing is far too complex for lay people to tackle on their own. For example, more than 100 mutations can cause cystic fibrosis, and most genetic tests only cover a small subset of these. But most lay people don't understand that negative results are not necessarily a clean bill of health.

The announcement last spring that Myriad Genetics would direct market cancer tests was particularly troubling, because such tests are only appropriate for a small number of individuals with certain familial cancers. The complex protocol includes taking a detailed family history, then finding and sequencing the gene from affected relatives (and there may be more than one gene).

Individuals who hear about such tests might approach their family physicians, but research shows that many are not trained in genetics. A paper in the NEJM, which looked at physicians ordering a test for a particular cancer, found that one third of these physicians interpreted the test incorrectly. One can assume that if physicians have trouble interpreting the results, patients will fare far worse. One survey found that fewer than 26% of Americans know what DNA is.

Geneticists believe testing should be preceded by informed consent. For consent to be informed, the patient must understand the disorder for which they are being tested, their current risk, how their risk might change as a result of testing, the ramifications of their being tested for their families and spouses, and the possibility of discrimination should third parties get their hands on this information.

Finally, no mechanism exists to review the accuracy of information provided with genetic tests to either health-care professionals or consumers. Individuals who get tested without professional counseling may buy trouble with their test results.

In our recent email interview of August 19, 2003, Howard Coleman added, "I agree with Dr. Quaid when the testing concerns the grave medical conditions caused by genetic disease and we don't offer those tests to the general public. The testing we do provide can help both the physician and the individual understand their ability to metabolize drugs which can help to prevent adverse drug reactions and how to optimize their diet."

***

6

Scientists and Physicians Comment on Pharmacogenetics

Why would the average consumer of health care want to have drug reaction testing, known as pharmacogenetic testing? The field is new, and still emerging. According to (reprinted with permission) Genelex's "Health and DNA" Web site (copyright 2003) at: http://www.healthanddna.com/, "The relatively new and emerging fields of pharmacogenetics studies how differences in individual genetic makeup affect the processing of drugs. We have known for about a half-century that individuals respond to the same drug and dosage in very different ways because of genetic variations called polymorphisms. Research shows that of all the clinical factors such as age, sex, weight, general health and liver function that alter a patient's response to drugs, genetic factors are the most important. Genelex is the first firm in the world to offer genetic drug reaction testing directly to the public." What I like about Genelex is that they also offer DNA testing for nutritional genetics, drug reaction testing, ancestry DNA testing, and DNA identity testing.

In the past, DNA traditionally had been used to identify people, mostly in forensic or paternity or relationship cases. Then a few years ago, companies began to test DNA for ancestry, which appeals also to genealogists, family historians, and oral historians. The use of DNA outside of court rooms and forensic laboratories and outside of government and university databases brought DNA to consumers of health care as well as to genealogists interested in molecular genealogy.

University laboratories and archaeology research institutes began using the science of archaeogenetics. Anthropologists who looked at HLA markers, those leukocytes or white blood cells and physicians or scientists who were interested in tissue typing for blood, marrow, and organ transplant donors worked with DNA to identify similar matches, people whose tissues and blood or marrow typing matched close enough for a transplant to take.

You had scientists who studied ancient DNA from fossils and mummies expand the science of archaeogenetics and population geneticists studying migrations of ancient peoples and the routes they took. When ancient and present day comparisons of genetic markers left trails that matched with archaeological relics, new branches of molecular genetics grew.

The science is still emerging, using DNA testing in more ways. Now, we have nutrition and genetics and drug testing and genetics. And there's so much more to arrive that will be available to the general consumer of healthcare, ancestry searching, and identity. From being a spectator in life watching new avenues of DNA testing unfold, the consumer now has the chance to become more involved in participating in and learning about how many ways DNA testing can benefit an individual at any age.

When I went back to Genelex's Web site to explore more material on DNA testing for a variety of purposes, I realized, the consumer has more choices than ever before and still more on the horizon. DNA testing is all about nourishment. From nutrigenomics to drug reaction testing, strategies for better health is being covered from all angles.

Perhaps you—as a consumer—want a healthier eating guideline customized to your genetic signature. Maybe you'd like a report on how your genes (and body) respond to certain chemicals, medicines, or substances. Whether it's a prescription drug or specific medicines or supplements that you buy over the counter, several genes are tested to see how your body metabolizes the substance. Not every gene in your body is tested to see how they respond to food or medicine—just specific genes and markers.

The science is new and changing, but the future is attractive to general consumers. You don't need a science background to begin your research on how your own body responds to what you put into it and what comes it from the environment. To begin, consumers need to become more involved in learning about what scientists and physicians are researching and why, and how all these new findings apply to you.

To become more involved as a consumer means being able to access information that evaluates the research, that finds credibility or flaws, and that helps you take more responsibility in your own maintenance. That's why I believe DNA testing can open doors.

Even with guarded caution, the benefits are to be explored and discussed. If you have already had your DNA tested for ancestry or identity, consider these new ways in which DNA testing can help you draw up a plan for eating and for any way you take care of your body—from exercise planning and nutrition to medicines and supplements if and when you need them, to making lifestyle changes for better health.

Consumers can learn a lot from news releases, including learning to understand where to begin to educate themselves about their genes. Some of these releases come from universities engaged in research, some from laboratories and genetic testing companies, some from the government, and from other scientific sources. Professionals in molecular genetics and in healthcare need to understand that an open door policy for public education, teaching oneself about DNA is good. To become more informed leads to becoming more involved in consumer's groups for understanding what "gene hunters" are doing.

Science has become so technical that most scientific journal articles are not understood by most consumers without science training. There has to be a midway point. So far, it's the media that translates scientific terminology into plain language for the consumer. Going one step farther, the medical journalist translates the scientific journals into articles and books whereby people with no science background can learn about their genes.

Finally, learning about how one's genes respond to food or drugs is another rung up on the ladder of self-education about how your body works. That's why there's the mass media for the consumer, the popular medical magazines that bring the consumer closer to healthcare professionals, and finally, the scientific and medical journal articles discussing the research.

At the top are the evaluators who let the consumers know which research studies were flawed in the past, and that usually filters down through the mass media. What some scientists call 'snake oil' may either be harmful or may be a forgotten remedy based on plants that worked in the past. For example, honey, cinnamon, and sesame oil. All three will resist bacteria and in the ancient past were used on minor wounds. A century ago colloidal silver was used on scratches to keep out bacteria.

In the Civil War days, it was used on wounds. Today, you can still make your own, inexpensively, but be careful buying brands containing aluminum. In the dentist's chair, colloidal silver mouthwash works as well as some of the more recent remedies such as washing your mouth with harsh substances that can cause ulcers in the mouth. The point is to check out the mechanisms that review accuracy in whatever you read. If you go the homeopathic or naturopathic route, make sure what you use does no harm for your individual genetic response. How do you metabolize what goes into your body? Genetic testing can help here.

When it comes to genetic testing, you can learn a lot from news releases—at least as a starting point in your research and education about your genes. Bioscience communicators have a role to play as interpreters between consumer and scientist. When I took my master's degree in English with a writing emphasis, it was through a graduate scholarship award in science writing. In genetics, being the communicator means bridging the gap between the growing body of knowledge in science and the consumer who might not even know what DNA is. If you have no science knowledge, start by reading press releases to open the first set of doors to understanding more about your genes, markers, and DNA.

What Consumers Can Learn From News Releases

NIGMS Awards $35 Million To UCSD-Led Consortium To Map Metabolic Pathways In Cells

According to a UCSD media release of August 11, 2003, the University of California, San Diego (UCSD) will lead an ambitious national effort to produce a detailed understanding of the structure and function of lipids—cellular fats and oils implicated in a wide range of diseases, including heart disease, stroke, cancer, diabetes and Alzheimer's disease.

The five-year, $35 million grant from the National Institute of General Medical Sciences (NIGMS) will support more than 30 researchers at 18 universities, medical research institutes, and companies across the United States, who will work together in a detailed analysis of the structure and function of lipids. The principal investigator of this collaboration is Edward Dennis, Ph.D, professor of chemistry and biochemistry in UCSD's Division of Physical Sciences and UCSD's School of Medicine.

Dennis notes that while sequencing the human genome was a scientific landmark, it is just the first step in understanding the diverse array of systems and processes within and among cells. Establishment of this consortium is a significant step in an emerging field called "metabolomics," or the study of metabolites, chemical compounds that "turn on or off cellular responses to food, friend, or foe," he explained.

Lipids are a water-insoluble subset of metabolites central to the regulation and control of normal cellular function, and to disease. Stored as an energy reserve for the cell, lipids are vital components of the cell membrane, and are involved in communication within and between cells. For example, one class of lipids, the sterols, includes estrogen and testosterone.

The initial phases of the project, known as Lipid Metabolites And Pathways Strategy (LIPID MAPS), will be aimed at characterizing all of the lipid metabolites in one type of cell. The term "Lipidomics" is used to describe the study of lipids and their complex changes and interactions within cells. Because this task is too extensive for a single laboratory to complete, researchers at participating centers will each focus on isolating and characterizing all of the lipids in a single class.

This information will then be combined into a database at <http://www.lipidmaps.org> to identify networks of interactions amongst lipid metabolites and to make this information available to other researchers. Shankar Subramaniam, Ph.D., professor of chemistry and bioengineering at UCSD's Jacobs School of Engineering and San Diego Supercomputer Center, will coordinate this aspect of the project.

The cell type selected for study is the macrophage, best known for its role in immune reactions, for example scavenging bacteria and other invaders in the body. Macrophage cells from mice will be used, rather than human cells, because there exists a "library" of mouse cells with specific genetic mutations. By studying cells missing certain genes, the research team will attempt to identify what genes code for those enzymes key in synthesis and processing of lipid metabolites. Christopher Glass, M.D., Ph.D., professor of cellular and molecular medicine at UCSD's School of Medicine, will coordinate the macrophage biology and genomics aspects of the consortium.

According to Dennis, one of the most difficult aspects of this project will be to ensure that all the participating research sites are using identical methods.

"There has never been an effort to detect all the lipid metabolites in a cell and to quantify the amounts of these lipids," says Dennis. "In order to be able take the data gathered by each lab and put them together to develop new ideas about interactions between lipid metabolites, it is essential to develop new technologies and methods and to standardize them so that they can be applied in the same way at each research site."

Once the researchers have developed these methods and used them to identify and quantify the lipid metabolites in the mouse macrophage, the methods can be applied to gather other information about lipid metabolites in cells.

The researchers plan to study the time and dose-dependent effects of lipopolysaccharide, or LPS, a component of the cell walls of many bacteria, on macrophages. Dennis' lab has studied the effects of LPS on macrophages for the last 20 years. A world authority on these effects, Dennis says that this large-scale, cross-institutional collaboration will create an understanding of LPS' effects on lipid metabolism in unprecedented detail, and set the stage for examining other macrophage effectors such as oxidized LDL, the so-called bad cholesterol which leads to atherosclerosis.

A detailed understanding of lipid metabolism will be valuable in drug design. Most people are already familiar with one class of drugs that interfere with lipid metabolism: the non-steroidal anti-inflammatory drugs (NSAIDs), which include aspirin and ibuprofen. These drugs block the synthesis of prostaglandins, a large group of chemicals secreted by macrophages that causes pain, inflammation and an immunological response.

Statins, which a large number of Americans take daily to lower their cholesterol levels, also dramatically alter lipid metabolites. With a detailed knowledge of each step in the lipid metabolic pathways, more effective drugs with fewer side effects can be designed to combat heart disease and a plethora of other diseases of lipid metabolism.

In addition to UCSD, participants in the LIPID MAPS consortium are: Avanti Polar Lipids, Inc.; Duke University Medical School; Georgia Institute of Technology; University of California, Irvine; University of Colorado School of Medicine; University of Texas Southwestern Medical School; and Vanderbilt University Medical School. Additional collaborators will include scientists from Applied Biosystems; Boston University School of Medicine; Harvard Medical School; Medical College of Georgia; Medical University of South Carolina; National Jewish Medical and Research Center; Scripps Research Institute; University of Michigan Medical School; University of Utah; and Virginia Commonwealth University.

For more information on the grant program, see the related release at http://www.nigms.nih.gov/news/releases/.

DNA testing isn't only for anthropologists. It includes but goes beyond forensic research in identity. It expands from population genetics to cellular nourishment. It's for the general consumer with no science training, the physician, the genetics counselor, and the nutritionist. DNA testing is for the genealogist and the oral historian. And there are many more applications of DNA testing/genetic testing on the horizon.

According to Genelex's Web site (copyright 2003) at: http://www.healthanddna.com/pharmacogenetics.html, (reprinted here with permission):

"The relatively new and emerging fields of pharmacogenetics and pharmacogenomics study how differences in individual genetic makeup affect the processing of drugs. Pharmacogenetics largely focuses on specific genes, such as drug-metabolizing enzymes, while pharmacogenomics deals with the entire human genome, including genes for numerous proteins in the body, such as transporters, receptors, and the entire signaling networks that respond to drugs and move them through the system.

- We have known for about a half-century that individuals respond to the same drug and dosage in very different ways because of genetic variations called polymorphisms. The first studies of what came to be pharmacogenetics were conducted in the early 1950s and examined drug metabolizing enzyme variants such as those in the Cytochrome P450 family. Research shows that of all the clinical factors such as age, sex, weight, general health and liver function that alter a patient's response to drugs, genetic factors are the most important.

- Although the genomes of individuals are 99.9% identical, the 0.1% difference means we have as many as 3 million polymorphisms, the most common being the single nucleotide polymorphism (SNP). Detecting polymorphisms is the foundation of pharmacogenetics.

- Misdosing of drugs in America costs more than $100 billion dollars a year, and is a leading cause of death. Pharmacogenetics helps to explain why some individuals respond to drugs and others don't. It can also help doctors identify individuals needing higher or lower doses and who will respond to a drug therapeutically and who might have an adverse drug reaction.

- Currently, the clinical use of pharmacogenetics is limited, with the most extensive use in Scandinavia to treat psychiatric illnesses. The technology is, however, steadily moving into other geographic and medical areas, especially cardiology and cancer treatment.

- The Human Genome Project has been an incredible boon to pharmacogenomics. As new data are decoded and more precise maps produced, we can begin to understand what specific genes are and how they function. We can also better appreciate the significance of DNA variation among individuals. Preventative medicine and pharmacogentics are the first disciplines to benefit from this vastly increased understanding of human diversity.

- Better understanding of the variation in the enzymes involved in drug metabolism and in the drug targets themselves will help explain why there is so much variation in drug response. We are now able to test for these variations before prescribing, and reduce the amount of human experimentation (and associated health care costs) needed to find the best therapy or dose. Drug companies can now develop drugs which are designed by choice, rather than happenstance, to be effective in wide range of the population.

- Predictive genetics is revolutionizing health care. DNA tests and personalized drug therapies are shifting the medical paradigm from detection and treatment to prediction and prevention and vastly increasing the efficacy, safety, and variety of therapeutic drugs. Predictive tests and preventative therapies reduce costs associated with hospitalization and lost productivity and will lead to a major redistribution of value in the health care industry."

According to the Genelex Web site at: http://www.healthanddna.com/drugreactiontest.html, (reprinted here with permission, copyright Genelex 2003.)

Drug Reaction Testing

Do not alter the dosage amount or schedule of any drug you are taking without first consulting your doctor or pharmacists.
"Research shows that of all the clinical factors such as age, sex, weight, general health and liver function that alter a patient's response to drugs, genetic factors are the most important. This information becomes even more crucial when you consider the fact that adverse reactions to prescription drugs are killing about 106,000 Americans each year—roughly three times as many as are killed by automobiles. This makes prescription drugs the fourth leading killer in the U.S., after heart disease, cancer, and stroke.

We currently offer CYP2D6, CYP2C9, and CYP2C19 screens that can help your physician or druggist predict your particular response to more than a quarter of all prescription drugs. These include such important medications as Coumadin (Warfarin), Prozac, Zoloft, Paxil, Effexor, Hydrocodone, Amitriptyline, Claritin, Cyclobenzaprine, Haldol, Metoprolol, Rythmol, Tagamet, Tamoxifen, Valium, Carisoprodol, Diazepam, Dilantin, Premarin, and Prevacid (and the over-the-counter drugs, Allegra, Dytuss and Tusstat). Click on the links at the Genelex Web site at: http://www.healthanddna.com/drugreactiontest.html to view a more complete list of drugs processed through these pathways.

Approximately half of all Americans have genetic defects that affect how they process these drugs. There are four different types of metabolizers, and we all fall into one of these categories for the variable pathways in Cytochrome P450 (this Cytochrome is responsible for creating the enzymes that process chemicals of all kinds through our bodies.) The easiest way to understand this is to picture a two lane highway.

- If you are the first type which is the norm, you would be an EXTENSIVE metabolizer. Both lanes of the highway are open and moving. Medications prescribed in normal doses will be metabolized by your body.

- If you are the second type, you would be an INTERMEDIATE metabolizer. This means that one lane of that highway is open and moving and the other lane is not, causing you to metabolize the medications more slowly. In this case you will need a lower dosage, and there is a chance of medications building up in your system causing adverse effects. It is especially important to monitor medications if you are in this category.

Intermediate metabolizers through the 2C9 pathway, for instance, have an increased risk of bleeding incidences when taking the common blood thinner Coumadin or Warfarin. For this reason, a recent article in the Journal of the American Medical Association, titled "*Association Between CYP2C9 Genetic Variants and Anticoagulation-Related Outcomes During Wayfarin Therapy,*" JAMA, April 3, 2002—Vol. 287, No. 13, recommends screening for CYP2C9 variants to reduce the risk of adverse drug reactions in these patients.

- The third type is a POOR metabolizer. In this case both lanes of the highway would be stopped. There is a possiblility that alternate routes can be found, but this type of metabolizer is potentially very dangerous, as there is a great chance for the medication to build up in your system making you very sick, or even killing you.

For example, a poor metabolizer of Phenytoin, a common antiepileptic would not be able to process the drug and would actually have an increased rather than decreased risk of seizure if prescribed this drug.

- The fourth type of metabolizer is ULTRA EXTENSIVE. This means you have additional lanes for processing, picture an Indy 500 speedway. In this instance, you literally burn through medications. If you were an Ultra extensive metabolizer through the 2D6 pathway and while in surgery and your doctor gave you codeine as a pain killer, you would receive no pain relief because the codeine would be metabolized so fast that it would have little or no effect on you.

The Testing Process

The process is simple. We send you a blood collection kit in the mail. You can either make an appointment with your doctor or we will provide you with the contact information for a phlebotomist in your area. Blood samples are overnighted to our laboratory and results are typically available in 15 business days.

Currently Available Tests

CYP2D6 (cytochrome P450 2D6) is the best studied of the DMEs and acts on one-fourth of all prescription drugs, including the selective serotonin reuptake inhibitors (SSRI), tricyclic antidepressants (TCA), betablockers such as Inderal and the Type 1A antiarrhythmics. Approximately 10% of the population has a slow acting form of this enzyme and 7% a super-fast acting form. Thirty-five percent are carriers of a non-functional 2D6 allele, especially elevating the risk of ADRs when these individuals are taking multiple drugs.

Drugs that CYP2D6 metabolizes include Prozac, Zoloft, Paxil, Effexor, Hydrocodone, Amitriptyline, Claritin, Cyclobenzaprine, Haldol, Metoprolol, Rythmol, Tagamet, Tamoxifen, and the over-the-counter diphenylhydramine drugs, Allegra, Dytuss, and Tusstat. CYP2D6 is responsible for activating the pro-drug codeine into its active form and the drug is therefore inactive in CYP2D6 slow metabolizers.

CYP2C9 (cytochrome P450 2C9) is the primary route of metabolism for Coumadin (Warfarin). Approximately 10% of the population are carriers of at least one allele for the slow-metabolizing form of CYP2C9 and may be treatable with 50% of the dose at which normal metabolizers are treated. Other drugs metabolized by CYP2C9 include Amaryl, Isoniazid, Sulfa, Ibuprofen, Amitriptyline, Dilantin, Hyzaar, THC (tetrahydrocannabinol), Naproxen, and Viagra.

CYP2C19 (cytochrome P450 2C19) is associated with the metabolism of Carisoprodol, Diazepam, Dilantin, Premarin, and Prevacid.

Other tests we are planning to provide include:

NAT2 (N-acetyltransferase 2) is a second-step DME that acts on Isoniazid, Procainamide, and Azulfidine. The frequency of the NAT2 "slow acetylator" in various worldwide populations ranges from 10% to more than 90%.

The advantages of Genelex's consumer genetic testing include:

- Safety. Decrease the chance that you will be the victim of an adverse drug reaction. More than 100,000 hospitalized Americans die of adverse drug reactions and two million outpatients have serious episodes each year. Knowledge of your DNA Drug Reaction Profile may help your physician or druggist prevent this from happening.

- Efficacy. Prescription drugs on the market today are usually prescribed by trial and error because they have been tested and approved in a "one size fits all" fashion. If you eliminate a particular drug more rapidly than the norm, taking the normal dose may be a complete waste of time. The drug simply will not work as prescribed, and it may take a while to discover this.

- Responsibility. Play a more active role in your health, your family's health, and in healthcare at large. DNA testing for drug reactions is just coming onto the market and you, the consumer, can help all of us take this major step toward better medicine."

Check out the common drugs processed by enzymes that Genelex tests. Genelex currently offers "DNA Prescription Drug Reaction Profiles that test 2D6, 2C9, and 2c19 functionality. The table below lists many commonly prescribed drugs that are metabolized through these pathways in the Cytochrome P-450 system. If your medication is not on the list, ask your pharmacist or physician about the metabolic pathway."

Common drugs prescribed by physicians or bought over-the-counter are processed by enzymes. Look at Genelex's Drug List by clicking on the PDF file to see Genelex's Drug list. See below, reprinted with permission, copyright Genelex, 2003. Or go to their Web site at: http://www.healthanddna.com/drugreactiontest.html and click on the PDF file to view the list of drugs at: http://www.healthanddna.com/drugchart.PDF. What you're looking for is prescription drug reaction testing. Your body metabolizes drugs differently than another person based on your individual genetic response. So look at the drug metabolization guide. Is your prescription drug listed there? What about drugs or supplements you buy over the counter? The typeset formatting is different on the PDF file. So look at the list below at: http://www.healthanddna.com/drugchart.PDF.

On the PDF file, you'll note that the formatting for numbers such as 2D6 is different in the PDF file than reprinted here due to formatting differences between a paperback book and the PDF file on the Web. So to get the correct formatting of the numbers such as 2D6 lined up with the correct drugs, use the PDF file to do your looking rather than the book, which is just a list of drugs tested. The PDF file has the correct numbers lined up with the correct names of the drugs.

Below is just a list of the drugs to give you an idea of what drug reaction testing is about and the names of the drugs listed. So use Genelex's PDF file, not the book if you want to see what numbers are lined up with what drugs. Below is to give you an idea of how a drug metabolization guide may be of benefit to a consumer when talking to your physician. The point is for you to learn how to become more involved in researching how your genes respond to what you put in your body.

Drug List

Prescription Drug Reaction Testing
Drug Metabolization Guide
(reprinted here with permission, copyright Genelex 2003.)

Substrates Metabolized through Cytochrome P-450

2D6
acetaminophen citalopram galanthamine norfluoxetine saquinavir
ajmaline clomipramine gallopamil nortriptyline sertinodole
alprenolol clozapine guanoxan octopamine setraline
amiflamine cobenzoline halofantrine olanzapine sildenafil
aminopyrine codeine haloperidol omeprazole S-metoprolol
amiodarone debrisoquine ibogaine ondansetron sparteine
amitriptyline delavirdine iloperidone opipramol tamoxifen

amphetamine deprenyl imipramine orto=Phebylphenol tamsulosin
amprenavir desipramine indinavir oxedrine tauromustine
aprinidine desmethylcitalopramindoramin oxycodone terfenadine
astemizole dexfenfluramine lidocaine parathion terguride
azelastine dextromethorphan lisuride paroxetine theophylline
benzydamine dextrromethorphan loratidine perhexiline thioridazine
bepranolol diazonin maprotline perphenazine timolol
bisoprolol diclofenac mequitazine phenacetin tolterodine
brofaramine diethyldithocarbamate methyl ester
methadone phenformin tramadol bromoperidol dihydrocodeine
methlyamphetamine pranidipine trifluperidol
buflomedil diltiazem methoxyamphetamine pregnenolone trimipramine
bufuralol diprafenone methoxyphenamine procainamide tropisetron
bunitrolol dolasetron metoprolol progesterone tryptamine
bupivacaine donepezil mexiletine promethazine tyramine
buthylamphetamine doxepin mianserin propafenone venlafaxine
captopril encainide milameline propranolol zolpidem
carteolol estradiol milameline propylajamaline zotepine
carvedilol estradiol minaprine prprofol zuclopenthixol
chlorophyros ethylmorphine mirtazapine quanoxan
chloropromazine ezlopitant nelfinavir mesylate quetiapine
chlorpheniramine flecainide nevirapine ranitidine
chlorzoxazone flunarizine nicardipine remoxipride
cilostazol fluoxetine nicotine retinol
cinnarizine fluperlapine nifedipine risperidone
fluphenazine norcodeine ritonavir
fluvastatin ropivacaine
fluvoxamine

Substrates Metabolized through Cytochrome P-450 2C19 2C9 ctd' 2C9 ctd'

amitriptyline aceclofenac tropisetron
citalopram acenocoumarol valproic acid
clominpramine acetaminophen venlafaxine
cyclophosphamide
alosetron verapamil
diazepam aminopyrine warfarin
imipramine amitriptyline zafirlukast

lansoprazole amitriptyline zidovudine
nelfinavir amprenavir zileuton
omeprazole antipyrine zolpidem
pantoprozole arachidonic acid
phenytoin artelinic acid
rabeprazole benzopyrene
bufuralol
butadiene monoxide
candesartan
carvedilol
celecoxib
chlorpyrifos
cinnarizine
cisapride
cloazapine
cyclophosphamide
dapsone
deprenyl
desmethyladinazolam
desogestrel
dextromethorphan
diallyl disulfide
diazepam
dibenzoanthracene
dibenzylfluorescein

Inhibitors Metabolized through Cytochrome P-450 2C19 2C9 2D6
cimetidine amiodarone amiodarone
flebamate fluconazole celecoxib
fluoxetine fluvastatin chlorpheniramine
fluvoxamine fluvoxamine cimetidine
ketoconazole
isoniazid clomipramine
lansoprazole lovastatin cocaine
omeprazole paroxetine doxorubicin?
paroxetine phenylbutazone fluoxetine

ticlopidine probenicid? halofantrine
sertraline levomepromazine
sulfaphenazole methadone
teniposide mibefradil
trimethoprim moclobemide
zafirlukast paroxetine
quinidine
ranitidine
red-haloperidol
ritonavir
terbinafine
chlorpheniramine
cimetidine
clomipramine
cocaine
doxorubicin?
fluoxetine
halofantrine
levomepromazine
methadone
mibefradil
moclobemide
paroxetine
quinidine
ranitidine
red-haloperidol
ritonavir
terbinafine

Inducers Metabolized through Cytochrome P-450
2C9 2C19 2D6
rifampin carbamazepine dexamethasone
secobarbital norethindrone rifampin?
rifampin

For immediate consultation Call 800-TEST-DNA (800-837-8362)
Hours 7:00 AM to 6:00 PM PST, 10:00 AM to 9:00 PM EST, fax 206-219-4000,
3000 First Avenue, Suite One, Seattle, WA 98121
E-mail: info@genelex.com Web: www.genelex.com
Se habla español. ©2003 Genelex Corporation

***

7

Scientists and Physicians Comment on Pharmacogenetics: Question: Where would consumers (with no science background) begin to search and learn about pharmacogenetics?

——-Original Message——-
From: Prakash Nadkarni
To: Anne Hart
Sent: Friday, August 15, 2003 9:19 AM
Subject: RE: query by writer

Hi Anne,

The world's biggest library, of course, is the Web (if you use a good tool like Google), and of course, you can walk to your nearest medical school if you need technical information-I know they shouldn't prevent use you from using the reference section unless they happen to be total A-holes (some of the Ivy league libraries are like that).

Re: interviews—E-mail is not a good medium for "interviews", though instant messaging probably is (I'm not sure how many scientists use it, though-I avoid it myself because I don't like to be interrupted while I'm in the middle of something.)

How it's going to help the average consumer—I don't know. My attitude is guarded optimism.

The cost of developing a new drug is frightfully high. If a new drug turns out to have an adverse effect that might result in its being pulled from the market, then, **if** it could be shown that this adverse effect has a clear-cut pharmacogenetic basis AND there is a very simple test that could detect susceptibility, then the drug could still be on the market.

—The flip side here : the really scary adverse effects are not all that common, they may occur in 1 of a 1000 individuals or less. Then you require a very large number of subjects to PROVE a pharmacogenetic basis for that adverse event. This costs a lot of money, and the FDA may not wait that long to pull the drug from the market if safer alternatives already exist. (There may be gray areas, though—troglitazone was one of the first of a new family of antidiabetic agents—the thiazolidinediones, or PPAR-gamma activators—that work by a unique mechanism. It was pulled from the market because of liver toxicity, but when this happened, many patients whose lives had been dramatically improved by the drug actually complained to the FDA that they were perfectly happy to continue taking it. Fortunately, newer and safer drugs from the same chemical family were developed, such as rosiglitazone and pioglitazone.

However, we don't know as yet whether there is a clearly and easily characterizable pharmacogenetic mechanism for thiazolidinediones-induced liver toxicity, so I'm really citing this as an example of a valuable drug that deserves and could get a second chance.

Similarly, if a genetic test indicates that a particular individual is potentially vulnerable to a dangerous adverse effect (for a potentially life-saving drug (e.g., an anticancer agent) which has dangerous side effects), then one could avoid this drug and seek an alternative drug to treat the same condition.

Re: Pharmacogenetics and diet, the lactose-intolerance of Han Chinese (as well as their greatly increased susceptibility to alcohol hangover with much more modest quantities of liquor) is well known, as is the fact that African-Americans require a larger quantity of eye drops to dilate the pupil prior to a retinal exam compared to people with light-colored eyes.

The complications here—it's not just a single gene for susceptibility, it's multiple genes (not to mention interaction of genes with environmental factors) that makes research in the field so damned difficult. Even when you have two people with the same susceptibility gene, the response to the same drug may vary tenfold,-there are numerous other genes that we haven't considered in the study that are responsible for the variation.

Genes and environment-repeated exposure to certain environmental agents may induce the expression of genes that result in greater tolerance to the agent. The ability of Russians to tolerate an astonishing amount of vodka (that they've been consuming since very young) is a case in point.

Also genes and age: study after study has shown that people who are susceptible to substance abuse (tobacco, booze etc.) are much more likely to get hooked if they're exposed to the substance during or before adolescence, which is why you have the laws about selling this stuff to minors. (We haven't studied brain plasticity enough—we don't know what's magical about the adolescence cut-off, but we do know that people who learn a foreign language by being immersed in that culture can learn speak it without an accent if exposed before adolescence, but not after.)

Hope this helps, and good luck with your book.

Regards,

Prakash

Prakash Nadkarni, MD
Associate Professor
Yale Center for Medical Informatics
New Haven, CT

To: 'Anne Hart'
Sent: Friday, August 15, 2003 12:07 PM
Subject: Re: query by journalist

Dear Ms. Hart:

Thank you for contacting the National Institute of General Medical Sciences, a component of the National Institutes of Health. NIGMS supports research in pharmacology, and more specifically, in pharmacogenetics. I know that you also contacted Dr. Rochelle Long at NIGMS, so I am responding on her behalf as well. NIGMS supports a network of pharmacogenetic researchers, listed at the following URL: http://www.nigms.nih.gov/pharmacogenetics/research_network.html. You could take a look at their research project summaries to get a sense for the kind of research that is going on. Also, you could try a Google News search on pharmacogenetics and/or pharmacogenomics to find published interviews/stories for context for your book.

Regarding consumer information, NIGMS has published two educational brochures on the topic. Online versions are posted at:
http://www.nigms.nih.gov/funding/medforyou.html

and
http://www.nigms.nih.gov/news/science_ed/genepop/

If you would like hard copies of either or both of these borchures, send an email request to pub_info@nigms.nih.gov.

Regarding the availability of genetic tests:

There are at least 2 pharmacogenetics tests that are widely available and offered by certified commercial clinical testing laboratories: TPMT and CYP2D6. There is no current standard for the widespread use of pharmacogenetics testing—yet. It is performed either as a difficult case referral to a tertiary care setting (e.g. Mayo Clinic), or due to patient interest in seeking it out (e.g. based on a web advertisement).

Physicians are divided on when adverse side effects that occur in rare patients should be dealt with: after they occur, or anticipated and prevented. Sometimes these events are uncomfortable (e.g. lack of pain control after taking codeine; once caught, a morphine based drug like vicodin can be used instead); sometimes these events are life-threatening (e.g. myelosuppression after an anticancer drug; once recognized, a lesser dose of the agent can be used).

The most modern research scientific approach is to tailor the drug choice and drug dose to the patient up front. But there are instances where this just simply isn't the prevailing standard of care, either for economic reasons, or for reasons of time, or for valid reasons (e.g. PTT testing in hospital after warfarin is performed, rather than pre-prescriptive genotyping makes sense, since genotype isn't the only influence on clotting time; diet matters too).

If you are interested in researching the industry angle:

Meeting report from last year's meeting of the NIH Pharmacogenetics Research Network (see especially the second keynote by Rebecca Eisenberg):
http://www.nigms.nih.gov/pharmacogenetics/2002meeting_report.html

Members of the NIH Pharmacogenetics Research Network Industry Liaison Group:
http://www.nigms.nih.gov/pharmacogenetics/
PRN_industry_liaison_roster.html

These folks, from small and big pharma and biotech, may be able to give you the view from the trenches.

A publication called The Scientist published a relatively savvy article on the topic last year; find the story at: http://www.the-scientist.com/yr2002/sep/research1_020930.html

(you have to register but it's free access to the full text)

I hope some of this information is helpful to you.

Alison Davis
Office of Communications and Public Liaison
NIGMS. NIH

-----Original Message-----
From: "Esteban Gonzalez Burchard, M.D."
To: "Anne Hart"
Sent: Thursday, August 14, 2003 9:33 PM
Subject: Re: Pharmacogenetics: What Can You Comment About Regarding Pharmacogenetics for the General Consumer?

Hi Anne:

I have attached a section that I wrote as part of a grant. You can use this as you please. As of today, there are not that many examples that are of use to the public. However there are examples developed at St. Jude Children's Hospital in Memphis in which they can test for mutations in the drug metabolizing enzyme that metabolizes specific chemotherapy drugs. In doing so, physicians can now identify who will be at risk for toxic levels of the chemotherapy drug.

The best publicly known example is alcohol and facial flushing in Asians. This is due to a genetic variant in the Alcohol Dehydrogenase Enzyme. This is a perfect example of a pharmacogenetic response.

Esteban

****

It is well established that among U.S. residents of similar socioeconomic status, there is greater cardiovascular morbidity and mortality among African Americans and Latino Americans than among Caucasians.[1-4] In addition, there is marked clinical heterogeneity in the prevalence and treatment of hypertension among specific ethnic and racial groups.

Specifically, there are marked differences in drug response to antihypertensive therapy with angiotensin converting enzyme inhibitors (ACE-I) among African Americans than among Caucasians.[5,6]. Although there are many potential explanations for this observation, including environmental and socioeconomic factors, one potential explanation is that the genetic predisposition to hypertension and to differences in drug response differs among racial and ethnic groups. In particular, we propose that genetic differences may explain, in part, the differences in response to ACE inhibitors between African American and Caucasian Americans.[7]

Hypertension among U.S. Populations: Cardiovascular disease is a substantial public health problem and a leading cause of death for all U.S. ethnic and racial minorities. The impact of premature morbidity from cardiovascular disease is devastating in terms of personal loss, pain, suffering, and effects on families and loved ones. The annual national economic impact of cardiovascular disease is estimated at $259 billion as measured in health care expenditures, medications, and lost productivity due to disability and death. <http://raceandhealth.hhs.gov/3rdpgBlue/Cardio/3pgGoalsCardio.htm>
A major modifiable risk factor for cardiovascular disease is high blood pressure (hypertension).

Disparities exist in the prevalence of risk factors for cardiovascular disease. Racial and ethnic minorities have higher rates of hypertension, tend to develop hypertension at an earlier age, and are less likely to respond to conventional antihypertensive therapies to control their high blood pressure.[6] For example, from 1988 to 1994, 35 percent of black males ages 20 to 74 had hypertension compared with 25 percent of all men. When age differences are taken into account, Mexican-American men and women also have elevated blood pressure rates. Among adult women, the age-adjusted prevalence of overweight continues to be higher for black women (53 percent) and Mexican-American women (52 percent) than for Caucasian women (34 percent). Although there are multiple potential explanations for the observed differences in the prevalence of hypertension, possible explanation include biologic or genetic differences.

Clinical and racial heterogeneity in response to treatment of hypertension:

Numerous large-scale clinical trials of therapy for heart failure and hypertension over the past decade have shown improvements in outcome with angiotensin converting enzyme inhibitors (ACE-I).[8–13] Interestingly, data from the second Vasodilator-Heart Failure Trial (V-HeFT II) indicated that although enalapril therapy was associated with significant reduction in the risk of death from any cause among white patients, no such benefit was observed among black patients.[14] In addition, the Beta-Blocker Evaluation of Survival Trail found significant benefits with the beta blocker bucindolol among white patients but not black patients.[15] A reduced response to ACE inhibitors in black patients as compared to white patients is well documented.

Decreases in blood pressure in black patients with hypertension have been shown to be less than white patients given similar doses of ACE inhibitors [16] and beta blockers.[17] Taken together, these data may suggest that there are racial/ethnic differences in therapeutic response to the treatment of heart failure and hypertension among different racial/ethnic groups. More importantly, these trials have raised the possibility that black patients with heart failure and hypertension may not benefit from commonly used doses of recommended therapies to the same extent as white patients.

Genetic differences among races: Naturally occurring genetic variants or polymorphisms of genes are an important source of genetic diversity. Single nucleotide polymorphisms (SNPs) account for over 90% of all human DNA polymorphism.[18] It is well established that there are differences in allele frequencies among various ethnic groups.[19,20] By determining the degree of disease risk associated with particular alleles, in conjunction with the allele frequency for a defined population, one can derive information about the population attributable risk, which is relevant to the public health impact for that population.

Genetics of Hypertension: In favor of genetic factors being an important in the pathogenesis and treatment of hypertension are observations from studies in families, twins, and clinical pharmacologic trials performed in ethnically diverse groups.6,21–28 Although there have been no systematic genome-wide screens performed across racial or ethnic populations, there have been individual investigations in African Americans, Mexican Americans, Chinese and Caucasians.29–33

All of these investigations found linkage between loci and hypertension. Interestingly, most of these loci differed by the ethnic population studied, suggesting genetic heterogeneity. This may suggest that there are "ethnic specific" genetic risk factors that are associated with hypertension in different racial and/or ethnic groups with resulting differences in drug response to anti-hypertensive drugs.

Genetic sequence variants among hypertension candidate genes: There is increasing evidence to suggest that drug metabolism alone does not account for the observed interindividual variability in drug disposition or response, but that other processes, including drug transport, are important determinants of drug disposition.[34,35] Drug transporters in the gastrointestinal tract are a major determinant of oral bioavailability of drugs whereas transporters in the kidney and liver are major determinants of renal and hepatic clearances.

Transporters in specific tissues (e.g., the blood brain barrier) control tissue specific distribution of drugs. The pharmacokinetics of the ACE inhibitors have been reviewed recently.[36,37] It is important to note that the absorption of several ACE inhibitors is low (approximately 40% of an oral dose is absorbed) and variable and appears to be controlled by the oligopeptide transporter, Pept1.[38]

Several ACE inhibitors are eliminated by active secretion in the kidney, and appear to interact with renal organic anion transporters (e.g., OAT1 and OAT3). Thus, genetic variation in transporters involved in absorption and elimination of ACE inhibitors will contribute to variation in drug response. It is possible that variants of these transporters present in different frequencies among various ethnic populations may explain (in part) ethnic differences in response to ACE inhibitors.

Summary and Significance: Currently available data indicate that cardiovascular morbidity and mortality disproportionately affect U.S. ethnic and racial minority groups. This proposal is the first phase of a systematic effort to identify biologic and genetic explanations for these differences in drug response to conventional antihypertensive therapies. We will recruit a local cohort of ethnically diverse subjects who will undergo phenotype analysis and genetic testing of candidate genes thought to be involved in the transport of anithypertensive drugs.

As part of the overall PMT project these genetic variants will be further tested with in vitro functional assays to determine the biologic relevancy of the identified genetic variants. As part of future studies, these subjects will be called back to participate in further pharmacokinetic testing. Subjects will be stratified by their respective genotypes and then undergo pharmacologic testing with conventional antihypertensive therapies.

These investigations will allow us to determine whether pharmacogenetic relationships exist. More importantly, these investigations will allow us to identify ethnic-specific differences in drug response by genotype. The overall aims of this proposal are to better understand the genetic factors that may contribute to the increased disparity from cardiovascular and hypertension morbidity and mortality seen among ethnically diverse populations. These aims are consistent with those established by Healthy People, which are to reduce disparities in health among different population groups.

Esteban González Burchard, M.D.
Pulmonary and Critical Care Medicine
Department of Medicine
University of California, San Francisco

Mailing Address:
    University of California, San Francisco
    Box 0833
    San Francisco, California 94143

Shipping Address:
    UCSF DNA Bank/Lung Biology Center
    San Francisco General Hospital
    Bldg 30, 5th Floor, Room 3501H
    1001 Potrero Avenue
    San Francisco, California 94110

From: Wein, Harrison (NIH/OD)
To: 'Anne Hart'
Sent: Friday, August 15, 2003 11:36 AM
Subject: RE: Thanks for replying

I just got the latest table of contents for *Environmental Health Perspectives*, and noticed that they feature pharmacogenomics in their new issue. There are a couple of articles in the issue. Here's the link: http://ehp.niehs.nih.gov/txg/docs/2003/111-11/toc.html.

Harrison

From: "Daniel Rubin"
To: "Anne Hart"
Subject: Re: query by writer
Date: Friday, August 15, 2003 5:15 PM

You can find educational information on our PharmGKB web site that should provide much of what you're looking for:
http://www.pharmgkb.org/do/serve?id=resources.education

Also, NIH as an educational brochure on pharmacogenetics for the public that may give you what you're looking for in terms of how this will benefit the public: http://www.nigms.nih.gov/funding/medforyou.html

I recall that Time magazine did a very nice piece in 2001 on pharmacogenomics directly addressing exactly what you're talking about, and that you should be able to find in a community public library.

Regards,

Daniel

UCSD Scientists Develop Novel Ways to Screen Molecules Using Conventional CDs and Compact Disk Players

Molecules attached to CDs in new techniques can screen for proteins. According to a UCSD August 18, 2003 press release, chemists at the University of California, San Diego have developed a novel method of detecting molecules with a conventional compact disk player that provides scientists with an inexpensive way to screen for molecular interactions and a potentially cheaper alternative to medical diagnostic tests.

A paper detailing their development will appear this week in an advance on-line edition of the journal Organic and Biomolecular Chemistry (http://xlink.rsc.org/?DOI=b306391G) and in the printed journal's September 21st issue.

"Our immediate goal is to use this new technology to solve basic scientific questions in the laboratory," says Michael Burkart, an assistant professor of chemistry and biochemistry at UCSD and a coauthor of the paper. "But our eventual hope is that there will be many other applications. Our intention is to make this new development as widely available as possible and to see where others take the technology."

"The CD is by far the most common media format in our society on which to store and read information," says La Clair. "It's portable, you can drop it on the floor and it doesn't break. It's easy to mass produce. And it's inexpensive."

Burkart and James La Clair, a visiting scholar in Burkart's laboratory who initially developed and patented the technique, said that since scientific laboratories often rely on laser light to detect molecules, it made sense to them to design a way to detect molecules using the most ubiquitous laser on the planet—the CD player.

Their technique takes advantage of the tendency for anything adhering to the CD surface to interfere with a laser's ability to read digital data burned onto the CD.

"We developed a method to identify biological interactions using traditional compact disk technology," explains La Clair, who provided the patent rights to the method to UCSD. "Using inkjet printing to attach molecules to the surface of a CD, we identified proteins adhering to these molecules by their interaction with the laser light when read by a CD player."

While usually anything, like a scratch on the CD surface, that would interfere with the detection of the bits of information encoded on a CD would be a drawback, the UCSD researchers actually exploited this error to detect molecules.

"That's the novelty of this," Burkart points out. "We are actually using the error to get our effect."

The typical CD consists of a layer of metal sandwiched between a layer of plastic and a protective lacquer coating. When a CD is burned, a laser creates pits in the metal layer. A CD player uses a laser to translate the series of pits and intervening smooth surface into the corresponding zeros and ones that make up the bits of digital information.

To do molecular screening, the researchers took a CD encoded with digital data, and enhanced the chemical reactivity of the plastic on the readable surface. They then added molecules they wanted to attach to this surface to the empty ink wells of an inkjet printer cartridge and used the printer to "print" the molecules onto the CD. This resulted in a CD with molecules bound to its readable surface in specific locations relative to the pits in the metal layer of the CD encoding the digital information. When the CD with these molecules attached is placed in a CD player, the laser detects a small error in the digital code relative to what is read from the CD without the molecules attached.

To detect proteins or other large molecules in a solution like a blood sample, the modified CD is allowed to react with the sample solution. Like a key that only fits in a certain lock, some proteins bind to specific target molecules. Thus, specific molecules on the surface of a CD can be used to "go fishing" for certain proteins in a sample. The attachment of these proteins will introduce further errors into the reading of the CD. Furthermore, since the molecules on the surface of the CD are at known locations relative to the bits of encoded information, the errors tell the researchers what molecules have attached to their target protein and, thus, whether or not that protein is present in the sample.

"James has even done this using CDs with music, like Beethoven's Fifth Symphony," says Burkart. "And you can actually hear the errors."

"How many people on this planet can actually hear a molecule attached to another molecule?" asks La Clair.

While a few bugs need to be ironed out before the technique can be used to accurately quantify the amount of a given protein in solution, Burkart plans to apply it immediately to help him screen for new compounds in his natural products chemistry research laboratory. Compared to the $100,000 price tag for a fluorescent protein chip reader, he points out, a CD player costs as little as $25.

The researchers envision many other potential applications for this technology outside the laboratory, particularly in the development of inexpensive medical diagnostic tests, now beyond the means of many people around the world, particularly in developing countries.

"In theory, anyone who has a computer with a CD drive could do medical tests in their own home," says La Clair. The researchers hope that by openly publishing their development in the scientific literature, others will customize the technology in a variety of ways, eventually leading to a wide range of inexpensive new diagnostic kits and other beneficial products.

"We plan to make this fully available and see what people come up with," says Burkart.

For additional information, see: http://discode.ucsd.edu/

***

8

DNA Testing for Nutritional Genomics and Ancestry

What does ancestry have to do with how your genes respond to food or medicine? Certain ethnic groups or 'races' respond in different ways to different medicines, dosages, foods, exercise, illnesses, lifestyle changes, and stress. Since most people have a mixture of genes from several races or ethnic groups going back to prehistoric times, only a DNA test of specific genetic markers or genes can reveal what you're are at risk for regardless of your dominant ethnicity. The genes respond in specific ways to chemicals in the environment and other factors.

That's why you can interpret a report, talk it over with your physician, and tailor what you eat, what drugs you take and the dosages for your condition, to what your genetic expression requires to be healthy. It's looking at you at the cellular level, the molecular level. Most people don't even know what DNA is or have no science background, so they need an easy to understand consumer's guide to nutritional genomics and pharmacogenetics. Consumers want to know how their individual, specific genes and possibly their ancestry respond to food—selected foods tailored or prescribed to nourish your individual genetic profile known as your *genetic expression*. Consumer involvement in nutritional genomics is important as it is in pharmacogenetics. How many products need to be tailored to your genes? Food? Medicines? Cosmetics? What else? Let's look at the issues.

Nutritional genomics, often abbreviated as 'nutrigenomics' is about increasing that success rate. How will science working together with the consumer tackle the issues confronting us as the population ages? Consumer involvement can democratize the science of nutritional genomics by improving diets for better health. You can ask to work on ethics boards or create your own. How is discovering deep ancestry through DNA testing related to the ways that food affects your health?

Ancestry and diet are linked by biology, culture, and choices. It's all about collaborating with your genes. Do you choose your food by habit or biology? Consumers need a guide to DNA testing for nutritional genomics as well as for ancestry and family history. Specific genetic variants *interact* and relate to nutrition.

Learn to interpret the expression of your genes before you count your calories. If you're supposed to eat 'bright' for your 'genotype,' then you begin by mapping your genetic expression and learning how the raw data applies in a practical way to what you consume. This means genetic testing, interpretation, and application to food.

Are you having your DNA tested to see how your genetic signature responds to the medicines you take? What's your genetic response to food, medicine, exercise, or nutraceuticals? If you're concerned about adverse reactions, are you having your genes screened, particularly CYP2D6, CYP2C9, and CYP2C19—to see what your particular response is to prescription medicines? What about the food you eat? Can you tailor what you eat to your genotype?

Does Your DNA Have a Core Identity That's Both Cultural and Biological? Not only can you trace your family history and ancestry with molecular genealogy or gene testing. You can create a DNA and genealogy time capsule. What are the relationships between your deep, ancient ancestry, your DNA, migrations, and how you customize what you eat to your genetic signature or how you tailor any medicine or supplements you take to specific genes and markers?

Are you interested in foods tailored to your genotype and have no science background? Curious about pharmacogenetics—exploring how your doctor tailors your medicine to your genes? Thinking about eating bright for your genotype? What about DNA and Ancestry? Or perhaps you would like to find out the definition of ***genomics, proteomics, metabolomics, or lipidomics***? How far can we take familiar genealogy and oral history techniques and link them to DNA testing for a variety of reasons from tailoring what you eat to your DNA to customizing what over-the-counter and prescription drugs your physician wants you to take? Before you take anything, ask your physician how your genes will respond to what you put into your body. What DNA and other tests can you take that will be responsible, credible, and appropriate measures for your present or future needs?

Genetic Testing When You're Unhappy With Your Doctor's Response to Questions?

Back in 1995, as a medical journalist, I felt unhappy with my doctor's response to my questions regarding different treatment methods for the symptoms of menopause. I was in my early fifties then, and before DNA testing was readily available to consumers, I decided to do my own research to find answers to questions. The commonly prescribed "one-size fits all" estrogen-progestin pills I received from my primary care physician at my HMO as I entered menopause was prescribed with the words, "If you go off your HRT, it would be like a diabetic going off insulin." I didn't quite believe those words, but that's what the doctor said when I asked about when I could go off the HRT. So I took my pills like he said, and within a few months, my cholesterol levels changed for the worse.

The pills worsened my genetic hypertension. Still, the doctor insisted I not stop taking the HRT. I went to another doctor who changed the routine. Instead of taking an estrogen and a progestin pill each day, he prescribed estrogen some days of the month and progestin other days of the month and warned that, "I'd better do something about my hypertension or I'll have to go on 'meds.'" And still, I felt worse on the estrogen and progestin the more months that past. After eleven months, I dumped the HRT and didn't go back to the HMO. It was time to do some research and become more involved in my own healthcare as a consumer.

At that time, around 1996, all those Web articles on the dark side of soy didn't yet come to my attention. The universities I visited were doing clinical trials with natural soy products for menopause, and magazine articles were telling me I had to have those 29 grams of soy each day for my bones and arteries. I didn't know who to believe. There were magazines warning of too much stimulation to the thyroid from too much soy isoflavones. I didn't know who to believe yet, but what I wanted was information.

Some nurses told me about natural yam-derived progesterone as an alternative to hormone replacement therapy (HRT). At first, not realizing the genetics of hypertension, I thought that the estrogen in HRT was the cause of my dangerously high blood pressure. Using the Internet to access newsgroups and WWW sites, I sought the alternative therapies and tried to find out whether they really worked. Was there anyone out there saying the studies had been flawed? And if so, how could I believe the critic?

What credentials or credibility did he or she have? And if there were no critics saying studies were flawed, perhaps no studies were ever performed? Who was right? So I turned to the medical journals on the shelves at the local Medical School library. I also turned to the Internet and to physicians online willing to answer my questions about these treatments. What I found were people who had the time to answer questions, but none were physicians.

Were there actual physicians online or available willing to answer questions? Who had the time and who had the credibility? Was it true that unproven holistic health and naturopathic or homeopathic remedies that worked a century ago came back now, and did they really work? Was it true that natural ingredients were better, but not used because the big drug companies could make a profit off of them? How much was snake oil and how much was plant-based cures? Who could I believe?

More important, what would work with my body, nasty genetic defects and all? And when my husband needed hernia surgery, how could the Internet answer questions? This was 1995, before genetic testing was available to consumers.

All I had to go by was the awful, scary reactions I had to dental anesthesia when having a compact wisdom tooth removed at age 26. The IV went in, and the next thing I started yelling I can't breathe. What were my genes telling me? Or the carbocaine for the root canal that caused my panic attack, tremors and convulsing, whereas the specific type of mepivacaine worked well and I remained calm because it didn't have vasoconstrictors to send my body into a panic response, unleashing catecholamine and other hormones that raise the blood pressure, heart rate, fear, tremors, and convulsing because of the special genetic response of my more or less defective autonomic nervous system to the anesthetic.

The popular archaeology magazine article clued me in that I might have a defect in my autonomic nervous system by reading a popular archaeology magazine article about ancient Egyptian mummies that mentioned an article in a medical journal. One eye pulled to the corner that shows up on photos or portraits, a withered left side of the face....It was all pointing to my panic disorder. Even an astrologer mentioned Neptune on the rising sign signaling my anxiety gene. I had inherited this gene from my father, who claimed he was too nervous to ever learn to drive. I never learned to drive either, and I passed the "anxiety genes" on to my son, the physician. What I was looking for was more than vague explanations of "anxiety genes." I had to pinpoint the response, that means in my situation, a hyperinsulinism response or insulin resistence, and to explore the possible need for what a decade later became known as "the Syndrome X diet." It all was genetic, or was it?

Of course, back in 1995 and 1996, I couldn't find a DNA test offered to consumers on the Internet. Instead I used the Web to search for alternative medicine and holistic health articles and supplements. Take a look at my article which appeared in **Internet World** 1996 February : 42–44, 46, 48. Reprinted with permission. It is also mentioned in archives online. See HEALTHNET NEWS VOL. XI, NO. 4 WINTER 1995, Lyman Maynard Stowe Library University of Connecticut Health Center: http://library.uchc.edu/departm/hnet/winter95.html

***

9

Menopause and Beyond Alternative Resources and Information Online

Alternative Resources Online for Menopause and Beyond

Through the Internet, you can obtain information your doctor hasn't got time to give you, and talk to people who share your problems.

By Anne Hart

I have found that doctors don't always have the time or breadth of knowledge to discuss alternative, customized solutions to my family's healthcare needs. That's why I turned to the Internet. After searching the Net for Web sites, support groups, or little-known health newsletters, I found dozens of physicians who offered to give me answers to my specific questions via e-mail.

Each member of my family required a different healthcare service and support group to address their particular problems. In my case, I am allergic to synthetic progestin, which eases the side effects of menopause. At the health-maintenance organization (HMO) to which I am a member, I asked my doctor for natural, yam-derived progesterone. He refused to prescribe it because he said it wasn't yet approved by the FDA, which left me with no alternatives to prevent osteoporosis, for which I'm at high risk. Instead I was put on hormone-replacement therapy, which involves taking oral estrogen. I believed that the oral estrogen was raising my blood pressure sky high. When I mentioned this to my doctor's nurse, she snapped, "Prove it."

So I set off on my journey to prove it. First I checked the **alt.support.menopause** newsgroup, which was extremely helpful. I posted the following question: *Is anyone else using high soy and vegetarian diets, the herb black cohosh, and natural, yam-derived, progesterone cream for menopause—to help prevent bone loss—instead of the usual estrogen and progestin? If so, what are your comments and experiences?*

The replies were practical, useful, and factual, providing medical references, titles of medical journal articles, and book bibliographies, as well as personal experiences and encouragement. For example, one person pointed me to an article entitled "Risks of Estrogens and Progestogens" in the December 1990 issue of *Maturitas*, an English-language European medical journal. The author, Dr. Marc L'Hermite, found that five to seven percent of women on conjugated equine estrogens could get severe high blood pressure and that they would return to normal when the hormone-replacement therapy was withdrawn. A bibliographic reference to this article also appeared in Dr. Lonnie Barbach's book The *Pause*. Not one physician at my HMO had mentioned these concerns.

To obtain more information about what alternative health solutions were available and how particular products would change my body or health, I searched the Web under the keywords "menopause," "alternative healthcare," "herbs," "homeopathy," and "naturopathy." I also looked under "natural progesterone."

Through this search I found the *MenoTimes*, a quarterly journal published in San Rafael, Calif. I subscribed because it had the information I sought on Dr. John Lee's book *Natural Progesterone: The Multiple Roles of a Remarkable Hormone*.

Also through my search I found a laboratory that would test my saliva and tell me whether my hormones were balanced. Most of all, I wanted to know how taking natural progesterone would affect my fast thyroid and adrenaline-drenched body, with its low blood sugar and excess insulin production. My HMO physicians did not answer these questions, but told me that going off conjugated equine estrogen and synthetic progestin was like a diabetic going off insulin. On the Net I found physicians who answered my letters, labs that sold natural progesterone cream that I could use to prevent bone loss after menopause, and other labs that could monitor my condition until I found a doctor in my community who would listen to alternative solutions to menopausal questions.

I even found the Menopause Matters page created by Susan Czernicka, who said she "learned about herbal treatment of menopause when there were too few resources available to help her with her many symptoms and too few medical providers with open minds." She will answer questions sent to her at susan270@world.std.com.

For those who don't have Web access, there is the Menopause mailing list. To subscribe, send e-mail to listserv@ psuhmc.hmc.psu.edu with **subscribe menopaus** *Your Name* in the message body.

The "black cohosh" that I mentioned in my letter to the **alt.support.menopause** newsgroup is an estrogenic herb and vasodilator, and I wanted to find out how safe it was and whether it was as good for menopause as the homeopaths and naturopaths claimed, as well as how much to use and what effects it would have. I found the **alt. folklore.herbs** newsgroup helpful, as well as articles by Anthony Brook entitled "Why Herbs?" and "Historic Uses of Herbs" at the Drum Holistic Herbs page.

I wanted to find out everything I could about natural progesterone, so I went to the Health and Science page at Polaris Network, http://www.polaris.net/~health/ which described itself as "A Guide to Understanding and Controlling PMS, Fertility, Menopause, and Osteoporosis." It contained information about natural progesterone and how it balances the side effects of unopposed estrogen, how it's required for proper thyroid function and progestin counterparts in the drug industry. It also offered a seemingly sound scientific and unbiased evaluation of how certain hormones affect the system, what the hormone's results are on various bodily functions, and side effects. And it had an excellent bibliography of books and medical journal articles on osteoporosis reversal using natural progesterone.

I wanted to query a physician about the high blood pressure resulting from the equine oral estrogen I was taking, and at the Atlanta Reproductive Health Centre page I was able to send e-mail to a doctor who answered my question quickly, providing information that I could consider when making my final decisions or in looking further. His answer was more to the point than the counseling I had received from my own physician.

Another doctor of mine had wanted to give me high blood pressure medicine on top of the estrogen and synthetic progestin. I asked him to consider the alternative—taking me off the hormones to see whether a low-salt diet and exercise could change things—because before I went on hormones my blood pressure wasn't high.

The Internet became one of my best alternative healthcare information resources after the last of three reproductive endocrinologists I saw (not affiliated with my HMO) told me to wait two months to see how I felt off the hormones.
Menopause is a mega-business. More than 100 books a year fill the store shelves on this subject and more than 3,500 baby boomers enter menopause daily. This captivated audience represents a huge market for makers of hormones, vitamins, and other specialty products, much of which is advertised on the Web. So a wealth of information has appeared on the Internet to meet the needs of the 38- to 55-year-olds undergoing menopause, as well as younger women with PMS, infertility, and contraception concerns.

Another resource I found informative was the
Women's Health Hot Line newsletter. Its topics include infertility, endometriosis, contraception, sexually transmitted diseases, stress, menopause, and PMS. Most of all, it doesn't close its pages to alternative therapies for women who can't tolerate standard hormone replacement therapy.

Hernia Hunt

This year my husband needed hernia repair surgery, but his busy HMO surgeon spent only a very short time briefing him a month before his surgery. Because it would take hours to describe in detail what is done during hernia repair surgery, he searched on the topic using Web search engines, which yielded a list of information about all kinds of hernias. The most thorough site discovered was the Hernia Information Home Page in England. It included articles that explained such things as the benefits of using mesh rather than stitches to close incisions. Information about hiatal hernias and diaphragmatic hernias could be found at the Collaborative Hypertext of Radiology. There was more information about hernias on the Net than my husband could possibly find time to read.

After seeing shark cartilage in many health food stores, my husband asked his surgeon about its ability to aid in faster wound-healing. The surgeon laughed, yet I found several references to articles discussing shark cartilage on the Internet. Some medical journal articles on the healing and other properties and uses of shark cartilage can be found on the <u>Simone Protective Pharmaceuticals</u> page. Also at the site I found health-style questionnaires, in-depth descriptions of a variety of nutrient products, and how each product affects the body. You'll also find information about where to order or buy shark cartilage from pharmacists.

Search Tips

From the many healthcare sites I visited, my three-ring binder is packed with more than 500 pages of answers to questions. Productive keyword searches can be made using terms such as "alternative health," "healthcare," "medical," "medicine," "nurses," "nutrition," "pharmacy," "physicians," and "smart drugs." I found the search word "healthcare" to be more specific for asking personal medical questions than trying to search under the word "health."

Some sites are best reached through links on dedicated healthcare Web pages. For example, Subhas Roy has a created a page with links to 25 other Internet health sites at <u>Health Info</u>. There also is a large collection of links at the <u>Internet Medical and Mental Health Resources</u> page. It's maintained by Jeanine Wade, Ph.D., a licensed psychologist in Austin, Texas. One particularly comprehensive directory of health and medical sites on the Web is <u>MedWeb</u>, which lists Web sites and mailing lists in 70 categories, from AIDS to toxicology.

One of the best medical referral Web sites I found was <u>Richard C. Bowyer's</u> page. (You'll find it when you scroll past all his genealogy information.) Bowyer's page has links to numerous sites, such as U.S. hospitals, medical resources, medical journals, medical schools, medical students, medicolegal resources, oncology, pathology, and different surgical disciplines, such as plastic surgery, general surgery, laparoscopic surgery, and telesurgery. Healthcare workers of all specializations also can find job opportunities on some of these sites.

My son, who recently became a physician, was interested in <u>Medline</u>, a collection of medical and scientific reports used by physicians, articles on the educational needs of physicians and the public, physicians' supplies, prescriptions, and advice by pharmacists about drugs. Clinical cancer information that is intended for physicians is useful to patients as well. There is a listing of surgeons according to the types of surgery they perform, the notes of the Physician Reliance Network, and a Gopher menu of physicians listed according to their specialty.

As for informative newsletters available on the Net, I highly recommend the University of California at Berkeley Wellness Letter at the Electronic Newsstand. It contains the latest news of preventive medicine and practical advice—including information on nutrition, weight control, self-care, prevention of cancer and heart disease, exercise, and dental care.

As a medical journalist, I found the Journal of the American Medical Association (JAMA) to be a reliable resource.

If you are looking for a description of a particular drug, the Physician's GenRx Web site provides a database of drugs you can search. You must register first.

In seeking answers to my health questions, I found the Internet to be a valuable source for a wide range of healthcare information. I'm sure your efforts will be rewarded, too.

Selected Health Sites

Anesthesiology & Surgery Center *Offers travel warnings and immunization, medical dictionary.*
Cancer Related Links *Facts, figures, prevention.*
Harvard Medical Gopher *Harvard publications, access to the Countway Library's online catalog (HOLLIS), and medicine.*
Health Letter on the CDC
Health on the Internet Newsletter *Links to NewsPages, CNN Food & Health, MEDwire, describes new health-related Web sites, topic of the month.*
HyperDOC *Sponsored by the National Library of Medicine.*
LifeNet *Positive thinking and right-to-life point of views. Discussion about euthanasia.*
The Mayo Clinic *tour of the Mayo clinic; description of programs.*
Med Help International *Provides medical information about many illnesses. Treatment is described in layman's terms.*
Mednews *A biweekly newsletter that welcomes submissions.*
Medical Information Resource Center *Lists and a referral directory.*
Medicine OnLine *Career-related educational content and discussion groups.*

Medscape *For health professionals and consumers. Bulletin boards and a quiz testing your knowledge of surgery. Registration required.*
The National Organization of Physicians Who Care *Nonprofit organization ensuring quality healthcare. Newsletters, articles about Medicare reform and HMOs.*
OncoLink, the University of Pennsylvania Cancer Resource *Thorough information about pediatric and adult cancers.*
Physicians Guide to the Internet *For new physicians.*
Robert Wood Johnson Foundation Gopher *Biological archives at Indiana university, molecular biology database, NYU Medical Gopher.*

***

10

Nutritional Genomics for the Consumer: How Are You Managing Your Gene Expression?

How are you managing your gene expression? In what direction are you moving? How do you make more intelligent *choices* of food to *nourish* your individual *genotype*? What is meant by *intelligent foods* that target and nourish *specific* genes?

Clinical dieticians and nutritionists, by allying with molecular geneticists, genetics counselors, physicians, molecular *genealogists*, family historians, phenomics professionals, nutritional and medical anthropologists, and archaeogeneticists are *collaborating* with consumers of genetics testing, but what are they really *sharing*?

If not so much raw materials such as DNA from donors, is shared, then how about *access* to information—databases and various discussion forums online and e-mailing lists equally open to consumers, licensed healthcare providers, and research scientists? Who controls access to new research—the consumers, the corporations, or the scientists?

Can the average consumer afford to find out what to eat for improved health and nourishment based upon tests of genetic expression? Can consumers *override* any inherited risks revealed in the genetic signature with foods and nutraceuticals individually tailored? What does it mean to eat 'smarter' foods that target specific genes compared to eating more intelligently regarding choice?

Scientists compare genetic distances between populations by comparing the frequencies of forms of genes called 'alleles.' Mutant alleles can be mapped as population genetics markers. Some, but not all mutations in genes may put you at risk for certain chronic diseases if you eat the wrong foods for your genotype. The solution is to eat more 'intelligent' foods customized to your individual genetic profile.

Research also looks at rare alleles. Their rarity gives them special power as markers of genetic similarity. There's a good chance two identical mutant alleles share a common origin. You can map genes for ancestral origin, migrations, or to reveal risks of disease depending upon which genes you map.

This book is for beginners with no science background. It's a consumer's guide book to nutritional genomics—genetic testing and profiling for foods tailored to your genotype and ancestry. The chapters also are about how to interpret DNA testing for family history and ancestry.

How do you as a consumer, not a scientist, choose the smartest food tailored to your genetic signature? How do you interpret your DNA test results for ancestry or family history? What is the *link* between tailoring your foods to your genetic expression and tracing your ancestry though DNA testing? And what genes are tested for either reason? How do you bridge the gap between nutritional genomics profiling and testing DNA for deep ancestral origins?

Does ethnicity play any role in tailoring your food and nutraceuticals, drug dosages, or healthcare? How much can the average consumer self-educate and/or start a private DNA bank for a consumer or patient group? How do you raise funds, contract with research scientists, and form or serve groups needing their DNA researched for specific reasons? How does learning how to interpret the results of your DNA tests for ancestry relate to understanding genetic tests for cardiovascular or other inheritable risks?

Start researching on your own what you need to know as a consumer to have more *choices* in customizing foods for your genetic signature—your genotype. What are some realistic applications of genetic testing and profiling?

This book will lead you to find out more about taking control of what happens to DNA that you may donate for research. You'll find out how to be in charge of your own nourishment and nutrition. Genetic profiling helps you to customize what you eat. How do you nourish your body? What can your genes reveal to you through genetic testing and profiling? It's your private information and should remain private. A good place to release it finally would be in a time capsule and history scrapbook for your heirs. Here are how some branches of human genetic history are linked to your nutrition, ancestry, and most of all nourishment.

Prosopography is all about human history and genes that travel because your genes have both a cultural and a biological component. The cultural component includes onomastics which is the study of the origin of a name and its geographical and historical utilization. Proteomics is about drug discovery. Pharmacogenetics researches how your genes respond to various drugs and dosages to avoid adverse reactions to medicines. Nutrigenomics (nutritional genomics) brings together nutrition and genetics.

Put all these branches of molecular genetics together with molecular genealogy. Add nutritional genomics—molecular nutrition, and what do you have? Knowledge of how every molecule in your body responds to certain foods, lifestyles or exercise can now be studied at the molecular level. New sciences such as pharmacogenetics open doors to learning how your genes respond to nutrition and nourishment. Maybe you want to know how your body responds to certain herbs, nutraceuticals (supplements), foods, or any chemical in your environment, even a skin lubricant or salve or a cosmetic. It's all within the sciences of pharmacogenetics and/or nutrigenomics. Today, it's not just about how your body responds. You look at the molecular level, the cellular level to study how your genes respond to nourishment, medicine, lifestyle, and the environment.

What if you take many prescription drugs and want to know how rapidly or slowly your body is metabolizing the medicine? You are concerned about the drugs building up in your body or interacting with one another. Pharmacogenetics tests several of your genes. With food menus, nutrigenomics tests other genes. At least you can find out whether you metabolize fast, slow, or like the majority of people. One size will never fit all people because genes recombine. They shuffle, and individuals have different responses to different drugs, cosmetics, or foods.

Multidisciplinary nutrition research and collaboration is necessary for nutritional genomics to bring together diverse expertise. Scientists working in the disciplines of nutrition, biochemistry, and genetics need to share, collaborate, and interface in this field. If scientists are more concerned about positioning themselves first in publishing their research and won't share DNA with all scientists, how can research ever move forward?

You might want to read "The Metabolic Basis of Inherited Disorders," 6th ed. McGraw-Hill, New York: 2649–2680, 1989. Then compare the latest research in nutritional genomics on how smart foods (foods tailored to your genetic signature) influence risk of chronic disease. The longer science studies the entire genome (rather than the specific SNPs for certain chronic diseases) the more information will be forthcoming on how food and lifestyle influence your health based on the genes you inherited.

According to the National Institutes of Health, (See their Web site at: http://www.nigms.nih.gov/funding/htm/yrgenes.html), "Your lifestyle, the food you eat, and where you live and work can all affect how you respond to medicines. But another key factor is your DNA, which contains your genes. Scientists are trying to figure out how the make-up of your DNA can contribute to the way you respond to medicines, including pain-killers with codeine like Tylenol®#3, antidepressants like Prozac®, and many blood pressure and asthma medicines. Scientific discoveries made through this research will provide information to guide doctors in prescribing the right amount of the right medicine for you.

"The National Institutes of Health aims to improve the health of all Americans through medical research that solves mysteries about how the human body normally works—and how and why it doesn't work, when disease occurs. One goal of this research is to help improve the good effects of medicines while preventing bad reactions."

Click on the Web site of the National Institutes of Health at: http://www.nigms.nih.gov/funding/htm/qanda.html to see their question and answer site. The point is that one size doesn't fit all when it comes to medicine or food or even cosmetics and skin products. According to the National Institutes of Health, here is the National Institutes of Health's answer to the question of "Aren't prescribed medicines already safe and effective?"

On their question and answer page, they reply, "For the most part, yes. But medicines are not 'one-size-fits-all.' While typical doses work pretty well for most people, some medicines don't work at all in certain people or the medicines can cause annoying, or even life-threatening, side effects." If you're wondering why one size doesn't fit all since our DNA is supposed to be so similar world-wide, it really varies due to some people's genetic variations, diversity, and mutations.

According to the National Institutes of Health (reprinted here with permission) from their Web site at: http://www.nigms.nih.gov/funding/htm/qanda.html#, "As medicines move through the body, they interact with thousands of molecules called proteins. Because each person is genetically unique, we all have tiny differences in these proteins, which can affect the way medicines do their jobs."

"The National Institutes of Health is providing money to scientists at universities and medical centers who come up with the best plans for carrying out research on how people respond differently to medicines. Curing and preventing disease is the National Institutes of Health's highest priority. Research on how people respond differently to medicines will make current and future treatments for diseases such as asthma, diabetes, heart disease, depression, and cancer safer and more effective. A bonus of this type of research will be a better understanding of the role many different genes play in causing or contributing to these and other diseases.

"The National Institutes of Health is providing money to scientists at universities and medical centers who come up with the best plans for carrying out research on how people respond differently to medicines.

"National Institutes of Health-funded scientists at universities and medical centers across the country will recruit volunteers from a wide variety of groups. Research of this type relies upon studying many different people with a broad range of genetic make-ups to find the small, but normal, genetic differences among them.

"Most of these research studies will involve simply rubbing the inside of a volunteer's cheek with a cotton swab. Scientists will then pull out DNA from inside the cheek cells they have collected. There are no health risks associated with this type of test."

"The first benefits to patients could come as soon as a few years from now. From then on, the knowledge gained through this type of research is expected to help doctors tailor the medicines they prescribe to best suit each patient's individual needs."

So what can the consumer do now to benefit as soon as possible from genetic testing? How can you apply the DNA test results to changing your lifestyle now in order to improve your well-being? Education is out there, and much of it is available free on the Web and in scientific journals available at most medical school libraries open to the public. How do you screen the information?

Foods Tailored to Your Genotype

How do your genes respond to what you eat? How many diet-by-DNA book titles are there? Books on smarter foods? Tailored menus? Extracts of plants? DNA tests for ancestry? Ancestry and eating? According to Dr. Fredric D. Abramson, Ph.D, S.M., President and CEO of **AlphaGenics, Inc.**, Genes are distributed, function, and work in such ways that nearly every reasonable diet could work well in about six percent of the population.

Are you eating smart foods—foods tailored to your genotype—DNA, your ancestry, and your entire genome of genes? Are you ready to get a picture of your response to your nutrition? How can you eat to nourish your genotype? According to Genomics 120, a science, nutrition, and health Web site at: http://g120.com/products/genomicscience.html, are you wondering why in the United States currently only 50,000 people out of some 280 million live to be even 100 years old, or that your body may be aging nearly twice the rate it should be because you're eating the wrong food for your genetic signature?

There is a strong connection between nutrition and genotype, especially in regards to your cardiovascular and central nervous system health. So you need to tailor foods intelligently to your genetic expression. The media buzz about 'intelligent' foods or 'smart' foods really means eating clean, safe, whole foods based on what your individual genes need to thrive. Not all your genes are tested. You might start your food research at the Web site of Food Resource, a source of science-based and business savvy information for the food industry at: http://food.oregonstate.edu/nutri.html.

What happens when diet books for your condition aren't working for you? Maybe salt restriction isn't working but exercise is for your condition. How do your genes respond to nutrition and *nourishment*? Are your genes intelligent, conscious, and communicating with you about their nutritional needs? If they are, so are the foods you eat. Your genes interact and collaborate as a team.

The language of communication is written in the human genome, in your individual genetic signature—in your DNA, in particular SNPs, and in all your genes and cellular material. Even your blood type is expressed in all the cells of your body. How does all this information signal you about what 'smart' foods and nutraceuticals to choose in order to help prevent or delay chronic disease for which your genes may put you at risk?

A slogan reads, "Smart foods for intelligent people." Nutritional genomics is a buzz word in the news. Testing DNA for ancestry also bridges gaps in regard to customizing smarter foods to your genotype. Phenomics is about customized healthcare and medicine tailored to your genetic profile. Prosopography is an independent science of social history embracing genealogy, onomastics and demography. If you're interested in metabolic typing, one Web site for Personal Metabolic Typing is at: http://www.wholebodyhealth.us/. Dentists may be interested to know that gum disease is genetic and may be caused by a genetic predisposition to diabetes, heart disease, or low birth weight. A genetic profile on patients with deep pockets of gum disease might be useful. Check out the Holistic Dental Network at: http://holisticdentalnetwork.com/products_services.php.

Cracking the human genome code is so new and tests so costly. Currently only certain genetic markers are tested. The genetic signatures tested include genes that tell you about risk for certain diseases. Nutritional genomics as a field of research also is abbreviated as a generic term to read 'nutrigenomics.'

Without testing all the genes, how can you know about all the diseases for which you may be at risk? And without knowing all the information that every one of your gene's reveals, how can you develop a plan to override your genetic risks by nourishing your genes with what they need to stay healthier? **Here is how some scientists answered these questions.**

According to Dr. Fredric D. Abramson, Ph.D, S.M., President and CEO of *AlphaGenics, Inc.*, "The key to using diet to manage genes and health lies in managing gene expression (which we call the Expressitype). Knowing your genotype merely tells you a starting point. Genotype is like knowing where the entrance ramps to an interstate can be found. They are important to know, but tell you absolutely nothing about what direction to travel or how the journey will go. That is why Expressitype must be the focus." You can contact AlphaGenics, Inc. at: http:// www.Alpha-Genics.com or write to: Maryland Technology Incubator, 9700 Great Seneca Highway, Rockville, MD 20850.

Alpha Genics, Inc. is a nutrigenomics science company. A sidebar on the company's Web site from Dr. Fredric Abramson, CEO reads (reprinted here with permission), "We are about to see a revolution in our concept of diet. Each of us is a unique organism and for the first time in human history, genetic research is confirming that one diet is not optimum for everyone.

Science is discovering that each individual's DNA processes food and nutritious supplements in a unique way. Through the development of a cutting-edge DNA analytical system and consumer guidance, Alpha Genics will be able to tune nutrition to meet the needs of each individual resulting in optimum health, peak performance, and enhanced creativity." What I also like about Alpha-Genics Inc. is that they have an independent, separate Ethics Board. Check out that Web site at: http://www.alpha-genics.com/ethics_board.php.

It is not part of the regular Board of Directors. It has five members: three outsiders, one representative from the Board of Directors, and one representative of the employees. The Ethics Board has no veto power, but has a seat on the Board of Directors. Compensation for the Ethics Board members comes through a blind trust, which means the Ethics Board has neither control nor knowledge of how the funds develop.

"I created this because I think companies need to have independent voices to provide reality checks," says Dr. Fredric Abramson. "It is something like that scene in Patton when he talks about the Roman conqueror returning home to glory, with someone standing just beside him reminding him that fame is fleeting. An independent Ethics Board helps us make better choices."

Consumers can bridge the gap between ethics boards and the media by acting as liaisons, ombudsmen, lobbyists, trustees, recruiters, communicators, independent board members, fee-for-service contractors, industry watchers, or volunteers. Get involved in the nutritional genomics industry.

You even can put together corporate gift baskets full of nutritional genomics products or samples. Throw a nutritional genomics party in any home, office, or meeting pace, in church basements, teacher's lounges after school, or at conventions. Make tape recordings about nutritional genomics and post the radio-length broadcasts to your Web site. So many news stories in the media give the impression that the average consumer will have to *wait a decade* for genomic testing to be applied to customized foods.

For example, the New York Times Magazine published an article by NY Times Magazine writer Bruce Grierson titled, "What Your Genes Want You to Eat: New Way to Look at Disease." The article also appeared in the Sacramento Bee, a daily newspaper, on Sunday July 13, 2003 on the Science page. Sundays are great for reading the Science page in the Forum section. There's time to read the cutting-edge science articles, many of which are reprinted from other, major urban publications. It's a family tradition for four decades.

I was impressed by the media buzz around the relatively new field of nutritional genomics. Last year the media buzz circled around testing DNA for ancient ancestry and genealogy, during which I took several DNA tests for ancestry and enjoyed the results. This year DNA and diet is fast-track news. DNA and foods also tie into food safety and security issues.

I like to know that when I go into a health food store and buy a package of imported powdered ginger under the title of herbs or botanicals that it's clean and contains no toxic pesticides, residues, bad bacteria, or unsafe chemicals. I often thought about who inspects these imports and do they rush through, take enough time, or have enough staff? If I buy fresh ginger root, I'm concerned whether it's organic or still full of pesticide residue. On the other hand, who has time to think these thoughts?

It all came rushing back when I went into a health food store last month, walked past an open bin where a child about twelve years old had just put her hand into the cous-cous, grabbed a fist full of the grain-like pasta, tossed it into her mouth, chewed it, spit it back into her palm and replaced the cous-cous that she had just spit into her hand, back into the bin with the rest of the grain. Her mother was busy looking at other products.

She didn't realize I was standing beside her ready to take a scoop myself of the oat groats in the next bin. The scoop hit the floor, and she picked it up and replaced it. The cashier was in the front of the store, the manager in the back room. Why couldn't the store change the bins so that nobody else could spit into the grain? When I brought this to the cashier's attention, she shrugged, looked down, and acted as if nothing had happened. I wondered at that instant, if this is how one consumer's word is received, how will the public perceive us?

There's power in numbers, in grouping together. I wished I had my video camera at that moment. There's also power in the media—the reputable, credible media that bridges gaps between science and the public. It's time for some consumers to become "media people." Let's watch the watchers and look at the media.

The article titled, "What Your Genes Want You to Eat: New Way to Look at Disease," had its first opening sentence beginning with a trip to the "'diet doc' circa 2013." I eagerly wanted this kind of testing to be available now, not in 2013. So I turned to Dr. Frederic Abramson, CEO of Alpha-Genics Incorporated to interview by email on his views. Dr. Abramson also teaches part-time at Johns Hopkins University, in the graduate program in Biotechnology.

I asked Dr. Frederic Abramson, CEO of Alpha-Genics Incorporated the following questions in an e-interview. Here are his answers.

1. When do you think genomic testing might be available to most consumers?

This is an important question. From a practical perspective, genomic testing—for part of a person's total genome—is possible today. We can test for several thousand genes right now. So this leads to two sub-questions. One, when will testing of an entire genome be possible? And two, when will low-cost testing be available.

Experts argue about whether we have 30,000 or 70,000 genes, or somewhere in between. Regardless of the number, we are five years away from a comprehensive full genotype test of all genes in a person. By genotype, I mean the identification of which genes a person has.

But it is not your genotype which determines things. It is the work that your specific genes do. Think of genotype as the location of exit ramps on an interstate. You need to know where these are, but they tell you nothing about where you are going and what the journey will be like.

To identify these, we must look at each gene's level of activity, called "gene expression." We call the gene expression the Expressitype. Gene expression changes over time for many genes. How do we know this? Because we age. We start as children, go though puberty, become adults, and then start declining. All of this is substantially under genetic control. Some people age faster than other. It's in their genes.

Right now testing costs a fair amount. And seldom is gene testing covered by insurance. But over time, the technology advances will enable very low cost tests. For example, measuring gene expression in several thousand genes can cost between $800 to $3,000, depending on who does it. But a Japanese company is working on a test that will end up costing less than $100 for 900 genes. Thus, one thing we can be sure of is that the costs will drop. Just like computers, VCR's and microwaves.

We are working to bring the test cost to be under $1,000, with a monthly follow-up of around $79. The monthly fee lets you contact us by phone or email anytime to ask whether something you want to eat might help you or not.

We are working with Carnegie Mellon University in Pittsburgh to develop a small implantable device that will measure vital chemicals in the blood, and send signals outside the body. This will let us track what is happening in a person around the clock, every day, with much more accuracy and less guesswork.

2. What do you think is the most important area of research in nutritional genomics today?

The most important area is to identify how the dietary system, which is composed of hundreds and thousands of chemicals in varying dosages, interacts with the thousands of genes in the genome to produce health, or illness. This is generally called "systems biology." Basically, we can no longer look at single things for easy answers. It won't be just a question of whether you eat blueberries or bananas or rice, but what balance of each of these you eat over time.

This points to the second area, to understand the dynamics of how changes in diet influence the work each person's genes is doing. The value of systems biology is that ultimately, we will be able to identify individualized responses to diet, based on genetic composition.

3. Do you have any advice for those who are looking for tailored diets for specific conditions—if genotype testing is not available today, what is the next best thing for the average consumer who has already had a DNA test of merely the mtDNA or Y chromosome for ancestry?

Ironically, many of the folk suggestions about diet weren't far wrong. So first look to your family history. If you have heart disease in your family, think about a diet that has a bit less fat and more antioxidants. There are similar observations about other conditions like arthritis and cancer.

Generally, if you have a DNA test at this point, it is for one or very few genes. This helps. But remember that most major health conditions are the result of many, many genes acting in concert, not just one gene.

I'll admit it can be confusing with so many different recommendations about diet. To me, this reflects the way in which our genetic diversity makes one diet work well for me and not well for you, while another diet has the exact opposite effect.

4. What is the area of research your company is focusing on now?

Our research focus is to understand how the specific ingredients of diet influence genetic activity. And by diet we include food, supplements, medications, chemicals in water, and even cosmetics, for all of these contain chemicals that influence genes.

The goal of our research is to translate the science into practical day-to-day advice for each person, based on his/her own genetic profile and genetic activity. We want to make genomics something that is useful for each of use every day instead of some industrial science.

This is the same thing that happened when Edison invented the light bulb. Suddenly, electricity was something that every home needed. For us, our success in delivering NutriGenomics to the consumer will make genomics something that everyone will want to use to live their lives a bit better.

5. Where can I refer readers today to learn more?

The amount of literature on the Internet is growing almost daily. Two weeks ago, the commission of the FDA mentioned NutriGenomics in his major speech at Harvard.

We welcome readers to contact us. We are assembling the world's first comprehensive NutriGenomic knowledge base, which we will use to help consumers make better choices. We build our research insights on the actual experience of what real people do.

Our current high priority is a totally new method to prevent viral infections using NutriGenomics. We discovered this in January, and have been working to get government support to conduct this important research. We believe we can develop a way to protect many people from the dangers of certain types of viruses, such as a weaponized flu.

6. Do you work with patients directly or only with their managing physicians or both?

We will work with patients directly and with physicians. When a physician is involved, we will be sure to include the physician in the information loop so the person continues to get the best care.

7. How many genes do you test? Do you prescribe a diet or nutraceuticals based on the results?

In the current stage, we will test about 2,000 genes, mainly for the cardiovascular cluster (cardiovascular disease, diabetes, obesity, hypertension, high cholesterol). The virus testing is planned to be a separate test.

The specific food/supplement recommendations are made directly to the person. These change as the person's genetic activity changes. It would be a mistake to prescribe one type of diet for life. Our genes and bodies don't work that way.

The procedure is simple. We get a sample from you, typically from your cheek lining or blood (if you go to a physician). We identify the genes (genotype) and the amount of genetic activity for each gene (expressitype). We provide you a report summarizing the results. Then, depending on your decision, we will provide you general dietary recommendations based on your genes, or will begin to work with you as often as daily to help you choose what you like to eat.

It is worth noting here that the so-called 'med diet' is based on a month of eating, whereas the USDA model is a daily model. We prefer the monthly approach for the evidence is that eating has a cumulative effect. Another way of saying this is "no one meal will ever hurt you. It is the combination of lots of bad meals that hurts. So knock yourself out."

Fredric D. Abramson, Ph.D., S.M.
President & CEO
AlphaGenics, Inc.
http://www.Alpha-Genics.com

According to Dr. Fredric D. Abramson, Ph.D, S.M., Esq. President and CEO of AlphaGenics, Inc., our genes determine how we respond to our environment. In the industry, nutritional genomics also is abbreviated as 'nutrigenomics.' Let's look at Dr. Abramson's article below titled: *"About NutriGenomics,"* Copyright 2002, 2003. Reprinted here with permission.

About NutriGenomics
Fredric Abramson, Ph.D., S.M., Esq.
Copyright 2002, 2003

Our genes determine who we are, how we develop and age, and how we respond to our environment. They are the blueprints that define our potential. But they do not act in a vacuum. They respond to and actually help shape our environment. That environment includes what we put into our body in food, water, supplements, nutraceuticals and pharmaceuticals.

Together, our genes and environment control our health and our susceptibility to disease. NutriGenomics involves decoding how the molecular composition of our diet, which includes food, supplements, pharmaceuticals and water) influences the work being done by our genes, and then defining personalized dietary strategies to tune each person's gene expression pattern, called the "Expressitype." This tuning process must be changed dynamically as the person's genetic activity changes.

"NutriGenomics lets people pick the foods they eat based on how well the foods make their genes function."

Because what we eat influences our health and the way our genes work, we have the opportunity to let people control their health destiny in a whole new way. This is what NutriGenomics is about: calibrating the mix and amounts of what ingredients we put into our body so that our genes work at their best, for our best health status. By building on each person's genetic uniqueness, NutriGenomics focuses on what that individual's genes are doing and how that person can pick and choose things from their environment that will make his/her genes work better or worse.

AlphaGenics approach is to identify what parts of your diet are making your genes work badly, and determine the best mixture in your diet to calibrate and tune the work your genes are doing. Our goal is to adjust your genetic activity, without changing your genes, by dynamically adjusting the mix of what you put into your body as your body changes over time.

> **Basic Terms**
>
> ▶ **Genotype**: The entire set of genes of an individual.
>
> ▶ **Expressitype**: The profile of gene expression activity in an individual at a moment in time.
>
> ▶ **Phenotype**: The observable characteristics of an individual.
>
> ▶ **NutriGenomics**: The science that relates how the molecular inputs from a person's environment influence and control gene expression.
>
> ▶ **Qwink**: A molecule or particle in the environment that can change gene expression directly or indirectly.
>
> ▶ **Expressitype Knowledge Transfer System**: A proprietary information/knowledge management system that integrates individual longitudinal data with a variety of scientifically verified data elements. Its outputs include personalized, dynamic dietary strategies as well as a variety of scientific and research-oriented solutions.

Our dietary environment influences how our genes work

Many different ingredients exist in a person's dietary environment. They exist as both natural molecules and artificial compounds and are found in food as well as in our water, and in dietary supplements, cosmetics and medications. Even in the air we breathe. Some are present by design; others as a by-product. So, for example, the common tomato is known to have over 300 different natural molecules. Some ingredients are man-made, but most are natural substances whose variety and health-effects are not even fully explored.

The key to NutriGenomics is identifying how specific compounds or chemicals in our diet influence the activity of one or more genes, and hence our health.

Our focus is on our genetic activity (Expressitype) and not just what genes we have (Genotype). Knowing your genotype is a lot like knowing where the entrance is to an interstate highway. It is very useful and important, but it doesn't tell you what direction to go in or when you will be able to stop for gas.

The term "Qwink" was coined to refer to any molecule in the dietary stream that influences gene expression. Qwinks are found in foods, supplements, cosmetics, water and pharmaceutical compounds. They include basic ingredients such as sugars, proteins and fat as well as very specialized substances such as vitamins or toxins. A Qwink may be a natural substance or synthetic.

A Qwink is neutral as to whether it has a positive or negative effect. If the gene expression moves in the direction to provide better health or reduce risk, it is good; if it moves to worsen a person's health status, it is bad. Thus, it is only the effects that are good or bad. For example, a pesticide residue that survives in the food chain and influences gene expression is a Qwink.

While pesticides are generally considered harmful, interestingly enough, there is some evidence that certain pesticides could actually reduce the risk of cancer, at least in model systems.[1]

NutriGenomics calibrates dietary inputs and patterns to provide different Qwink types and concentrations to the genetic machinery.

The impact of a Qwink typically depends on its concentration and dosage frequency.

One of the amazing aspects of human biology is that it responds differently to varied Qwink concentrations. Take arsenic, for example. Most people know that arsenic in high doses can kill. What is less commonly known, however, is that at very low concentrations, arsenic is considered a nutrient.

The same will be found for how Qwinks work. At very low concentrations, one set of genes may be affected while a much higher concentration may change expression in an entirely different set of genes.

NutriGenomics is personalized genomics

The way a person's dietary environment influences gene expression and health is called NutriGenomics. NutriGenomics is a scientific platform that permits focused, personalized adjustments to a person's dietary environment. An advantage of NutriGenomics is that it is a unifying standard that includes the effect of dietary supplements, nutraceuticals, pharmaceuticals and cosmetics in addition to

---

1. John Milner, National Cancer Institute, personal communication, January 2003.

what a person normally eats, as well as environmental factors such as toxins, contaminants and infectious agents, such as viruses and bacteria. The scientific goal is to identify molecular levers coming from a person's environment that can move the person's gene expression patterns in an appropriate direction, and to further identify environmental adjustments that can help improve the person's health status through changes in gene expression.

Our core scientific proposition is that evaluating and modulating a person's Expressitype as it dynamically changes over time is a more potent and acceptable way to prevent and control disease than working with genotype alone. The practical difference between focusing on genotype versus expressitype is comparable to the difference between buying a car based on what Consumer Reports says is its repair frequency and listening to your car coughing and sputtering when you are driving down the highway.

The cycle begins with a person's environment, which interacts with the person's genotype and can up-regulate or down-regulate various genes. The gene expression pattern, the expressitype, in turn, works to produce proteins and otherwise control the person's metabolism and physiology to produce the phenotype. It is a person's phenotype at which we observe health and illness.

Examples of phenotype include eye color, blood groups, and various chronic and acute diseases. A person's phenotype in turn helps shape the environment. For example, phenotype can influence what a person chooses to eat. A person who doesn't feel well may eat differently; a person who changes their physical activity may also change how they eat; and so on.

AlphaGenics designed its research and intervention around this cycle. By capturing data and information from each individual, including serial measures of expressitype, and blending this data with scientific knowledge from genomics, nutrition, pharmacology, toxicology, and medicine, we can focus on unraveling the systems biology map of how the complex environment interacts with and influences the comparably complex genome.

Our genome works as a system that takes us through our life cycle

Our genetic apparatus is a complex system designed to sustain our lives in intimate communication with the environment, from birth through puberty and the gradual aging process leading to the end of our lives. It is this complexity that has made it so difficult to find cures or treatments for so many diseases like cancer and heart disease. It is this same complexity that explains why a drug will work in one person, be ineffective in another, and harm a third. It is this complexity that confirms why different people need entirely different diets to lose weight, for example.

The fact is also that almost all of the major health conditions that concerns society today—cancer, heart disease, diabetes, obesity, neurologic disorders and so on are all multifactorial and polygenic. Multifactorial means that it is both environment and genetics in combination that explains when, why and where different diseases occur. Polygenic means that multiple genes are involved, not just one.

Three specific aspects of this complexity are worth exploring. First, the number of different genes involved with each of the diseases that plague us can be in the hundreds or even thousands. Second, virtually every chemical that we put into our body influences some genes. And third, our genes typically work collaboratively, where one influences another. Our genes work as an integrated, cohesive system.

A change in gene expression is not always a benefit

The goal of NutriGenomics is to move gene expression patterns to some "best" state. However, because the human genome is complex, an important issue is that an "increase or a decrease" in gene expression is not "always" linked with a benefit. A lot may depend on what the other genes involved are doing as well.

Many different genes are involved in the common health conditions

Most of the common conditions that concern our health are polygenic, i.e. involve many different genes as mentioned above. The goal of the Human Genome Project was to identify thousands of genes and to link these genes to specific health effects. Scientists are also pinpointing exactly where each of these genes are located, or "mapped," on our chromosomes.[2]

Some examples include breast cancer (over 200 genes, 39 of which are mapped to 17 different chromosomes), obesity (over 200 genes; 15 mapped to 11 chromosomes); Diabetes (60 mapped genes to 17 different chromosomes); Lymphoma (200 mapped genes on 22 chromosomes); and hypertension (19 genes on 10 chromosomes).[3]

---

2. Humans have 22 pairs of chromosomes plus the X and Y.

Qwinks impact multiple genes, not just one

What we put into our body typically influences more than one gene. For example, the natural compound retinoic acid can change expression in more than 500 different genes[4] while the enzyme Cofactor Q10 will up-or down-regulate over 100 different genes.[5]

Genes work together as a coherent system, not as independent actions.

For a polygenic condition, it is important to realize that the genes involved do not work by themselves but work in a coherent, structured system. We can take as an example the more than 200 different genes implicated in obesity. If each of these genes worked independently of one another, there would be 2 to the 200th power different combinations. This is 1.6 times ten to the 60th power; that is, 1.6 followed by 60 zeros. This number, 1.6 times ten to the 60th, is very large. In fact, it is larger than all the humans who have ever lived on the earth. It is larger than all the organisms, including viruses that have ever lived on the earth.

So clearly, just by observing that different groups or types exist in obesity, one realizes that these 200 genes don't work independently. In fact, what must happen is that when one gene turns on, it in turn regulates ten or twenty others. This interaction among our genes leads to a relatively small, countable number of combinations of gene activity. Instead of the astronomic number calculated above, we have only 100's or 1000's of combinations for this example of obesity. These combinations are the "typical" patterns of gene activation, so to speak the molecular bar codes, each of which could be associated with a form of the disease.

So it is possible to imagine two type of obese people, one with five obesity genes turned on, say numbers 1, 50, 95, 150 and 200, and the rest turned off; the second with twelve different obesity genes turned on. These two different combinations could translate into different obesity effects, where one type of person gains weight easier than the other on the same diet.

---

3. Obtained from the Human Genome Project web site.
4. Balmer and Blomhoff, Gene Expression Regulation by Retinoic Acid, Journal of Lipid Research, Vol. 43, 1773–1808 (November 2002).
5. Linnane AW, et al. Cellular redox activity of CoQ10: effect of Q10 supple. On human skel. muscle. Free Rad Res 2002; 36(4): 445–453.

While our genes are fixed at birth, the work they do changes as we age

Each person is born with genes they inherited from their parents. The identity of your genes is called the "Genotype." The Genotype is a person's unique collection of genes. Classic genetics described genotypes for hair and eye color, and blood types. For all practical purposes, the genotype is fixed when your parents' sperm and egg unite.

Genes are dynamic engines. Their work varies depending on circumstances and time. The entire genomic system changes its work, as illustrated by the aging process that begins with infancy, moves through childhood, puberty and adulthood, and then to senescence. These genetic mechanisms are starting to be understood.

An example of gradual, almost invisible changes in genetic activity is observed in how menstrual cycles shorten as a woman ages. Research done at the University of Minnesota in the 40's shows that a 40 year old woman has one more menstrual period a year than a 20 year old, and that the one-day per year change in cycle length is linear from 20 to 40. This gradual change suggests that key genetic components to target in NutriGenomics are genes whose activity undergoes very gradual changes over a long time period. This further suggests that many key NutriGenomic interventions will be based on cumulative effects.

It's Gene Expression (Expressitype) Which is Key

The accomplishments of the Human Genome Project mean we will soon be able to identify every gene in a person, in effect, that person's genetic blueprint. Some of these genes will indicate greater risks for certain diseases; other genes will mean less risk than average. An example is the BRCA1 gene for breast cancer. Women with BRCA1 have a significantly higher risk of developing breast cancer than those without it.

The genotype, however, is only a starting place to understand what our genetic destiny might be. This is because the work a gene does is likely to change over time, both according to preprogrammed rules, and in response to environmental influences, notably our food. Some genes are turned on early in life, say during pregnancy, and are switched off permanently. Other genes may just sit there waiting to be activated or to become phenotypically manifest. This appears to be the case with certain diseases that arise later in life, such as Huntington's chorea, a single dominant gene disorder that shows itself in a person's 50's.

Further complicating the situation is that when a gene is "on" it may be operating at one activity level one day and a higher or lower level the next day. A gene's activity is exhibited as "genetic expression," a term that refers to the gene producing RNA and otherwise interacting with a person's metabolism and physiology.

Let's call, in analogy to Genotype, the entire profile of expression of each gene throughout the genome, the "Expressitype." Thus, for any health condition, each person has his or her own Expressitype. The Expressitype is the actual amount of genetic activity produced by each gene that is implicated in or related to a specific health status or condition.

Since for any particular gene that is turned on, the expression level may be changing over time, and because the gene itself may be turned on or off at a particular point in time, a person's Expressitypes is likely to change.

A person's NutriGenomic profile, then, measures what each gene is doing at a point in time. By stringing together a series of NutriGenomic profiles, it is possible to see whether a gene is changing, the direction a gene is going and how fast it is getting there.

The gene-environment interaction is very powerful

The interaction between genes and environment is undisputed. Part of the environment is dietary—what we eat. Another part is lifestyle—how we live our everyday lives. Figure 1 is a general overview of NutriGenomics, starting with the dietary inputs at the base and moving through the genome to the end result of health or illness.

Even well defined genetic disorders respond to environment differences. For example, if identical twins each have the single dominant gene for Huntington's chorea, and the twins have different dietary and exercise programs, the onset of Huntington's chorea can be delayed up to eight years in one of the twins.

Diet has been shown to alter gene expression in several ways Qwinks can act on how DNA is transcribed into RNA They can be involved in various metabolic pathways and can increase or decrease the concentrations of other materials needed by the genome. In other words, Qwink effects can be direct, by changing gene activity itself, or indirect by shifting what is available for the gene to use.

Foods, supplements and other chemical sources are the foundation of NutriGenomics. The ingredients that actively regulate the genome, the Qwinks, can turn genes on and off, and can change the amount of activity for each turned-on gene. The Qwinks help determine the Expressitype, which is the profile of what each gene is doing at a moment in time. Gene expression becomes translated into health, wellness or illness through a person's metabolism and the many proteins in the proteome. Moreover, a person's health status, in turn, influences gene activity and can even change what might be acceptable to eat. An example would be creating an immune response to a certain food (i.e., strawberries) that makes the food dangerous. This system is dynamic, and changes over time as we age and our health changes. Tying these all together are the many biochemical, physiological and metabolic processes with cells and tissues.

◆ ◆ ◆

Let's profile another company that helps you eat better by revealing disease risks, so your physician can concentrate on focused prevention and treatment tailored to your individual genetic signature. Today, there are companies such as **Genovations™**.

According to Marketing Communications at the North-Carolina-based Great Smokies Diagnostic Laboratory, the company approaches preventive health care by using genetic analysis to provide individuals and their physicians with critical information to more effectively control present and future health. Genovations uses information stored in each person's genes to reveal disease risks and to help each individual lead a healthier life though focused prevention and individualized treatment.

Genes provide the key to a wealth of information unique to each individual that, once unlocked, can serve as a comprehensive guide to a healthier life. By identifying specific genetic risk factors and changes in dietary, lifestyle, nutritional supplements and medications that are most likely to improve each patient's health, Genovations enables physicians to practice truly individualized preventive medicine.

Most consumers will ask about what happens when you go for a genetic test. It's more than the DNA test you might have taken to look at your mtDNA or Y-chromosome for deep ancestry. Instead, the kind of genetic testing you'd take at Genovation, for example is to evaluate specific portions of your genetic code.

Genovations™ testing evaluates specific portions of the genetic code that vary from person to person. These variations are called single nucleotide polymorphisms, or SNPs (pronounced "snips") for short. Everyone has SNPs—they're what make people different from one another—our hair color, our height, our voice, even key aspects of our personality. SNPs are the very seat of our individuality. SNPs also affect our health.

Some SNPs can make people more susceptible or more resistant to common diseases. Others may make the body respond differently to certain diets or lifestyle habits. Read the *Complete Blood Type Diet Encyclopedia*. What nutritional genomics needs is a complete human genome diet dictionary. With 40,000 or more genes in each person shuffling with each generation, individuals inherit different genotypes.

At the same time, some abnormal genes are inherited and make members of some families more at risk than other families. **Here is an e-letter from Kay Patrick, Product Manager, Genovations.**

From: Kay_Patrick
To: 'Anne Hart'
Sent: Tuesday, July 22, 2003 6:46 AM
Subject: RE: Thanks. Just a few questions

Dear Anne:

I have attempted to answer your questions. Please let me know if you have additional questions.

Why are Genovations tests only available through a physician?

Genomic test results are most meaningful and best understood when they are clinically interpreted within the context of a patient's complete health history. For this reason, genomic testing is best managed by a licensed health care professional.

Genovations tests are available only through trained, licensed health care practitioners. Not all practitioners are adequately trained to interpret and use the genomic information revealed in these tests, so Genovations has taken a leading role in providing genomic medical education to health care practitioners.

What does Genovations ™ testing do?

Genovations testing focuses only on SNPs in the genetic code that are associated with common health conditions, such as heart disease, osteoporosis, allergies and asthma, for which simple treatment strategies exist to reduce risk. Testing also evaluates SNPs that affect how each person's body is likely to respond to specific diets, supplements, and medications.

Therapeutic recommendations based upon the genetic results are provided for the practitioner to develop a treatment protocol. In this way, test results can provide the physician with a "road map" for developing a comprehensive, personalized health action plan for each patient.

Example:

Heart Disease

MYTH: All persons with a family history of heart disease can minimize their cardiovascular risk by reducing their dietary intake of fat and salt, exercising more, and taking a cholesterol lowering medication.

REALITY: Heart disease is not simply a condition caused by excess fat and cholesterol. Research has revealed that there are many other modifiable risk factors, some genetically influenced that can predispose a person to heart disease. For example, some people have a genetic inability to properly metabolize folic acid in the body. This can lead to a build-up of homocysteine in the bloodstream, causing increased risk of blood clots and atherosclerosis. Yet none of the "one-size-fits-all" conventional therapies listed above would reduce this risk. For a person with this genetic variation, the only way to reduce risk is to take the active form of folic acid, which is not found in common vitamin supplements.

High blood pressure is typically treated by restricting an individual's salt intake. However, not all people have the genetic variation (a single nucleotide polymorphism, or SNP-pronounced "snip") that allows them to respond effectively to a salt-restricted diet. Based on their genetic make-up, they may respond better to aerobic exercise. By testing genetic variations, the physician can better identify which therapy is likely to be the most effective for lowering blood pressure in each patient.

Apo E is a protein in the body that affects cholesterol levels. There are three major genetic variations of the Apo E gene that can affect how each person's body breaks down fat and cholesterol in the diet. These three variations can lead to an increased, average, or decreased risk of heart disease. By knowing an individual's Apo E genetic variation, the physician can prescribe the dietary and lifestyle changes, nutritional supplements, or prescription medications most likely to lower cholesterol levels effectively.

What is the process?

Practitioners incorporate genomic testing into their complete work-up along with other phenotypic diagnostic tests. Blood samples are collected in the office, shipped to our laboratory, tested here in our genomics lab and the reported results are shipped back to the ordering practitioner who schedules a follow-up with the patient.

How much does a Genovations test cost?

The cost of Genovations testing depends on the testing option chosen by the patient and the practitioner. The comprehensive testing program, which assesses health risks for a wide range of conditions (including heart disease, osteoporosis, and immune dysfunction) costs about $1,000-$2,000. Focused testing for health risks associated with a single area or condition costs about $300-$500.

Hope that helps.

Sincerely,

Kay Patrick

Product Manager, Genovations
www.genovations.com

◆   ◆   ◆

An overview of what genes are, how they are inherited, and how many are in the human body is presented on the Web site at: http://www.genovations.com/patient_overview.html.

What Can SNPs Tell You About Your Genotype?

SNPs can even affect whether certain drugs are likely to have severe side effects or not. By evaluating these health-related SNPs, Genovations™ testing allows physicians to develop preventive strategies that are tailored to each patient's unique genetic makeup and health risks. This helps to reduce the guesswork that results from using "one-size-fits-all" approach to preventive medicine.

Genovations™ is a line of tests that is only available for a licensed health care provider (doctor) to order for a patient. The test results are provided directly to the health care provider, who consults with his or her patient. Genovations does not interpret test results for the patient or the health care provider. A patient's health care provider uses the test result information to determine appropriate treatment protocols for the patient. Genovations™ does not provide specific nutriceuticals or supplements recommendations. That is up to the health care provider.

According to marketing communications, "Genovations does not test for SNPs. We currently have 4 profiles that identify specific SNPs related to heart disease, detoxification problems, bone health, and immunology related problems. Each profile identifies between 7 and 15 SNPs." If you want more information about Genovations, write to them, email them, or telephone. Genovations' phone number and email address are listed at its Web site. Since books hang around libraries for years, but phone numbers may or may not change, you can reach Great Smokies Diagnostic Laboratory or Genovations™ as follows: Corporate Headquarters, Great Smokies Diagnostic Laboratory/Genovations™ 63 Zillicoa Street, Asheville, NC 28801, USA

Or go to their Web sites at: http://www.genovations.com/ and/or at: www.gsdl.com. See the Web site at: http://www.gsdl.com/assessments/finddisease/cardiovascular/metabolic_glycemia.html

Read about metabolic glycemia. If you have insulin resistance, or too much insulin pouring out every time you eat certain carbohydrates, you'd want to read more about this and how it changes or affects your arteries and other parts inside your body.

There are excellent sites on the Web to educate yourself before you decide to get tested. Check out your biomarkers. Before you see your doctor, know what questions to ask and what tests to ask for. Then talk it over with your physician.

When you take control over what you eat you are educating yourself as to your body's nutritional requirements to maintain health, stay fit, and prevent or delay chronic illness at any age. Your genetic profile tells you how your genes are responding to the food you eat.

With 30,000, 40,000, or 70,000 genes in each individual and each person inheriting a different genotype, you'd need a customized diet book for each person on Earth.

Each individual's genetic signature is different. Nobody responds exactly the same way to certain foods. That's why genetic testing helps to *customize* food plans for the individual. Some people have allergies. Others respond to foods gradually and silently inside their arteries and organs.

Another buzz word in the media is 'smart' or 'intelligent' foods. It's not that the food is smart or genetically changed. It's that the food in a relatively unprocessed, natural, clean state is free from vermin, full of life and enzymes. That prescribed food according to the person's gene expression is then healthier to eat for that particular person. Another person could be allergic to it and become sicker.

The food itself is freer of toxins. Additionally, the 'smart' or 'intelligent' food terminology really means the act of tailoring or customizing the food to the person's genetic profile. It does not mean the food is genetically altered and may or may not be dangerous to humans. 'Intelligent' foods refer to the results of your genetic testing.

You're the person being smarter by eating what the results of genetic tests reveal. The results are interpreted by a professional. In addition, you need to find out for yourself how to apply the test results to what happens to you when you eat a particular food combination or take a certain group of nutraceuticals. In making a food such as raw soy milk, for example, you need to know that raw soy can take certain nutrients out of your body if the soy milk is not cooked for a certain length of time before it is bottled, cooled and then consumed.

Blue, purple, and red-colored fruit or vegetables are excellent for your health, if your genotype says so. Eating raw blue berries also will take out certain vitamins from your body unless the berries are frozen or cooked.

You need to know how to process raw vegetables and fruits either by cooking and/or freezing before they are consumed so that eating them day after day in certain amounts won't lead to the leaching out of various vitamins, enzymes, or minerals from your body. Knowledge is important here. When you are prescribed certain foods, make sure you know how to prepare them for the maximum nutrient benefits.

Diets need to be tailored to individuals. There are only four blood types, but individual genotypes are different for each person. You'd need a customized diet book for each person on Earth. Each individual's genetic signature is different. Nobody responds exactly the same way to certain foods. That's why genetic testing helps to customize food plans for the individual. Some people have allergies. Others respond to foods gradually and silently inside their arteries and organs.

Not only do people of different ethnicities react differently to the same doses of medicines, but people of different ethnicities react differently to certain foods. It's touchy to bring in the term "race," but studies have found that people of a certain race react differently to certain drugs for certain conditions, and that people of another race react differently to the same dosage given to another race. You can read in medical journal articles for example, on how African Americans respond to glaucoma drugs compared to Caucasian people. You can read about how the same dose of a drug given to an East Asian and a non-East Asian will be too high a dose for the East Asian.

You can read about foods—how people from Northern Europe tolerate milk better than people from Southern Europe, and how people from Southeast Asia hardly tolerate milk at all in many cases. Then, there is always the exception. People are of mixed ancestry. What you need to learn about is the way intelligent foods work with your genes.

You don't know who will react in which ways at the genetic level to the food or medicine or whether the dosage of the drug or the amount of the food will be tolerated or too high or low. Even with varying dosages of nutraceuticals—supplements such as vitamins, food extracts, and minerals, some people benefit and others show no change. Read the results of conflicting studies in medical articles. If the dosage of vitamins is hotly debated, drug dosages and ethnicity is another topic in the research arena.

Some people have inherited risks for certain chronic diseases. Those people need to eat foods and perhaps take nutritional supplements that will prevent or delay the onset of those problems. Nutritional genomics fills an important need in maintaining health and quality of life. From the *Institute of Food Research (IFR)* in Norwich, UK, **Dr. Ruan Elliot, lead IFR scientist,** sent me this e-letter in response to my questions about what is being researched there. Here's the reply.

Dear Anne:

My main research interests are in using so-called functional genomic techniques to define mechanisms by which diet and specific components of the diet promote human health. These powerful techniques are set to revolutionise the way we approach fundamental nutrition research. On top of this, as you will appreciate, there is also the aspect of inter-individual genetic variation, the impact that this has on health and the potential variations in optimal dietary requirements.

To my mind, these two areas are locked together. We need to properly understand the processes by which components of the diet (nutrients, and micronutrients) work individually and together to keep us healthy so as to be able properly to define optimal nutrition for sub-populations or individuals properly based on their genetics.

You can find descriptions of my work and research interests at the following URLs;
http://www.ifr.ac.uk/public/FoodInfoSheets/EDPgenomics.html,
http://bmj.com/cgi/reprint/324/7351/1438.pdf
I hope this is helpful.

Best regards,

Ruan Elliott

For further information, contact the Institute of Food Research, Norwich

Research Park, Colney, Norwich NR4 7UA, UK. To view their Web site, phone and fax numbers or email address, on the Internet, go to: http://www.ifr.ac.uk/about/.

My philosophy about genetic testing is to remember a quote by Richard Feynman, Nobel Laureate: "The best way to predict the future is to invent it." Your genes are hard-wired for certain foods, but not all foods make compatible software. Consumers need a guide book to nutritional genomics. You'll hear terms such as gene expression, genetic signatures, risk, intelligent foods, and tailoring the food to your genotype. What it all means is that your body is looking for customized nourishment.

If you really want to take charge of your own health and nutrition, learn all that you can possibly find out about how to apply the results of your DNA testing, genotype testing, metabolic, blood, ancestry, racial percentages testing, and body chemistry or allergy testing to what you eat, the nutraceuticals and supplements you select, your exercise style and lifestyle. Everything starts at the cellular level, even the way your body reacts to stress and exercise with cortisol or with relaxation. Individuals react to certain types of exercise, foods, or perceived stress situations differently.

Some nutrients, foods, herbal compounds or other supplements cause relaxation in some people and panic attacks in others. Then there are allergies to consider. It starts with an expression of your genes in reaction to the environment. Some people get an increase in ocular pressure from sleeping on pillows containing a stuffing to which they are allergic. One sip of caffeine can start a panic reaction in one person and relaxation in another. Your reaction is in your genes. It's about body type, another genetic expression, whether you have inherited genes for anxiety of certain lengths, and the whole interplay and interface of one team of genes with another group in your body. I refer to this interplay of genes as a "rhumba of rattlesnakes" at the molecular level.

You have a part to play in all this, and it isn't always as the passive patient or recipient. You have the *genomer* and the *genomee*....Be the *genomer* for a change—the person in the driver's seat. Take control. Take charge of how to interpret your DNA tests for risk and diet changes or for ancestry and family history or for any other purpose for which you want to test your genome or any region of your genetic profile. To be able to understand how to read and interpret a DNA test and apply it to foods and supplements and to know how the foods will actually effect your genetic expression—that kind of knowledge is power.

The information is publicly available in medical school libraries, on the Internet's scientific databases, and in various journals in the nutrition and genetics fields. Most of these sources are open to the public without you having to be a scientist to learn about how your body responds to food at the cellular level. Start by joining various online groups, visiting your library, and reading the latest medical and scientific journal articles in the field of nutritional genomics. Network with patient support groups that use genetic testing. Start your own email list message board for consumers who want to learn and listen in addition to sharing resources of information.

The consumer's role is to compare, review, and find out who is best qualified to work with an individual's genes. How does a consumer discern between snake-oil and reputable companies in this growing field? Who is qualified? Consumers need an explanation in plain language what is healthy to eat, not for the world, but for the individual. So it's up to the consumer to do some research and learn a lot more about nutritional genomics.

Beyond food what does an individual's gene expression require in terms of exercise and lifestyle? It's time to educate your body about nutritional genomics and about DNA and ancestry. What foods should you eat and what nutritional supplements (nutraceuticals) would benefit your health? One way to tell is to test your genetic markers. Which genetic markers? The entire genome? Or specific SNPs that signal risk of certain chronic diseases? Nutritional genomics should be available to all consumers, not only those with money to pay for expensive testing. Sure, in the future the price of genetic testing will come down, but senior citizens and parents today want to know what they can eat that will agree with them and their families.

Researching the Web under "nutritional genomics" I found a company in the United Kingdom called Sciona. According to Sciona's Web site at: http://www.sciona.com/coresite/index.asp?p=1, Sciona is a venture capital backed company that researches and develops tests for common variations in genes which affect your individual response to medicines, food and the environment.

There are around 40,000 genes in the human genome. Sciona identifies those genes that influence a certain function such as cardiovascular status and tests for these as a set or 'panel'. This information is then used by appropriately qualified practitioners to provide you with health advice.

Testing of specific genes rather than the entire genome usually is done by various companies at this time. In the future, the number of genes tested may increase, or science may find which particular genes interact with other genes to put you at risk if you eat certain foods, or whether your specific genes work together in such a way that you can eat almost anything without developing chronic diseases.

According to Sciona's Web site testing specific genes for certain chronic illnesses is useful in guiding aspects of the treatment of diseases such as heart disease and osteoporosis, which are influenced by your genes, lifestyle and environment. For the consumer, knowing which foods to eat to influence your gene expression is important.

According to Sciona's Web site, Sciona's team of geneticists, molecular biologists, medical doctors and dieticians work with universities and other companies to identify the significant genes underlying a particular effect. The effects of these, and other factors such as your diet, are then analyzed to give you specific advice on courses of action tailored to your own genetic makeup and circumstances.

Filling out questionnaires allow consumers to think in terms of focus sheets. Besides DNA testing, food choices, and consultations with your physician, you need to focus on your habits and think how realistically you answer a questionnaire. Think about from which direction you want to participate in nutritional genomics—marketing, research, science, consumer awareness, forming support groups, media, or other. What are your basic interests and how can you apply them to this field? You can contact Market America, Inc. at 1302 Pleasant Ridge Road, Greensboro, NC 27409. Their Web site is at: http://www.marketamerica.com. Here is an e-letter from Andy Aldridge, Public Relations Director, Market America, Inc.

Dear Anne:

After extensive market surveying and testing, Market America and Cellf, a division of Sciona, have partnered together to offer the Nutri-Physical™ Gene SNP DNA Screening Analysis program. This product allows consumers to submit a sample of their DNA to have it analyzed. Once the analysis is complete, consumers receive a report that outlines possible deficiencies along with lifestyle and diet changes that can be carried out to address possible vulnerabilities. To accompany the DNA analysis, Market America also developed a questionnaire that, when analyzed in conjunction with the Gene SNP product, results in a suggested list of customized vitamins and supplements, available in one formula, if desired by the consumer.

The company's Isotonix® Custom Formula makes choosing the correct nutritional supplementation a simple and efficient process. From a Distributor Custom Web Portal, a customer submits answers to a dietary and lifestyle questionnaire and has the option of purchasing a unique custom formula nutraceutical that specifically addresses their individual needs.

Many companies are talking about NutriGenomics. Through our partnership with Cellf for DNA analysis and Garden State Nutritionals for manufacturing and customization, we are actually doing it.

Regards,

Andy Aldridge
Public Relations Director
Market America, Inc.

***

11

Consumers Need to Be Involved in Quality Control

You need a voice in quality control. What the consumer needs to understand are the roles of genes in healthcare, and how the *roles of genes* interact when you take in nutrition. Consumers, corporations, venture capitalists the government, taxpayers, and research institutions invest billions of dollars each year to develop this understanding. One example of a consumer group involved in quality control is when parents group together to form their own DNA bank, recruit people to donate DNA for research, and develop databases and Web sites disseminating information on a particular genetic condition.

Consumers can don few or many hats in nutritional genomics. There are avenues to explore varying from watchdog, marketing, research, public relations, parenting, safety, event planning, publishing, gerontology, videography, genealogy, healthcare, to broadcasting.

If you do your research, you'll find that venture capitalists who in the last decade invested heavily in the computer industry are now looking to invest in biotechnology. The power of gene technology drums up business and communication also for patent attorneys, journalists, and inventors.

To participate in nutritional genomics in a variety of capacities as a consumer, you can write to the United States Food & Drug Administration (FDA), Department of Agriculture (USDA), and National Institutes of Health (NIH), as well as university laboratories, pharmaceutical manufacturers, and government agencies worldwide. Get involved in the power of genetic technology at some level. The FDA is the agency that's responsible for 80 percent of the United States food supply, according to a July 1st 2003 speech given by the Commissioner, Food and Drug Administration.

In that July 1, 2003 speech before Harvard School of Public Health, Mark B. McClellan, MD, PhD, Commissioner, Food and Drug Administration, said, "All of you—consumer advocates, representatives of the food industry, nutrition scientists, and other food experts—have a collective commitment to the issues we face at FDA that is integral to our ability us fulfill our mission. And your help is needed more than ever. Now more than ever, we all must work together to find better solutions."

It's important to read this speech and to look at materials on the Web at the Harvard School of Public Health. Click on the Harvard School of Public Health's Web site at: http://www.hsph.harvard.edu/now/jul11/conference.html. Read the important facts there and check out the forums. The current headline at the Web site notes that "A July 2003 conference at the Harvard School of Public Health 'spurred' dialogue with the nation's food industry on the subject of "'Changing the American Diet' To Improve Health." Read the materials there and think for yourself about how the food you eat is processed and marketed.

In the speech, you'll find key words such as "consumer advocates" and "collective commitment." Your help indeed is needed to work together as a team, to share, with a purpose of finding better solutions. That's why it's important to listen and learn all you can about the future of nutritional genomics. The consumer's involvement is important. The theme emphasized "collaborating to improve the American diet." Collaborating with consumers also means working to decrease obesity and epidemics of diabetes in children.

You can look at the American diet from the point of view of those who work in DNA testing, from those who work with family history and ancestry research, or from those who work in food packaging and processing. It really hits home when you look from the point of view of the consumer or from marketing and product management. Everybody has to eat.

◆ ◆ ◆

Consumer Surveillance

There's another branch of nutritional genomics that instead of only testing your DNA to find out which foods are healthiest for your genes, focuses on *manipulating plant micronutrients* to improve human health. See the article on the Internet, a PDF file on a Web site at: **http://www.ipef.br/melhoramento/genoma/pdfs/dellapenna99.pdf**.

The volume of imported food is growing each year. Consumers have a field cut out for them—surveillance. As FDA increases its examinations and sampling at borders, consumers can work together to research information about food imports and inspection.

A laboratory can only sample so many products. Consumers can take a role in food security, perhaps looking at industry to identify problems or threats. What the consumer's role entails is better information and collaboration. Everyone needs to keep costs down.

Plant biotechnology of food and feed is another area of consumer interest. If you buy food that comes from overseas, do you ever wonder who oversees the packaging and shipping of those products? Are there really enough inspectors to go around? Consumers worry about the widespread use of sugar in soft drinks. In addition to having your DNA tested, you need to understand how what you eat influences your health at all ages.

Another way consumers can oversee quality control is by forming public interest research groups funded by grant money, private donors, institutions, or the government. You can become a volunteer in nutritional genomics, an ombudsman, a lobbyist, or start your own consumer research interest group.

You can turn a hobby of nutritional genomics or DNA for ancestry and genealogy into a business by affiliating yourself with a university lab which you contract to do testing from your DNA testing clients. There are open doors for consumer involvement depending upon your skills and interests. Nutritional genomics needs public speakers and technical writers to relay to the public what innovations the experts are bringing to healthcare and food systems design.

From running a summer camp for teens interested in nutritional genomics internships or learning experiences to recruiting DNA donors to create a DNA bank or in researching and writing about genomics, there are a variety of doors. Consumers have power in numbers. You can even enter as a venture capitalist with a goal of raising funds even if you have no funds of your own and plenty of determination to learn to ropes.

Don't overlook nutritional genomics for the pet care industry from foods to medicine. Contact the veterinary schools about their research on how foods affect genetic signatures of pets or race horses. Check out the Web site for Research Diets at: http://www.researchdiets.com/.

Research Diets, a New Jersey company since 1984 has formulated more than 6,000 distinct *laboratory* animal diets for research in all areas of biology and related fields at hundreds of pharmaceutical, university, and government laboratories around the world. Nutritional genomics isn't only for humans or laboratory animals. Did you ever think about how your dog or cat could benefit by genetic testing to determine which foods are healthiest?

Talk to your veterinarian to see who is researching how nutraceuticals and better food can help your pet's health, especially when the pet is older. What about nutritional genomics for farm animals or pets? Find out who is doing what kind of nutrition research for better health.

Or you can tap the venture capitalist watering holes and open your own nutritional genomics business. Consumers can hire, outsource, or contract fee-for-service with licensed healthcare staff and geneticists specializing in nutritional genomics research. It's not difficult to raise funds. Billions already are invested in the gene technology industry. Sometimes parents of children with genetic conditions are the first to band together to form DNA banks for medical research. Don't overlook agri-nutrition. It's about what vegetables and fruits you eat that in turn impact your genes.

Medical schools and research institutions often seek federal grant money. Someone is investing billions for this research. Find out who are the investors. Who benefits from this knowledge first? Who has access? Is it the consumer who pays for testing, healthcare, nutraceuticals, and selected foods? It's up to the consumer to take a look at quality-control of the entire nutritional genomics arena.

Everyone is concerned about cell degeneration from a lack of critical nutrition. Unless you know how your genes are impacted by various foods and nutraceuticals, how can you can decide what's good or bad for your body other than looking at family history?

The consumer is concerned with how nutritional genomics as an area of research is applied to healthcare with a goal of disease prevention. For the consumer with no science background seeking a beginner's guide book to the power of gene technology, the first step is to realize that genomics has applications for many more areas of exploration than nutrition, healthcare, archaeology, genealogy, oral history, and population genetics.

So should you get screened? With the enormous variety of diet books on the market, healthy-eating advice abounds. How do you know what works to keep your arteries unclogged and your organs nourished for the combination of genes you inherited? People are different, but similar advice often is given to everyone.

What you need is a personalized report, time capsule, scrap book, or profile that has not only your genetic test results, but everything about your lifestyle, stress reactions, exercise. For example, if your body reacts to exercise by secreting way too much cortisol, should you be taking vitamin C? Will the vitamin restore to a healthier state or further damage your organs? If the vitamin is prescribed, what amounts of it will work best with your genetic expression or genotype? How do you find out whether this regimen is good for you or not? It is to these types of questions consumers want answers.

Information on food groups, vitamins, minerals, extracts, and nutraceuticals in an easy-to-understand format and binder are what consumers need. You also have to learn how your DNA test was interpreted. If someone interprets the test for you, how can you apply the test results to eating smarter foods for your genotype or selecting the right dosages of nutraceuticals? You'd have to keep paying for advice on every aspect of your nutrition as another branch of healthcare.

If you know the language of your genes or can learn it yourself, it makes your road to a healthier self clearer. Consumers need access to and knowledge of quality control. Who is performing the tests? What kind of medical supervision is officiating at the marriage of healthcare to nutritional genomics? Who is at the top overseeing the quality control—the consumer? It should be. You pay to fund the research in tax dollars. So form your consumer groups and look into quality control in the field of nutritional genomics, DNA testing, and related businesses that will open to serve the consumer and the licensed heath-care professional.

As the power of gene technology reaches the masses, all types of offerings will find a way to apply DNA test results to what you eat, how you exercise, what you wear, your prescribed medicines, therapies of various kinds, what music will change your physiological responses based on your genotype, and even what career, lifestyle, mate, childbearing plans, or hobby you choose.

DNA samples usually are obtained by rubbing a brush swab on the inside of your cheek, completing a questionnaire, and mailing the brush to a testing laboratory. Sometimes in the case of a rare genetic disease, blood samples are taken, but most DNA testing is done with a cotton or felt swab or mouthwash. After several weeks, your physician would get a report to interpret for you for nutritional genomics. In testing DNA for ancestry, the report would be sent directly to you.

How specifically is the report customized for you? Even tiny differences in your genes influence the way your body metabolizes foods and excretes toxins. These reports look for *variations*. Your feedback consists of dietary information. Some companies assess your present eating habits and foods with your genetic profile. Advice is offered. For example, some tests look at the type of meat you eat, whether your meat and fish are smoked.

The tests look at the types of vegetables. How many cruciferous vegetables do you eat—such as cabbage, cauliflower, Brussels sprouts, and broccoli? How many raw vegetables do you eat? How many sprouted legumes or whole grains not processed into flakes? Did you know that cauliflower is a low-carbohydarate vegetable, but broccoli is higher in carbohydrates? Did the sprouts you ate have bacteria on them that affected you?

This is important if you are insulin resistant or have too much insulin pouring outing each time you eat a high-carbohydrate vegetable. For example, pectin consumption results in an insulin release. What kind of antioxidants do you eat? How much folate are you taking in, and how does that folate impact your arteries—neutral, good, or worse?

What types of whole grains are you eating? Are you eating your whole oat groats or eating foods high on a high-glycemic index that rapidly turn to sugar in your blood and require more insulin to be secreted which can lead to other problems in excessive amounts.

Does your genetic profile reveal a need for a Syndrome X diet or other measures to halt the excessive insulin secretion each time you eat carbohydrates or proteins? What kind of weight problems or control do you have?

How much sugar do you eat? Do you drink liquid candy such as soda pop? How do diet drinks affect your genetic expression compared to sugared drinks? What kind of saturated fats do you eat and how does your body react to that kind of fat? Do you smoke? What damage has smoking caused your body? What kind of exercise do you get?

Walking moderately may be enough for your body, and too much exercise could bring about high cortisol levels—the stress hormone. Does your body react to exercise as stressful or relaxing? What allergies do you have? All these questions need to be compared against your genetic signature. What genes are assessed by the testing company? Do you have a genetic propensity to alcoholism? Which gene predisposes you to alcoholism or drug addiction when you are under stress? Did you inherit the anxiety gene? If so, what is the right career for you? Do you have the genes that predispose you to panic disorder? Find out how many genes are studied and which genes relate to what physical aspects of your health?

Look for a company with excellent quality control. Recommendations should be practical and proven. Your goal is to improve short and long-term health and prevent chronic diseases or at least delay them as long as possible. If the food you are prescribed don't make you feel better, find out why. The tests you take and the food or nutraceuticals prescribed should be scientifically proven in reputable studies. You can check out the studies in medical journal articles. Make sure they were not flawed studies or studies so old that new studies have proven opposite conclusions.

How can you tell for sure what risks you have and what needs your genes have for certain foods or nutraceuticals without testing *all* the genes in the human genome of an individual? If all the genes aren't tested, how can science really know everything there is to know about your individual gene expression? What's important to know now—going with what you have available? When will the research teams be ready to reach out to the average consumer of nutritional products and healthcare? Today there is a divide between what's going on in research and what's available to consumers. That's understandable because the research is still going on.

Without knowledge of all your genetic markers, how can you develop a plan for maximum health through nourishment? Why must this plan be in the hands of a licensed professional instead of an informed, self-taught consumer? If the consumer is armed with knowledge of what the individual's genes reveal and how it relates to certain foods and supplements, the consumer than can take back control and power over his or her own health.

Knowing your genes is only a beginning. How do you apply that knowledge to practical applications as in how certain foods affect your body chemistry and metabolism? If you have a plan A and a plan B, and you know the unexpected can always kick in, can you still have that feeling of a little more control over the way your body treats you? In a world so out of control, consumers of healthcare are looking for a semblance more of control and power over their sense of well-being and mood.

Diet books written for the masses may not work well with your particular gene expression. Science no longer tells you that you have inherited some gene mutations or defects. So your gene expression causes certain chemicals in your body not to work normally. Instead, you are told that you have an individual gene expression requiring you to eat this instead of that.

If you eat this combination of food in the morning, that excessive insulin pouring out each time you eat high carbohydrate vegetables or fruits won't narrow your arteries so quickly. It sounds so positive. It's practical, and tells you specifically what you should eat and when and how the food affects particular organs or metabolic and chemical reactions in your body as a result of consuming a particular food or nutraceutical. At last, there's a positive solution.

You can use intelligent food to override your defective genes that put you at risk and nourish your gene expression by fulfilling your genetic signature's deepest nutritional needs. How fast are the professional dietitians and nutritionists tuning in to nutritional genomics?

The key for the average consumer without degrees in nutrition or genetics is to learn how to look at a printout of your genetic markers and be able to interpret what that means in terms of which foods and nutraceuticals to consume for your health based on what you are at risk or pre-disposed to come down with given the right interplay of environment, lifestyle, food, attitude, and perception of stress.

It's not as hard as you think to find this information in libraries and in online databases. Any subject that takes money away from professionals making a living giving advice in any field is going to require effort to learn. The key is to find sources willing to share information. You bet they are around, and a lot of scientists are willing to share. Some scientists are concerned about their careers and reputations. They should be. Some are cautious about with whom they share information.

On the other hand, some scientists do not share. It may not be the fault of the scientists, though. If you look in the archives of the Los Angeles Times, you'll perhaps find an article dated July 18, 2003 titled, "Whose DNA Is it, Anyway?" Read that article. It's about a person whose mother has Alzheimer's disease and who, according to the article, was "trying to coax Alzheimer's patients and their families to donate DNA" to a university. It's not only about one person because more than ten thousand people donated DNA for the cause.

How about you starting a nutritional genomics DNA bank, new support groups, non-profit organizations, online discussion groups, or Web sites such as "Moms for Genome-Tailored Meals?" Think about how many consumers, not scientists, whose families have a condition or who are interested in a condition arrange events to encourage people to donate DNA to research.

How many consumers go out to assisted living complexes, senior centers, adult education classes, gerontology workshops, nursing homes and churches or family meetings to recruit people who have relatives with a certain condition such as Alzheimer's? How many consumers reach out to the actual Alzheimer's sufferers to recruit people to give DNA to science—to research, often located at or connected to various university laboratories? This is one way consumers get involved in science without necessarily having credentials in science.

All it takes is a deep interest to take action and learn more about the subject. Some research companies are independent, founded by scientists, but work closely with a variety of universities. Consumers are involved every day in science, usually on the end of recruiting people for donating DNA, planning events, fund-raising, writing about in the mass media, encouraging children to take an interest in science, acting as ombudsmen in nursing homes, running consumer awareness classes for seniors or parents, working with genealogy societies for various ethnic groups whose DNA is being studied by scientists for diversity, geography, or genetic diseases, and philanthropy are but a few ways to get involved in scientific research.

According to the article, you have a situation where DNA was used to search for Alzheimer's genes. Science itself has a great mission, to someday find a cure. The DNA is like a pointer. What the consumer is at risk for is that the research will stop. When the research stop for whatever reason, who controls the DNA you've donated? Read the Los Angeles Times article, "Whose DNA Is it, Anyway?" Think about your own DNA and thousands of others who have donated for testing. Who controls the DNA archive when the research is halted in midstream?

Who owns all the DNA samples—including yours? Who will inherit and use the samples and for what type of research? What if you have a group of family members suffering from a condition that inheritable and you don't know whether you'll get the disease or not?

What if you donate DNA along with others hoping it will point toward a cure, but the project ends? It's like the old question of who owns the living embryos in vitro when the project is over? Almost anyone with a nursing, social work, or other allied healthcare background can open a home-based or other business part time online or fulltime recruiting people for medical trials.

Around the world, people give their DNA for medical research. Why do people donate DNA? It's not always to find their ancestry or ethnic origins, because in many cases, results become statistical and no feedback is given to any individual or even to members of an entire group. What shows up are anonymous statistics in scientific journal articles or books. People often give DNA in the hope that science will find a cure for their or their family member's genetic disorder or inheritable disease.

People donate DNA to see whether a new test will work better for them, such as a test of racial percentages, or in the case of a disease, a test that will reveal risk that could be overcome with certain foods or medicines or even gene therapy. People hope that the medical field has a way of injecting normal genes into a child or person that will take hold and correct a defect in the existing gene expression. Part of gene therapy is about the introduction of new genes that could fix the old genetic problem, and often it works in certain cases.

What consumers can do for themselves is to take part in creating lending libraries for DNA. You need to have lending libraries open not only for scientists, but for the consumer to at least read about research. Of course, scientists who actually work in laboratories with DNA could enjoy the free and open access to the DNA managed by special DNA librarians. It helps research move forward. As a consumer you are a taxpayer.

Do you know your tax dollars are paying the cost of much of this scientific research, especially the research going on at state universities? If you are paying for this research, then the scientists are public servants to you, the consumer. You have the right to this information as much as any scientist, since you are paying the bills for the research.

The problem is that not all scientists freely share DNA all the time hoping some other scientist will find a breakthrough. It reminds me of competing journalists on different papers vying for the news scoop of the day. The DNA is collected and sometimes not shared at all.

The reason is that a scientist's career is at stake. In research, a scientist's reputation and job depends too often upon breakthroughs that help the career. Will sharing lift a scientist's career by his or her bootstraps? Sharing could mean the other guy finds the breakthrough first. Then in walk the patent attorneys. Scientists can file patents on genes.

What if your child is autistic? What scientists will share DNA with other scientists in research on autism or Alzheimer's or any other condition? What's your role as a consumer? It's to create a situation for cooperation. Consumers tend to form patient support groups and to build their own DNA banks. That's only one way of taking consumer action in a positive way to move research forward and point toward solutions and cures for the benefit of the health-care recipient—you, the consumer.

What can you do with nutritional genomics that is in your interest? You can continue to build DNA banks backed by consumers with an interest in nutritional genetics or DNA ancestry, or gene testing for disease markers. Share among yourselves. After all, you are not all scientists, and therefore have no career in science about which to be concerned. I speak as a consumer here rather than as a science communicator.

What you can start as a consumer is a movement to pool DNA samples into a managed library and DNA bank open to all scientists. Consumers could have access to reading about the research and learning to interpret their own DNA and other genetic tests of SNPs, since most consumers don't have access to laboratories to perform research. The time for the consumer's role in moving research forward is here, and consumers must learn all they can about how to interpret the entire human genome.

Your first step would be to make a list of the various institutes that study particular conditions of interest to you or written in your genes as your risk condition. Start with the ones funded by you as a taxpayer. Learn more about what your genes show you. When you talk to scientists, you'll find out how many will or will not share anything from information to DNA. Talk to the scientists. Get on emailing lists where scientists talk among themselves. If it really were true that scientists never share information, there wouldn't be any libraries or databases. Once anything is published and is not classified by the government as secret, it's shared through journals and libraries. The idea is to do the same with the raw material, the DNA through gene banks and DNA libraries as well as databases.

You already know some scientists don't like to share raw materials such as DNA with other scientists. Just ask most biology educators at universities to share their experiences with geneticists with consumers, let alone the media. If you think your DNA is a gift to the public like your donation dollars, think patents. What if you're a consumer who works with donors? What if your dream is to see children with certain diseases tested for little cost?

What if there are battles between scientists and hospitals for patents? What if a hospital takes out a patent on your DNA or those of a bunch of donors you as a consumer recruited from your efforts, support group, or fund-raising events? What if the hospital is awarded the patent and starts to charge fees?

What if that results in some testing programs closing? It happens over and over. What if the costs of developing certain types of DNA or other genetic tests are or remain too high? These are all the questions that consumers want answered by scientists, hospitals, research institutions, universities, and licensed healthcare professionals.

When a consumer so enthusiastic about learning all about genetics, DNA for ancestry, or nutritional genomics put effort into raising money and providing DNA donors, the consumer wants to be sure that what's used for research doesn't always go to pay for patents that result in fees and costs to the consumer or anyone else becoming unaffordable just to create or invent the new genetic tests.

Media has created videos romanticizing "gene hunters" poised on the cutting-edge of a new frontier—inner space. Gene hunting is entertainment. Yet there are little controls or rules—not even a law that requires scientists to share DNA with other scientists. If gene hunting is another Wild West frontier focused on individuality of gene expression, where are the public interest research groups here?

Back in 1975 I was a book author employed in a temporary job full time for ten months writing the consumer manuals and handling all incoming calls to the consumer complaint center of a public interest research group. My job was to write a book that ended up in the Attorney General's office. The book was a consumer manual of how to effectively complain.

Today, the consumer needs a guidebook or manual to learn how to effectively organize what should be in the public interest—research. That research could be anything from the frontiers of nutritional genomics to the study of food on a disease or the disease itself. Sometimes research is shared with the public long after everyone else knows about it. If it's archaeology, it's shared quickly. If it's gene research, that's another story. Sometimes patents get in the way, sometimes costs, and sometimes egos.

Databases are open to the public. The bioinformatics profession manages databases of bioscience information stored in computer software. You can get a certificate in bioinformatics at many community colleges or through extended study programs at various colleges and get your foot in the door of bioinformatics through an internship course.

On the other hand, you have scientists, universities, private companies, and research institutions vying for money and/or information being returned to them for their work in giving information to the databases that hold DNA information. It's all about your taxpayer dollars.

As consumers, you pay your taxes to the government. In return, the government takes your tax dollars and dangles that big carrot of federal grant money in front of the institutions employing the gene hunters. And if those scientists refuse to share DNA with other, competing scientists, well, they won't get the federal grant money that's really your tax dollars.

So who's at the root here who should be in control? It's you, the consumer because you supply the government with your tax money that makes up the federal grant money that goes to the many of the places that employ scientists who test DNA for research.

Write to the National Institutes of Health and ask who is sharing what with whom. How are your questions answered? You could get yourself a career as a grants writer and work your way up to become a grants administrator. That's a long career route where you never know whether or not you'll be chosen for the job. The highest you might get is as a serious student of grants administration research. Another route is to ask scientists why sharing DNA could destroy their careers. It's all a matter of who gets to publish first.

In my quest to find people to chat with, a few scientists turned me down not because I don't have a science degree, and not because I don't work for a major newspaper or magazine, but because I'm *the first journalist* to write a book on genomic nutrition before the scientists have had a crack at it. And the book isn't for scientists speaking to other scientists, but a consumer's guide to nutritional genomics and ancestry by DNA testing.

Consumers don't need to be slapped in the face with technical jargon unless the genetics terminology is defined in a glossary. The purpose of this book is to inspire consumers to look into the subject of DNA testing not only for deep ancestry, molecular genealogy, or archaeogenetics, but for tailoring intelligent foods to their genetic signatures. If the buzz words in the news are nutritional genomics, then now is the time to organize. The entire area of DNA testing needs direction, quality control, and applications to what consumers perceive as critical needs.

You see dozens of books published on population genetics, archaeogenetics, the peopling of various continents, mitochondrial Eve, ethnic DNA, diseases and DNA, ethnic genetic diversity, food intolerance, allergies, inherited illnesses, aging, and DNA for ancestry. How many general-reader type books are published on smart foods based on genetic signatures and SNPs rather than chemical, blood, or metabolic signatures? So positioning yourself first matters if you're a scientist who must publish and patent. If you're an educator with an academic institution, positioning yourself first in publishing is crucial to your career, even factoring in getting tenure.

A consumer's world is based on sharing. A scientist's world is based on fear of competition. It should not be that way, but it is. And few will admit it. What competition? There's no scarcity here. There's room for everyone? Then how come so many excellent PhDs in various sciences can't find jobs in their field?

If you want to check this out, talk to recruiters in the sciences and find out how many people are out of work with PhDs in a variety of sciences. Is it age discrimination? Ask the recruiters for their experiences. They will share with you in most cases. Try writing a book on careers in genetics. See what the recruiters have to say to you. Some scientists at some research institutions also will tell you that before they can share DNA, they have to get consent from everyone who donated DNA.

What do you do as a consumer? You don't work alone. Power is in numbers. You organize, get people together and pitch in as a growing group to start your own DNA bank, perhaps consisting of DNA donors. Then you'll have the problem of contacting universities and researchers at a wide variety of places to see whether they are interested.

For the average consumer who doesn't want to take any action, just find out what foods are healthier, the problem is simple. Just research your own genes yourself. All your genes interact as a team. And as all these genes work together, you usually can't point to one gene as being responsible for one action or disease. You can have a defect in one gene that causes a problem or only a risk.

Both consumers and geneticists are interested in finding out what mutation happens in a gene that leads to chronic illness at any age. When a gene oozes really bad protein or the wrong amount of protein—ah, oh.

The interaction between the rest of your genes and the chemical and metabolic systems go wild at the cellular level. You get sick. Gene hunters who study the mutations in your genes can study mutations for ancestry or for sickness. Different genes are studied for ancestry, and muations there have little to do with illness. So where does the consumer go to start? That depends on what you want from your DNA—ancestry, or a prescription for intelligent food?

If you're talking about illness, you start with patient support groups. For ancestry, it's the genealogy groups, ethnic groups, and DNA mailing lists online or start an archaeogenetics club or e-mailing list. You can start your own academic journal. For nutritional genomics, the field is wide open for you to start your own DNA collection banks. Who controls your DNA—you or the research 'industry'? You, of course. You control the direction of any research performed on your DNA, and it's time to take DNA by the reigns and direct it—yes, you the consumer. Your tax dollars fund research. So take control of your own DNA.

In the field of nutritional genomics, you, the consumer must take charge of your DNA and begin to build collections in DNA banks where the DNA donors are in control of the direction of the research. Nutritional genomics is not only for parents of children with genetic defects that need the research so urgently. It's also for anyone who wants to know what foods to eat to stay healthier for longer periods of time.

You can start organizing family networks to review and compare the nutritional genomics industry. You can build collections of DNA in your own DNA banks and create databases and DNA libraries. Instead of re-inventing the wheel by duplicating the existing databases, fill the gaps where your efforts are needed and rewarded. Go to a variety of genetics, foods, and DNA-related conferences and let people know you are starting your own nutritional genomics DNA bank and/or intelligent food bank.

The skill you need is not a PhD in genetics, but the desire to persuade people to come together. You need to be a catalyst. If you're one of those unemployed people with science training and are good at public speaking, organizing people, and fund-raising, it could be the career you're waiting for.

If you are a parent with no particular skill other than public speaking and an interest to learn all you can about nutritional genomics from self-education, it's a way to take action. Be prepared to outwit Mr. or Ms. Medical Research Politics just as you have outwitted corporate politics and the games your mother never taught you. If you are lucky enough to be a homemaker, this is the perfect part time career—finding loopholes in medical research politics. It works best if you are not a doctor's wife. Being a doctor's mother is even better for this career. And if you're a patent attorney or work for one, this is another ball game.

The first loophole is to form your own group. Work as a group and put your power in the numbers. Work with scientists in a group also. If you have DNA to donate, give it to a group of scientists that you can organize as a group to work together. There's the problem of individual scientists sharing DNA with one another. If there's no group of scientists, ask them to form a group to work with your group. Organize for nutritional genomics.

Pool the scientists, not the DNA. Otherwise, you'll have a bunch of DNA on your hands and no place to store it. You've got to organize scientists together in a group to work with your organization. First, you need to start an organization. It doesn't take a lot of money. It's about people working together. Consumers aren't afraid of competition. Scientists are afraid of competition because positioning oneself first is the rule of the game if they are to keep their jobs or get ahead.

Consumers are parents, kids, or retirees, not scientists or medical researchers whose careers will be harmed by grouping together to share information or DNA. Most scientists will not share DNA with other scientists. You can go to scientists in other countries and talk to them. See if they will pool and share DNA for research. There's a far better solution. Start your own DNA bank. You organize parents—especially ones with a little money or prestige in their non-science careers. These elite parents have enough smarts to do fund-raising and hire people to collect the DNA to put in your own DNA bank.

What kind of people perform the best fund-raising and organizing? Public relations people—extroverts, people-people, public speakers, entertainers, and the media, for starters. Don't bother the introverted scientists who need to hold their jobs by positioning themselves first in publishing the fruits of their research. Contact public relations and marketing people such as fund-raisers and marketing communications managers.

Hire people not afraid to speak out and not afraid to learn about nutritional genomics. Start a DNA bank, a nutritional genomics resource center made up of many families. Collect a lot of DNA. Ask for private donations. Share your DNA bank's DNA with any scientist. Let the medical schools know about you. Again, you don't need any type of academic degree or credentials to organize a DNA bank. You just need to hire a staff to collect the DNA and make it available to medical schools and any other scientist for research. Who owns the DNA? The donors.

If you want to network with other parents and consumers who have done this, contact people who have organized their own DNA banks, such as the Autism Genetic Resource Exchange. They are on the Web at: http://www.agre.org/.

Talk to the medical schools that use the DNA banks. Read the Los Angeles Times article, "Whose DNA Is It, Anyway?" July 18th 2003. The article also mentions how the Autism Genetic Resource Exchange got started. You can do the same for researching nutritional genomics, a field that got started only back in 2000, and you don't have to have a background in any science to organize people, hire staff, and get started recruiting families to join you in creating your own DNA bank. Do you know how much you are needed by the medical schools?

When you speak to scientists, you may get cooperation, but you'll also come across someone who will tell you to scram because you don't have credentials, or in my case, because I make up stories for a living as a novelist, overlooking 35 of my non-fiction books. Don't let anyone distract your attention from your goal by trying to focus your attention on perceived shortcomings. Anyone who turns you away probably wants first position. True love is about making sure the loved one is positioned first. What nutritional genomics needs is a little tender loving care.

Who do you contact as your first customer? Try the *research-oriented* medical schools. They don't have anywhere near the resources to collect as much DNA as they need. Contact the heads of the molecular genetics departments at the medical schools.

Not everyone interested in nutritional genomics will want to start a DNA bank devoted exclusively to nutritional genomics. There's a place for all levels of participation. You can hire a DNA collector or learn how to collect DNA from people if you're a people-person. You can raise money from donations or ask your state for money. Have each donor sign a statement allowing the DNA to be shared by other research scientists.

Find someone who is associated with a reputable laboratory or with a laboratory in a medical college to store the DNA samples. Talk to biostatistics professionals. The more DNA you collect, the better it is for the numbers crunchers. Not much is learned from only a few DNA samples unless you're comparing Neanderthals to modern humans.

Nutritional genomics as an industry and area of science needs its own support groups. Most diseases have support groups. It's time support groups were formed for preventing or delaying genetic risks some people are pre-disposed to from occurring through learning more about smart foods tailored to individual genetic signatures.

What's your solution? Since scientific research is for the benefit of the consumer of healthcare and nutrition, start by going to other consumers and getting feedback by reviewing, comparing, and polling consumers on their experiences with testing. Are the consumers satisfied? What would they like to see improved? Today's consumers of genetic testing have enough money to spend on tests that help them make decisions about food, health, or ancestry research. You've seen Michael Moore's TV Nation. How about creating a video called Nutritional Genomics Nation? If you don't want to produce a video or a media presentation, how about visiting a database and a Web site?

Still feeling a bit out of control? No need to. Just talk to the professionals in the field and make friends. Aside from professional competition that makes some scientists want to position themselves first in publishing and patenting, scientists really are friendly, welcoming people—especially when you are positioned to give them the kind of publicity in the media they want at this time. Nutritional genomics is a buzz word in the media. It's a hot topic this year. Scientists are finding this one-chance shot at getting positioned first in the media by publications of the highest repute.

Librarians at medical school libraries and publishers really like it when you read magazines that would otherwise be read by only a few. I used to spend months reading journals at the UCSD medical library to get ideas for topics that would result in books—either novels or nonfiction how-to series. So you, the consumer, also can learn the important connection between how to read the results of a genetic test and how to apply it to choosing the right food.

Right now, that key is in the hands of licensed healthcare professionals and scientists. It's time the consumer learned how to interpret the results of genetic testing and how to apply the results to a better way of eating. Self-education is the answer. And even if you put your life in your doctor's hands, how do you know whether he or she is qualified to make the connection between your genes and recommended foods or whether those foods will work with your specific gene expression?

Find out what shows up in your relatives and what you may or may not have inherited at the molecular level. A test of your biomarkers will at least give you that handle on your own healthcare, foods, and lifestyle. The point is to educate yourself about your individual genetic signature regarding risk and the implications genetic markers have on your health now and in your own future.

Keep your profile private, and use the information to choose the best possible combinations of foods to help prevent or delay any chronic illness for which you may be at risk. You don't have to give your genetic profile to anybody who could use it against you such as insurance companies and employers. What you can use your profile for is to choose the foods that your body needs. Let your gene expression show you what to eat, how much, and when.

When you look for a diagnostic lab, keep in mind that you want to deal with companies that provide their testing to physicians or similar licensed healthcare providers and not to anyone from the general public. The reason is that you need to know how to interpret these tests and how to apply what you learned.

For example, if genetic testing reveals your at risk for a certain chronic disease, how would you know that unless you can look at the genetic markers, see the risk, and be able to judge which foods and supplements would cut that risk. What I'd like to see in the future is that the consumer would directly be able to get that information from publicly available sources while maintaining privacy.

What I like about the company is that they have a Web site with clinician support where you can educate yourself about what to look out for in your own body so you can at least ask important questions when you see your doctor for testing. And for physicians and clinicians, it's an excellent site to become more informed in specific areas.

The ideal situation would be to see the results of your testing and be told what risks you have. Then you'd be able to buy a book or look up enough medical journal articles. In the future, as a consumer I would like to see more consumer education so that your preventive care wouldn't always have to be solely in the hands of your managing physician.

You need to have more control over what you eat to lessen your disease risks. For you as a consumer to concentrate on focused prevention and treatment tailored to your individual genetic signature, you need knowledge. If knowledge is power, than those with the knowledge you need and don't have leaves you pretty powerless. So you have to educate yourself via books and medical journal articles, the Internet's Web sites and asking questions.

When you see your doctor, often you have only a few minutes to ask questions as doctors are pressed for time. Make a list of questions you want to ask. You might want to compare opinions and answers between alternative health care licensed professionals and your usual primary care physicians or HMO nurse practitioners.

Then check out with medical journal articles and news or even consumer feedback and reviews of what you're seeking. If you don't see enough Web sites that review genetic testing companies or research institutions and if you don't see reviews of biotech companies doing research, start your own Web site to review the many companies springing up.

Compare the nutritional genomics companies that deal with licensed healthcare professionals, and get answers, opinions, reviews, comparisons, feedback, polls, and consumer reports. You have consumer reports on cars, mattresses, and washing machines, why not a consumer report site or publication on genetic profiling for nutrition and health? Start a public interest research group on companies that look at genes or look at disease in various ways. You'll learn a lot from consumers just as teachers learn more from their students than from many of their books. Ask for feedback. It's your genes.

Learn the terminology. Find out how nutraceuticals affect your body before you decide what nutraceuticals to take—vitamins, minerals, antioxidants. What works for people with your risk, may not work with your specific genetic signature.

What foods can you as an individual eat to lessen the risk or delay the onset of the chronic ailment for which you are at risk? I'd like to see the entire genome tested, not only a few SNPs for the major chronic diseases, but all the genes.

Since cracking the human genetic code in its entirety is so new, science may not have the full impact of which genes react with what foods to build a healthier you. Does anybody really know today which genes are responsible for the way your body reacts to certain foods, medicine, or lifestyles?

Science can tell you the specific SNPs for certain chronic, degenerative diseases. Think about it, is that all there is—as far as all the genes responsible for those chronic diseases? What about the rest of your genetic signature? Are there hidden files at the cellular level?

Intelligent foods are attracting the attention of the big, international food design systems firms. Who else is interested in nutritional genomics? Rushing into the buzz about nutritional genomics include patent attorneys, genetics counselors, universities, pharmaceutical manufacturers, and physicians. Nutritional genomics also attracts the interest of naturopaths, the alternative health and health food industry, and merchants of nutraceuticals. Then there are the big, international food systems design companies. Anybody interested in health and healing or food and nutrition is paying attention to research in nutritional genomics. It's time for the consumer to do some research.

What is your stance on nutritional genomics? I visualize peering inside my genes every morning to know how my body responds to carbohydrates. I can feel too much insulin being released, hitting me like a bomb when I used to eat donuts and coffee the first thing in the morning. A few minutes later I was in tremors, caught between the sugar and caffeine. Needless to say, the whole process led to hardening of my arteries and more.

Only when I switched to a diet that calmed my body type, intelligent foods for my biomarkers, did I realize the connection between what I ate, how I exercised, and what changes were happening in my body in the past sixty-plus years. What I needed was foods tailored to my genotype. My relatives eating processed carbohydrates and trans-fatty acids didn't last very long. As the only survivor, I had to listen to what my body was communicating to me at the molecular level, to be aware. The best way to be aware is to listen to how your genes respond to what you eat.

Genes communicate beyond a chemical and metabolic level of consciousness through the language of biomarkers. They alert you to risk or no risk in a measurable, physiological way. And the language shows up not as words but as risks on genetic tests. Sometimes genes mutate or make mistakes in copying. There also are other reasons for defects in specific genes. Or you may not have any defects. Your biomarkers evolve over millenniums based on what your ancestors ate and the climate in which they lived for the past ten, twenty, or forty thousand years.

You can do something about risk such as lower your homocysteine levels with nutraceuticals and vitamins if the levels are too high. Look for clues in family history, your own blood tests and physical exams...and most important, your genetic risk profile.

That's why knowledge of healing foods tailored for your specific genetic profile should be your right. There are food and nutrient solutions to most risks that show up on tests. It's better to know how to overcome the obstacle than to remain in the dark about what foods will delay or prevent future or present chronic illness. Information about your genetic profile should always be private.

When should it not be private? What if you have a contagious disease or apply for a commercial pilot's license or drive trucks or busses, trains or boats? Then regular medical exams will reveal the expression of your genes as you age. What is your opinion on privacy? Have you looked at the polls on the subject of nutritional genomics? What do you think about genetic testing for nutrition and nutraceuticals? There are allied fields to explore as a consumer.

Research the publications and articles online or in print on pharmacogenetics and phenomics—the sciences of tailoring your medicine and healthcare to your genetic profile. Explore proteomics (drug discovery). Read about bioinformatics (managing bioscience information and statistics in computer databases and using computer science technologies with genetics research. You work with both computer programming and bioscience information.) The information is either on the Web or in medical and science journals you, the consumer can find on the shelves of university and medical school libraries.

You can purchase a library card and make use of medical and science libraries at universities. As a senior citizen in lifelong learning, I was able to purchase a library card through a retirement-age group at my nearest university. There are medical journals you can read in the library or subscribe to. Take notes. Use the Internet to read medical and scientific journals online. Use the library photocopy machine for articles you want to take home if you aren't online. Do your homework and teach yourself about the wonders that are out there in the burgeoning field of nutritional genomics.

It's true that a little knowledge can do a lot of harm, but what you're interested in is for the consumer to be able to self-educate at least to the level of the media, and to have the same access as the media to see what is evolving in fields that look at disease in new ways and in looking at health and preventive nutrition in new ways.

For example, when I used green tea extract to cure my gum problems, my dentist was happy for me. Right now nutritional genomics has buzz appeal. It's in the news. At the root is change. The way scientists look at your health today is that nutrients play your body like a violin. Your entire system is an orchestra, and you have to listen to the music.

How does an orchestra of musicians play together in sync? By being interactive. Your body is part a team. And the systems biology that nutritional genomics is all about works interactively. First science looked at the parts, and then the whole person. Now science looks at the orchestra playing together in sync, the systems biology at the genetic level. It all works together, and what you eat may be able to keep the symphony in sync, playing a healthier, more stable rhythm. Nutritional genomics is more about your individual genes that your ethnic group as a whole entity. What if you didn't inherit the gene to digest milk properly?

In the book **Archaeogenetics**, McDonald Research Institute Monographs, 2000, there's an excellent article based on a study of lactase diversity titled, "*Lactase Haplotype Diversity in the Old World*," (chapter 36) page 305, by Edward J. Hollox, Mark Poulter and Dalls M. Swallow. Science knows that lactase persistence is a genetic trait. It shows up in various frequencies in different populations.

The point is you can't point to genetic markers to distinguish one ethnic group or race from another in order to prescribe a certain dosage of a drug or food because there is diversity in individuals. You have to look at the genes, the genetic profile first to see whether the individual inherited a specific genetic marker.

On the other hand, some members of some races react differently to some drugs and foods, and you can't prescribe for one person based on how the majority of his or her ethnic group or race reacts to the drug or food without seeing the genetic profile to see what the individual inherited. I say before you prescribe a dosage consider how the person's ethnic group reacts to that dosage....but know what genes the individual inherited first. That's at the root of nutritional genetics. Because it's such a hot buzz word in the news, there is lot of discussion about the changes going on.

Scientists are becoming mighty particular about the reputation of who they open up to because of the old adage that says your own scientific reputation in a relatively new field in the throes of change depends upon not only on who you talk to, with whom you're friends, but also on the reputation of the publication for whom the media person works that you open up to.

That's why it's nearly impossible for a freelance book author who works for no publication and is not under assignment by any editor or publisher to get an interview with a scientist willing to talk about nutritional genomics. I found *almost*, but not all doors closed as far as getting scientists to chat with me online about what's new and what's news in nutritional genomics. Yet, as you see here, some excellent scientists in the field did give me information at no cost to me for my book on their wonderful, new and changing field.

On the other hand, scientists working with DNA testing for ancestry or archaeogenetics, including those in Europe, greeted me warmly. They readily provided comments and quotes freely at no cost to me and with the friendliest of attitudes. So for those scientists who treated me equally well as any other media professional, thank you, I deserved that. I don't write for the tabloids. I don't put down anyone, ever. I write books for the general reader about choices.

I read scientific and medical journals and news and go to conventions and seminars with other media people writing about the sciences and belong to professional associations for journalists and authors who write about science. The scientists mentioned in this book from the nutritional genomics industry were most courteous and responded quickly to my email. They are friendly people who gave me the facts I needed for this book, responding immediately to my email. For busy people, that is awesome. Thank you all again.

Scientists need to protect their careers and reputations, but they also need to remember that the consumer is the bread and butter of the nutritional genomics industry. What the consumer wants most from the nutritional genomics industry besides the profile and the prescribed foods and nutraceuticals are respect and privacy. Grandma knows best here. It's not a matter of preaching to the choir, of scientists selling to physicians, licensed health care professionals, or genetics counselors. The consumer's body as a recipient of healthcare is involved in the product, and the product is more than a genetic profile in a database. The key word is choice.

What's abuzz about nutritional genomics is that one size doesn't fill all ethnic groups when it comes to prescribing dosages of medicine or certain foods based on one's race or country of origin. For example, in the past, many people who lived in certain parts of northern or north central Europe for thousands of years mutated a gene to digest milk without symptoms such as gas, bloating, diarrhea, and cramps.

Many people who lived for thousands of years in Southeast Asia did not inherit a mutation to digest milk without symptoms. You can't assume the ethnic group reacts as a whole because there is diversity. Not all Northern Europeans can digest the lactose in milk, and not all Southeast Asians get the runs and gas from drinking milk because of lactase intolerance.

That's why nutritional genomics (nutrigenomics) speaks out in the same way that archaeogenetics by DNA testing speaks about deep ancestry and population expansions. My all-vegetarian diet of soup and salad followed by a lot of muffins and frozen yogurt desserts eaten in my forties and fifties was making my insulin resistance/hyperinsulinism worse. The burgers and fries I ate up to the end of my thirties didn't help.

Midway into my sixties decade I decided to eat smart foods, but which foods were right for me, now in my mid-sixties, with a history of parents and sibling dying young of hardened arteries, chronic anxiety, and genetic hypertension? Would my vegetarian diet be appropriate or would it be best to add salmon, and if so, how many times per week? What about green tea extract with polyphenols? Would the caffeine in it be harmful, or could I find a brand that was truly decaffeinated?

I've heard while chatting with people interested in DNA and nutrition, phrases such as "*eat 'bright' for your genotype.*" It seems to be a word play on a NY Times best-selling book that I've read a few years ago titled *Eat Right 4 Your Type* by Dr. Peter D'Adamo. The Web site is at: http://www.dadamo.com/. There are lists of medical journal articles in the book, and I was able to read the medical articles referenced explaining the scientific basis of how lectins work. I followed the Blood Type Diet. I enjoyed the Web site designed to educate anyone about the scientific links between blood types and nutrition. Dr. Peter D'Adamo, is the author of *Eat Right 4 Your Type, Cook Right 4 Your Type*, and the *Complete Blood Type Diet Encyclopedia.*

The popularity of these diets recommended for the various blood types—O, A, AB, and B are backed by medical journal articles of scientific studies referenced. Eating according to your blood type is a theory, but backed by scientific studies. The entire concept of genetically individualized nutrition seems to have evolved starting with blood type. For example, there are scientific studies of how various lectins agglutinate the blood depending on one's blood type. You can see reviews of many diets at the Diet Reviews and Information Web site at: http://www.chasefreedom.com/eatrightforyourtype2.html.

It's as if the sciences of genetics and nutrition had its roots beginning with blood testing regarding how certain foods affect people with certain blood types. Before the entire human genome code was known, scientists used a metabolic and chemical approach to study how one's food intake influenced one's health. Blood was tested, physiological responses, metabolism, glucose, and other tests to measure anything from allergies to whether one had diabetes. Today we have the whole genome at our fingertips. Tests are costly, but the science is here now.

We don't have to wait another ten years to know how our genes react to food, exercise, environment, relationships, stress, or lifestyle. For example, if you look on the Web you may come across something like this: some people who prescribe nutraceuticals based on blood type might suggest that if you have blood type A, and your body is full of cortisol after heavy exercise, change the exercise or take some vitamin C. Is this medical education or medical advice? It's a thin line. And the person who puts this suggestion on the Web may or may not have an M.D. Naturopaths and nutritionists can make suggestions on the Web. How should you react?

You can ask your own physician when it comes to medical advice. On the other hand, is your physician trained in nutrition or nutritional genomics? Will you be referred to another specialist? Would that specialist be a nutritional genomics consultant, a naturopath, a genetics counselor, a nutritionist, a nurse practitioner, a nutraceuticals company, a testing lab, or who? That's why the consumer needs to take control of his own research about healthcare.

Scientists look at proteins in food and allergies, or whether a certain blood type became agglutinated when a certain food was consumed. Back in the era of World War One blood testing was linked to population genetics. Today, research on the entire human genome code has revealed a more comprehensive way of studying health and nutrition through nutritional genomics. Instead of only looking at blood type and the way food affects you—because your blood type is reflected in how every cell in your body reacts to a certain food—science looks at all your genes.

If you're a consumer with no science background, how can you make money in nutritional genomics? You can put up an informational Web site. Compare, review, and evaluate the genetic testing companies that have contact with consumers. You can compare, review, and evaluate the research companies and the university programs. Basically, you don't need a degree to develop a consumer-oriented information Web site and/or database or an annual book of facts published for consumers. What you can do is compare what's available to consumers in the field of genetic and genome testing. It can include nutrition, ancestry, pharmacogenetics, nutraceuticals, and related sciences such as phenomics.

You can provide services in a variety of categories even with no science background to start with by listing information on companies as a type of consumer report on the nutritional genomics industry, on the DNA testing for ancestry industry, and related health and nutrition or nutraceutical firms. If you have no money, begin by asking the reputable companies you list for funding.

Talk to experts who know which companies are reputable who can be asked to fund your online business. Ask the food industry for funding. Talk to the patent attorneys. If the universities have little money for research and can't help you, talk to those where the universities go for funding. What government agencies do the universities contact for funding their research in nutritional genomics? What large corporations? Which philanthropists? Start with the largest, international food companies and the healthcare industries.

Then work down to the nutrition companies and the people in alternative health care. If you only want to give out information, that's helpful to the consumer. Develop a database. You want the companies to sponsor your efforts. The first step is to write up a plan and keep knocking on doors. Do a bit of fund raising. If nothing happens, create an informational Web site anyway. I did at www.newswriting.net. My purpose was to give information on genetics, broadcast my talks, and promote my books with excerpts and articles.

Yours may be to compare and review companies involved in any aspect of the DNA testing or genomic profiling arena and possibly to review books and publications. You could include feedback from customers of DNA and genome profiling and testing firms, research organizations and institutions, and any company dealing with the public or the healthcare systems and food systems design corporations or the alternative health markets.

Suppose you weren't interested in any business but wanted to check out the reviews and comparisons of DNA testing companies at a Web site. So the possibilities are limited only by your creativity. If someone tells you that you haven't the credentials to do something, start a Web site reviewing and comparing the kind of companies in which you are interested, and let the consumers give you feedback as well as the professionals and experts in that field.

There are two voices to be heard—the consumer's and the expert from the most reputable companies in any field. That's what informational sites are about—selling facts to nontraditional markets, reviewing, and comparing. There's room for more than a few of these informational sites.

Certain people have different responses to nutrition than the majority of people. Twenty percent of people respond one way, twenty percent another way, and sixty percent still another. Do you respond to eating fruits by getting a bulging belly and too much insulin in your blood? Your DNA wants smarter food. Here's the big picture: Your DNA has cultural, biological, and nutritional core identities. Are you ready to look for the smile in your genes? Do mothers know best what foods their children will respond to according to the rules of nutritional genomics? Are the rules in place yet?

Your genes express their biological and cultural components in health or disease based upon the type of foods you eat. Your genes express their needs based on the type of exercise you do to stay fit as well as the way you perceive the stress in your environment. Your DNA is alive. It's conscious. It's a whole you in miniature. I'm a medical journalist interested in finding out my own response to nutrition. So I began by questioning scientists and other experts in genetics.

How about your response to what you eat? It's okay to want to eat smarter by learning to interpret the how-why-when-where-who-what-where of your entire genome and apply it to practical use in eating healing foods and choosing if needed, helpful nutraceuticals. The goal is to get consumers interested in looking at their own picture of health at the molecular level. It's a fantastic subject to learn. Question all authority and think for yourself. Eating for your genotype is here at last to free you from food cravings. Or is it available to all equally—to poor and rich alike? I say it is if you look deep enough into the research available to the public.

If you could get a printout of your entire DNA genome, what would it tell you about smart foods? How far would you go, how much would you pay to stay healthier and to prevent or delay chronic disease? What would you do to preserve your privacy and keep your employer, your HMO, or your primary care physician from looking at your entire genetic profile? The details are in the DNA.

12

Intelligent Nutrition or Smart Foods? Who Makes The Rules in Nutritional Genomics?

When a mystery and romance novelist turns to writing nonfiction books on DNA, from nutritional genomics to archaeogenetics, most scientists with doctorates do a double-take. What's encouraging is that some scientists in nutritional genomics and in other areas of genetics are speaking out to me, as you can see in the e-letter below, reprinted with permission:

From: David Crawford
To: Anne Hart
Sent: Thursday, July 17, 2003 9:00 AM
Dear Anne:

Kudos to you for standing up to the profit warlords, politics and professional jealousy!

I am also a "full-time volunteer who writes for the love of science."

Albert Swietzer said "a person who is truly happy is the one who has diligently searched and sought out how to serve others". That is our founding truth at Interface Medical Research, Inc.

We work as consultants to the nutragenomic and nutraceutical community. From basic research, product design and development and clinical studies, we are committed to serve the scientific community and public.

We would be tickled to be listed as a service provider in you book. Thank you!

Please use my letter as you see fit.

David S. Crawford, PhD
Research Director

Interface Medical Research, Inc.
545 Farr Avenue
Wadsworth, OH 44281

Thank you, scientists who were willing to talk to me about how foods, genes, and healthcare are linked. It's as if there's a pyramid with your genes at the top and nutrition—foods and nutraceuticals—at one end of the triangle linked to your healthcare at the other angle. It's a triumvirate of intelligent nutrition tailored to your genes.

What I like about the science of bioinformatics is that the field is about managing databases of biological data. The future of bioinformatics helps to ensure that the growing body of information from molecular biology and genome research is placed in the public domain.

Information should be available equally to self-taught freelance writers as well as staff journalists and to the potential consumer of genome testing and research. Information should be accessible freely to all facets of the scientific community in ways that promote scientific progress.

If you need to do some fundraising to establish a DNA bank or support group project, look to celebrities, retired celebrities, media and film producers, and entertainers. Each year movie stars frequently headline Hollywood fundraisers and similar events that bring in money, sometimes in the millions, to donate to causes, most often for various campaign committees or health causes.

You might make some phone calls to see whether celebrities might want to get involved in planning events that would bring people together to raise funds for nutritional genomics-related projects, perhaps on healing foods and individual genomes. You might develop a slogan such as "Don't let your *genomee* become your enemy." Think in terms of foods for the majority in the midst of individual genetic differences.

What foods are best for the majority, 60 percent of the population who usually follow food guidelines recommended by healthcare professionals, dieticians and nutritionists based on general physical exams? How do you respond individually to certain foods and diets? Does your immune system go down when you fast or consume sugary foods such as fruit juice, when you exercise or travel? What foods contribute to your well-being?

When you research a study, find out who has done the study and whether it has held up to the critics or was found to be flawed by someone credible. If the study cannot be shown to be flawed, consider it as evidence to explore further. For example, on July 22, 2003, the Bee News Services of the Sacramento Bee, a daily newspaper published an article titled, "Fish Diet May Help Seniors, Study Says." The sub-title read, "Weekly helpings may cut Alzheimer's risk, doctors argue." The article listed Chicago as the origin of the news article.

What's missing in the article is any mention of what study the news piece referred to. The subject of the news reads: "Older people who eat fish at least once a week may cut their risk of Alzheimer's disease by more than half, a study suggests." The only problem is that the study is not named.

The news article proceeds to mention that the study adds to the evidence that what you eat might influence your risk of developing the chronic illness. What study? It refers to "a growing body of scientific evidence." I would have liked to see the name of the study so I could read the details in a scientific journal that usually publishes abstracts of and articles connected with studies. Where can I find this "body of evidence?"

I want to read the research studies linking various foods to cutting risk of various illnesses for myself. How about you? Is the recommended food good for my genetic expression and everyone else's or only for the twenty percent of the population who can eat most diets and still remain healthy into old age? What about the other twenty percent who can't eat certain foods without damage to their arteries and organs?

What the article does discuss is that the evidence adds to accumulated information on the subject of reducing risk of developing several chronic illnesses such as cancer, Alzheimer's, or heart disease, if people eat a diet rich in fish, fruits, and vegetables and low in saturated fats from red meat.

There are statistics from the Alzheimer's Association in the article. Approximately four million people in the USA are afflicted with Alzheimer's now, and Alzheimer's cases are expected to rise to 14 million by 2050. The evidence presented in the news article was that 'researchers' found "that people 65 and older who had fish once a week had a 60 percent lower risk of Alzheimer's than those who never or rarely ate fish."

Who did the study? Who are the researchers? When was it done? No mention was given of the amount of mercury in the fish eaten. The article mentions that the meals included fish sticks and tuna sandwiches. No mention of whether the tuna was canned, canned with salt or no salt added, canned with oil or water packed, or whether it was fresh tuna or how it was prepared—grilled, fried, or baked, or boiled? Was the research done by the fish industry? What about the warning labels in supermarket's fish departments about the mercury in certain species of fish?

My neighborhood supermarket has a big sign posted at the fish counter for women of child-bearing age or pregnant women to avoid certain types of seafood with the names of the species mentioned due to the mercury levels. I see an article in the August 2003 issue of *Reader's Digest*, a cover story titled, "Hidden Dangers in Healthy Foods," that connects fish eating with individual reports of nervous system problems, illness, and hair loss due to mercury residues in fish.

Science writers should be accepted as media professionals by the science community without discrimination as to their credentials and should have equal access with all staff media of large newspapers, magazines, or broadcast networks to the growing body of information in genome research or any other area of molecular biology, especially areas related to preventive health and nutrition, foods, and vitamins and minerals.

The bioscience communicator whether self employed or staff employed acts as a go-between, a liaison between the consumer and the healthcare system or research laboratory, a type of educator and ombudsman. The communicator should be respected and allowed to access databases of biological data. Making the complex easier to understand for the consumer is news.

The scientific community should include journalists who specialize in writing about specific areas of science such as genomics without requiring writers to have science degrees. We attend enough genomics seminars, read enough journal articles, observe conventions, and read enough monographs and books to know what questions to ask the experts. In addition to looking toward the media as a liaison and ombudsman, the consumer of genetic testing is concerned about privacy issues.

For nutrition purposes only, results of tests for genetic risks should be private. The whole idea is to lessen the risk before a chronic disease develops. You can change how your genes express themselves based on food and nutraceuticals, but you can't swap your genes for some other genes.

Nutritional genomics is all about effects of intelligent nutrition—smart foods—on how my genes express themselves. Don't laugh at me. Smile. I need to know how specific dietary chemicals will nourish me at the molecular level. You also need to know about the effects of food on your own health.

Your genetic information—a printout of your entire genome—along with a profile interpreting it in terms of risk, disease, and recommendations for foods and/or supplements, should be kept private to be shared only by you and your healthcare professionals, and perhaps your heirs in a time capsule about DNA and medical history to be passed on to future generations. It's about what makes you tick at the molecular level, about DNA and health, behavior, and mood...If you are what you eat, why should you eat that way, and what will happen to your health at the molecular level when you stray?

Do you metabolize your meal fast or slow—burn your calories quick or over hours? Your genetic profile will give you clues, but you need to learn how to interpret it. And for the time being, you'll find it difficult to learn for yourself and take charge of your healthcare, because getting your entire genome tested is done in reputable places under the supervision of a physician managing your healthcare. You'll have to work hard to take control of your diet and your genes.

You have to work towards taking control of your genetic testing. The only way to do that for now is to learn how to interpret a genetic test of the entire genome and how to apply that to a diet that works for you. You need to get control of your genes. They are yours. The food you eat is yours. Why should putting diet and genes together cost a fortune now and be cheap in a decade? Knowledge is not only power. It saves you a bundle.

Okay, so the doctors need to eat, too. I should know. My son and son-in law are physicians. Just ask their wives about eating smart foods. The point is there are ways to learn how to look at a DNA test and look at a diet and understand how your genes respond to what you eat. Books are online and in the medical libraries. Read them, you autodidactic learners. Educate yourself. If a social worker can take a two-year masters degree program and become a genetics counselor, and if a nutritionist and a physician can learn to talk to people about what diets are healthiest for them, then you, too can learn to interpret a DNA test of your entire genome.

The trick is finding someone who comes highly recommended to give you your genomic testing. You do need to send your DNA to a lab. Where do you start? Talk to professionals in the field and find out what books and journals they read in order to understand how to interpret the DNA tests. Find out how your genes respond to what you eat. Very few people have the time to be interviewed, but keep asking around.

Find out whether anyone in the field or recently retired will teach an extended studies course for the public. Go to conventions and conferences where professionals in nutritional genomics congregate. Make friends. Read my romantic intrigue novel series on DNA. Or my books on tracing your ethnic DNA. Cook for your genes.

Have a nutritional genome feast/party. Invite people who can help you match your DNA to your diet or apply for a job in the field. Whatever you do, get a handle on, control of what you eat and how what you eat influences how your genes respond to that food.

Write about your dieting experiences. Go on the radio. Talk to people. Read professional publications. Sign up to receive the news of the industry. Ask yourself whether you want to get into an allied business from another angle where you'd rub elbows with people in the field. However you choose, get your entire genome tested, and find out how your genes respond to recommended diets compared to what you eat now. The whole industry is moving toward this becoming routine in about ten or more years. If you're as old as I, don't wait, start asking questions now, and find out what foods are healthiest for your genotype.

How does what you eat affect the expression of your genes? Nutritional genomics is causing a rhumba of rattlesnakes in the media. Research the company before you spend money on testing your DNA. Is the company focusing on DNA-driven genealogy in its DNA testing? Or is the firm targeting DNA tests for nutritional genomics answers? Is the company testing DNA markers and gene mutations to see whether there are pharmaogenomics reactions to specific medicines? Is your physician able to interpret correctly the DNA test results before prescribing anything?

If you don't know how to interpret your DNA test for ancestry, you'd consult a DNA-driven genealogy firm that works with DNA testing laboratories, usually at universities and/or private labs. If you are testing DNA for medical reactions, skin care product reactions, or to find the diet that works for your health using customized nutrition for your metabolism, genetic profile or condition, you'll need to find a physician and possibly a nutritionist/dietician who works with nutritional genomics professionals in the medical field and has had training in nutritional genomics and special diets. If you're looking for a diet for insulin resistance, have your genes tested to see how you respond to proteins and carbohydrates. You can also go to a specialist in metabolic-based diets, but look at the cellular level also to see how your particular genes react to anything from the environment, foods, medicines, or cosmetics.

For example, if the physician doesn't have a working contract with nutritional genomics professionals from reputable companies, how will the individual interpret a printout of your entire genome in relation to tailoring your foods? How many physicians today are trained in nutrition let alone nutritional genomics, a field that began around the year 2000?

So check out your healthcare professionals and the companies doing the research to make sure what they offer is what you want. Read the medical journals connected with nutritional genomics. You can find them in local university medical school libraries open to the public. Read some of the latest journals in the field. Go to conventions and attend meetings of associations of professionals who are connected with nutritional genomics and ask for reputable referrals. You can network with people in the field. Read about what research is being done by various companies.

When I was single and in my twenties, circa 1962, I joined clubs for single professionals where the scientists and technical people congregated so I could talk to them and enter the awe of their world in order to write about the aura of enthusiasm that engulfed them in such important work. I majored in professional writing/English and did get an M.A. degree, working my way through from the age of seventeen with a full-time day job, but since the age of eleven I had a deep interest in reading and daydreaming about science.

In 1962 my best female friend majored in chemistry and math, and I felt stifled by my day job typing important men's manuscripts and finishing my degree at night. Through chatting with the experts—the scientists, I found the inspiration to play with words and channel creative expression to novels featuring exciting, learned characters as scientists. I mean, they actually talked to me, a retired janitor's kid whose mom was a maid. (No—none of them proposed marriage, but I got to write romance and suspense novels about fictional scientists.) I needed something more—science news.

It was the high point of my life. Back in 1962 I ask each scientist who asked me to dance whether there was a new way to look at disease. I wanted to know then what my genes really wanted me to eat. What if genes could talk? What would they whisper to each of us? The DNA helix was big news with Watson and Crick visualizing the double helix. I ended up visualizing a double helix myself—writing the romance novel titled, *The Bride Wore a Double Helix*. The outcome was I wasn't left on my own. We talked about foods, as in how a quarter of a fried potato eaten each day for a year could cause someone with "specific genes" or "fat cells" to gain 'x' number of pounds.

Our conversations on the ballroom floor were about consumers getting their proteins analyzed. More than forty years later, science has spectrometers, and you can find bioinformatics databases that manage nutrition-related information. You can now have a genetic profile compared and cross-referenced to your blood chemistry, metabolic tests, or any other areas of your physical exam or genetic profile.

You can look at genetic risks and find the foods and nutraceuticals that will help you. If you know eating fruit all day will make your blood sugar problems worse, because too much insulin is released into your bloodstream due to a genetic situation, you have some more control. That's what it's all about—control. It's you armed with ways to override your genes as much as you can with healing foods and nutraceuticals, if needed.

Don't only get tested and then left on your own on the dancefloor. You need some kind of medical supervision so you can learn how your test was interpreted and why those foods were prescribed as well as the effects those foods will have on the expression of your genes. You need to look at yourself at the molecular level. When you've self-educated yourself, you can take more control over what you eat knowing why the food affects you the way it does—at the cellular levels. You'll know why you're salt sensitive or not, and what happens inside your cells when you eat foods that make react the way you do.

Everyone takes nourishment expecting to feel well after eating. Don't let the caricature of yourself hold you back from exploring how you react to dietary chemicals in food. Look at the nuances, the changes in concentration of the nutrients you take in. Look at how the chemicals in foods make you feel. Eat certain foods and take your pulse and blood pressure. What makes your physiological responses healthier?

How long does it take between eating and seeing the results in your health? You need to compare your genotype, that genetic information such as your genome, your DNA, and look at the way you respond to a particular food or meal. That's why a lot of diet books don't consider your individuality at the molecular level. Many stop at the metabolic or chemical level or just look at blood types. That's important too, but so is your entire genome, all your genes and markers and DNA. What nourishment do your genes require for you to become healthier, delay the onset of chronic diseases, and feel well?

Your genes shuffle each generation. They recombine. Food, like DNA moves beyond family history, ancestry, and molecular genealogy. You need to realize the cultural, biological, and nutritional components of your genome in order to eat smarter for your genotype.

Here's how to eat bright for your genotype. The new field, blooming since about 2000 is called nutritional genomics. Nutritional genomics link what you eat to your physician-supervised DNA test of your entire genome and your healthcare. Nutritional genomics is about food plus supplements when needed and any medicine and/or supplements termed nutraceuticals.

Your genes will give out the signals of what whole foods are needed to delay or prevent infirmity. Who is going to test that DNA, and who will teach you to interpret the test when you can no longer afford a supervising physician for healthcare and have to feed yourself on your own on what money or lack of it may come?

Tailoring your food to your genome reminds me of what I heard for the past thirty years in the holistic food fanfare world of eating brighter for your individual metabolic self. Only today it goes beyond the metabolism and moves to your entire genome—your genotype. That's why a lot of diets don't work for everyone, because food affects people in different ways. What's one person's nourishment is another's allergy.

I've been touting eating healthy since 1959. It's what the holistic food and health conferences and conventions were spouting for at least three decades. Now it's arrived: eating bright according to your genotype. You have to eat smart, that is intelligent diets tailored just for you if you want to be healthy. That is, unless you're lucky enough to be part of the 20 percent of the population who can eat any diet and stay healthy longer.

Ever since I avidly read, tried, and was successful at eating for my blood type after reading about it a few years ago, and later for my personality type. More than a decade ago I tried to eat right and take the right supplements for my menopause.

When my genes acted differently to hormone therapy, I stopped the therapy back in the early nineties when my healthcare professional insisted that going off of estrogen and progestin was tantamount to a diabetic going off insulin.

Nevertheless, after looking at my physiological responses to progestin, I dumped the pills and took up exercise and raw veggies. What works well for another's genome, didn't work for mine. After reading a book on body type and diet, I cut out the sugar, being a "thyroid body type," and my large, over-stimulated thyroid felt great again without the excess sugar. And since I began to eat whole grains and raw veggies, I tried to take charge of my own diet.

Before nutritional genomics, I had to read medical articles on blood type and diet or body type and diet, and figure out whether I was eating according to my genotype. Then I turned to books on eating for syndrome X, hyperinsulinism, and tried diets of 45% carbohydrates, 35–40% good oils, and 15% proteins to stop the excess insulin and insulin resistance after sixty from making my big tummy flatter on my thin, 124 pound frame. Without knowing my genotype, it was all based on metabolism, blood chemicals or blood type, ancestry, or body shape. Now with nutritional genomics, I can get to the molecular level, my entire genome…as soon as the price of testing becomes affordable to me…and available to consumers.

You can lobby for nutritional genomics to be available to everyone equally. How about the right to have your genome matched to a healthy diet? I want a good diet today that works for me. Why not match your genome to a diet that works to make you healthier? What's happening in this industry?

Consumer, be alert. I can't wait another ten years for nutritional genomics to be available to everyone on demand at little cost. I want it now. My solution is to write about it. My eight grandchildren will no doubt have it as their dads are physicians, and it will be a part of their children's lives. So what can you do now with your DNA as far as practical applications of DNA testing not only for ancestry or population genetics, but for nutritional genomics?

Scientists use the word 'interface' a lot as do business executives and people with day jobs. I have no day job. I'm a white-haired bag lady, so I don't have to interface. I interlace instead. What I do is greet you with a smile and hug. I don't interface with you. Only in nutritional genomics you have to study how to interface between your diet and your genes. Yes, that's you in there in the middle right between your diet and your genes. Now interface. Take a picture of your gene expression. Now smile. Snap. You've just developed a picture of your gene expression. It's profiled and digitally filed. You now have a picture of how you respond to what you eat. I'll say it again scientifically:

Nutritional genomics researches the interface between what you eat and your genetic processes. Scientists analyze your single nucleotide polymorphisms (SNPs)[1]. Your gene expression is profiled to get a picture of your response to your nutrition.

Individual nutrition is emphasized in the new field of nutritional genomics where you tailor what you eat, the nutritional supplements you take, and your entire health program of diet and exercise, work and lifestyles according to your own genetic profile—your genome. Forget about diets for large numbers of people.

Maybe those diets fit the sixty percent of us who can eat those things, but what about the twenty percent who need tailored menus according to our genes? Or what about the other twenty percent who can eat almost any food and still not develop the main diseases that could stem from eating the foods not right for our genotype—heart disease, hypertension, asthma, hardened arteries, cancer, loss of memory, and more? And are these degenerative diseases linked to changes in our genes? How do the different ethnic groups react to different foods?

---

1. SNPs see http://www.ornl.gov/TechResources/Human_Genome/faq/snps.html. SNPs are "DNA sequence variations that occur when a single nucleotide (A, T, C, or G) in the genome sequence is altered." The letters are pronounced as "snips."

All this research shines under the umbrella of nutritional genomics. This year it's tracing ancestry by looking at DNA test results, and next year the focus will be on nutritional genomics, how to tailor what you eat according to your genetic markers. That's where the focus of my next book is. Most people aren't aware of the research being done at universities or at companies targeting research on nutrition and your DNA.

There are new fields within nutritional genomics such as the study of how people's DNA change according to their diets. There's the study of pharmacodiagnostic drugs, and phenomics, tailoring your medicine, therapy, and healthcare or exercise and lifestyle to your genes.

Books abound on eating right for your chemical, blood type, and metabolic systems. At the molecular level, you'll see books popping up on eating bright for your genotype, smart diets for your genetic markers. One diet book will not fit all. You have books on eating for Syndrome X, for diabetes, and other illnesses that appeal to groups, but what about eating bright for your genotype? Studies compare the immune systems of people from different ethnic groups or geographic areas.

This country faces a diabetes epidemic in children. Sugar is added to health foods such as soy milk or almond milk and other foods or beverages to bring people back by taste, even to 'addict' people to the sugar so they should buy more of the health food product in some cases.

Try finding an unsweetened almond, rice, or oat milk. You'll have to make your own from water and pureed grain. You'll have a hard time finding unsweetened soy milk, too but it's around if you look long enough at various shelves in health stores. Some health food products are drenched in salt so salt-sensitive seniors can't easily find a decent meat substitute that isn't loaded with salt, sugar, or fat. All this—in the midst of a revolution of research focusing on tailoring your food to your DNA to prevent or slow down premature degenerative diseases.

Make sure you're working with a reputable company that arranges to work with a managing physician. Beware of companies that test different aspects of your body to predict what to eat as the testing isn't of the complete genome and may not be done with a managing physician to interpret the results of the tests for prescribing your diet.

Don't waste money on incomplete or inaccurate testing. Watch out for testing companies that offer nutritional consulting based on genetic tests for a variety of complex diet-disease-or diet-health associations as they may not be reputable. Find out who is reputable before you spend any money. If you are testing for single gene defects such as PKU or hypolactasia, these can be tested for and diets arranged based on the test results.

Check out reputable companies that offer tests which are supervised. Some companies research the linkages between genes, diet and health. Don't go to a company that offers unsupervised tests. Learn the difference between a research company looking at how genes, diet, and health are linked, and a genetic testing service company. Beware of the unsupervised test or the company selling snake oil when it comes to finding out how your diet is linked to your genes and your health.

How do you know the difference? Talk to professionals in the research field before you think of any testing, and make sure everything done is supervised by a physician managing your personal healthcare who also is participating in research of how health, diet and genes are linked.

Think about all the Japanese who eat a dairy-filled American diet and land up with diseases not found in Japan when they kept the ethnic diet. What happens when other indigenous people take on a processed-food Western diet high in whatever disagrees with the nutritional expression of their genetic markers. It's a horizontal expression of a vertical desire. The remedy is tailoring your nutrition—foods and supplements—to your genotype as well as customizing your type of exercise and other lifestyle events.

Perhaps it's time to tailor our DNA to our multitude of ethnic diets and first find out which one we inherited or which diet works for our genetic expression. For me it's Omega 3 oils and low-carbohydrate vegetables, B complex vitamins, whole grains, and the Syndrome X diet, but how can we know for sure until our entire genome is passed through the test, and for now, the test is still expensive?

So we have to guess, perhaps at the mixture of ethnicities, family members and their lifelong genetic expression of what they ate. Look at what your families ate and what it did to them. Did you inherit any of those gene mutations? How has what you ate influenced your own health? The answer may lie in the SNPs, the genetic markers. Stay tuned for reporting on new research. These sciences include not only 'nutrigenomics,' but also tailored diets and understanding your genotype.

Related sciences that look at genetic variation to customize healthcare strategies include proteomics, metabolomics, bioinformatics, biocomputation, and phenomics. It's all about getting the details and the big picture of your nutritional status, requirements, and genotype. You can eat better according to your genotype without having to wait another ten or twenty years before testing your entire genome is affordable.

Start by observing your body's reaction to the nutritional and exercise environment. It's about personalized nutrition and medicine. Nutrients are not just nutrients. You have macronutrients, micronutrients, and antinutrients. The details come out in gene expression. Nutrients can alter your gene expression without changing your genes. Look at your metabolism, DNA, and gene expression the way you use your DNA test results to trace your ancestry. The door is open to personalize diets and customize your medical care. Under the umbrella of smarter nutrition, DNA plays a role that treats you as an individual rather than a member of a special group.

If you're an older individual, it's not possible to wait another decade to look for a customized smart diet to eat "bright" for your genotype. You can have your DNA tested now. Only be on the alert and research the company you're working with so you don't get scammed. A DNA test of your mtDNA or Y chromosome for deep ancestry is not enough of a test. Neither is a test of your racial percentages.

You need more clues—metabolic, chemical, and genetic clues. There are plenty of books on how to eat according to your metabolism or chemical clues, but you need more genetic information. What you need is your entire individual genetic makeup. If you can get a company to give your entire genome a pass-through and genetic printout, it could be helpful in preparing customized diets to help prevent and ease chronic disease.

The only obstacle is that you'd need either a professional to prepare the diet by interpreting your DNA test of your entire genome, and not all physicians can look at your genes and prescribe a diet. What you'd need would be a qualified, accredited, and experienced nutritional genomic specialist to consult with you, look at your genetic profile, and prescribe a diet. You'd need someone with enough experience in nutrition and genetics to know what foods would be best for you as a person. No diet fits all people, not even all ethnic groups. It must be individualized.

That's why it's called "intelligent nutrition." Smart menus are customized to your genomic profile. You eat at the molecular level. Eating smart for your genotype is science-driven nutrition. You can write a diet book for a specific individual, who's a member of a specific ethnic group of whom you can't prescribe a one-size-fits-all menu for that person within any group. There's too much individual diversity.

Where do you start? Research is the first step for any smart nutritional genomics consumer. You start by reading and going to conventions of food technologists and nutritional genomics professionals and/or students. Do your own homework. Start with reading about the metabolic diets based on an individual's chemistry. A decade ago, it was eating according to your blood type or metabolic type.

Now it's eating according to your entire genetic profile, your genome. The trend is becoming molecular, eating down to the atomic level in your molecules. That's because your genes express themselves at the molecular level, within and from the cells. The "eat according to your individual profile" movement began with alternative healing movements that always are keyed into scientific research at the genetic, molecular, and chemical/metabolic levels.

In most every alternative and holistic health magazine you find, the footnotes contain references to studies and medical journal articles. Check those out. It's a good use of time to learn how to read articles in medical journals and look up the terminology. If a study is flawed and is reviewed in another publication, read it. If a study holds up with time, keep a scrapbook of the study or articles for your reference.

Only now it's beyond looking at blood type or the lectins that agglutinate your blood from harsh reactions to foods, individual reactions. It's beyond chemical and metabolic, it's now genetic. You inherit and pass on recombination of genes that express themselves in various, individual ways, even within the same family. How you react to food is genetically determined and expressed not only in and through the genes, but also in behavior, mood, and sense of well-being.

Think of the potential for this science-driven food-for-health industry. You must explore the innovations and find professionals to network with to explore and evaluate the forthcoming innovations. Get on the mailing lists of the nutritional genomic research institutions. These research institutions may not do genetic testing of individuals for diets, but are engaged in the type of research you want to learn about before you put your health in the hands of a managing physician who must not only interpret your genome but prescribe diets and/or nutraceuticals.

You've heard of pharmaceuticals. Well, think nutrition and nutrition supplements. Research those nutraceuticals. Make sure everything is supervised, but that you still have an opportunity to learn about why and how your genes respond to what's prescribed, that is, you still have control over what goes into your body and knowledge of how it will affect your health. After you've finished being supervised for your own health, learn all you can about how genes and the diet work together to make you healthier. Then take control of your diet and celebrate your knowledge of a new language—a way to communicate with the expression coming from your genes.

Nutritional genomics is a science dedicated to researching smart diets customized for individuals with a purpose of creating a healthier population. Ask yourself as a consumer, how come it's so expensive right now to test the entire genome of a human to get a prescribed diet and so cheap to test the entire genome of a dog or race horse for breeding or diet? Lobby for costs to come down so diets can be prescribed according to one's genotype.

Besides the research scientists working or studying the field of nutritional genomics in academic or consumer consulting capacities, you have food industry leaders flocking to hear the scientists specializing in genetics and nutrition because the field is potentially a money-making enterprise for nutritional genomics-based food industries.

The whole idea of nutritional genomics got its wings around the year 2000 when studies revealed that a "dumb diet" as opposed to a "smart diet" (based on your gene expression) can fan the flames of your chronic disease risk. So to prevent or delay the chronic diseases for which your genes may be at risk, a diet prescribed only for your genes would help delay or prevent those chronic illnesses. You would need a printout of your genes based on DNA testing to find out for what diseases you may be at risk. Then the diet would be prescribed to prevent or delay those diseases.

Not only the diet, but the exercises and lifestyle and any other nutritional supplements would be recommended. The idea of an intelligent diet would be to nourish your genes at the molecular level based on what the genes needed to express themselves in the healthiest way possible for you as an individual.

So why wait a decade for science to come up with prescriptive diets? There's a lot you can do today with DNA testing and biotechnology before the impact on medical care, on the way foods are processed or not processed, and on your own health becomes influenced by the bottom line—profit.

Science has mapped the human genome. Back in 2001, you have the beginnings of the "marriage" between genomics—mapping your genes, and medicine. You have alternative medicine clawing at the door for decades demanding this knowledge and using it long before the medical fields ever thought of offering courses in nutrition beyond an introduction.

Now medicine and science is working to identify how genes work to change your health. In the meantime, the alternative health books have been emphasizing this all along—how to eat according to your body type, your blood type, your metabolic type, your chemistry, for years. Now medicine and the food industry is finally listening. The turning point came when the functions of various genes were identified with the aim of finding out how these functions affect your health, your risk of disease, and how the genes interplay with your lifestyle, environment, stress level, and even how you spend your day, let alone what you eat.

What you want to know now, especially if you're an older person who can't afford to eat the wrong foods for the next decade, is how and why your genes predispose you to sickness, premature aging, or obesity. While you're out there fighting for organic food, or worried about how genetically engineered crops will make you ill, here's one more concern: think how food is processed and how the process will express itself through your genes. Now customize your diet for your specific needs based on your genes—the expression of your genome based on the foods you eat and the lifestyle and exercise you do. In other words, eat "bright" for your genotype.

What is nutritional genomics? It's a market. It has legal and industrial implications for the food industry. We anthropology and bioscience communicators have been writing about alternative health for decades, exploring the whole oat grains and raw vegetables diets for years, looking at whether fish diets give us lower blood pressure or just a dose of mercury and writing about all sides of the whole foods spectrum, the details and the big picture.

Looking out toward the next decade, nutritional genomics professionals want to attract the baby boomers because of the large size of its population. However, the need is great right now among us parents of baby boomers, the senior silent generation. We are now over sixty and want to reverse our age-related conditions that pop up in the sixties and seventies decade of our lifestyles. We want nutritional genomics now, and we won't wait a decade. So how do we start, our little mtDNA or Y-chromosome DNA tests in hand that we sought to search for our deep maternal or paternal ancestry?

Don't you dare wait for boomers to demand smart diets prescribed for individuals according to their genome. We won't be scammed by diagnostics or counseling aimed at seniors by less than professional firms. We want the nutritional genomics professionals to be our entrepreneurs. To ensure this, we stick to the whole organic foods, but wonder which foods are best to nourish our genetic needs? Cooked, or raw? What specific foods are good for us as individuals, and who will prescribe them? How much will it cost, and how scientific is the information?

Could the genomics revolution actually be the alternative health movement and the whole foods movement joining together? Its purpose would be to look at the molecular expression of specific foods on the body. During the past 30 years, I've been attending holistic health conventions hearing lectures on the benefits of Omega 3 fatty acids, flax seed oils, fish oils, whole grains, low-glycemic diets for hyperinsulinism, syndrome X diets, raw vegetables, juicing with the pulp, the affects of too much fruit on triglycerides, how to flatten big bellies caused by eating sugar in people with hyperinsulinism, enzymes, food supplements, vitamins, and minerals, and what's needed to absorb calcium or magnesium, B-vitamins and TMG to reduce high homocysteine.

All of these are reactions expressed through specific genetic problems. And what I had to do 30 years ago was to take a specific, detailed look at how my body reacted physiologically, and change the food, customizing the food based on how that food influenced my physiology, that is the way my genes expressed themselves and reacted.

It always worked for me. Cut out the sugar, and the belly goes down. Exercise, cut the salt, and the symptoms change. Eat those whole grains that have a lower glycemic level, and the hyperinsulinism subsides until the next meal. Your genes are not my genes. What will work for you? Tailor eating to your own molecular level.

Know all about your genes and how food influences the way they express themselves. It is easy to guess at what the genes needed based on the symptoms and bodily response and also by looking at what the wrong diet did to family members who thrived on a Japanese, vegetarian, or Greek diet and succumbed on a high dairy, sugar, bread, coffee, canned tuna, hamburger, fries, too much yogurt, and chicken diet. The clue was to cut the calorie intake in half, and the disease retracted. So you got a handle on what the genes expressed.

A decade ago, I read books on the metabolic approach. I was a fast burner (metabolic), so I was supposed to look younger and thrived on eating protein and fat in the morning. Sugary cereals caused too much insulin and the shakes in the morning and weakness. Eating fish such as cooked salmon fried in olive oil mixed with egg whites in a patty with no salt, but celery seed, garlic, and onion powder felt healthy for breakfast. Oatmeal was too high glycemic and made me weak and shaky.

By working backwards from the symptoms, I could guess at my genes by what strengthened me after eating and gave me energy. Now I can work backgrounds from the general to the specific, from the effects of food on my feelings to the specific genes that put me at risk for the effects of too much insulin or insulin resistance and work backwards and forwards to prescribe myself low-glycemic foods that flatten my tummy and make me feel stronger and full of energy throughout the day.

I attended these holistic health fairs since 1970 when they arose out of the natural foods industry and the alternative medicine movement of naturopaths and nutritionists who were trying to get the medical industry to listen to them. Now they are listening as both sides—the alternative medicine and natural foods industries plug into the nutritional genomics research field from one end, and the medical and food industry from the other end. One group is visionary—looking toward the future. The other is benchmarking, based on what sold successfully in the past.

Prehistoric people ate berries in larger quantities than we eat today, when the berries were in season. Nutritional genomics researches diverse health effects for diverse genes. Flavinoid's effects on inflammation and on the nervous system are being researched. Ask yourself what can you eat that will protect your genes? Did prehistoric peoples have better nutrition then than the standard Western diets encouraged today? Did prehistoric peoples have similar genes to what we have today? Or did the slow-mutating regions of our DNA signal reactions to the change in diet and climate?

From the alternative food and health industries I learned about polyphenols. You find it in green tea, red wine extract, and other vegetable tannins. Anthocyanins are a type of polyphenol. According to Polyphenols Laboratories at: http://www.polyphenols.com/main.php?PHPSESSID=976bfb699f4c396cfb032a25fa7877fd on the Web, "Anthocyanins is a large water-soluble pigment group found in a large number of fruits, vegetables and flowers.

These are the pigments which give plants their brilliant colors ranging from pink through scarlet, purple and blue. Some pharmaceutical effects of anthocyanins have been suggested, for example in treatment of cardiovascular diseases and in ophtamology. The antioxidant potentials of anthocyanins are high."

According to Marilyn Sterling, R.D.'s article in the December 2001 Issue of *Nutrition Science News*, "Eaten in large amounts by primitive humans, anthocyanins are antioxidant flavonoids that protect many body systems. They have some of the strongest physiological effects of any plant compounds, and they are also things of beauty: anthocyanins provide pigment for pansies, petunias, and plums." (Anthocyanins are a separate class of flavonoids from proanthocyanidins, discussed in *NSN* 2000;5(6):231–4.)

Back in folk medicine of the 12th century, bilberry, a bioflavinoid, was fed to young women to induce menstruation. Bilberry (*Vaccinium myrtillus*) is one of several anthocyanins. British pilots during World War II took Bilberry to improve their night vision. So if your genomic profile calls for anthocyanins, at least you'll know what plant pigments in fruit juices or certain wines or in nutraceuticals such as bilberry extract will be of help.

The genes of the plants, such as the genes in the plant pigments feed your genes. The work goes on at the molecular level. In the past, bilberry was used to treat ulcers. If it's found that bilberry may increase the production of stomach mucus to protect the bacteria that causes ulcers from attacking as bilberry was a traditional treatment for ulcers in the past, then folklore herbal medicines may have come full circle to play a role in genomic profiling.

See the Web site at http://www.medpalett.no/index.php?lang=en.

Then there are the procyanidins, found in grape seed extract and other plant extracts. I take grape seed extract in a capsule. I heard about polyphenols and procyanidins from the alternative health market and sometimes on radio shows that most people would call the "woo-woo" industry.

I heard all about grape seed capsules on the Art Bell radio talk show, the show that showcases UFOs along with holistic health speakers and topics of past life and so on. Sometimes you hear about these products in mainstream media and sometimes in alternative health media. The benefits of nutritional genomics has made it to mainstream mass media with articles in daily newspapers and the NY Times news magazine.

What you as consumers need to do is to read the guidelines of the National Research Council. They are the group that offer guidelines of recommended dietary allowances (RDA) based on the results research. So with advent of nutritional genomics, research may open doors for those guidelines to shift.

Now biotechnology has joined up with talking business applications. The polyphenols found in berries is being researched to relieve arthritis suffering. For the last three decades the holistic food conventions have been touting the same blueberries, grapes, raspberries, blackberries, cherries, to help relieve the pain of arthritis. Grape seed extract which contain procyanidins have been said to help strengthen capillaries.

Now science finds out that eating compounds such as polyphenols found in dark berries inhibits the expression of a particular gene. You still have the arthritis, but the pain lessens or goes away. Does it work? Try it and see how the expression of your particular gene reacts to polyphenols.

The procyanidins in my grape seed extract and the polyphenols in the berries or green tea I consume daily not only strengthened my capillaries, it got rid of my 'rhoids. Then there's body shapes. In Ayurvedic medicine of ancient India, body shapes are divided into vatta, thin, pitta, athletic/muscular, and kapha, rounded, with a layer of fat under the skin. Vatta is less prone to arthritis and more prone to anxiety. I'm a vatta.

So just by looking at body shapes you gain a handle on something about the expression of your genes. In the health food stores there are recommended food and teas for vatta, pitta, and kapha types. Just read Deepak Chopra's books, and you'll learn.

The wisdom of the ancients around the world knew something about genetic expression. The body shapes are genetic and react differently to foods. A vatta sipping coffee will sometimes feel anxiety from the caffeine, whereas a kapha may need the jolt to the nervous system to get going. Even without a DNA test, you can look back at reactions to food from the various body shapes and see what foods hit you like a nerve-shattering bomb of hyperinsulinism after you eat, and what food calms you or makes you more alert.

Before you try anything, read the studies and find out which, if any, are flawed, and which stand up to science. Does it work? Only your individual gene can tell you. It speaks not in words, but in the expression of pain or no pain or in other symptoms of strength and energy or weakness and fatigue or chronic illness symptoms. Will your physician work with you? Who will? What about training physicians in nutrigenetics consulting? Will other professionals fill the gap?

What nutritional genomics needs are companies that will sell food that is good for our body organs. Right now I walk into a health food store and find the healthy alternatives to the milk that makes me sick filled with sugar, rice sweeteners, fructose, corn syrup, or other additions, including salt that makes me sicker than drinking milk. Soy milk marked 'plain' is not plain. It's sweetened. You have to look for unsweetened, and most supermarkets don't carry it because they answer when you ask, "Will it sell if it doesn't taste sweet?"

Why are food companies addicting us to sugar knowing we'll come back and buy more of the product because of sugar cravings not because we want the healthier, alternative products. So you have to make your own to leave out the sugar, salt, barley malt, rice flour, or other additives that sell the product by addicting you to the taste rather than selling the healthy alternative point.

A few physicians sometimes also produce and sell vitamins, minerals, or vegetarian food products. Are these products best for your particular genome? Does one size fit the masses or only a percentage of people? My last resort was to make my own soy, oat, rice, and almond milk by cooking the grain in water and pureeing it in a blender. That way I can drink some almond milk by blanching the almonds and removing the bitter skin, then putting it in the blender with water. I soak soy beans overnight, puree them in a blender with water, then cook for 20 minutes and strain into a bottle. Soy milk has to be cooked. Foods have to be researched. Some types of sprouts may contain salmonella or other bacteria on the seeds.

Food companies need to join forces with nutritional genomics specialists who need to join with physicians, HMO-employed nutritionists, and other health care professionals as well as with genetics counselors. If the food industry doesn't promote food good for specific parts of the body, the people who produce food supplements and other health and diet tools will step in and plug into the genomics field.

If all would work together, food could be produced to normalize cholesterol. However, there is a need to create foods for those who need special foods such as low-salt, low-sugar, and low-fat foods that don't take out one and put in the other. For example, low-fat foods often contain a lot more sugar and/or salt than regular food products. Again, taste is being sold, not health. Look at soy cheese.

Most of it contains 290 mg. of salt. Regular Swiss cheese often contains only 40 mg of salt. Or look at labels of some fat-free cookies with lots more sugar and salt added than you'd find in those unhealthier cookies full of trans-fatty acids, true, but less salt and sugar. My answer is to bake my own cookies.

I take a cup of oat bran mix it with a half cup of flax seed meal and some cooked whole oat groats, add a few almonds, a handful of millet soaked in carrot juice overnight, and sweeten with fruit juice or a banana. I add two or three egg whites and a quarter cup of lecithin granules and a little soy milk if needed. Then I roll out the cookies and bake until light brown at 350 degrees F.

Before you boy low-fat organic alternative food products such as soy milk, look at the label and see whether the sweeteners added will send your blood sugar up. If you are genetically at risk for diabetes II in your mature years, try making your own alternative milk products from soy or almonds or grains such as oat or rice. Or look for unsweetened products when you can find them. Talk to supermarkets and convince them that unsweetened products are not marked 'plain' in many cases, and that they will sell, especially to the senior citizens who often don't know where to look for unsweetened or no salt added products.

Know your genes. The old adage "know thyself" means know how your genes express yourself. There are not enough specialized physicians to go around to all the baby boomers who will be knocking down doors in a decade for healthcare. Don't wait for the boomers. Senior citizens must take nutrition into their own hands and do the research on their genes now, not in a decade.

If you're retired, it's time to make nutritional genomics your main hobby as far as researching the effect of food and exercise on your genes. Nutritional genomics is linked to many other scientific areas of study. It's the most important tool for self identity ahead of personality tests and DNA tests for ancestry.

Ally yourself with others researching all the disciplines and branches of the life sciences. Food industry people need to read up on nutritional genomics. You don't need a special degree to know that food changes the way your genes express themselves. The prescribed appropriate and specific food combinations won't change your genes, but it will allow what genes you have to express themselves in the healthiest way they can. Some people may need to change their food or nutritional supplements based on their gene expression.

You could be taking vitamins and minerals in the wrong amounts or emphasizing the wrong supplements. Find out which ones you need and how much according to your genetic needs. Again, you don't have to wait another ten years until this is available to the masses. Do your research homework.

There are ways to find out what works with your genes. Join organizations that focus on allying people in the life sciences. Join the various life sciences alliances. People in many disciplines need to join together. For example the food industry, genomic scientists, healthcare professionals, journalists in bioscience communications, and the alternative health care markets need to plug into one another's research and resources. What's missing from these alliances? It's the consumer. Take charge of your search for self-identity at the genetic level.

If you don't see a new development and you're a consumer, perhaps you can develop it yourself. You don't need a degree in nutrition to make your own soy milk in a blender if you see the food products on the market are adding too much sugar and salt or barley flour to the milk you see on the shelves. Write to the companies. If nothing happens, make your own nutritious food.

The future of nutritional genomics is based on whether it becomes profitable. If it doesn't become profitable, no it won't be shelved. It will move to the holistic health markets and the alternative medical markets. The food industry is quick to move in with its own strategies, but it will always be profit as long as there is profit in healthy food. The mission is to convince the buyers in the supermarkets that people will buy food that isn't loaded with sugar, salt, and flour when the fat is taken out.

Nutritional genomics can plug into the phenomics industry—the science of customizing and tailoring your medicine to your genome. The U.S. diagnostics industry is a $2.3 billion business. You want your genotype? Your key-words in your consumer search for companies are "science-based." Contact science-based companies. Most do research, not individual genetic testing, unless it's for single-gene problems. So do your homework and don't bother the research companies if you want a genetic test. What you want to do is search for a managing physician who works with the reputable, scientific-based companies to look at your entire genome before any diet is prescribed.

You want to talk with people who are supervised under an umbrella and find out what research and testing is being done by whom, why, and how, where and when. Don't get taken in by snake oil-type unsupervised testing. Know the company and get recommendations from scientists and attorneys in the genome research business that work with supervising physicians concerning your healthcare.

You can look at a printout of your genes, your genome and conference with health professionals in consultation with you and your nutritional genomics consultants to find out whether a diet alone will help or whether a 'nutraceutical' supplement, medicine, or drug will help you along with the diet. The first place to convince and check out is your HMO and your insurance company. Find out whether you'll be reimbursed for the expense of having your DNA tested and for any sessions with genetic counselors. Again, don't wait until this becomes common and genetic counselors are flooded with boomers demanding tailored diets and/or nutraceuticals.

What will happen is either the price of the testing will come down or the number of people demanding it will make the price of health care premiums go up, or testing may or may not be covered by insurance in the future when the demand becomes high. You can help yourself now by forming alliances with genetic counselors and research scientists in nutritional genomics. Or at least read their research and join the message boards in nutritional genomics on the Internet. If you don't ask questions now, the answers later could become expensive or scarce, if the price goes down.

Make business alliances with people in nutritional genomics fields. What you're doing is taking research off the shelves of medical school libraries and bringing science to yourself—as a consumer. Again, you don't need a special degree to research this field for your own use. You can become a collector of published research in the field. Clip articles in the news or save Web-based articles for your scrapbook on this subject. Contact people in the news to answer your questions. Get your DNA tested—the whole genome. It's worth the thousand bucks or so now, even if the price goes down to a hundred later. If you get the right diet, ask yourself would it be worth it to your health?

What you want to find out about yourself are the links between your genes and your HMO or other health care systems. If you understand yourself, identify yourself at the genetic, molecular level, you'll think differently about your own health. You'll think differently about the health of your children and grandchildren if you think in terms of nutritional genomics as preventative medicine. It could delay the onset of what your genes predispose you to get. If you're at risk, instead of worrying, the dietary change could delay the onset or prevent it.

Time is on your side when you know how your genes express themselves. It's just the molecular you expressing yourself creatively. And since the body is hardwired to try and heal itself, you could lend a helping hand with the specific diet tailored to your genes. A diet book can't fit a whole population any more than one shoe size will fit a whole group of people.

Think about your health at the molecular, cellular level, at the genetic level. It's all about using your DNA for ancestry in a new way, from genealogy to nutritional genomics. Now you can find out whether your vitamins and minerals or other supplements are actually being absorbed and whether you need most of them. If you do, then you'll know which ones are working and which are not.

Food industries, the agribusiness, genomics research, healthcare, and allied life sciences all are working together. The goal is to bring this to the consumer level now and not in the future. You have industries contributing millions to biotechnology centers. Many of these centers are university-based. It's time to bring the university ivory tower lab to the consumer. Food and drug industries are pouring millions into centers to further research. You have only a few scientists working on projects that bring the fields of genomics, nutrition, and health together.

You need more catalysts like this to bring people together to do more than 'interface.' You need consumer action to 'interact' with the scientists who are working in projects that 'interface' the genomics people with the nutrition researchers and the healthcare professionals and genetics counselors. You need consumer involvement and consumer-based research.

You need bioscience communicators to make the complex easy to understand by the consumer with little or no science background. You need the consumer to take charge of his or her own health, privacy, nutrition, and research. Extended studies courses would help bring the consumer into taking charge by learning what research is being done and comparing his own DNA genome test to various interpretations of the test by different professionals such as his or her physician, genetics counselor, nutritionist, and nutritional genomics scientist.

All these fields need to be brought together under an umbrella accessible to the consumer…not in the future, but now, and with the privacy issues intact for the consumer, and the research open to the public in publications or online to be read by consumers as well as by allied science professionals and the media.

Who will write the regulations? It should be up to the consumer to have privacy in individual genetic tests. Genetic profiles are the business of only the person taking the DNA test to be used for planning a diet or lifestyle change. And unless the tests are totally anonymous as to name and address, collections of genomes would benefit scientists studying ethnic groups, but only if there was total anonymity. Nobody wants to be labeled and have somebody else know the label can't be altered. It's too controlling. Privacy should come first. Privacy should be respected so people numbered not by name and address, but by anonymous numbers or letters would ensure the privacy of large groups of people.

Scientists would be eager to collect large numbers of genomes for databases. Your genes are private. They are your time capsule. Your genome test would be something you'd pass on in a family time capsule scrap book to your heirs or future generations to learn about what your genes expressed, what was in the family, and what may or may not have been inherited by heirs.

Each individual has a different genetic pattern due to recombination. Some genes are inherited, but not all. So a time capsule or book binder with your DNA test may be of interest for medical history to your grandchildren, but how much of this they have inherited wouldn't be known unless they had a test themselves and compared genes. The consumer could take an extension course or read about DNA and learn about what, how, and why, when and where the genes express themselves.

Check out the genetic testing products. Do your research so you won't get scammed by companies that could spring up anytime in the future anywhere and target certain customers such as senior citizens or Boomers. Learn which companies are reputable and check them out by satisfied customers and by the reputations of the people who work there. Make sure the companies have professionally trained and experienced people. Check with the universities and the genomic associations and centers to make sure you are being tested by a company they would use themselves.

Most people have no clue as to how to apply genetics to their personal life. Many fear privacy invasions. Few are told what risks and benefits surround eating bright according to your genotype. As regulations come into focus, ask yourself how to validate the testing. When you are able to validate the test, then you know you've learned enough to get tested. Always ask professionals how they validate the genetic test and learn from their answers. Then check out their answers with research and other opinions from various professionals in the field.

What can you teach yourself about the applications of nutritional genomics testing for your individual health concern? How do you apply the information you've read? How do you validate that information? If a company makes a health claim, find out how to prove it and test it before you pay for testing. Pass the word to parents and children about getting their children tested so genetic profiles will be used to prescribe better diets and exercise programs for children as well as parents. Go into the schools and speak at PTA groups on getting children tested.

With the epidemic of diabetes in children increasing, perhaps a genetic profile might help parents to plan healthier diets, beverages, and snacks for children. It might also help the fare in school cafeterias. The idea is to get the word out at the consumer level as to what research they can do as consumers regarding applications of testing and validation of the tests. Then what follows are the smart diets and diaries of changes in health for all ages. The time for tailored eating and lifestyle changes is now.

By the time we wait for cheaper DNA tests of our whole genome and the chance to apply the test results to a practical menu, in what shape will we be? Ask yourself, should what we eat need medical prescription? Learn for yourself. Powerful knowledge is out there for the public and the media. Take control of your genes, your food, and your nutraceuticals. Your destiny will be in the hands of nutritional genomic scientists until you learn the science of nutritional genomics and research how you are impacted at the molecular level by how and what you eat. Reactions to lifestyle, environment, relationships, and exercise, also are expressed through your genes and your body chemistry.

13

What Products are Available Now for the Consumer?

Research has to be applied to thrive, to interest business. Somebody has to fund genomics research for convergence to bring together and link genomics to food systems design and to healthcare. Will it be the largest international food technology companies that funds nutritional genomics as a research tool for better healthcare?

We go from the general—the body—the specific—the genes, from whole foods to nutraceuticals such as concentrated extracts. You end up healthier and with your medical privacy, a special kind of virginity, intact. You also end up with whole foods tailored to your genes instead of tailored foods for the masses in a one-size fits all planet. Foods need to respect your diversity, including extracts and nutraceuticals. That's why exploring what your genes require for health is an innovation you need to learn.

According to Minneapolis-based Cargill's Web site at http://www.cargill.com/about/index.htm, "Cargill, Incorporated is an international marketer, processor and distributor of agricultural, food, financial and industrial products and services with 97,000 employees in 59 countries. The company provides distinctive customer solutions in supply chain management, food applications, and health and nutrition."

Cargill's media release of July 14th 2003 reports that food and pharmaceutical specialties are part of Cargill's Food System Design (FSD) platform, a group of specialty ingredient businesses that formulate integrated food solutions for customers. Cargill Health & Food Technologies, is a leading developer, processor and marketer of science-based, healthy ingredients for food and dietary supplements worldwide. H&FT is part of Cargill's Food System Design initiative in which Cargill businesses work with customers to produce ingredient solutions for affordable, nutritious, convenient and appetizing consumer products. H&FT is a business unit of Cargill, Incorporated.

FSD works closely with its customers to produce innovative products and services that deliver nutritious, convenient and good tasting food. It is a unit of Cargill, Incorporated. The idea of new product concept is the result of extensive consumer research. For example, Cargill announced a new product concept in its July 14th 2003 press release, a flavorful raspberry tea that supports bone health. It's one of the concept products unveiled by Cargill Health & Food Technologies (H&FT) at the Institute of Food Technologists' Annual Meeting and Food Expo in Chicago, that ran from July 13th–16th, 2003.

Suppose you had a DNA test of your entire genome for the purposes of nutritional genomics analysis and food prescription as part of a health program of diet, exercise, nutrition, lifestyles changes—the works. Assume you have just been counseled by your managing physician and nutritional genomics counselor that the way your genes expressed themselves required certain types of nutrients.

If a particular food or nutrient taken at a certain or any time is what would make your genes respond in a healthier way, delaying or preventing chronic disease now or later in life, then you might reach for what your individual genes might need. Perhaps it would be the refreshing prototype beverage, aptly named Bone Appetit. It contains AdvantaSoy™ isoflavones, Oliggo-Fiber™ inulin and calcium. Emerging science indicates that soy isoflavones may help maintain healthy bones and that inulin may help boost calcium absorption.

Let your genes speak to you by learning the language of your genes—your DNA. Learn or teach yourself what your genes are saying to you. They are alive. They speak. They are you encased in a molecular universe and make up every part of you. Listen for their message. What do your genes whisper in your ear? Is it a need for a prototype beverage? If so, what nutrients are in the beverage that you need?

Visualize your genes expressing themselves in the language of genomics showing you what they require. Up until now, you had to wait for your genes to communicate by your symptoms, by how you felt. Now listen to your genes by how they think. That's what intelligent nutrition is about, not only eating prescribed smart foods, but listening to your smart genes as well.

What nutritional genomics would do in your case is determine whether for you as an individual with a diet tailored to your genes would need a beverage containing soy isoflavones, inulin, and calcium. First, such a DNA test could help you determine whether your body would react healthier or not so healthy to those products. Let's say, it's just what your individual body needs. That's you.

Someone else may show different needs and different reactions to various products. The point is that the food industries have an open door to participating in the whole nutrition genomics revolution. It's like the Internet, with connecting links from your genes to the nutrients your body needs to operate at maximum efficiency. See the potential for a lot more convergence here of food industries, healthcare, and genomics stemming from cracking the human genome code a few years ago?

It's like the Internet making instant contact between your genes, your food, your healthcare systems, and the food industry. Convergence also works by going from macronutrition to micronutrition, from the holistic you to you at your molecular level, your cellular level. By making right what's inside your cells, the outer you works better. Food system design and your genes are linked through a burgeoning industry and research arena called nutritional genomics. It starts with you having your DNA tested, not just for ancestry, but the whole genome, for links with the healthcare industry.

Now you have food system design, the healthcare industry, your genes, and the nutritional genomics research area linked, webbed, as compared to the development of the Internet. Convergence. You are now a catalyst, bringing various sciences and industries closer together. Business becomes science and science brings research in a more informed well of knowledge.

According to Cargill's July 14th press release, "The functional beverage segment is one of the most innovative in the industry," said Steve Snyder, Cargill H&FT director of sales & marketing. "Based on our consumer research, we know that successful products in this category will be those that taste great and meet consumer demand for products that positively impact health. Clearly, flavored teas fit well into today's healthy consumer's lifestyle. Our goal in creating Bone Appetit and debuting it during IFT is to spur exciting new product ideas with our partners in the food and beverage industry," he said.

Bone Appetit is made with AdvantaSoy™ Clear isoflavones. Proprietary processing technology results in beverages that retain their traditional flavor, color and consistency. "AdvantaSoy™ Clear isoflavones allow us to meet the challenge of creating new functional beverages that promote health and retain the delicious flavor and aroma which made them popular in the first place," said Snyder.

Cargill H&FT provides proprietary science and technology for food applications, new product development, product prototype development and regulatory support. The International Food Technology (IFT) convention attendees were invited to learn more about these products and partnering with Cargill Health & Food Technologies, when they visited booth #1748 at IFT.

Cargill's H&FT business is part of a larger Cargill initiative called Food System Design. As such, Cargill Health & Food Technologies works in concert with customers to produce ingredient breakthroughs and other food system solutions that result in appetizing, nutritious, and convenient consumer products.

Do you eat lecithin granules? I do. I need that particular nutrient for my health. What about you? The food industry also knows you sprinkle lecithin granules on your food for specific reasons related to your health. In Minneapolis, Minnesota and Stuttgart, Arkansas on July 9, 2003, Cargill and Riceland Foods, Inc., announced their intention to form a strategic alliance to manufacture, market and sell innovative lecithin products to food, pharmaceutical, and technical customers worldwide. Under terms of the agreement, lecithin produced by Riceland at its Arkansas (USA) facility will be marketed and distributed by Cargill. The agreement took effect by Sept. 1, 2003.

According to Cargill's July 9th 2003 press release, Riceland, a farmer-owned cooperative based in Stuttgart, Arkansas, has been a leading manufacturer of deoiled lecithin for 25 years. Riceland supplies high quality powdered and granulated deoiled lecithin products globally. Cargill Lecithin has a global sales organization and modern production facilities in Europe and South America.

"The Riceland alliance establishes Cargill as a lecithin supplier in North America," said Jens Heiser, president of Cargill's lecithin product line. "It also supports our goal of growing the lecithin business globally by serving our customers with premium products and the application properties they desire. Well-known brands like LECIGRAN™, LECIPRIME™ and LECIPERSE™ open up exciting development possibilities and are excellent additions to our existing business."

Cargill is a proven leader in lecithin marketing, logistics and identity preserved supply chain solutions. "This alliance is a major growth initiative for Cargill Lecithin and will foster innovation in the deoiled lecithin market," said R. Creager Simpson, president of Cargill's Food & Pharma Specialties North America, where the lecithin product line resides. "It supports Cargill's ability to provide breakthrough solutions with new, value-added products such as identity-preserved, non-genetically modified deoiled lecithin."

Riceland President and CEO Richard E. Bell said the alliance will allow Riceland to operate its lecithin manufacturing facilities at more efficient rates. "In recent years we've been handicapped by not having a source of conventional (non-GM) crude lecithin for processing. This new arrangement with Cargill helps us fill that gap."

According to the press release, lecithin is derived from soybean processing. In its refined liquid form it is used as a natural emulsifier. The deoiling process results in powdered and granular lecithin. Lecithin is used in a variety of foods as a blending agent, dough stabilizer, egg replacer and instantizing aid. It is also an important ingredient in pharmaceuticals, dietary supplements and cosmetics.

Riceland Foods, Inc., a farmer-owned cooperative headquartered in Stuttgart, Ark., has been a leading manufacturer and supplier of liquid and deoiled lecithin products for 25 years. It also operates soybean processing and vegetable oil refining plants and is the world's largest miller and marketer of rice. Its products are marketed globally under the Riceland and Chef-way labels, private labels, as ingredients and in bulk.

So you see the convergence here, the big, international food industries that design food technology systems are *listening* at food conventions to research done by the companies and universities engaged in research in nutritional genomics. Twenty years ago eating health foods was the butt of jokes. You've heard on mass media TV the clichés about how vegetarians and people who walk into health food stores look sickly. You've seen the look of disgust on some people's faces when invited to a vegetarian restaurant. Often they responded with another cliché, "Oh, you're not going to eat those twigs and sprouts?"

The tide has turned. No one makes fun of people who go into health food stores anymore. No one smirks at vegetarians who eat lunch at soup and salad bars instead of taking power lunches with three martinis, steak, fries, and a sedentary afternoon. Yet for some the soup and salad may not work well, and an hour later they are shaking with excessive insulin. On the other hand, the person eating the fatty steak and potatoes might have clogged arteries. So how you react to a meal is determined by your genes—your entire genome, with all its mutations. Some people can eat any type of meal and stay healthy. It's all in your genes as if your genome is a deck of cards shuffling with each generation, excerpt for your mtDNA and Y-chromosomes that pass on your remote ancestry key.

Even as more young people today demand natural food, those type of stores may also sell red meat because some people eat it. There's the competition in those types of stores. And with diets focusing on animal protein versus vegetable protein, the question remains: Will your genes express themselves in excessive insulin both with the high animal protein and high carbohydrate meals? And if so, what's the solution? The Syndrome X diet was helpful to me. What about you? The answer lies in going deeper, that is to the molecular level and seeing how your own genes react to different food combinations.

You'll hear warnings about avoiding the snake-oil companies popping up in the future or here already, but no one dared name them in print. You'll be told to research reputable companies. You'll find out about research companies and DNA testing companies, and learn the difference. Maybe you'll enroll in university programs of research or study and find internships or careers in nutritional genomics. The door is open, but where are the jobs? You don't always have to approach from a research stance into nutritional genomics careers. There are the patent attorneys.

The search for funding for research is real. Who has the money? In the meantime, know your genes. Are you ready for a preventive health profile not only at the chemical or metabolic level, but at the level of your entire genome? Don't forget your pet's genes, either. What food is best for your dog, cat, or race horse? Have you thought about what laboratories are researching in pet-related nutritional genomics? What genomic innovation is next? Let's take a look at ancestry and DNA.

If you're going to learn about how and what to eat for your genotype—your genetic signature—let's consider your ethnic origins. Food is a cultural component. And genes express themselves through cultural components. Genes have more than a biological signature. There's ancestry, ethnicity, and culture to consider when customizing foods. When the media mentions smart foods or intelligent foods, it's more than a perceived consciousness of your genes and the living food nutrients. What is really implied is tailoring or customizing food as you would configure software to your computer's operating system.

You have to tailor the food, choose the food based on what your genes require to function optimally. Since people are more or less diverse within their ethnic group, they may have inherited or not inherited certain genes that express themselves differently with certain foods. One example is milk/lactase tolerance or intolerance. If thirty-six percent of Southern Europeans can tolerate the lactase in milk and the rest cannot, that's one way ethnicity, genes, and food may interact.

If more Northern Europeans can tolerate the lactase in milk without digestive symptoms because ten thousand years ago a gene mutation in Northern Europe (Finland) allowed the people there and subsequently in most of Northern Europe to drink milk—tolerate the lactase in milk—it's a gene expression. If about sixty-eight percent of North American Ashkenazim and approximately sixty-six percent of Greeks can't tolerate the lactase in milk, it's a genetic signature.

How does your body handle certain foods? Find out what foods to eat from genetic testing and profiling. Tailored foods prescribed or recommended should be based on you as one individual rather than the ethnic group to which you belong. People of any one ethnic group are diverse enough to require individual testing. Let *your own genes* point the direction of your nutrition, exercise, and lifestyle plans, not merely the statistical averages of your ethnic group.

Your donated DNA may end up in statistical tables. Are you getting feedback for healthcare, or only a certificate printing out some of your markers? What practical applications can consumers find from DNA testing for nutritional genomics and also from testing DNA for ancestry? If you want to go beyond looking at DNA for deep ancestry or current DNA matches, what's out there for the consumer interested in human genomics?

***

14

DNA Testing DNA for Ancestry

Are you Curious about Your Ethnic Family Origins from Last Century Back 10,000 Years And How Their Genes Responded to What they Ate?

What your ancestors ate is part of your genetic history as a *cultural* component. How food they ate long-term affected their health is part of a *biological* component you may have inherited. People have different eating customs, choices, and habits. Lifestyle, environment, and exercise also play roles. What did your ancestry and food history look like if you had to draw a map? How does your ancestry relate to the effects food, nutraceuticals, or medicine have on your genes?

Genealogists are now using molecular genealogy—comparing and matching people by matrilineal DNA lineages—mtDNA or patrilineal Y-chromosome ancestry and/or racial percentages tests. People interested in ancestry now look at genetic markers to trace the migrations of the human species. Here's how to trace your genealogy by DNA from your grandparents back 10,000 or more years. Where did they wander and camp?

Anyone can be interested in DNA for ancestry research, but of interest to Jews from Eastern Europe is to see how different populations from a mosaic of communities reached their current locations. From who are you descended? What markers will shed light on your deepest ancestry? You can study DNA for medical reasons or to discover the geographic travels and dwelling places of some of your ancestors.

What you're studying is non-randomness. You use DNA as a tool to study ancestry and the history of your ancestors as part of a larger population. You look for similar patterns.

Ashkenazim and Sephardim separated about 1,000–1,500 years ago. What happened genetically to each branch since that time? This book is about researching the possible origins of Eastern European Jewish genetic markers and DNA such as mtDNA and Y-chromosomes. It's a book that addresses the questions beginners have when studying how to interpret DNA test results for family history and ancestry.

DNA Ancestry Tests for the Consumer

With whom in the world do you share a common ancestry? With whom do you share a common ancestry and history? Looking at DNA to study ancestry or to study medical issues is all about tracing patterns—from patterns in the genes to patterns in the migrations. Some of the best genetic feedback material I received was from Roots for Real, a DNA testing firm that lets you see the geographic area of your DNA matches for ancestry from their Roots for Real mtRadius database. The map they sent me shows where in the world I have matches for my mtDNA, my matrilineal lineage. The idea is to see where my estimated maternal geographic ancestral roots would be on a map.

They enclosed an article by Dr. Peter Foster titled, "Mitochondrial DNA, the Peopling of the World and Your Own Haplogroup's Story." This article gave me some background in depth of how the world was populated. More important to me, it pointed out where my own specific genetic group or haplotype originated. The article is an excellent introduction for beginners trying to interpret the results of their DNA tests for ancestral roots study. For men, mtDNA or Y-chromosome can be tested for male and female ancestry. The fascinating marriage of genetics with archaeology provided an open door to view where my ancestors might have been 10,000 years ago—namely, in what today is the town of Bar sur Aube, France or the geographic center of my origins at 48.30N, 4.65E to the nearest location.

How could that be? I don't know anyone who is related to me that came from that area. I had been reared by my auburn-haired, silver-eyed American family who had ancestors generations ago living in a balmy, southern Mediterranean family eating the typical Mediterranean fanfare from grape leaves to feta cheese pizza, from olive oil to parsley and grain salads, and mostly vegetarian foods. My listening in music focused on Eastern Mediterranean quarter tones and Greek bouzouki.

My fair, freckled complexion and hazel eyes gave me a clue to a wider geographic range of genes. Sure enough, the DNA test revealed that my deep, ancient, maternal line ends up mostly in Northern Europe. So what I discovered from that point, the tracing of geography began as I mapped the route Mitochondrial Eve, the mother of all humans on Earth today took when she left eastern

Africa more than 80,000 years ago, fanned out to Asia, turned around, and landed in Europe by 21,000 years ago.

As a member of the leisure, quite-well-over-62 set with my time free all day, I have dedicated my working hours to tracing the geographic path of travel mitochondrial Eve took and researching publications about ancient and modern nutrition as I follow her descendants leaving genetic prints or mutations in their mtDNA as they moved from campsite to campsite from one end of the world to the other.

That's the awe about studying the peopling of the world. Because scientists now have genetic dating and DNA sequencing, you can trace your own ancestry from origin to present day migrations at least statistically. That's to Dr. Frederick Sanger, Nobel Prize winner, 1980, you can trace and understand your family tree by DNA mutations and genetic markers. It's a lot like looking at fingerprints people made as they touched the world to gain balance. It's also watching their footprints trot around the globe. In it, you'll find your own identity, or at least a few good possible places your ancestors might have been.

What you'll see are matches—people living today whose exact mtDNA sequences match your own, showing that somewhere back in the deep maternal ancestry you and they shared a single lineage connected from mother to daughter. The same can be said for looking at the male Y-chromosome about sharing a single lineage connected from father to son. Once you find your matches, you can see what geographic location they are at today and where the probable origin spot was geographically in the distant past.

After you find your matches, you are free to find out whether they want to be contacted or whether you want to start a group for people with the same matching mtDNA or Y-chromosome. Of course, privacy is usually asked for, but some people might want to meet their mtDNA or Y-chromosome matches. Somewhere in time, you might have shared an ancestor, maybe 10,000years ago or maybe 250 years ago or in-between. It's like looking at living archaeology joining genetics as a new field of archaeogenetics.

What science finds today may all be changed tomorrow. The field is so new, the new technology evolving so fast that data has to be updated frequently. You learn starting with genealogy and oral history and moving onto genetics that each haplogroup split at an early stage into small groups. The pioneers colonized the world and expanded demographically and geographically. In some tropical areas of the world, original DNA types didn't die out. If the population was large enough, and food was plentiful, the people thrived and retained an overall similarity to the original African ancestors. In Europe and Northern Asia, people were confronted with cold climate and Neanderthals.

Conditions were fierce, almost arctic with tundra and penguins in the Mediterranean. By 20,000 years ago only small groups could pioneer into Europe and Asia. Most took refuge in Southern Europe, in Spain and the Balkans. These areas became the homelands of haplogroup H and V. At the height of the Ice Age of 20,000 years ago, Native Americans were closed off from Asia by glaciers, leaving only a small group in Alaska until 11,400 years ago. Genetic variation continued to evolve in different areas of the world due to climatic changes. People followed the herds or the places where food and water were available as part of their migrations.

◆　◆　◆

Do You Eat for Your Paleolithic Homo Sapien Genes?

Did you ever wonder what diets the oldest anatomically modern people ate based on their DNA? Are prehistoric people's DNA from 20,000 to 50,000 years ago similar to your DNA in the present? Does the prehistoric hunter's DNA differ enough to require a change in food and lifestyle to nourish your genes? How about the Neolithic grain farmer's diet? Does it agree or upset your own DNA? Would you prefer a carbohydrate-based diet, a protein-based diet, or a mixed diet? How about a balanced diet of 45% carbohydrates, 15% proteins, and the rest good oils such as olive oil, fish oils, and flax seed oils providing the Omega 3 fatty acids? Which diet is for you? Are there companies out there who can tell you what metabolic or genetically-based foods and nutrition in general is best for your individual type? Yes. The question remains, when hundreds of genes work together to cause a condition, how do you single out the gene to test before you can prescribe? You look at the symptoms, but you need to get to the root cause, the nutrition at the cellular level.

Can genetic testing reveal whether your genes need a Paleolithic or Neolithic style of eating? If you're not into "fad diets" what do your genes really need to stay healthy in the way of food or supplements? I recommend the book titled, *Complete Food and Nutrition Guide* from the American Dietetic Association John Wiley & Sons, NY 2002 (or later editions).

In addition, you need to compare alternative nutrition books, and as a result of DNA testing, find out what your genes really need for nourishment. It's still too early to tell whether the genetic testing is accurate as disease may be caused by mutations in over a hundred different genes. How do you know which genes to research for what need or ailment?

That's why a broad reading approach covers more territory for the beginner until you get enough feedback to make wise choices. Eating is about choices, and taking prescribed or over-the-counter medicines is about knowing how your genes respond to the chemical or the dosage. In the old days, you'd talk to your allergist. Now you take a DNA test and look at the molecular/cellular level interactions.

What if you have the genes or at least the maternal lineages or mtDNA of the oldest human group that first came to Europe or anywhere else? You're only looking at two or three percent of your entire genome, but for deep, ancient, ancestry purposes let's have a look. Genealogy buffs and anthropologists enjoy mapping population genetics, particularly the prehistoric migrations.

Since it has been said by Canadian geneticist and physician, Dr. Dr. Charles Scriver, genetics has both a cultural component as well as a biological, where do your genes appear geographically in the ancient past, the recent past, and the present? Do you descend from a matrilineal line in the Middle East, in West Asia, in Northern Europe, or in Southern Europe? Here is a sample personal matches list from an mtDNA database. All the people noted on the list by a number have the same mtDNA sequences and match my mtDNA sequences. They may have been tested for different numbers of markers, but they are all in the database because they have the same mtDNA sequences that showed up on the number of sequences tested.

◆ ◆ ◆

Here's a letter from Daniel F. Hayes, MD, Clinical Director, Breast Oncology Program, University of Michigan Comprehensive Cancer Center. He is one of several physicians I asked to comment on the topic of 'pharmacogenetics'—tailoring your medicines to your genetic signature. Other physicians and scientists I've asked to speak out and explain about nutrigenomics for the consumer.

The point is consumers need to become more involved in both nutrigenomics and pharmacogentics. They need to become as involved as they are in learning to interpret their DNA tests for family history and ancestry or to make time capsules. Check out how your genes respond to skin care products or supplements and food interactions. Where do you start? As Dr. Hayes notes, pharmacogenetics is in its earliest stages. What's the consumer's role right now? Let's look at the doctor's letter on how consumers may be genotyped and analyzed for patterns in the DNA.

———-Original Message———
From: Daniel Hayes
To: Anne Hart
Sent: Sunday, August 24, 2003 1:07 PM
Subject: Pharmacogenetics.

Pharmacogenomics is in its infancy. However, I believe that in the future every potential patient may be "genotyped" by examining his or her DNA, which can easily be collected from either a simple blood specimen or a swab of the inside of your mouth. Developing technology will permit analysis of the patterns of the genes that control response to and metabolism of different drugs and medicines. These patterns differ from one person to another, according to what one inherited from one's parents, much like hair and eye color, etc.

Although it will take an enormous amount of careful and sophisticated clinical research, I believe doctors will then be able to prescribe either different drugs, or the same drugs at different doses and schedules, based on these inherited patterns. Currently we do so, but crudely, based mostly on a patient's size (kids take smaller doses than adults), or obvious organ dysfunction (people with kidney or liver problems might need less or more of an agent than those with normal function).

We and others have begun preliminary studies addressing these issues, and we hope that they will give us important insights not only into where we are now, but where our investigations should proceed in the future.

dfh

Daniel F. Hayes, M.D.
Clinical Director, Breast Oncology Program
University of Michigan Comprehensive Cancer Center
6312 Cancer Center
1500 E. Medical Center Drive
Ann Arbor MI 48109–0942

Here's how to tailor your food, medicines, and skin-care products to your DNA. This is your consumer's guide to genetic testing kits. Your DNA, including your ancient ancestry and ethnicity has a lot to do with how your body responds to food, medicine, illness, exercise, and lifestyle, but just how much? And how do you know which DNA kits and gene testing are reliable and recognized?

Learning about DNA to understand and improve your health is now interactive and available to the average consumer, not limited to students and teachers, but to anyone else. In the last few years genealogy buffs, parents, and anyone interested in DNA without a science background took an interest in DNA tests rests that reveal deep maternal and paternal ancestry. Currently consumers with little or no science background are interested in learning about drug metabolism—pharmacogenetics. How your body metabolizes medicine is as important as how your body metabolizes food. Nutrigenomics is about how your genes respond to food and how to tailor what you eat to your DNA. Consumer DNA interest ranges from forensics and anthropology to nutrition, caregiving, family scrapbooking and healthcare knowledge. Nurses are becoming more interested in DNA.

The DNA consumer revolution began when media broadcasts revealed to the public that fast computers had revealed the human gene code. Once more TV opened doors. Suddenly, a gap between science and consumers had to be bridged by available interactive education.

A proliferation of products relating to DNA emerged. The internet shows DNA summer day camps for students and teachers. DNA testing companies and books emerged geared to the average consumer. Genealogists tried to interpret DNA for ancestry. People left other non-science-related businesses to open up DNA testing companies for ancestry research, contracting out to university research laboratories to do the DNA testing. Again, opportunities opened doors to the public.

Nutrigenomics product marketers sought those who wanted a diet tailored to their genetic signature. Pharmacogenetics reports customized medicines in order to prevent adverse drug reactions. Pharmacogenomics studies the entire genome in relation to chemicals and drugs, whereas pharmacogenetics researches specific genes and markers to look for adverse drug reactions for individual clients or patients. Finally, DNA testing products emerged offering to tailor skin care products such as creams and cosmetics to your individual genetic signature.

If you've had an interest in learning about how to interpret your DNA test results for ancestry, you now can see the links to understanding how to tailor your food, lifestyle, exercise, medicines, supplements, and skin care products—in fact numerous environmental chemicals—to your genetic expression. It's not only about food anymore or ancestry alone, or medicine.

DNA testing also is about kits sent to you directly or to your physician. It's about tailoring to your DNA skin products, cosmetics and anything you put into or on your body that gets absorbed. It's about what chemicals are in your water and home-grown vegetables.

No science background? Don't worry. There's a DNA summer camp near you, or an educational experience in learning about DNA now available to the average consumer. Educators, scientists, and multimedia producers have teamed up to teach you the wonders of DNA, your genes and your lifestyle.

What's left? Physicians and genetic research scientists need to talk more to each other because most family doctors don't have time to read the proliferation of publications reporting new advances in genetics or other areas of science that directly affect consumers. It looks like it's the consumer's job to bring people together through the media and through consumer's watchdog organizations, professional associations, and support groups. Key words: action and public education about DNA through multimedia and consumer involvement. I highly recommend the DNA Interactive Web site at: http://www.dnalc.org/. Consumers need to know more about how to interpret DNA test results for whatever purpose they seek—tailoring diets or drugs, skin care products, or seeking out ancient family history or ancestral lineages. The science is new enough to have many more applications on the horizon that consumers can digest in the future. What can the average consumer with no science background learn about applications of DNA testing currently? Start with the publications and the interactive DNA learning sites.

You'll soon become familiar with the DNA terminology. The goal is to bridge the gap between science and the consumer, let alone the gap between science and medicine. So you'll have to invite your busy family doctor to join you in creating consumer groups. It'll work fine. Doctors and scientists usually are found conversing together at parties.

The triangle now includes the consumer. If you have children, bring them into the fold. There are now wonderful DNA summer camps. So include your children's teachers. Learning about DNA as a consumer might make you wish you had majored in genetics. Link your beginning self-taught DNA studies to your special field of interest such as your healthcare or your ancestry.

For history buffs, there's always molecular genealogy for family history. History buffs can follow population genetics. Anthropology enthusiasts can read about archaeogenetics. Bring archaeology and DNA together.

Nutrigenomics links DNA research to nutrition for the diet-conscious. Pharmacogenetics helps you to tailor specific medicines to your genes. DNA gets into all walks of life and work. Start with population genetics and physical anthropology if you wish. I like the popular book titled: *The Real Eve*, by Stephen Oppenheimer. Carroll & Graf Publishers, NY 2003. Or start with nutrition and genetics. First find out what field you like best—nutrition, pharmacy, genealogy, anthropology, or heathcare—and read a book for beginners on DNA related to understanding your particular area of interest, such as nutrition or family history. It's about discoveries.

- A blizzard of discoveries are published monthly in recognized scientific journals found in local medical school libraries open to the public. Only a few consumers ever look at them, and still fewer physicians. Doctors are busy with so many patients and paperwork or bureaucracy. Consumers may not know information is accessible to them. And few can keep up with the proliferation of material in science publications.

- For information, resources, the research network, and references on pharmacogenetics (education) see the Pharmacogenetics and Pharmacogenomics Knowledge Base Web site at: https://preview.pharmgkb.org/resources/education.jsp. As far as education, the Web site features links and articles on the following subjects: What is Pharmacogenetics? Asthma Case Study, CYP2D6 Case Study, The National Institutes of General Medical Sciences (NIGMS), Medicines For You, Minority Pharmacogenomics, The Importance of Genetic Variation in Drug Development, Publications, and News Clippings.

- Click on the Dolan DNA Learning Center at: http://www.dnalc.org/. The Dolan DNA Learning Center at Cold Spring Harbor is entirely devoted to public genetics education. The gene almanac is an online resource that provides timely information about genes in education. The Dolan DNA Learning Center is the world's first science museum and educational facility promoting DNA literacy.

- Dolan's Saturday DNA program is designed to offer children, teens and adults the opportunity to perform hands-on DNA experiments and learn about the latest developments in the biological sciences.

- If you're interested in student "DNA Camps" see the Web site at: http://www.dnalc.org/programs/workshops.html. Student summer day camps have fun with DNA and enzymes and study DNA science or genetic biology. Students and high-school teachers can participate. There are a lot of ways to become involved in learning more on these topics. I wish there were DNA day camps for senior citizens newly retired with time free at last, or for parents, children, and teachers to participate and learn together.

The student summer day camp workshops feature such wonderful learning experiences as the genomic biology and PCR workshop. This new workshop is based on lab and computer technology developed at the DNALC in the past year. The workshop focuses on the use of the polymerase chain reaction (PCR) to analyze the genetic complement (genome) of humans and plants. DNA educational centers bridge gaps between scientists and communicators.

Many physicians have not yet been trained in nutritional genomics or in pharmacogenetics—how to correctly interpret a DNA test for consumers. Who advises consumers how to tailor their food or drugs to their genes in ways that consumers can immediately use? Who instructs them how to make time capsules out of their ancestry from DNA reports?

For the time being, it's the bioscience journalist who acts as a communicator, the liaison, the publicist, the broadcaster, the bearer of news, the turner of complex terms into plain language, the reporter and the publisher, showing consumers, physicians, and scientists how to bridge the gap between science and healthcare.

Who acts as the middle person between genetics and medicine or between nutrition and healthcare? Who bridges the gap between the dietician and the geneticist? For now, it's the media, the science writer writing for the mass media or an audience of general readers—consumers like you and I. Who should join the science media team? Consumer watchdog groups, concerned physicians, genetics researchers—scientists, DNA testing companies, and bioscience publishers.

Instead of worrying about physician apathy or consumers turning to alternative medicine, let's turn to scientific knowledge available to all consumers if you know where to look and what is accessible. If you want to take charge over how your body responds to food or medicine or lifestyle changes, you need to take control by finding out how your genes respond to what you put into your body from what you eat to the drugs you take to the chemicals in your environment.

Has a local or national newspaper reported on rocket fuel or other specific chemicals in your water supply that went into your home-grown vegetables that made your thyroid go wild, find out what research is being done on the situation. Or if your body responds to various foods in certain ways or various drugs in other ways, you can take control by learning what new advances are available to you. Check out what is credible and what works for you. Minorities may have genes that respond differently to various dosages of medicines. Check out the Web site for research on minority pharmacogenetics at: <http://www.sph.uth.tmc.edu:8057/gdr/default.htm>. The Minority Pharmacogenetics website is devoted to issues of minorities, populations and pharmacogenomics.

Have you ever met a doctor who keeps up with all the latest advances? Who has time? Consumers do. Look at the breakthroughs in the journals published monthly. Why worry that medical discoveries aren't being delivered fast enough to patients or clients if you can see what's happening in the journals?

Consumers need to form watchdog groups and to link up with the media. You need to become involved in accessing scientific pioneers. As a journalist, my job is to convey information to the public to bridge the communication gap. As a consumer, your job is to take care of your body. Nobody should walk in medical or nutritional ignorance when the information you might need "is out there." You don't need any science degree or license to read public information. Let's all bridge the communication gap between consumers, scientists, and physicians.

Consumers want applied knowledge. Here's where to start. On the Web, look at: http://www.nigms.nih.gov/funding/medforyou.html. It's the page for The National Institutes of General Medical Sciences (NIGMS). It's the home page for the National Institute of General Medical Sciences, a component of the National Institutes of Health, the principle biomedical research agency of the United States Government. "NIGMS supports basic biomedical research that is not targeted to specific diseases, but that increases understanding of life processes and lays the foundation for advances in disease diagnosis, treatment, and prevention," according to the Pharmacogenetics and Pharmactogenomics Knowledge Base at: http://www.nigms.nih.gov/funding/medforyou.html.

If your doctor hasn't the time to take advantage of current treatment findings, you as a consumer can read what your physician may not get to read for a while. Picture one future fantastic scenario regarding DNA testing where spouses are chosen based on DNA reports. If this sounds too bizarre, visualize choosing foods based on DNA reports. That sounds credible. The point made here is that you could help to bridge the gap between research and conventional medicine. You and I have our work cut out for us, speaking as a consumer and as a journalist who loves reading scientific articles.

The scene at the "speed dating" table opens with a young woman waving her latest DNA test result under the nose of her blind date. "I'll show you mine, if you'll show me yours," she says, smiling. The fellow opens his DNA test result folder or database. They exchange DNA test result printouts and interpretations.

"Hmm…regarding ancestry," he says, "Looks like we're both members of the same mtDNA matrilineal clan."

That could be haplogroup H, L, M, or with whatever worldwide mtDNA haplogroup letter or matrilineal 'clan' they match. "Now let's look at your specific genetic markers and genes tested for adverse reactions to this list of medicines or foods or…." She gazes up at him smiling. "I like your lack of risk for diseases," he says. "And I like the way your gene mutations put you at risk for all these inheritable diseases," she answers.

"Do you also have any inheritable real estate and cash? I'd like to be beneficiary." He looks at her askance and holds up a little green bottle. "With this gene equalizer, my risk is greatly lowered." They exchange DNA printouts once more and move to different tables to begin another Dating by DNA interview.

Science fiction version of "blind dating by DNA?" Maybe, but consumers want practical results from DNA testing that they can apply to real-life situations. What kind of personalized medicine can consumers find today regarding genetic testing? How many ways can you use DNA tests? Tailoring your food, customizing your medicines by avoiding adverse drug reactions, looking at your ancient ancestry, family history, or genealogy by DNA through "molecular genetics" and perhaps, choosing your spouse by DNA?

Genetic screening of couples hoping to marry or couples seeking answers to fertility questions have been screened for inheritable diseases in several countries. How many ways can you market DNA tests? For what purposes can you use DNA tests? Who will review the products? What consumers need is a guide to genetic testing kits sold directly to the public and a guide to those sold only to physicians and healthcare professionals.

***

15

Personalized Medicine from DNA Testing Companies

How do consumers or consumer's watchdog groups regulate doctors' services? Are DNA test kits sold to consumers really doctor's services or are they medical devices? Consumers want their healthcare and nutrition tailored to their individual genetic signatures, and they want it to be affordable and available to everyone. The FDA regulates medical devices, not physician's services. Where does a DNA kit sent directly to consumers fit into a definition? Scientists and physicians study haplotypes when they map complex-disease genes. Genetic researchers look at the abundance of single-nucleotide polymorphisms (SNPs). They use terms such as "single-locus analyses."

Consumers want those terms translated into plain language by bioscience communicators, since most scientists and physicians are too busy to continuously write about genes for the public. That's why public education programs about DNA exist, including the summer day camps for students and teachers. We need the training programs and camps also for parents—for entire families from great grandma and pa to elementary school children and teens.

When it comes to studying how your genes respond to food or drugs, scientists avoid analyses techniques with limited power and instead focus on economical alternatives to molecular-haplotyping methods. So you, as a consumer, presumably with little or no science background, are in a position to learn about your own genes and about DNA in general.

You can start with some of the DNA public education learning programs open to anyone. Since the gene hunters began to explore the entire human genome, they opened a door to the public for knowledge accessible to all. At first consumers were interested in learning how to interpret their DNA test for ancestry and family history. Next, consumers wanted to know how to tailor their drugs and food to their genes—their genetic expression. Finally, consumers found they could customize skin care products to their DNA reports as well.

Consumers ask important questions. What's waiting for you in a DNA kit that will be valuable to your health? How will you understand and apply the scientific reports to your daily lifestyle? How do you find out whether the various DNA testing companies are credible and recognized by which group? Who are the watchdogs here? Who are the experts? What do you do with the information you receive after a DNA test? Should you test for deep ancestry, adverse response to drugs, or tailoring your food to your DNA? How do you apply the information in a practical way?

Should the consumer only deal with companies that sell their DNA test kits to physicians and similar healthcare professions such as dieticians and genetics counselors? Or should the average consumer buy a DNA testing kit directly from a company that also markets to consumers, bypassing physicians? What do you think as a consumer? Which stance are you taking? For resources and articles, see the Web site at: http://syndromexmenu.tripod.com.

What happens when a DNA testing company sells its DNA tests or kits to companies that turn around and sell personalized nutritional or cosmetics products so that consumers can buy skin products or nutritional supplements customized to their genetic signatures? Is this good for consumers? What do you think? I like DNA testing. The interpretation of the tests is tricky. Results can have a lot more than one interpretation, even for physicians and scientists. So how does the consumer learn to screen the screeners?

Should a company tell consumers which specific genes are tested? Some companies do and others don't. How do you know whether the dose of vitamins you take is beneficial or harmful? Do you need to cut back or add more? What can DNA testing do for you, the consumer, presumably starting with no science background? As a bioscience journalist, I'm not taking sides here, but I really like the variety of DNA tests.

Some types of genetic test kits are sold directly to the public, and others are marketed only to physicians and similar healthcare professionals. From the consumer's point of view, there is always the question of whether the tests are needed and if they are, how accurate and valuable are the tests, and who is trained and experienced enough to interpret the answers? If you're going to take a DNA test, it could be for ancestry, for nutrition planning, or to see whether any prescribed or over-the-counter drug, chemical, supplement, nutraceutical, food, or herb you take will have adverse effects on your body. In the past you could test for allergies to some substances and foods, some chemicals in the environment, but now there's genetic testing.

Specific genes are tested for risk or reaction. The type of consumer most likely to order a genetic test for drug response or a "pharmacogenetic test" would be an individual with medical conditions taking several drugs and concerned how his or her body handled the mix of medicines, food, and lifestyle. Each person has his or her own genetic signature. You rinse your mouth with a type of mouthwash and send the contents in a small tube to a laboratory, hospital, office, or company to be analyzed. Or you swab the inside of your cheek with a felt tip and send the swab to a hospital or genetic testing company so your DNA—specific genetic markers—can be analyzed.

In a few weeks, you get a report. The genetic testing company looks at specific genes and markers. If you're concerned about drugs, you want to know how slowly your body is breaking down any of the drugs you take or might take in the future. You don't want to gulp down a prescription or over-the-counter drug and have it build up in your body to the point you land in the emergency room of a hospital. The whole point of a genetic test for adverse drug reactions is to alert your physician to adjust the dosage of your medicine or change the type of medicine.

On one side, the consumer sees the physician and scientist diagnosing rare diseases by looking at particular genes. On the other, the consumer wants information on how genes work. Firms that sell gene tests directly to consumers need to fill an education gap. Certain tests can be sold directly to consumers, such as DNA testing for ancestry, genealogy, family, oral history, DNA matches, and surname projects where people with the same or similar last names are matched by DNA to see whether there's a relationship in the past.

What DNA testing companies can do for the consumer is to widely teach the consumer about population genetics—the peopling of the world. How about for disease? Genetic disease specialists have been in charge of this new science. Lately, companies that test DNA now market directly to consumers as well as to their physicians. Here's a chance to educate consumers as well as physicians in everything they need to know about their genes. Physicians want to know how to interpret DNA tests for disease markers. Consumers want this information directly. So there needs to be a level for education for both physician and consumer.

If a DNA testing firm markets directly to the consumer, the test should be interpreted for the consumer and the physician. Consumers are paying to learn what medications to avoid. In the past allergists told people what foods to avoid.

What consumers want is to know what foods, vitamins, minerals, supplements, skin creams, herbs, medicines, and other nutraceuticals will work best with their individual genetic signatures. Who will teach them? If the physician has only a few minutes to consult with each patient, who else has the time? Either the DNA testing companies must educate the consumer, or the consumer as an autodidact, must teach himself from reliable sources.

Today anyone with a genealogy hobby an open a DNA testing company and contract out to a laboratory in a university to test DNA for clients. When it comes to test DNA for diseases or drug reactions, it's different than sending a printout regarding the client's ancient general ancestry such as the results of a DNA test for mtDNA (matrilineal ancestry) or Y-chromosome (patrilineal ancestry) in the deep past. Those tests are general and would be of interest to people matching ancestors or learning about population genetics an ancient migrations.

When it comes to tailoring food or drugs to your genotype, how far can DNA testing companies bypass the genetic disease specialists and market directly to the consumer? The answer is as far as it takes to teach the consumer about his or her own genetic expression. When you look for a DNA testing company, ask yourself whether the company diagnoses diseases? Does the company help you choose your medicines, food, or cosmetics according to your genetic signature? Or does the company work with your physician and you. Does the company pass you over entirely and only send the results to your personal physician?

Look at the marketing efforts of the company. Investigate gene research for yourself by the many articles available online and in the medical libraries and popular magazines that go to physicians. If you buy a test, is it affordable? Do you need the particular DNA test?

If one of your relatives has a specific condition you might have inherited that will show symptoms when you reach a certain age or presently, you need to know what you can do to delay or prevent the situation. You need to know that the results of DNA testing is for a particular purpose, such as testing adverse reactions of your body to certain prescription drugs. Find out the value of the tests from reliable sources. If you start with your personal physician, make sure he or she can correctly interpret a genetics test. Some physicians can't because they were never trained to do so.

Consult the various professional associations for referrals to genetic disease specialists rather than rely on a generalist when it comes to interpreting your test while you learn how to interpret your tests yourself. The purpose is to make sure your physician's interpretation and your interpretation agree. If you have any doubt, you need more information and a second opinion.

What you want to avoid is confusion. Are the tests available today reliable? Find out from several opinions of specialists. To reach these specialists, the professional associations for the various genetic disease specialists would be of help as would the publications. You can talk to the research labs at universities specializing in genetic testing. The institutes are online and have their own Web sites.

Talk to federal health officials about what's on the consumer market in your country and in other countries. Some people send their DNA swabs to companies and/or labs in other countries. Currently genetic tests are sold without a special 'watch' organization that reviews them. So look at any publications put out by the U.S. Food and Drug Administration (FDA). The FDA wants consumers to become more involved in their own food safety and security monitoring. Keep a folder on DNA testing company marketing claims and check them out.

Companies can tell you which mix of vitamins works best or what food to eat based on genetic testing and/or health questionnaires. For example, check out GeneLink Inc., New Jersey. Their Web site is at: http://www.bankdna.com/breakthrough_profiling.html.

According to Multex, an investments site at: http://biz.yahoo.com/p/g/gnlk.ob.html, "GeneLink Inc. is a bioscience company that offers the safe collection and preservation of a family's DNA material for later use by the family to identify and potentially prevent inherited diseases. GeneLink has created a new methodology for single nucleotide polymorphisms (SNPs)-based on genetic profiling. The Company plans to license these proprietary assessments to companies that manufacture or market to the nutraceutical, personal care, skin care and weight-loss industries. GeneLink's operations can be divided into two segments, the biosciences business, which includes three SNP-based, proprietary genetic indicator tests, and DNA Banking, which involves the use of its proprietary DNA Collection Kit." Contact GeneLink at: 100 S. Thurlow Street,Margate, NJ 08402. See the Web site at: http://biz.yahoo.com/p/g/gnlk.ob.html.

GeneLink also offers gene testing for skin care products. According to a March 6, 2003 media release posted at GeneLink's "In the News" Web page at: http://www.bankdna.com/news_articles/03_06_03.html, GeneLink, Inc. (OTCBB:GNLK) "entered into a collaborative agreement with DNAPrint Genomics, Inc. (OTCBB:DNAP) whereby the companies will combine certain scientific and intellectual property resources to develop and market "next generation" genetic profile tests to the $100 Billion plus personal care and cosmetics industry. Further details were not disclosed.

"GeneLink invented the first genetically-designed patentable DNA test for customized skin-care products, and DNAPrint brings its ultra-high throughput genotype capability and ADMIXMAP platform to the partnership. The companies anticipate screening millions of candidate markers to broaden proprietary product offerings.

"Tests are designed to assess genetic risks for certain skin and nutritional deficiencies and provide a basis for recommending formulations that have been specifically designed to compensate for these deficiencies."

I like this idea. I'm allergic to hair tint and aloe vera, and it would b wonderful to find skin care products that didn't turn my skin red and make it itch. Think of all the allergy warnings and patch tests on hair products. If everyone walked into a beauty salon with a list of products that can be safely put on the scalp or other skin without an adverse response, that would be my definition of comfort. Also, check out GeneLink's test which tells clients which mix of vitamins is appropriate for an individual's genes without getting an adverse reaction from the vitamin combination. I know certain supplements over-stimulate my thyroid. I'd love to have these kinds of tests.

How about you? Still, you need to do your homework on the usefulness of the tests or the claims. Perhaps you need to deal with a company in another country. Sciona Ltd. in England gives nutritional advice based on your DNA and a questionnaire related to your health. **The company sells tests only through doctors and dieticians.** How do you know which defect is associated with any specific gene when the gene is defective? With so many interpretations, consumers can become confused from test results. Sometimes scientists in various companies don't know which results are accurate predictions. That's why consumer watchdog groups are necessary.

The only problem is that the average consumer usually can't afford to hire genetics counselors unless they have a specific genetic disease. If you have a reduced ability to process something, it's not a disease, but unless you take action, your body could suffer in other ways resulting from the reduced ability to process an essential nutrient from your diet that your body needs to work with other processes. It's like a chain reaction.

That's why DNA testing is helpful. At the same time, you can't assume some people need more of one vitamin just because their body processes a nutrient at a reduced rate. The science is still very new. A lot of data is still in the works. You don't know whether certain genetic profiles need special diets. Consumers want to see rare clinical data. Consumers most of all want to know the value of what the DNA tests predict. And no one wants to take a test and then worry for a decade.

What consumers want are not only tests of prescription drugs and food as to whether adverse reactions happen with individuals based on their genetic testing, but personalized anti-aging formulas, cosmetics, creams, and other products customized to the individual's genetic markers are needed.

These tests must be reliable, consumers say. Feedback is helpful. Customer reviews of the companies are needed. Who will publish customer reviews of the products and the DNA testing firms? These consumer reviews could also include input from scientists and physicians. Physicians should take the DNA testing themselves after they are properly trained to interpret the tests before looking at their patient's results.

Nutritionists and dieticians show concern that medical schools only give very brief and shallow courses in nutrition to graduating physicians. What about a course in interpreting genetics tests? Will this be left to naturopaths and homeopaths? Or can medical schools include courses in DNA test interpretation to physicians other than genetic disease specialists? Healthcare consumers should ask for DNA testing from their HMOs before taking prescribed medicines. Will insurance companies pay for testing?

Look at the various DNA testing companies. Also look at health screening companies. One would be HealthcheckUSA, San Antonio, Texas for example, sells a mail-order test for iron buildup in the blood known as hereditary hemochromatosis. Results go to the consumer. If you need a test for cystic fibrosis or a blood clotting disorder known as factor V Leiden, check out this company. Their Web site is at: http://www.healthcheckusa.com/media.html, or you can write to them at: **HealthcheckUSA,** 8700 Crownhill Rd., Suite 110, San Antonio, TX 78209.

Practicing physicians started HealthCheck USA in 1987 to provide health awareness screening to customers throughout the USA. The tests provided to their customers are the same as those ordered by physicians across the country. Many physicians refer their patients to Healthcheck USA. Since 1987, HealthCheck USA has provided over 500,000 health awareness screening tests to satisfied customers nationwide.

The company is associated with the country's major fully accredited medical reference laboratories, to ensure quality, accurate test results. HealthcheckUSA has partnered with Virtual Medical Group, Inc., to offer physician interpretation of results. According to HealthcheckUSA's Web site, "You can have your results reviewed and interpreted by a Board Certified Physician. By selecting the "Physician Interpretation of Results" option, you will be mailed instructions with the hard copy of your results. These instructions will give you a toll free number along with your username and password to access your interpretation 72 hours after your blood draw. This process is completely confidential, private, and secure."

Here the consumer has the best of both worlds. What tests are provided? Check out the list of tests provided at the Web site: http://www.healthcheckusa.com/testsweoffer.php.

When any company does health screening, find out whether it is DNA testing, blood testing, comprehensive tests, or whether you are sent test kits such as for colon cancer screening, prothrombin (factor KII) DNA test, cystic fibrosis (DNA test), Factor V R2 (DNA test) or other. Are the results sent back to you directly as a consumer? Or do you purchase additional services such as a physician's interpretation of the results?

What consumers want are genetic tests for metabolism and health. In the past, genetic testing was used by doctors when the patients were concerned about specific diseases diagnosed by genetic testing. Today, the general consumer of healthcare and healthcare alternative medicine wants genetic testing for general healthcare in the absence of a specific disease. Consumers want to know how to handle the risks they might have inherited to delay or prevent what happened to their grandparents. Also parents of children and couples worried about fertility issues also want these types of tests. Genetic testing shouldn't be used to screen people out of insurance, employment, or anything else.

Health screening should be focused on matching people to the best possible foods, nutraceuticals, and medicines when and if needed. It should be an inclusive not an exclusive process. Like personnel departments that focus on screening out applicants, genetic testing should not be used to screen people out or to exclude based on genetic risks. Instead, it should be used to draw people into learning about taking responsibility for their own health habits such as choosing the right foods. It's all about choice, like the science of nutrition.

In the past, genetic tests were used on people who suffered from rare genetic disease. Consumers are worried about claims by companies that sell DNA tests to the public. Presently, consumers are asking the FDA to review tests before they are sold to the public or to physicians. The FDA reviews drugs and other medical appliances, why not DNA tests or other health screening tests sold to consumers? Should the government be reviewing these tests? What do you think? Or should the tests be monitored by the companies offering them? Or should the approach be to the tests similar to vitamins and minerals found in health food stores? What side will you take? Do you have enough information to even begin to take sides? Who runs and regulates the DNA testing companies that may or may not be headed by physicians and geneticists?

Testing DNA for ancestry would not have the same impact on someone's health as testing DNA for a specific disease marker. Still, the people doing the testing should have the same qualifications. What about the people doing the interpreting? What stance do you take? Then there's the question of privacy which is essential.

If you're testing for DNA ancestry, you might want to meet your DNA match to correspond with online. If you're testing for a disease marker, you want privacy. You may want to join a support group of families with similar DNA markers for the specific risk or disease you've inherited. You want meetings held in a hospital or some other private, medical setting. Personal issues come up.

Research the various associations of physicians, such as the American Academy of Family Physicians and other similar groups. You can find opinions there. You can bring in lawyers who want to make sure laws are put into effect to keep employers and insurers from using your results to terminate your job or kick you off of health or life insurance plans if your genes put you at risk.

Set up groups made up of advisories of experts that include consumer groups. What consumers with no science background can do is to work with the US Department of Health and Human Services to monitor various panels of experts. Some people want regulations to change. Others don't. There are hundreds of groups out there who advise the US Dept. of Health on what policies they 'should' adopt.

Consumers ask for reviews. Experts ask for reviews. Consumers don't want their vitamins taken away from health food stores. In the midst of it all are the wonderful DNA tests. Talk to pharmacists as well. If you're having adverse reactions to drugs, you eagerly want a DNA test you can take at home as a consumer. What I like about Genelex Corporation's drug metabolism test is that the test can tell you whether your body is taking too long to break down the drugs you're consuming. Consumers need that kind of calmness that comes from knowing personal physicians aren't prescribing too much of what you may or may not need. Maybe you're worrying about your pharmacist giving you the wrong medicine or dosage. Whatever you're concern, a home DNA test is a tool for consumers.

Monitoring your own health is your responsibility, not merely your doctor's. You can get DNA tests sent to your home or scans from clinics. Prescription drugs are available on the Internet. How your body will metabolize what you put into it is important. Are you becoming your own doctor? Is the average consumer becoming more knowledgeable? Or is it more accurate to say most people haven't an inkling of what DNA is?

Pharamcogenetics and nutrigenomics offer tools to consumers. The movement is towards taking charge of your own healthcare. Are doctors worried consumers are taking money away from them? Do tests create more problems than they solve? What happens when scans or other types of tests show non-existent problems?

Research what comes out of the National Institutes of Health. Consumers want to know what the results of their tests mean and how they can apply the information in practical ways. Who interprets the tests? Who is reviewing genetic tests sent directly to consumers and physicians? Right now, it's the consumer's 'job' to ask these questions. If you don't know what vitamins to take, perhaps you need a DNA test to tell you what vitamins work best with your individual genes.

DNA tests seek out mutations in the genes. If a mutation is only remotely associated with a risk, do you ignore it or customize your food and vitamins to the mutation that is only remotely connected to the risk? What action do you take as a consumer? You talk to a physician who knows about genes and your particular risk or you go to a genetics counselor. If you have certain forms of a gene that results in cancer of a certain area of your body by a certain age, you take action right away. Some mutations are associated with disease and some are not, such as certain ancestry markers.

Some genes account for only a tiny percentage of certain cancers. Not everyone needs to be tested. If you have a family reason to be tested, then get tested. If one of your siblings or parents has an inheritable disease, get tested. You can check out Myriad Genetics based in Utah regarding their breast and ovarian cancer tests, available only through doctors. Their Web site is at: http://www.myriad.com/. Or contact them at: Myriad Genetic Laboratories, Inc. | 320 Wakara Way, SLC UT 84108-1214.

The Controversy over Whether Genetic Tests should be Sold to the Public

Professionals working in molecular genetics may be divided among those who think genetic tests should be sold to the general consumer and those who think they should not. The professionals who think genetic tests should not be sold to the public may underestimate the intelligence of the public's drive to learn as much as possible about their own genes—the DNA, ancestry, health response to medicine and chemicals in their environment, their response to prescription and over-the-counter drugs, and their response to food, nutraceuticals and other supplements.

Those who are against selling genetic tests to the public fear that genetic testing is way too complex for the average consumer to understand. The average consumer consists of the person who says "it's way over my head," or "my eyes are glazing over" when you mention anything scientific, and the individual who wants to learn as much as possible about his or her own DNA, health, ancestry, and genetic response to food. Most consumers only need to be told that more than a hundred mutations can cause a specific disorder or disease. Average consumers can understand that. Those who underestimate the intelligence and desire to learn of an individual about his or her own health or the health of children and parents usually are the ones who are first to speak out in censorship of a consumer's right to learn all that is possible to learn at the moment about his own genes.

When a consumer is told that negative results don't always signal health, he or she is intelligent to understand that plain English or any language statement. It's a learning process. Some genetics companies are in direct marketing. Some test DNA only for ancestry. Other genetics companies test for genetic responses to drugs or for various genetic health risks that can be helped by changes in diet. The aim of these companies is cooperation. That means cooperation between patient, physician, genetics counselor, dietician or nutritionist and anyone else involved in the healthcare team working with the patient.

In fact in genetics testing, patients are really clients seeking information, feedback, and recommendations for changes in lifestyle and diet. In drug response genetic testing, the client wants to know how his or her body metabolizes the prescription or over-the-counter medicines used. The whole idea of genetic testing is to bring the healthcare team closer to the healthcare consumer. Instead of a one-size-fits all attitude with drugs, foods, or nutraceuticals, or the usual five to fifteen minute consultation with a doctor, the consumer can have the chance to learn everything publicly available about how his or her genes respond to food, drugs, chemicals in the environment, or lifestyle changes.

If a genetics testing company markets cancer tests, it should have the responsibility to let the patient know that the tests are meant for specific people with specific familial cancers and not for the entire consumer population. When a genetics company takes a family history, it opens doors to the individual not only to explore a familial inherited disease or risk, but the entire history and ancestry of the person's family and the DNA of the family members.

Individuals need to learn that there could be more than one gene involved. So teaching consumers about their DNA and genes is important. Classes could be springing up to train people what to expect before they undergo testing. It's a matter of learning and teaching. This is one more way to make use of retired professionals and other scientists and physicians or genetics counselors working in molecular genetics—helping consumers learn what they can expect from testing.

Consumers need to learn about how important it is to sequence certain genes from many family members. Consumers need to get in touch with affected family members and form a support group for DNA testing, perhaps creating a time capsule for future generations with genetic information resulting from testing.

Most consumers go to their family doctors first. Yet how many family physicians are interested in, let alone trained in molecular genetics? Not so many. Most physicians may not interpret a genetic test correctly because they haven't been trained to do so. In may be that the consumer is the first person to request training be set up to show physicians how to interpret genetic tests, especially for such diseases as cancer. Consumers have more power as a group. So consumers need to become involved, work together as an organization, and make sure not only the physicians are trained, but also themselves as consumers of genetic testing. Perhaps the physician and consumer can learn together in groups set up for both, where at a point, physician and consumer meet together. Will this present the physician differently in the eyes of the patient? Not really. The science is new enough so that consumer and patient could learn together to unravel the mysteries of genetics not unlike checking the clues in a mystery novel.

Consumers need to read more medical journals such as the New England Journal of Medicine. Spend some time in the library of your local medical school reading about how physicians interpret genetic testing. Look at surveys. As of now, not too many general consumers from the public at large even know what DNA is. So you have to educate yourself, perhaps as a new hobby. First find out what DNA is. Look it up in the dictionary.

Then start reading magazine articles about genetic testing. Look on the Web at the testing companies. Then you can graduate to reading articles in medical journals. The gap between the physician and the consumer needs to be narrowed at least when it comes to your own DNA and genetic testing for the things consumers put into their body such as food and prescribed or over-the-counter medicines, supplements such as herbs, and other nutraceuticals.

Everyone talks about the consumer's consent to testing. The problem is that the consumer needs to be given information. If you're going to eat 'smarter' you have to become informed first. If you have an inheritable risk or disease, you need to understand everything you can find that is publicly accessible about the risk, the disease, or the genes involved. Material is online, but is it credible? There are always the articles in medical journals, but can you understand the terminology? If not, look up the terms in glossaries and dictionaries. Make a list of terms and learn how to make the journal articles understandable in plain language.

Think about risk for a moment. Your risk could change. How are you going to receive the information from genetic testing? What are you going to do about it as far as changing your lifestyle, diet, or medicines? If you are concerned how your family might react to genetic testing, ask them whether they would change their food choices or other lifestyle changes based on the information. You can always keep the information private, but it could save the lives of family members to get tested and to change what can be changed, such as food choices or exercise routines.

Most consumers worry more about discrimination in employment and health insurance due to companies finding out their genetic risks. Privacy must be kept, and information only given to the consumer of the testing, not his employer or health insurance company. It's nobody's business but your own as far as your genetic information. So, measures to keep out strangers from using your genes against you must be in order. That's why numbers instead of names work better with testing.

The big issue is accuracy. If so many physicians cannot correctly interpret a genetic test and advise their patients, more training is needed for the physicians and the patients. It's a case of whose watching the watchers. Who is going to review the genetic testing itself as to accuracy? What if the tests are not accurate? Right now nobody is reviewing the testing companies other than themselves. Nobody is reporting directly to the consumer after checking the accuracy of the genetic tests. And nobody is reporting directly to the physician with that same information.

If you're going to get tested, at least learn as much as most professional genetic counselors know about interpreting your tests. It can be done without going back to school for a graduate degree in molecular genetics. The information is available to the public from libraries, the Internet, databases, and medical school libraries, journals, and professional articles. Many are on the Web in PDF files. Subscribe to the journals or read them in libraries and learn how to interpret your own tests. Then form a consumer's group to watch the watchers. Learn how to check the accuracy of tests done by the testing companies.

You have to become involved in your own healthcare and nutrition. Take charge and take responsibility. Consumers can't be passive. You won't know how your body responds to food or medicine or lifestyle changes until symptoms appear. Start by looking at the health of your family members and realizing what is inheritable, and what you can do to help yourself if you've inherited what they have or will get.

The first step for consumers is to start learning about DNA and genetic testing just as they have learned about genealogy and family history. Classes online or in adult education can be set up as well as support groups and organizations of consumers.

Scientists and Physicians Comment on Pharmacogenetics

Why would the average consumer of health care want to have drug reaction testing, known as pharmacogenetic testing? The field is new, and still emerging. According to (reprinted with permission) Genelex's "Health and DNA" Web site (copyright 2003) at: http://www.healthanddna.com/, "The relatively new and emerging fields of pharmacogenetics studies how differences in individual genetic makeup affect the processing of drugs. We have known for about a half-century that individuals respond to the same drug and dosage in very different ways because of genetic variations called polymorphisms. Research shows that of all the clinical factors such as age, sex, weight, general health and liver function that alter a patient's response to drugs, genetic factors are the most important. Genelex is the first firm in the world to offer genetic drug reaction testing directly to the public." What I like about Genelex is that they also offer DNA testing for nutritional genetics, drug reaction testing, ancestry DNA testing, and DNA identity testing.

In the past, DNA traditionally had been used to identify people, mostly in forensic or paternity or relationship cases. Then a few years ago, companies began to test DNA for ancestry, which appeals also to genealogists, family historians, and oral historians. The use of DNA outside of court rooms and forensic laboratories and outside of government and university databases brought DNA to consumers of health care as well as to genealogists interested in molecular genealogy. University laboratories and archaeology research institutes began using the science of archaeogenetics. Anthropologists who looked at HLA markers, those leukocytes or white blood cells and physicians or scientists who were interested in tissue typing for blood, marrow, and organ transplant donors worked with DNA to identify similar matches, people whose tissues and blood or marrow typing matched close enough for a transplant to take.

You had scientists who studied ancient DNA from fossils and mummies expand the science of archaeogenetics and population geneticists studying migrations of ancient peoples and the routes they took. When ancient and present day comparisons of genetic markers left trails that matched with archaeological relics, new branches of molecular genetics grew. The science is still emerging, using DNA testing in more ways. Now, we have nutrition and genetics and drug testing and genetics. And there's so much more to arrive that will be available to the general consumer of healthcare, ancestry searching, and identity. From being a spectator in life watching new avenues of DNA testing unfold, the consumer now has the chance to become more involved in participating in and learning about how many ways DNA testing can benefit an individual at any age.

When I went back to Genelex's Web site to explore more material on DNA testing for a variety of purposes, I realized, the consumer has more choices than ever before and still more on the horizon. DNA testing is all about nourishment. From nutrigenomics to drug reaction testing, strategies for better health is being covered from all angles. Perhaps you—as a consumer—want a healthier eating guideline customized to your genetic signature. Maybe you'd like a report on how your genes (and body) respond to certain chemicals, medicines, or substances. Whether it's a prescription drug or specific medicines or supplements that you buy over the counter, several genes are tested to see how your body metabolizes the substance. Not every gene in your body is tested to see how they respond to food or medicine—just specific genes and markers.

The science is new and changing, but the future is attractive to general consumers. You don't need a science background to begin your research on how your own body responds to what you put into it and what comes it from the environment.

To begin, consumers need to become more involved in learning about what scientists and physicians are researching and why, and how all these new findings apply to you. To become more involved as a consumer means being able to access information that evaluates the research, that finds credibility or flaws, and that helps you take more responsibility in your own maintenance.

That's why I believe DNA testing can open doors. Even with guarded caution, the benefits are to be explored and discussed. If you have already had your DNA tested for ancestry or identity, consider these new ways in which DNA testing can help you draw up a plan for eating and for any way you take care of your body—from exercise planning and nutrition to medicines and supplements if and when you need them, to making lifestyle changes for better health.

Consumers can learn a lot from news releases, including learning to understand where to begin to educate themselves about their genes. Some of these releases come from universities engaged in research, some from laboratories and genetic testing companies, some from the government, and from other scientific sources. Professionals in molecular genetics and in healthcare need to understand that an open door policy for public education, teaching oneself about DNA is good. To become more informed leads to becoming more involved in consumer's groups for understanding what "gene hunters" are doing.

Science has become so technical that most scientific journal articles are not understood by most consumers without science training. There has to be a midway point. So far, it's the media that translates scientific terminology into plain language for the consumer. Going one step farther, the medical journalist translates the scientific journals into articles and books whereby people with no science background can learn about their genes.

Finally, learning about how one's genes respond to food or drugs is another rung up on the ladder of self-education about how your body works. That's why there's the mass media for the consumer, the popular medical magazines that bring the consumer closer to healthcare professionals, and finally, the scientific and medical journal articles discussing the research. At the top are the evaluators who let the consumers know which research studies were flawed in the past, and that usually filters down through the mass media. What some scientists call 'snake oil' may either be harmful or may be a forgotten remedy based on plants that worked in the past. For example, honey, cinnamon, and sesame oil. All three will resist bacteria and in the ancient past were used on minor wounds. A century ago colloidal silver was used on scratches to keep out bacteria.

In the Civil War days, it was used on wounds. Today, you can still make your own, inexpensively, but be careful buying brands containing aluminum. In the dentist's chair, colloidal silver mouthwash works as well as some of the more recent remedies such as washing your mouth with harsh substances that can cause ulcers in the mouth. The point is to check out the mechanisms that review accuracy in whatever you read. If you go the homeopathic or naturopathic route, make sure what you use does no harm for your individual genetic response. How do you metabolize what goes into your body? Genetic testing can help here.

When it comes to genetic testing, you can learn a lot from news releases—at least as a starting point in your research and education about your genes. Bioscience communicators have a role to play as interpreters between consumer and scientist. When I took my master's degree in English with a writing emphasis, it was through a graduate scholarship award in science writing. In genetics, being the communicator means bridging the gap between the growing body of knowledge in science and the consumer who might not even know what DNA is. If you have no science knowledge, start by reading press releases to open the first set of doors to understanding more about your genes, markers, and DNA.

What Consumers Can Learn From News Releases

NIGMS Awards $35 Million To UCSD-Led Consortium To Map Metabolic Pathways In Cells

According to a UCSD media release of August 11, 2003, the University of California, San Diego (UCSD) will lead an ambitious national effort to produce a detailed understanding of the structure and function of lipids—cellular fats and oils implicated in a wide range of diseases, including heart disease, stroke, cancer, diabetes and Alzheimer's disease.

The five-year, $35 million grant from the National Institute of General Medical Sciences (NIGMS) will support more than 30 researchers at 18 universities, medical research institutes, and companies across the United States, who will work together in a detailed analysis of the structure and function of lipids. The principal investigator of this collaboration is Edward Dennis, Ph.D, professor of chemistry and biochemistry in UCSD's Division of Physical Sciences and UCSD's School of Medicine.

Dennis notes that while sequencing the human genome was a scientific landmark, it is just the first step in understanding the diverse array of systems and processes within and among cells. Establishment of this consortium is a significant step in an emerging field called "metabolomics," or the study of metabolites, chemical compounds that "turn on or off cellular responses to food, friend, or foe," he explained.

Lipids are a water-insoluble subset of metabolites central to the regulation and control of normal cellular function, and to disease. Stored as an energy reserve for the cell, lipids are vital components of the cell membrane, and are involved in communication within and between cells. For example, one class of lipids, the sterols, includes estrogen and testosterone.

The initial phases of the project, known as Lipid Metabolites And Pathways Strategy (LIPID MAPS), will be aimed at characterizing all of the lipid metabolites in one type of cell. The term "Lipidomics" is used to describe the study of lipids and their complex changes and interactions within cells. Because this task is too extensive for a single laboratory to complete, researchers at participating centers will each focus on isolating and characterizing all of the lipids in a single class.

This information will then be combined into a database at <http://www.lipidmaps.org> to identify networks of interactions amongst lipid metabolites and to make this information available to other researchers. Shankar Subramaniam, Ph.D., professor of chemistry and bioengineering at UCSD's Jacobs School of Engineering and San Diego Supercomputer Center, will coordinate this aspect of the project.

The cell type selected for study is the macrophage, best known for its role in immune reactions, for example scavenging bacteria and other invaders in the body. Macrophage cells from mice will be used, rather than human cells, because there exists a "library" of mouse cells with specific genetic mutations. By studying cells missing certain genes, the research team will attempt to identify what genes code for those enzymes key in synthesis and processing of lipid metabolites. Christopher Glass, M.D., Ph.D., professor of cellular and molecular medicine at UCSD's School of Medicine, will coordinate the macrophage biology and genomics aspects of the consortium.

According to Dennis, one of the most difficult aspects of this project will be to ensure that all the participating research sites are using identical methods.

"There has never been an effort to detect all the lipid metabolites in a cell and to quantify the amounts of these lipids," says Dennis. "In order to be able take the data gathered by each lab and put them together to develop new ideas about interactions between lipid metabolites, it is essential to develop new technologies and methods and to standardize them so that they can be applied in the same way at each research site."

Once the researchers have developed these methods and used them to identify and quantify the lipid metabolites in the mouse macrophage, the methods can be applied to gather other information about lipid metabolites in cells. The researchers plan to study the time and dose-dependent effects of lipopolysaccharide, or

LPS, a component of the cell walls of many bacteria, on macrophages. Dennis' lab has studied the effects of LPS on macrophages for the last 20 years. A world authority on these effects, Dennis says that this large-scale, cross-institutional collaboration will create an understanding of LPS' effects on lipid metabolism in unprecedented detail, and set the stage for examining other macrophage effectors such as oxidized LDL, the so-called bad cholesterol which leads to atherosclerosis.

A detailed understanding of lipid metabolism will be valuable in drug design. Most people are already familiar with one class of drugs that interfere with lipid metabolism: the non-steroidal anti-inflammatory drugs (NSAIDs), which include aspirin and ibuprofen. These drugs block the synthesis of prostaglandins, a large group of chemicals secreted by macrophages that causes pain, inflammation and an immunological response.

Statins, which a large number of Americans take daily to lower their cholesterol levels, also dramatically alter lipid metabolites. With a detailed knowledge of each step in the lipid metabolic pathways, more effective drugs with fewer side effects can be designed to combat heart disease and a plethora of other diseases of lipid metabolism.

In addition to UCSD, participants in the LIPID MAPS consortium are: Avanti Polar Lipids, Inc.; Duke University Medical School; Georgia Institute of Technology; University of California, Irvine; University of Colorado School of Medicine; University of Texas Southwestern Medical School; and Vanderbilt University Medical School. Additional collaborators will include scientists from Applied Biosystems; Boston University School of Medicine; Harvard Medical School; Medical College of Georgia; Medical University of South Carolina; National Jewish Medical and Research Center; Scripps Research Institute; University of Michigan Medical School; University of Utah; and Virginia Commonwealth University.

For more information on the grant program, see the related release at http://www.nigms.nih.gov/news/releases/.

DNA testing isn't only for anthropologists. It includes but goes beyond forensic research in identity. It expands from population genetics to cellular nourishment. It's for the general consumer with no science training, the physician, the genetics counselor, and the nutritionist. DNA testing is for the genealogist and the oral historian. And there are many more applications of DNA testing/genetic testing on the horizon.

According to Genelex's Web site (copyright 2003) at: http://www.healthanddna.com/pharmacogenetics.html, (reprinted here with permission):

"The relatively new and emerging fields of pharmacogenetics and pharmacogenomics study how differences in individual genetic makeup affect the processing of drugs. Pharmacogenetics largely focuses on specific genes, such as drug-metabolizing enzymes, while pharmacogenomics deals with the entire human genome, including genes for numerous proteins in the body, such as transporters, receptors, and the entire signaling networks that respond to drugs and move them through the system.

- We have known for about a half-century that individuals respond to the same drug and dosage in very different ways because of genetic variations called polymorphisms. The first studies of what came to be pharmacogenetics were conducted in the early 1950s and examined drug metabolizing enzyme variants such as those in the Cytochrome P450 family. Research shows that of all the clinical factors such as age, sex, weight, general health and liver function that alter a patient's response to drugs, genetic factors are the most important.

- Although the genomes of individuals are 99.9% identical, the 0.1% difference means we have as many as 3 million polymorphisms, the most common being the single nucleotide polymorphism (SNP). Detecting polymorphisms is the foundation of pharmacogenetics.

- Misdosing of drugs in America costs more than $100 billion dollars a year, and is a leading cause of death. Pharmacogenetics helps to explain why some individuals respond to drugs and others don't. It can also help doctors identify individuals needing higher or lower doses and who will respond to a drug therapeutically and who might have an adverse drug reaction.

- Currently, the clinical use of pharmacogenetics is limited, with the most extensive use in Scandinavia to treat psychiatric illnesses. The technology is, however, steadily moving into other geographic and medical areas, especially cardiology and cancer treatment.

- The Human Genome Project has been an incredible boon to pharmacogenomics. As new data are decoded and more precise maps produced, we can begin to understand what specific genes are and how they function. We can also better appreciate the significance of DNA variation among individuals.

Preventative medicine and pharmacogentics are the first disciplines to benefit from this vastly increased understanding of human diversity.

- Better understanding of the variation in the enzymes involved in drug metabolism and in the drug targets themselves will help explain why there is so much variation in drug response. We are now able to test for these variations before prescribing, and reduce the amount of human experimentation (and associated health care costs) needed to find the best therapy or dose. Drug companies can now develop drugs which are designed by choice, rather than happenstance, to be effective in wide range of the population.

- Predictive genetics is revolutionizing health care. DNA tests and personalized drug therapies are shifting the medical paradigm from detection and treatment to prediction and prevention and vastly increasing the efficacy, safety, and variety of therapeutic drugs. Predictive tests and preventative therapies reduce costs associated with hospitalization and lost productivity and will lead to a major redistribution of value in the health care industry."

According to the Genelex Web site at: http://www.healthanddna.com/drugreactiontest.html, (reprinted here with permission, copyright Genelex 2003.)

Drug Reaction Testing

Do not alter the dosage amount or schedule of any drug you are taking without first consulting your doctor or pharmacists.

"Research shows that of all the clinical factors such as age, sex, weight, general health and liver function that alter a patient's response to drugs, genetic factors are the most important. This information becomes even more crucial when you consider the fact that adverse reactions to prescription drugs are killing about 106,000 Americans each year—roughly three times as many as are killed by automobiles. This makes prescription drugs the fourth leading killer in the U.S., after heart disease, cancer, and stroke.

We currently offer CYP2D6, CYP2C9, and CYP2C19 screens that can help your physician or druggist predict your particular response to more than a quarter of all prescription drugs. These include such important medications as Coumadin (Warfarin), Prozac, Zoloft, Paxil, Effexor, Hydrocodone, Amitriptyline, Claritin, Cyclobenzaprine, Haldol, Metoprolol, Rythmol, Tagamet, Tamoxifen, Valium,

Carisoprodol, Diazepam, Dilantin, Premarin, and Prevacid (and the over-the-counter drugs, Allegra, Dytuss and Tusstat). Click here to view a more complete list of drugs processed through these pathways.

Approximately half of all Americans have genetic defects that affect how they process these drugs. There are four different types of metabolizers, and we all fall into one of these categories for the variable pathways in Cytochrome P450 (this Cytochrome is responsible for creating the enzymes that process chemicals of all kinds through our bodies.) The easiest way to understand this is to picture a two lane highway.

- If you are the first type which is the norm, you would be an EXTENSIVE metabolizer. Both lanes of the highway are open and moving. Medications prescribed in normal doses will be metabolized by your body.

- If you are the second type, you would be an INTERMEDIATE metabolizer. This means that one lane of that highway is open and moving and the other lane is not, causing you to metabolize the medications more slowly. In this case you will need a lower dosage, and there is a chance of medications building up in your system causing adverse effects. It is especially important to monitor medications if you are in this category.

Intermediate metabolizers through the 2C9 pathway, for instance, have an increased risk of bleeding incidences when taking the common blood thinner Coumadin or Warfarin. For this reason, a recent article in the Journal of the American Medical Association, titled "*Association Between CYP2C9 Genetic Variants and Anticoagulation-Related Outcomes During Wayfarin Therapy,*" JAMA, April 3, 2002—Vol. 287, No. 13, recommends screening for CYP2C9 variants to reduce the risk of adverse drug reactions in these patients.

- The third type is a POOR metabolizer. In this case both lanes of the highway would be stopped. There is a possiblility that alternate routes can be found, but this type of metabolizer is potentially very dangerous, as there is a great chance for the medication to build up in your system making you very sick, or even killing you.

For example, a poor metabolizer of Phenytoin, a common antiepileptic would not be able to process the drug and would actually have an increased rather than decreased risk of seizure if prescribed this drug.

- The fourth type of metabolizer is ULTRA EXTENSIVE. This means you have additional lanes for processing, picture an Indy 500 speedway. In this instance, you literally burn through medications. If you were an Ultra extensive metabolizer through the 2D6 pathway and while in surgery and your doctor gave you codeine as a pain killer, you would receive no pain relief because the codeine would be metabolized so fast that it would have little or no effect on you.

The Testing Process

The process is simple. We send you a blood collection kit in the mail. You can either make an appointment with your doctor or we will provide you with the contact information for a phlebotomist in your area. Blood samples are overnighted to our laboratory and results are typically available in 15 business days.

Currently Available Tests

CYP2D6 (cytochrome P450 2D6) is the best studied of the DMEs and acts on one-fourth of all prescription drugs, including the selective serotonin reuptake inhibitors (SSRI), tricyclic antidepressants (TCA), betablockers such as Inderal and the Type 1A antiarrhythmics. Approximately 10% of the population has a slow acting form of this enzyme and 7% a super-fast acting form. Thirty-five percent are carriers of a non-functional 2D6 allele, especially elevating the risk of ADRs when these individuals are taking multiple drugs.

Drugs that CYP2D6 metabolizes include Prozac, Zoloft, Paxil, Effexor, Hydrocodone, Amitriptyline, Claritin, Cyclobenzaprine, Haldol, Metoprolol, Rythmol, Tagamet, Tamoxifen, and the over-the-counter diphenhydramine drugs, Allegra, Dytuss, and Tusstat. CYP2D6 is responsible for activating the pro-drug codeine into its active form and the drug is therefore inactive in CYP2D6 slow metabolizers.

CYP2C9 (cytochrome P450 2C9) is the primary route of metabolism for Coumadin (Warfarin). Approximately 10% of the population are carriers of at least one allele for the slow-metabolizing form of CYP2C9 and may be treatable with 50% of the dose at which normal metabolizers are treated. Other drugs metabolized by CYP2C9 include Amaryl, Isoniazid, Sulfa, Ibuprofen, Amitriptyline, Dilantin, Hyzaar, THC (tetrahydrocannabinol), Naproxen, and Viagra.

CYP2C19 (cytochrome P450 2C19) is associated with the metabolism of Carisoprodol, Diazepam, Dilantin, Premarin, and Prevacid.

Other tests we are planning to provide include:
NAT2 (N-acetyltransferase 2) is a second-step DME that acts on Isoniazid, Procainamide, and Azulfidine. The frequency of the NAT2 "slow acetylator" in various worldwide populations ranges from 10% to more than 90%.

The advantages of Genelex's consumer genetic testing include:

- Safety. Decrease the chance that you will be the victim of an adverse drug reaction. More than 100,000 hospitalized Americans die of adverse drug reactions and two million outpatients have serious episodes each year. Knowledge of your DNA Drug Reaction Profile may help your physician or druggist prevent this from happening.

- Efficacy. Prescription drugs on the market today are usually prescribed by trial and error because they have been tested and approved in a "one size fits all" fashion. If you eliminate a particular drug more rapidly than the norm, taking the normal dose may be a complete waste of time. The drug simply will not work as prescribed, and it may take a while to discover this.

- Responsibility. Play a more active role in your health, your family's health, and in healthcare at large. DNA testing for drug reactions is just coming onto the market and you, the consumer, can help all of us take this major step toward better medicine."

Check out the common drugs processed by enzymes that Genelex tests. Genelex currently offers "DNA Prescription Drug Reaction Profiles that test 2D6, 2C9, and 2c19 functionality. The table below lists many commonly prescribed drugs that are metabolized through these pathways in the Cytochrome P-450 system. If your medication is not on the list, ask your pharmacist or physician about the metabolic pathway."

Common drugs prescribed by physicians or bought over-the-counter are processed by enzymes. Look at Genelex's Drug List by clicking on the PDF file to see Genelex's Drug list. See below, reprinted with permission, copyright Genelex, 2003. Or go to their Web site at: http://www.healthanddna.com/drugreactiontest.html and click on the PDF file to view the list of drugs at: http://www.healthanddna.com/drugchart.PDF. What you're looking for is prescription drug reaction testing. Your body metabolizes drugs differently than another person based on your individual genetic response. So look at the drug metabolization guide. Is your prescription drug listed there? What about drugs or supplements you buy over the counter? The typeset formatting is different on the PDF file. So look at the list below at: http://www.healthanddna.com/drugchart.PDF.

On the PDF file, you'll note that the formatting for numbers such as 2D6 is different in the PDF file than reprinted here due to formatting differences between a paperback book and the PDF file on the Web. So to get the correct formatting of the numbers such as 2D6 lined up with the correct drugs, use the PDF file to do your looking rather than the book, which is just a list of drugs tested. The PDF file has the correct numbers lined up with the correct names of the drugs.

Below is just a list of the drugs to give you an idea of what drug reaction testing is about and the names of the drugs listed. So use Genelex's PDF file, not the book if you want to see what numbers are lined up with what drugs. Below is to give you an idea of how a drug metabolization guide may be of benefit to a consumer when talking to your physician. The point is for you to learn how to become more involved in researching how your genes respond to what you put in your body.

Drug List

Prescription Drug Reaction Testing
Drug Metabolization Guide
(reprinted here with permission, copyright Genelex 2003.)

Substrates Metabolized through Cytochrome P-450

2D6
acetaminophen citalopram galanthamine norfluoxetine saquinavir
ajmaline clomipramine gallopamil nortriptyline sertinodole
alprenolol clozapine guanoxan octopamine setraline
amiflamine cobenzoline halofantrine olanzapine sildenafil
aminopyrine codeine haloperidol omeprazole S-metoprolol
amiodarone debrisoquine ibogaine ondansetron sparteine
amitriptyline delavirdine iloperidone opipramol tamoxifen
amphetamine deprenyl imipramine orto=Phebylphenol tamsulosin

amprenavir desipramine indinavir oxedrine tauromustine
aprinidine desmethylcitalopramindoramin oxycodone terfenadine
astemizole dexfenfluramine lidocaine parathion terguride
azelastine dextromethorphan lisuride paroxetine theophylline
benzydamine dextrromethorphan loratidine perhexiline thioridazine
bepranolol diazonin maprotline perphenazine timolol
bisoprolol diclofenac mequitazine phenacetin tolterodine
brofaramine diethyldithocarbamate methyl ester
methadone phenformin tramadol bromoperidol dihydrocodeine
methlyamphetamine pranidipine trifluperidol
buflomedil diltiazem methoxyamphetamine pregnenolone trimipramine
bufuralol diprafenone methoxyphenamine procainamide tropisetron
bunitrolol dolasetron metoprolol progesterone tryptamine
bupivacaine donepezil mexiletine promethazine tyramine
buthylamphetamine doxepin mianserin propafenone venlafaxine
captopril encainide milameline propranolol zolpidem
carteolol estradiol milameline propylajamaline zotepine
carvedilol estradiol minaprine prprofol zuclopenthixol
chlorophyros ethylmorphine mirtazapine quanoxan
chloropromazine ezlopitant nelfinavir mesylate quetiapine
chlorpheniramine flecainide nevirapine ranitidine
chlorzoxazone flunarizine nicardipine remoxipride
cilostazol fluoxetine nicotine retinol
cinnarizine fluperlapine nifedipine risperidone
fluphenazine norcodeine ritonavir
fluvastatin ropivacaine
fluvoxamine

Substrates Metabolized through Cytochrome P-450 2C19 2C9 ctd' 2C9 ctd'

amitriptyline aceclofenac tropisetron
citalopram acenocoumarol valproic acid
clominpramine acetaminophen venlafaxine
cyclophosphamide
alosetron verapamil
diazepam aminopyrine warfarin
imipramine amitriptyline zafirlukast
lansoprazole amitriptyline zidovudine

nelfinavir amprenavir zileuton
omeprazole antipyrine zolpidem
pantoprozole arachidonic acid
phenytoin artelinic acid
rabeprazole benzopyrene
bufuralol
butadiene monoxide
candesartan
carvedilol
celecoxib
chlorpyrifos
cinnarizine
cisapride
cloazapine
cyclophosphamide
dapsone
deprenyl
desmethyladinazolam
desogestrel
dextromethorphan
diallyl disulfide
diazepam
dibenzoanthracene
dibenzylfluorescein

Inhibitors Metabolized through Cytochrome P-450 2C19 2C9 2D6
cimetidine amiodarone amiodarone
flebamate fluconazole celecoxib
fluoxetine fluvastatin chlorpheniramine
fluvoxamine fluvoxamine cimetidine
ketoconazole
isoniazid clomipramine
lansoprazole lovastatin cocaine
omeprazole paroxetine doxorubicin?
paroxetine phenylbutazone fluoxetine
ticlopidine probenicid? halofantrine

sertraline levomepromazine
sulfaphenazole methadone
teniposide mibefradil
trimethoprim moclobemide
zafirlukast paroxetine
quinidine
ranitidine
red-haloperidol
ritonavir
terbinafine
chlorpheniramine
cimetidine
clomipramine
cocaine
doxorubicin?
fluoxetine
halofantrine
levomepromazine
methadone
mibefradil
moclobemide
paroxetine
quinidine
ranitidine
red-haloperidol
ritonavir
terbinafine

Inducers Metabolized through Cytochrome P-450
2C9 2C19 2D6
rifampin carbamazepine dexamethasone
secobarbital norethindrone rifampin?
rifampin

For immediate consultation Call 800-TEST-DNA (800-837-8362)
Hours 7:00 AM to 6:00 PM PST, 10:00 AM to 9:00 PM EST, fax 206-219-4000,

3000 First Avenue, Suite One, Seattle, WA 98121
E-mail: info@genelex.com Web: www.genelex.com
Se habla español. ©2003 Genelex Corporation

16

Effects

-----Original Message-----
From: Prakash Nadkarni
To: Anne Hart
Sent: Friday, August 15, 2003 9:19 AM
Subject: RE: query by writer

Hi Anne,

The world's biggest library, of course, is the Web (if you use a good tool like Google), and of course, you can walk to your nearest medical school if you need technical information-I know they shouldn't prevent use you from using the reference section unless they happen to be total A-holes (some of the Ivy league libraries are like that).

Re: interviews—E-mail is not a good medium for "interviews", though instant messaging probably is (I'm not sure how many scientists use it, though-I avoid it myself because I don't like to be interrupted while I'm in the middle of something.)

How it's going to help the average consumer—I don't know. My attitude is guarded optimism.

The cost of developing a new drug is frightfully high. If a new drug turns out to have an adverse effect that might result in its being pulled from the market, then, **if** it could be shown that this adverse effect has a clear-cut pharmacogenetic basis AND there is a very simple test that could detect susceptibility, then the drug could still be on the market.

-The flip side here: the really scary adverse effects are not all that common, they may occur in 1 of a 1000 individuals or less. Then you require a very large number of subjects to PROVE a pharmacogenetic basis for that adverse event. This costs a lot of money, and the FDA may not wait that long to pull the drug from the market if safer alternatives already exist. (There may be gray areas, though—troglitazone was one of the first of a new family of antidiabetic agents—the thiazolidinediones, or PPAR-gamma activators—that work by a unique mechanism.

—It was pulled from the market because of liver toxicity, but when this happened, many patients whose lives had been dramatically improved by the drug actually complained to the FDA that they were perfectly happy to continue taking it. Fortunately, newer and safer drugs from the same chemical family were developed, such as rosiglitazone and pioglitazone.

However, we don't know as yet whether there is a clearly and easily characterizable pharmacogenetic mechanism for thiazolidinediones-induced liver toxicity, so I'm really citing this as an example of a valuable drug that deserves and could get a second chance.

Similarly, if a genetic test indicates that a particular individual is potentially vulnerable to a dangerous adverse effect (for a potentially life-saving drug (e.g., an anticancer agent) which has dangerous side effects), then one could avoid this drug and seek an alternative drug to treat the same condition.

Re: Pharmacogenetics and diet, the lactose-intolerance of Han Chinese (as well as their greatly increased susceptibility to alcohol hangover with much more modest quantities of liquor) is well known, as is the fact that African-Americans require a larger quantity of eye drops to dilate the pupil prior to a retinal exam compared to people with light-colored eyes.

The complications here—it's not just a single gene for susceptibility, it's multiple genes (not to mention interaction of genes with environmental factors) that makes research in the field so damned difficult. Even when you have two people with the same susceptibility gene, the response to the same drug may vary tenfold,-there are numerous other genes that we haven't considered in the study that are responsible for the variation.

Genes and environment-repeated exposure to certain environmental agents may induce the expression of genes that result in greater tolerance to the agent. The ability of Russians to tolerate an astonishing amount of vodka (that they've been consuming since very young) is a case in point.

Also genes and age: study after study has shown that people who are susceptible to substance abuse (tobacco, booze etc.) are much more likely to get hooked if they're exposed to the substance during or before adolescence, which is why you have the laws about selling this stuff to minors. (We haven't studied brain plasticity enough—we don't know what's magical about the adolescence cut-off, but we do know that people who learn a foreign language by being immersed in that culture can learn speak it without an accent if exposed before adolescence, but not after.)

Hope this helps, and good luck with your book.

Regards,

Prakash
———

Prakash Nadkarni, MD
Associate Professor
Yale Center for Medical Informatics
New Haven, CT

To: 'Anne Hart'
Sent: Friday, August 15, 2003 12:07 PM
Subject: Re: query by journalist

Dear Ms. Hart:

Thank you for contacting the National Institute of General Medical Sciences, a component of the National Institutes of Health. NIGMS supports research in pharmacology, and more specifically, in pharmacogenetics. I know that you also contacted Dr. Rochelle Long at NIGMS, so I am responding on her behalf as well. NIGMS supports a network of pharmacogenetic researchers, listed at the following URL: http://www.nigms.nih.gov/pharmacogenetics/research_network.html. You could take a look at their research project summaries to get a sense for the kind of research that is going on. Also, you could try a Google News search on pharmacogenetics and/or pharmacogenomics to find published interviews/stories for context for your book.

Regarding consumer information, NIGMS has published two educational brochures on the topic. Online versions are posted at:
http://www.nigms.nih.gov/funding/medforyou.html
and
http://www.nigms.nih.gov/news/science_ed/genepop/

If you would like hard copies of either or both of these borchures, send an email request to pub_info@nigms.nih.gov.

Regarding the availability of genetic tests:
There are at least 2 pharmacogenetics tests that are widely available and offered by certified commercial clinical testing laboratories: TPMT and CYP2D6. There is no current standard for the widespread use of pharmacogenetics testing—yet. It is performed either as a difficult case referral to a tertiary care set-

ting (e.g. Mayo Clinic), or due to patient interest in seeking it out (e.g. based on a web advertisement).

Physicians are divided on when adverse side effects that occur in rare patients should be dealt with: after they occur, or anticipated and prevented. Sometimes these events are uncomfortable (e.g. lack of pain control after taking codeine; once caught, a morphine based drug like vicodin can be used instead); sometimes these events are life-threatening (e.g. myelosuppression after an anticancer drug; once recognized, a lesser dose of the agent can be used).

The most modern research scientific approach is to tailor the drug choice and drug dose to the patient up front. But there are instances where this just simply isn't the prevailing standard of care, either for economic reasons, or for reasons of time, or for valid reasons (e.g. PTT testing in hospital after warfarin is performed, rather than pre-prescriptive genotyping makes sense, since genotype isn't the only influence on clotting time; diet matters too).

If you are interested in researching the industry angle:

Meeting report from last year's meeting of the NIH Pharmacogenetics Research Network (see especially the second keynote by Rebecca Eisenberg):
http://www.nigms.nih.gov/pharmacogenetics/2002meeting_report.html

Members of the NIH Pharmacogenetics Research Network Industry Liaison Group:
http://www.nigms.nih.gov/pharmacogenetics/
PRN_industry_liaison_roster.html

These folks, from small and big pharma and biotech, may be able to give you the view from the trenches.

A publication called The Scientist published a relatively savvy article on the topic last year; find the story at: http://www.the-scientist.com/yr2002/sep/research1_020930.html

(you have to register but it's free access to the full text)

I hope some of this information is helpful to you.

Alison Davis
Office of Communications and Public Liaison
NIGMS. NIH

———-Original Message———
From: "Esteban Gonzalez Burchard, M.D."
To: "Anne Hart"
Sent: Thursday, August 14, 2003 9:33 PM

Subject: Re: Pharmacogenetics: What Can You Comment About Regarding Pharmacogenetics for the General Consumer?

Hi Anne:

I have attached a section that I wrote as part of a grant. You can use this as you please. As of today, there are not that many examples that are of use to the public. However there are examples developed at St. Jude Children's Hospital in Memphis in which they can test for mutations in the drug metabolizing enzyme that metabolizes specific chemotherapy drugs. In doing so, physicians can now identify who will be at risk for toxic levels of the chemotherapy drug.

The best publicly known example is alcohol and facial flushing in Asians. This is due to a genetic variant in the Alcohol Dehydrogenase Enzyme. This is a perfect example of a pharmacogenetic response.

Esteban

****

It is well established that among U.S. residents of similar socioeconomic status, there is greater cardiovascular morbidity and mortality among African Americans and Latino Americans than among Caucasians.[1-4] In addition, there is marked clinical heterogeneity in the prevalence and treatment of hypertension among specific ethnic and racial groups.

Specifically, there are marked differences in drug response to antihypertensive therapy with angiotensin converting enzyme inhibitors (ACE-I) among African Americans than among Caucasians.[5,6]. Although there are many potential explanations for this observation, including environmental and socioeconomic factors, one potential explanation is that the genetic predisposition to hypertension and to differences in drug response differs among racial and ethnic groups. In particular, we propose that genetic differences may explain, in part, the differences in response to ACE inhibitors between African American and Caucasian Americans.[7]

Hypertension among U.S. Populations: Cardiovascular disease is a substantial public health problem and a leading cause of death for all U.S. ethnic and racial minorities. The impact of premature morbidity from cardiovascular disease is devastating in terms of personal loss, pain, suffering, and effects on families and

loved ones. The annual national economic impact of cardiovascular disease is estimated at $259 billion as measured in health care expenditures, medications, and lost productivity due to disability and death. <http://raceandhealth.hhs.gov/3rdpgBlue/Cardio/3pgGoalsCardio.htm>
A major modifiable risk factor for cardiovascular disease is high blood pressure (hypertension).

Disparities exist in the prevalence of risk factors for cardiovascular disease. Racial and ethnic minorities have higher rates of hypertension, tend to develop hypertension at an earlier age, and are less likely to respond to conventional antihypertensive therapies to control their high blood pressure.[6] For example, from 1988 to 1994, 35 percent of black males ages 20 to 74 had hypertension compared with 25 percent of all men. When age differences are taken into account, Mexican-American men and women also have elevated blood pressure rates. Among adult women, the age-adjusted prevalence of overweight continues to be higher for black women (53 percent) and Mexican-American women (52 percent) than for Caucasian women (34 percent). Although there are multiple potential explanations for the observed differences in the prevalence of hypertension, possible explanation include biologic or genetic differences.

Clinical and racial heterogeneity in response to treatment of hypertension:

Numerous large-scale clinical trials of therapy for heart failure and hypertension over the past decade have shown improvements in outcome with angiotensin converting enzyme inhibitors (ACE-I).[8–13] Interestingly, data from the second Vasodilator-Heart Failure Trial (V-HeFT II) indicated that although enalapril therapy was associated with significant reduction in the risk of death from any cause among white patients, no such benefit was observed among black patients.[14] In addition, the Beta-Blocker Evaluation of Survival Trail found significant benefits with the beta blocker bucindolol among white patients but not black patients.[15] A reduced response to ACE inhibitors in black patients as compared to white patients is well documented.

Decreases in blood pressure in black patients with hypertension have been shown to be less than white patients given similar doses of ACE inhibitors [16] and beta blockers.[17] Taken together, these data may suggest that there are racial/ethnic differences in therapeutic response to the treatment of heart failure and hypertension among different racial/ethnic groups. More importantly, these trials have raised the possibility that black patients with heart failure and hypertension may not benefit from commonly used doses of recommended therapies to the same extent as white patients.

Genetic differences among races: Naturally occurring genetic variants or polymorphisms of genes are an important source of genetic diversity. Single nucleotide polymorphisms (SNPs) account for over 90% of all human DNA polymorphism.[18] It is well established that there are differences in allele frequencies among various ethnic groups.[19,20] By determining the degree of disease risk associated with particular alleles, in conjunction with the allele frequency for a defined population, one can derive information about the population attributable risk, which is relevant to the public health impact for that population.

Genetics of Hypertension: In favor of genetic factors being an important in the pathogenesis and treatment of hypertension are observations from studies in families, twins, and clinical pharmacologic trials performed in ethnically diverse groups.[6,21–28] Although there have been no systematic genome-wide screens performed across racial or ethnic populations, there have been individual investigations in African Americans, Mexican Americans, Chinese and Caucasians.[29–33] All of these investigations found linkage between loci and hypertension. Interestingly, most of these loci differed by the ethnic population studied, suggesting genetic heterogeneity. This may suggest that there are "ethnic specific" genetic risk factors that are associated with hypertension in different racial and/or ethnic groups with resulting differences in drug response to anti-hypertensive drugs.

Genetic sequence variants among hypertension candidate genes: There is increasing evidence to suggest that drug metabolism alone does not account for the observed interindividual variability in drug disposition or response, but that other processes, including drug transport, are important determinants of drug disposition.[34,35] Drug transporters in the gastrointestinal tract are a major determinant of oral bioavailability of drugs whereas transporters in the kidney and liver are major determinants of renal and hepatic clearances.

Transporters in specific tissues (e.g., the blood brain barrier) control tissue specific distribution of drugs. The pharmacokinetics of the ACE inhibitors have been reviewed recently.[36,37] It is important to note that the absorption of several

ACE inhibitors is low (approximately 40% of an oral dose is absorbed) and variable and appears to be controlled by the oligopeptide transporter, Pept1.[38]

Several ACE inhibitors are eliminated by active secretion in the kidney, and appear to interact with renal organic anion transporters (e.g., OAT1 and OAT3). Thus, genetic variation in transporters involved in absorption and elimination of ACE inhibitors will contribute to variation in drug response. It is possible that variants of these transporters present in different frequencies among various ethnic populations may explain (in part) ethnic differences in response to ACE inhibitors.

Summary and Significance: Currently available data indicate that cardiovascular morbidity and mortality disproportionately affect U.S. ethnic and racial minority groups. This proposal is the first phase of a systematic effort to identify biologic and genetic explanations for these differences in drug response to conventional antihypertensive therapies. We will recruit a local cohort of ethnically diverse subjects who will undergo phenotype analysis and genetic testing of candidate genes thought to be involved in the transport of anithypertensive drugs.

As part of the overall PMT project these genetic variants will be further tested with in vitro functional assays to determine the biologic relevancy of the identified genetic variants. As part of future studies, these subjects will be called back to participate in further pharmacokinetic testing. Subjects will be stratified by their respective genotypes and then undergo pharmacologic testing with conventional antihypertensive therapies.

These investigations will allow us to determine whether pharmacogenetic relationships exist. More importantly, these investigations will allow us to identify ethnic-specific differences in drug response by genotype. The overall aims of this proposal are to better understand the genetic factors that may contribute to the increased disparity from cardiovascular and hypertension morbidity and mortality seen among ethnically diverse populations. These aims are consistent with those established by Healthy People, which are to reduce disparities in health among different population groups.

Esteban González Burchard, M.D.
Pulmonary and Critical Care Medicine
Department of Medicine
University of California, San Francisco

Mailing Address:
 University of California, San Francisco
 Box 0833
 San Francisco, California 94143

Shipping Address:
 UCSF DNA Bank/Lung Biology Center
 San Francisco General Hospital
 Bldg 30, 5th Floor, Room 3501H
 1001 Potrero Avenue
 San Francisco, California 94110

From: Wein, Harrison (NIH/OD)
To: 'Anne Hart'
Sent: Friday, August 15, 2003 11:36 AM
Subject: RE: Thanks for replying

I just got the latest table of contents for Environmental Health Perspectives, and noticed that they feature pharmacogenomics in their new issue. There are a couple of articles in the issue. Here's the link: http://ehp.niehs.nih.gov/txg/docs/2003/111-11/toc.html.

Harrison

From: "Daniel Rubin"
To: "Anne Hart"
Subject: Re: query by writer
Date: Friday, August 15, 2003 5:15 PM

You can find educational information on our PharmGKB web site that should provide much of what you're looking for:
http://www.pharmgkb.org/do/serve?id=resources.education

Also, NIH as an educational brochure on pharmacogenetics for the public that may give you what you're looking for in terms of how this will benefit the public: http://www.nigms.nih.gov/funding/medforyou.html

I recall that Time magazine did a very nice piece in 2001 on pharmacogenomics directly addressing exactly what you're talking about, and that you should be able to find in a community public library.

Regards,

Daniel

UCSD Scientists Develop Novel Ways to Screen Molecules Using Conventional CDs and Compact Disk Players

Molecules attached to CDs in new techniques can screen for proteins. According to a UCSD August 18, 2003 press release, chemists at the University of California, San Diego have developed a novel method of detecting molecules with a conventional compact disk player that provides scientists with an inexpensive way to screen for molecular interactions and a potentially cheaper alternative to medical diagnostic tests.

A paper detailing their development will appear this week in an advance on-line edition of the journal Organic and Biomolecular Chemistry (http://xlink.rsc.org/?DOI=b306391G) and in the printed journal's September 21st issue.

"Our immediate goal is to use this new technology to solve basic scientific questions in the laboratory," says Michael Burkart, an assistant professor of chemistry and biochemistry at UCSD and a coauthor of the paper. "But our eventual hope is that there will be many other applications. Our intention is to make this new development as widely available as possible and to see where others take the technology."

"The CD is by far the most common media format in our society on which to store and read information," says La Clair. "It's portable, you can drop it on the floor and it doesn't break. It's easy to mass produce. And it's inexpensive."

Burkart and James La Clair, a visiting scholar in Burkart's laboratory who initially developed and patented the technique, said that since scientific laboratories often rely on laser light to detect molecules, it made sense to them to design a way to detect molecules using the most ubiquitous laser on the planet—the CD player.

Their technique takes advantage of the tendency for anything adhering to the CD surface to interfere with a laser's ability to read digital data burned onto the CD.

"We developed a method to identify biological interactions using traditional compact disk technology," explains La Clair, who provided the patent rights to the method to UCSD. "Using inkjet printing to attach molecules to the surface of a CD, we identified proteins adhering to these molecules by their interaction with the laser light when read by a CD player."

While usually anything, like a scratch on the CD surface, that would interfere with the detection of the bits of information encoded on a CD would be a drawback, the UCSD researchers actually exploited this error to detect molecules.

"That's the novelty of this," Burkart points out. "We are actually using the error to get our effect."

The typical CD consists of a layer of metal sandwiched between a layer of plastic and a protective lacquer coating. When a CD is burned, a laser creates pits in the metal layer. A CD player uses a laser to translate the series of pits and intervening smooth surface into the corresponding zeros and ones that make up the bits of digital information.

To do molecular screening, the researchers took a CD encoded with digital data, and enhanced the chemical reactivity of the plastic on the readable surface. They then added molecules they wanted to attach to this surface to the empty ink wells of an inkjet printer cartridge and used the printer to "print" the molecules onto the CD. This resulted in a CD with molecules bound to its readable surface in specific locations relative to the pits in the metal layer of the CD encoding the digital information. When the CD with these molecules attached is placed in a CD player, the laser detects a small error in the digital code relative to what is read from the CD without the molecules attached.

To detect proteins or other large molecules in a solution like a blood sample, the modified CD is allowed to react with the sample solution. Like a key that only fits in a certain lock, some proteins bind to specific target molecules. Thus, specific molecules on the surface of a CD can be used to "go fishing" for certain proteins in a sample. The attachment of these proteins will introduce further errors into the reading of the CD. Furthermore, since the molecules on the surface of the CD are at known locations relative to the bits of encoded information, the errors tell the researchers what molecules have attached to their target protein and, thus, whether or not that protein is present in the sample.

"James has even done this using CDs with music, like Beethoven's Fifth Symphony," says Burkart. "And you can actually hear the errors."

"How many people on this planet can actually hear a molecule attached to another molecule?" asks La Clair.

While a few bugs need to be ironed out before the technique can be used to accurately quantify the amount of a given protein in solution, Burkart plans to apply it immediately to help him screen for new compounds in his natural products chemistry research laboratory. Compared to the $100,000 price tag for a fluorescent protein chip reader, he points out, a CD player costs as little as $25.

The researchers envision many other potential applications for this technology outside the laboratory, particularly in the development of inexpensive medical diagnostic tests, now beyond the means of many people around the world, particularly in developing countries.

"In theory, anyone who has a computer with a CD drive could do medical tests in their own home," says La Clair.

The researchers hope that by openly publishing their development in the scientific literature, others will customize the technology in a variety of ways, eventually leading to a wide range of inexpensive new diagnostic kits and other beneficial products.

"We plan to make this fully available and see what people come up with," says Burkart.

For additional information, see: http://discode.ucsd.edu/

***

17

How Do Your Genes Respond?

What does ancestry have to do with how your genes respond to food or medicine? Certain ethnic groups or 'races' respond in different ways to different medicines, dosages, foods, exercise, illnesses, lifestyle changes, and stress. Since most people have a mixture of genes from several races or ethnic groups going back to prehistoric times, only a DNA test of specific genetic markers or genes can reveal what you're are at risk for regardless of your dominant ethnicity. The genes respond in specific ways to chemicals in the environment and other factors.

That's why you can interpret a report, talk it over with your physician, and tailor what you eat, what drugs you take and the dosages for your condition, to what your genetic expression requires to be healthy. It's looking at you at the cellular level, the molecular level. Most people don't even know what DNA is or have no science background, so they need an easy to understand consumer's guide to nutritional genomics and pharmacogenetics. Consumers want to know how their individual, specific genes and possibly their ancestry respond to food—selected foods tailored or prescribed to nourish your individual genetic profile known as your *genetic expression*. Consumer involvement in nutritional genomics is important as it is in pharmacogenetics. How many products need to be tailored to your genes? Food? Medicines? Cosmetics? What else? Let's look at the issues.

Nutritional genomics, often abbreviated as 'nutrigenomics' is about increasing that success rate. How will science working together with the consumer tackle the issues confronting us as the population ages? Consumer involvement can democratize the science of nutritional genomics by improving diets for better health. You can ask to work on ethics boards or create your own. How is discovering deep ancestry through DNA testing related to the ways that food affects your health?

Ancestry and diet are linked by biology, culture, and choices. It's all about collaborating with your genes. Do you choose your food by habit or biology? Consumers need a guide to DNA testing for nutritional genomics as well as for ancestry and family history. Specific genetic variants *interact* and relate to nutrition.

Learn to interpret the expression of your genes before you count your calories. If you're supposed to eat 'bright' for your 'genotype,' then you begin by mapping your genetic expression and learning how the raw data applies in a practical way to what you consume. This means genetic testing, interpretation, and application to food.

Are you having your DNA tested to see how your genetic signature responds to the medicines you take? What's your genetic response to food, medicine, exercise, or nutraceuticals? If you're concerned about adverse reactions, are you having your genes screened, particularly CYP2D6, CYP2C9, and CYP2C19—to see what your particular response is to prescription medicines? What about the food you eat? Can you tailor what you eat to your genotype?

Does Your DNA Have a Core Identity That's Both Cultural and Biological?

Not only can you trace your family history and ancestry with molecular genealogy or gene testing. You can create a DNA and genealogy time capsule. What are the relationships between your deep, ancient ancestry, your DNA, migrations, and how you customize what you eat to your genetic signature or how you tailor any medicine or supplements you take to specific genes and markers?

Are you interested in foods tailored to your genotype and have no science background? Curious about pharmacogenetics—exploring how your doctor tailors your medicine to your genes? Thinking about eating bright for your genotype? What about DNA and Ancestry? Or perhaps you would like to find out the definition of ***genomics, proteomics, metabolomics, or lipidomics***? How far can we take familiar genealogy and oral history techniques and link them to DNA testing for a variety of reasons from tailoring what you eat to your DNA to customizing what over-the-counter and prescription drugs your physician wants you to take? Before you take anything, ask your physician how your genes will respond to what you put into your body. What DNA and other tests can you take that will be responsible, credible, and appropriate measures for your present or future needs?

Is Genetic Testing Right if You are Unhappy with Your Doctor's Response to Questions?

Back in 1995, as a medical journalist, I felt unhappy with my doctor's response to my questions regarding different treatment methods for the symptoms of menopause. I was in my early fifties then, and before DNA testing was readily available to consumers, I decided to do my own research to find answers to questions. The commonly prescribed "one-size fits all" estrogen-progestin pills I received from my primary care physician at my HMO as I entered menopause was prescribed with the words, "If you go off your HRT, it would be like a diabetic going off insulin." I didn't quite believe those words, but that's what the doctor said when I asked about when I could go off the HRT. So I took my pills like he said, and within a few months, my cholesterol levels changed for the worse.

The pills worsened my genetic hypertension. Still, the doctor insisted I not stop taking the HRT. I went to another doctor who changed the routine. Instead of taking an estrogen and a progestin pill each day, he prescribed estrogen some days of the month and progestin other days of the month and warned that, "I'd better do something about my hypertension or I'll have to go on 'meds.'" And still, I felt worse on the estrogen and progestin the more months that past. After eleven months, I dumped the HRT and didn't go back to the HMO. It was time to do some research and become more involved in my own healthcare as a consumer.

At that time, around 1996, all those Web articles on the dark side of soy didn't yet come to my attention. The universities I visited were doing clinical trials with natural soy products for menopause, and magazine articles were telling me I had to have those 29 grams of soy each day for my bones and arteries. I didn't know who to believe. There were magazines warning of too much stimulation to the thyroid from too much soy isoflavones. I didn't know who to believe yet, but what I wanted was information.

Some nurses told me about natural yam-derived progesterone as an alternative to hormone replacement therapy (HRT). At first, not realizing the genetics of hypertension, I thought that the estrogen in HRT was the cause of my dangerously high blood pressure. Using the Internet to access newsgroups and WWW sites, I sought the alternative therapies and tried to find out whether they really worked. Was there anyone out there saying the studies had been flawed? And if so, how could I believe the critic?

What credentials or credibility did he or she have? And if there were no critics saying studies were flawed, perhaps no studies were ever performed? Who was right? So I turned to the medical journals on the shelves at the local Medical School library. I also turned to the Internet and to physicians online willing to answer my questions about these treatments. What I found were people who had the time to answer questions, but none were physicians.

Were there actual physicians online or available willing to answer questions? Who had the time and who had the credibility? Was it true that unproven holistic health and naturopathic or homeopathic remedies that worked a century ago came back now, and did they really work? Was it true that natural ingredients were better, but not used because the big drug companies could make a profit off of them? How much was snake oil and how much was plant-based cures? Who could I believe?

More important, what would work with my body, nasty genetic defects and all? And when my husband needed hernia surgery, how could the Internet answer questions? This was 1995, before genetic testing was available to consumers.

All I had to go by was the awful, scary reactions I had to dental anesthesia when having a compact wisdom tooth removed at age 26. The IV went in, and the next thing I started yelling I can't breathe. What were my genes telling me? Or the carbocaine for the root canal that caused my panic attack, tremors and convulsing, whereas the specific type of mepivacaine worked well and I remained calm because it didn't have vasoconstrictors to send my body into a panic response, unleashing catecholamine and other hormones that raise the blood pressure, heart rate, fear, tremors, and convulsing because of the special genetic response of my more or less defective autonomic nervous system to the anesthetic.

The popular archaeology magazine article clued me in that I might have a defect in my autonomic nervous system by reading a popular archaeology magazine article about ancient Egyptian mummies that mentioned an article in a medical journal. One eye pulled to the corner that shows up on photos or portraits, a withered left side of the face....It was all pointing to my panic disorder. Even an astrologer mentioned Neptune on the rising sign signaling my anxiety gene. I had inherited this gene from my father, who claimed he was too nervous to ever learn to drive. I never learned to drive either, and I passed the "anxiety genes" on to my son, the physician. What I was looking for was more than vague explanations of "anxiety genes." I had to pinpoint the response, that means in my situation, a hyperinsulinism response or insulin resistance, and to explore the possible need for what a decade later became known as "the Syndrome X diet." It all was genetic, or was it?

Of course, back in 1995 and 1996, I couldn't find a DNA test offered to consumers on the Internet. Instead I used the Web to search for alternative medicine and holistic health articles and supplements. Take a look at my article which appeared in **Internet World** 1996 February : 42–44, 46, 48. Reprinted with permission. It is also mentioned in archives online. See HEALTHNET NEWS VOL. XI, NO. 4 WINTER 1995, Lyman Maynard Stowe Library

University of Connecticut Health Center: http://library.uchc.edu/departm/hnet/winter95.html

Menopause and Beyond Alternative Resources And Information Online

Alternative Resources Online for Menopause and Beyond

Through the Internet, you can obtain information your doctor hasn't got time to give you, and talk to people who share your problems.

By Anne Hart

I have found that doctors don't always have the time or breadth of knowledge to discuss alternative, customized solutions to my family's healthcare needs. That's why I turned to the Internet. After searching the Net for Web sites, support groups, or little-known health newsletters, I found dozens of physicians who offered to give me answers to my specific questions via e-mail.

Each member of my family required a different healthcare service and support group to address their particular problems. In my case, I am allergic to synthetic progestin, which eases the side effects of menopause. At the health-maintenance organization (HMO) to which I am a member, I asked my doctor for natural, yam-derived progesterone. He refused to prescribe it because he said it wasn't yet approved by the FDA, which left me with no alternatives to prevent osteoporosis, for which I'm at high risk. Instead I was put on hormone-replacement therapy, which involves taking oral estrogen. I believed that the oral estrogen was raising my blood pressure sky high. When I mentioned this to my doctor's nurse, she snapped, "Prove it."

So I set off on my journey to prove it. First I checked the **alt.support.menopause** newsgroup, which was extremely helpful. I posted the following question: *Is anyone else using a high soy and vegetarian diet, the herb black cohosh, and natural, yam-derived, progesterone cream for menopause—to help prevent bone loss—instead of the usual estrogen and progestin? If so, what are your comments and experiences?*

The replies were practical, useful, and factual, providing medical references, titles of medical journal articles, and book bibliographies, as well as personal experiences and encouragement. For example, one person pointed me to an article entitled "Risks of Estrogens and Progestogens" in the December 1990 issue of *Maturitas*, an English-language European medical journal. The author, Dr. Marc L'Hermite, found that five to seven percent of women on conjugated equine estrogens could get severe high blood pressure and that they would return to normal when the hormone-replacement therapy was withdrawn. A bibliographic reference to this article also appeared in Dr. Lonnie Barbach's book *The Pause*. Not one physician at my HMO had mentioned these concerns.

To obtain more information about what alternative health solutions were available and how particular products would change my body or health, I searched the Web under the keywords "menopause," "alternative healthcare," "herbs," "homeopathy," and "naturopathy." I also looked under "natural progesterone."

Through this search I found the *MenoTimes*, a quarterly journal published in San Rafael, Calif. I subscribed because it had the information I sought on Dr. John Lee's book *Natural Progesterone: The Multiple Roles of a Remarkable Hormone*.

Also through my search I found a laboratory that would test my saliva and tell me whether my hormones were balanced. Most of all, I wanted to know how taking natural progesterone would affect my fast thyroid and adrenaline-drenched body, with its low blood sugar and excess insulin production. My HMO physicians did not answer these questions, but told me that going off conjugated equine estrogen and synthetic progestin was like a diabetic going off insulin. On the Net I found physicians who answered my letters, labs that sold natural progesterone cream that I could use to prevent bone loss after menopause, and other labs that could monitor my condition until I found a doctor in my community who would listen to alternative solutions to menopausal questions.

I even found the Menopause Matters page created by Susan Czernicka, who said she "learned about herbal treatment of menopause when there were too few resources available to help her with her many symptoms and too few medical providers with open minds." She will answer questions sent to her at susan270@world.std.com.

For those who don't have Web access, there is the Menopause mailing list. To subscribe, send e-mail to listserv@ psuhmc.hmc.psu.edu with **subscribe menopaus** *Your Name* in the message body.

The "black cohosh" that I mentioned in my letter to the **alt.support.menopause** newsgroup is an estrogenic herb and vasodilator, and I wanted to find out how safe it was and whether it was as good for menopause as the homeopaths and naturopaths claimed, as well as how much to use and what effects it would have. I found the **alt. folklore.herbs** newsgroup helpful, as well as articles by Anthony Brook entitled "Why Herbs?" and "Historic Uses of Herbs" at the Drum Holistic Herbs page.

I wanted to find out everything I could about natural progesterone, so I went to the Health and Science page at Polaris Network, http://www.polaris.net/~health/ which described itself as "A Guide to Understanding and Controlling PMS, Fertility, Menopause, and Osteoporosis." It contained information about natural progesterone and how it balances the side effects of unopposed estrogen, how it's required for proper thyroid function and progestin counterparts in the drug industry. It also offered a seemingly sound scientific and unbiased evaluation of how certain hormones affect the system, what the hormone's results are on various bodily functions, and side effects. And it had an excellent bibliography of books and medical journal articles on osteoporosis reversal using natural progesterone.

I wanted to query a physician about the high blood pressure resulting from the equine oral estrogen I was taking, and at the Atlanta Reproductive Health Centre page I was able to send e-mail to a doctor who answered my question quickly, providing information that I could consider when making my final decisions or in looking further. His answer was more to the point than the counseling I had received from my own physician.

Another doctor of mine had wanted to give me high blood pressure medicine on top of the estrogen and synthetic progestin. I asked him to consider the alternative—taking me off the hormones to see whether a low-salt diet and exercise could change things—because before I went on hormones my blood pressure wasn't high.

The Internet became one of my best alternative healthcare information resources after the last of three reproductive endocrinologists I saw (not affiliated with my HMO) told me to wait two months to see how I felt off the hormones.

Menopause is a mega-business. More than 100 books a year fill the store shelves on this subject and more than 3,500 baby boomers enter menopause daily. This captivated audience represents a huge market for makers of hormones, vitamins, and other specialty products, much of which is advertised on the Web. So a wealth of information has appeared on the Internet to meet the needs of the 38- to 55-year-olds undergoing menopause, as well as younger women with PMS, infertility, and contraception concerns.

Another resource I found informative was the Women's Health Hot Line newsletter. Its topics include infertility, endometriosis, contraception, sexually transmitted diseases, stress, menopause, and PMS. Most of all, it doesn't close its pages to alternative therapies for women who can't tolerate standard hormone replacement therapy.

Hernia Hunt

This year my husband needed hernia repair surgery, but his busy HMO surgeon spent only a very short time briefing him a month before his surgery. Because it would take hours to describe in detail what is done during hernia repair surgery, he searched on the topic using Web search engines, which yielded a list of information about all kinds of hernias. The most thorough site discovered was the Hernia Information Home Page in England. It included articles that explained such things as the benefits of using mesh rather than stitches to close incisions. Information about hiatal hernias and diaphragmatic hernias could be found at the Collaborative Hypertext of Radiology. There was more information about hernias on the Net than my husband could possibly find time to read.

After seeing shark cartilage in many health food stores, my husband asked his surgeon about its ability to aid in faster wound-healing. The surgeon laughed, yet I found several references to articles discussing shark cartilage on the Internet. Some medical journal articles on the healing and other properties and uses of shark cartilage can be found on the <u>Simone Protective Pharmaceuticals</u> page. Also at the site I found health-style questionnaires, in-depth descriptions of a variety of nutrient products, and how each product affects the body. You'll also find information about where to order or buy shark cartilage from pharmacists.

Search Tips

From the many healthcare sites I visited, my three-ring binder is packed with more than 500 pages of answers to questions. Productive keyword searches can be made using terms such as "alternative health," "healthcare," "medical," "medicine," "nurses," "nutrition," "pharmacy," "physicians," and "smart drugs." I found the search word "healthcare" to be more specific for asking personal medical questions than trying to search under the word "health."

Some sites are best reached through links on dedicated healthcare Web pages. For example, Subhas Roy has a created a page with links to 25 other Internet health sites at <u>Health Info</u>. There also is a large collection of links at the <u>Internet Medical and Mental Health Resources</u> page. It's maintained by Jeanine Wade, Ph.D., a licensed psychologist in Austin, Texas. One particularly comprehensive directory of health and medical sites on the Web is <u>MedWeb</u>, which lists Web sites and mailing lists in 70 categories, from AIDS to toxicology.

One of the best medical referral Web sites I found was <u>Richard C. Bowyer's</u> page. (You'll find it when you scroll past all his genealogy information.) Bowyer's page has links to numerous sites, such as U.S. hospitals, medical resources, medical journals, medical schools, medical students, medicolegal resources, oncology, pathology, and different surgical disciplines, such as plastic surgery, general surgery, laparoscopic surgery, and telesurgery. Healthcare workers of all specializations also can find job opportunities on some of these sites.

My son, who recently became a physician, was interested in <u>Medline</u>, a collection of medical and scientific reports used by physicians, articles on the educational needs of physicians and the public, physicians' supplies, prescriptions, and advice by pharmacists about drugs. Clinical cancer information that is intended for physicians is useful to patients as well. There is a listing of surgeons according to the types of surgery they perform, the notes of the Physician Reliance Network, and a Gopher menu of physicians listed according to their specialty.

As for informative newsletters available on the Net, I highly recommend the University of California at Berkeley Wellness Letter at the Electronic Newsstand. It contains the latest news of preventive medicine and practical advice—including information on nutrition, weight control, self-care, prevention of cancer and heart disease, exercise, and dental care.

As a medical journalist, I found the Journal of the American Medical Association (JAMA) to be a reliable resource.

If you are looking for a description of a particular drug, the Physician's GenRx Web site provides a database of drugs you can search. You must register first.

In seeking answers to my health questions, I found the Internet to be a valuable source for a wide range of healthcare information. I'm sure your efforts will be rewarded, too.

Selected Health Sites

Anesthesiology & Surgery Center *Offers travel warnings and immunization, medical dictionary.*
Cancer Related Links *Facts, figures, prevention.*
Harvard Medical Gopher *Harvard publications, access to the Countway Library's online catalog (HOLLIS), and medicine.*
Health Letter on the CDC
Health on the Internet Newsletter *Links to NewsPages, CNN Food & Health, MEDwire, describes new health-related Web sites, topic of the month.*
HyperDOC *Sponsored by the National Library of Medicine.*
LifeNet *Positive thinking and right-to-life point of views. Discussion about euthanasia.*
The Mayo Clinic *tour of the Mayo clinic; description of programs.*
Med Help International *Provides medical information about many illnesses. Treatment is described in layman's terms.*
Mednews *A biweekly newsletter that welcomes submissions.*
Medical Information Resource Center *Lists and a referral directory.*

Medicine OnLine *Career-related educational content and discussion groups.*
Medscape *For health professionals and consumers. Bulletin boards and a quiz testing your knowledge of surgery. Registration required.*
The National Organization of Physicians Who Care *Nonprofit organization ensuring quality healthcare. Newsletters, articles about Medicare reform and HMOs.*
OncoLink, the University of Pennsylvania Cancer Resource *Thorough information about pediatric and adult cancers.*
Physicians Guide to the Internet *For new physicians.*
Robert Wood Johnson Foundation Gopher *Biological archives at Indiana university, molecular biology database, NYU Medical Gopher.*

***

18

Nutritional Genomics for the Consumer

How are you managing your gene expression? In what direction are you moving? How do you make more intelligent *choices* of food to *nourish* your individual *genotype*? What is meant by *intelligent foods* that target and nourish *specific* genes?

Clinical dieticians and nutritionists, by allying with molecular geneticists, genetics counselors, physicians, molecular *genealogists*, family historians, phenomics professionals, nutritional and medical anthropologists, and archaeogeneticists are *collaborating* with consumers of genetics testing, but what are they really *sharing*?

If not so much raw materials such as DNA from donors, is shared, then how about *access* to information—databases and various discussion forums online and e-mailing lists equally open to consumers, licensed healthcare providers, and research scientists? Who controls access to new research—the consumers, the corporations, or the scientists?

Can the average consumer afford to find out what to eat for improved health and nourishment based upon tests of genetic expression? Can consumers *override* any inherited risks revealed in the genetic signature with foods and nutraceuticals individually tailored? What does it mean to eat 'smarter' foods that target specific genes compared to eating more intelligently regarding choice?

Scientists compare genetic distances between populations by comparing the frequencies of forms of genes called 'alleles.' Mutant alleles can be mapped as population genetics markers. Some, but not all mutations in genes may put you at risk for certain chronic diseases if you eat the wrong foods for your genotype. The solution is to eat more 'intelligent' foods customized to your individual genetic profile.

Research also looks at rare alleles. Their rarity gives them special power as markers of genetic similarity. There's a good chance two identical mutant alleles share a common origin. You can map genes for ancestral origin, migrations, or to reveal risks of disease depending upon which genes you map.

This book is for beginners with no science background. It's a consumer's guide book to nutritional genomics—genetic testing and profiling for foods tailored to your genotype and ancestry. The chapters also are about how to interpret DNA testing for family history and ancestry.

How do you as a consumer, not a scientist, choose the smartest food tailored to your genetic signature? How do you interpret your DNA test results for ancestry or family history? What is the *link* between tailoring your foods to your genetic expression and tracing your ancestry though DNA testing? And what genes are tested for either reason? How do you bridge the gap between nutritional genomics profiling and testing DNA for deep ancestral origins?

Does ethnicity play any role in tailoring your food and nutraceuticals, drug dosages, or healthcare? How much can the average consumer self-educate and/or start a private DNA bank for a consumer or patient group? How do you raise funds, contract with research scientists, and form or serve groups needing their DNA researched for specific reasons? How does learning how to interpret the results of your DNA tests for ancestry relate to understanding genetic tests for cardiovascular or other inheritable risks?

Start researching on your own what you need to know as a consumer to have more *choices* in customizing foods for your genetic signature—your genotype. What are some realistic applications of genetic testing and profiling?

This book will lead you to find out more about taking control of what happens to DNA that you may donate for research. You'll find out how to be in charge of your own nourishment and nutrition. Genetic profiling helps you to customize what you eat. How do you nourish your body? What can your genes reveal to you through genetic testing and profiling? It's your private information and should remain private. A good place to release it finally would be in a time capsule and history scrapbook for your heirs. Here are how some branches of human genetic history are linked to your nutrition, ancestry, and most of all nourishment.

Prosopography is all about human history and genes that travel because your genes have both a cultural and a biological component. The cultural component includes onomastics which is the study of the origin of a name and its geographical and historical utilization. Proteomics is about drug discovery. Pharmacogenetics researches how your genes respond to various drugs and dosages to avoid adverse reactions to medicines. Nutrigenomics (nutritional genomics) brings together nutrition and genetics.

Put all these branches of molecular genetics together with molecular genealogy. Add nutritional genomics—molecular nutrition, and what do you have? Knowledge of how every molecule in your body responds to certain foods, lifestyles or exercise can now be studied at the molecular level. New sciences such as pharmacogenetics open doors to learning how your genes respond to nutrition and nourishment. Maybe you want to know how your body responds to certain herbs, nutraceuticals (supplements), foods, or any chemical in your environment, even a skin lubricant or salve or a cosmetic. It's all within the sciences of pharmacogenetics and/or nutrigenomics. Today, it's not just about how your body responds. You look at the molecular level, the cellular level to study how your genes respond to nourishment, medicine, lifestyle, and the environment.

What if you take many prescription drugs and want to know how rapidly or slowly your body is metabolizing the medicine? You are concerned about the drugs building up in your body or interacting with one another. Pharmacogenetics tests several of your genes. With food menus, nutrigenomics tests other genes. At least you can find out whether you metabolize fast, slow, or like the majority of people. One size will never fit all people because genes recombine. They shuffle, and individuals have different responses to different drugs, cosmetics, or foods.

Multidisciplinary nutrition research and collaboration is necessary for nutritional genomics to bring together diverse expertise. Scientists working in the disciplines of nutrition, biochemistry, and genetics need to share, collaborate, and interface in this field. If scientists are more concerned about positioning themselves first in publishing their research and won't share DNA with all scientists, how can research ever move forward?

You might want to read "The Metabolic Basis of Inherited Disorders," 6th ed. McGraw-Hill, New York: 2649–2680, 1989. Then compare the latest research in nutritional genomics on how smart foods (foods tailored to your genetic signature) influence risk of chronic disease. The longer science studies the entire genome (rather than the specific SNPs for certain chronic diseases) the more information will be forthcoming on how food and lifestyle influence your health based on the genes you inherited.

According to the National Institutes of Health, (See their Web site at: http://www.nigms.nih.gov/funding/htm/yrgenes.html), "Your lifestyle, the food you eat, and where you live and work can all affect how you respond to medicines. But another key factor is your DNA, which contains your genes. Scientists are trying to figure out how the make-up of your DNA can contribute to the way you respond to medicines, including pain-killers with codeine like Tylenol®#3, antidepressants like Prozac®, and many blood pressure and asthma medicines. Scientific discoveries made through this research will provide information to guide doctors in prescribing the right amount of the right medicine for you.

"The National Institutes of Health aims to improve the health of all Americans through medical research that solves mysteries about how the human body normally works—and how and why it doesn't work, when disease occurs. One goal of this research is to help improve the good effects of medicines while preventing bad reactions."

Click on the Web site of the National Institutes of Health at: http://www.nigms.nih.gov/funding/htm/qanda.html to see their question and answer site. The point is that one size doesn't fit all when it comes to medicine or food or even cosmetics and skin products. According to the National Institutes of Health, here is the National Institutes of Health's answer to the question of "Aren't prescribed medicines already safe and effective?"

On their question and answer page, they reply, "For the most part, yes. But medicines are not 'one-size-fits-all.' While typical doses work pretty well for most people, some medicines don't work at all in certain people or the medicines can cause annoying, or even life-threatening, side effects." If you're wondering why one size doesn't fit all since our DNA is supposed to be so similar world-wide, it really varies due to some people's genetic variations, diversity, and mutations.

According to the National Institutes of Health (reprinted here with permission) from their Web site at: http://www.nigms.nih.gov/funding/htm/qanda.html#, "As medicines move through the body, they interact with thousands of molecules called proteins. Because each person is genetically unique, we all have tiny differences in these proteins, which can affect the way medicines do their jobs."

"The National Institutes of Health is providing money to scientists at universities and medical centers who come up with the best plans for carrying out research on how people respond differently to medicines. Curing and preventing disease is the National Institutes of Health's highest priority. Research on how people respond differently to medicines will make current and future treatments for diseases such as asthma, diabetes, heart disease, depression, and cancer safer and more effective. A bonus of this type of research will be a better understanding of the role many different genes play in causing or contributing to these and other diseases.

"The National Institutes of Health is providing money to scientists at universities and medical centers who come up with the best plans for carrying out research on how people respond differently to medicines.

"National Institutes of Health-funded scientists at universities and medical centers across the country will recruit volunteers from a wide variety of groups. Research of this type relies upon studying many different people with a broad range of genetic make-ups to find the small, but normal, genetic differences among them.

"Most of these research studies will involve simply rubbing the inside of a volunteer's cheek with a cotton swab. Scientists will then pull out DNA from inside the cheek cells they have collected. There are no health risks associated with this type of test."

"The first benefits to patients could come as soon as a few years from now. From then on, the knowledge gained through this type of research is expected to help doctors tailor the medicines they prescribe to best suit each patient's individual needs."

So what can the consumer do now to benefit as soon as possible from genetic testing? How can you apply the DNA test results to changing your lifestyle now in order to improve your well-being? Education is out there, and much of it is available free on the Web and in scientific journals available at most medical school libraries open to the public. How do you screen the information?

You don't know who will react in which ways at the genetic level to the food or medicine or whether the dosage of the drug or the amount of the food will be tolerated or too high or low. Even with varying dosages of nutraceuticals—supplements such as vitamins, food extracts, and minerals, some people benefit and others show no change. Read the results of conflicting studies in medical articles. If the dosage of vitamins is hotly debated, drug dosages and ethnicity is another topic in the research arena.

Some people have inherited risks for certain chronic diseases. Those people need to eat foods and perhaps take nutritional supplements that will prevent or delay the onset of those problems. Nutritional genomics fills an important need in maintaining health and quality of life. From the ***Institute of Food Research (IFR)*** in Norwich, UK, **Dr. Ruan Elliot, lead IFR scientist,** sent me this e-letter in response to my questions about what is being researched there. Here's the reply.

Dear Anne:

My main research interests are in using so-called functional genomic techniques to define mechanisms by which diet and specific components of the diet promote human health. These powerful techniques are set to revolutionise the way we approach fundamental nutrition research. On top of this, as you will appreciate, there is also the aspect of inter-individual genetic variation, the impact that this has on health and the potential variations in optimal dietary requirements.

To my mind, these two areas are locked together. We need to properly understand the processes by which components of the diet (nutrients, and micronutrients) work individually and together to keep us healthy so as to be able properly to define optimal nutrition for sub-populations or individuals properly based on their genetics.

You can find descriptions of my work and research interests at the following URLs;
http://www.ifr.ac.uk/public/FoodInfoSheets/EDPgenomics.html,
http://bmj.com/cgi/reprint/324/7351/1438.pdf

I hope this is helpful.

Best regards,

Ruan Elliott

For further information, contact the Institute of Food Research, Norwich Research Park, Colney, Norwich NR4 7UA, UK. To view their Web site, phone and fax numbers or email address, on the Internet, go to: http://www.ifr.ac.uk/about/.

My philosophy about genetic testing is to remember a quote by Richard Feynman, Nobel Laureate: "The best way to predict the future is to invent it." Your genes are hard-wired for certain foods, but not all foods make compatible software. Consumers need a guide book to nutritional genomics. You'll hear terms such as gene expression, genetic signatures, risk, intelligent foods, and tailoring the food to your genotype. What it all means is that your body is looking for customized nourishment.

If you really want to take charge of your own health and nutrition, learn all that you can possibly find out about how to apply the results of your DNA testing, genotype testing, metabolic, blood, ancestry, racial percentages testing, and body chemistry or allergy testing to what you eat, the nutraceuticals and supplements you select, your exercise style and lifestyle. Everything starts at the cellular level, even the way your body reacts to stress and exercise with cortisol or with relaxation. Individuals react to certain types of exercise, foods, or perceived stress situations differently.

Some nutrients, foods, herbal compounds or other supplements cause relaxation in some people and panic attacks in others. Then there are allergies to consider. It starts with an expression of your genes in reaction to the environment. Some people get an increase in ocular pressure from sleeping on pillows containing a stuffing to which they are allergic. One sip of caffeine can start a panic reaction in one person and relaxation in another. Your reaction is in your genes. It's about body type, another genetic expression, whether you have inherited genes for anxiety of certain lengths, and the whole interplay and interface of one team of genes with another group in your body. I refer to this interplay of genes as a "rhumba of rattlesnakes" at the molecular level.

You have a part to play in all this, and it isn't always as the passive patient or recipient. You have the *genomer* and the *genomee*....Be the *genomer* for a change—the person in the driver's seat. Take control. Take charge of how to interpret your DNA tests for risk and diet changes or for ancestry and family history or for any other purpose for which you want to test your genome or any region of your genetic profile. To be able to understand how to read and interpret a DNA test and apply it to foods and supplements and to know how the foods will actually effect your genetic expression—that kind of knowledge is power.

The information is publicly available in medical school libraries, on the Internet's scientific databases, and in various journals in the nutrition and genetics fields. Most of these sources are open to the public without you having to be a scientist to learn about how your body responds to food at the cellular level. Start by joining various online groups, visiting your library, and reading the latest medical and scientific journal articles in the field of nutritional genomics. Network with patient support groups that use genetic testing. Start your own email list message board for consumers who want to learn and listen in addition to sharing resources of information.

The consumer's role is to compare, review, and find out who is best qualified to work with an individual's genes. How does a consumer discern between snake-oil and reputable companies in this growing field? Who is qualified? Consumers need an explanation in plain language what is healthy to eat, not for the world, but for the individual. So it's up to the consumer to do some research and learn a lot more about nutritional genomics.

Beyond food what does an individual's gene expression require in terms of exercise and lifestyle? It's time to educate your body about nutritional genomics and about DNA and ancestry. What foods should you eat and what nutritional supplements (nutraceuticals) would benefit your health? One way to tell is to test your genetic markers. Which genetic markers? The entire genome? Or specific SNPs that signal risk of certain chronic diseases? Nutritional genomics should be available to all consumers, not only those with money to pay for expensive testing. Sure, in the future the price of genetic testing will come down, but senior citizens and parents today want to know what they can eat that will agree with them and their families.

Researching the Web under "nutritional genomics" I found a company in the United Kingdom called Sciona. According to Sciona's Web site at: http://www.sciona.com/coresite/index.asp?p=1, Sciona is a venture capital backed company that researches and develops tests for common variations in genes which affect your individual response to medicines, food and the environment.

There are around 40,000 genes in the human genome. Sciona identifies those genes that influence a certain function such as cardiovascular status and tests for these as a set or 'panel'. This information is then used by appropriately qualified practitioners to provide you with health advice.

Testing of specific genes rather than the entire genome usually is done by various companies at this time. In the future, the number of genes tested may increase, or science may find which particular genes interact with other genes to put you at risk if you eat certain foods, or whether your specific genes work together in such a way that you can eat almost anything without developing chronic diseases.

According to Sciona's Web site testing specific genes for certain chronic illnesses is useful in guiding aspects of the treatment of diseases such as heart disease and osteoporosis, which are influenced by your genes, lifestyle and environment. For the consumer, knowing which foods to eat to influence your gene expression is important.

According to Sciona's Web site, Sciona's team of geneticists, molecular biologists, medical doctors and dieticians work with universities and other companies to identify the significant genes underlying a particular effect. The effects of these, and other factors such as your diet, are then analyzed to give you specific advice on courses of action tailored to your own genetic makeup and circumstances.

Filling out questionnaires allow consumers to think in terms of focus sheets. Besides DNA testing, food choices, and consultations with your physician, you need to focus on your habits and think how realistically you answer a questionnaire. Think about from which direction you want to participate in nutritional genomics—marketing, research, science, consumer awareness, forming support groups, media, or other. What are your basic interests and how can you apply them to this field? You can contact Market America, Inc. at 1302 Pleasant Ridge Road, Greensboro, NC 27409. Their Web site is at: http://www.marketamerica.com. Here is an e-letter from Andy Aldridge, Public Relations Director, Market America, Inc.

Dear Anne:

After extensive market surveying and testing, Market America and Cellf, a division of Sciona, have partnered together to offer the Nutri-Physical™ Gene SNP DNA Screening Analysis program. This product allows consumers to submit a sample of their DNA to have it analyzed. Once the analysis is complete, consumers receive a report that outlines possible deficiencies along with lifestyle and diet changes that can be carried out to address possible vulnerabilities. To accompany the DNA analysis, Market America also developed a questionnaire that, when analyzed in conjunction with the Gene SNP product, results in a suggested list of customized vitamins and supplements, available in one formula, if desired by the consumer.

The company's Isotonix® Custom Formula makes choosing the correct nutritional supplementation a simple and efficient process. From a Distributor Custom Web Portal, a customer submits answers to a dietary and lifestyle questionnaire and has the option of purchasing a unique custom formula nutraceutical that specifically addresses their individual needs.

Many companies are talking about NutriGenomics. Through our partnership with Cellf for DNA analysis and Garden State Nutritionals for manufacturing and customization, we are actually doing it.

Regards,

Andy Aldridge
Public Relations Director
Market America, Inc.

Consumers Need to Be Involved in Quality Control

You need a voice in quality control. What the consumer needs to understand are the roles of genes in healthcare, and how the *roles of genes* interact when you take in nutrition. Consumers, corporations, venture capitalists the government, taxpayers, and research institutions invest billions of dollars each year to develop this understanding. One example of a consumer group involved in quality control is when parents group together to form their own DNA bank, recruit people to donate DNA for research, and develop databases and Web sites disseminating information on a particular genetic condition.

Consumers can don few or many hats in nutritional genomics. There are avenues to explore varying from watchdog, marketing, research, public relations, parenting, safety, event planning, publishing, gerontology, videography, genealogy, healthcare, to broadcasting.

If you do your research, you'll find that venture capitalists who in the last decade invested heavily in the computer industry are now looking to invest in biotechnology. The power of gene technology drums up business and communication also for patent attorneys, journalists, and inventors.

To participate in nutritional genomics in a variety of capacities as a consumer, you can write to the United States Food & Drug Administration (FDA), Department of Agriculture (USDA), and National Institutes of Health (NIH), as well as university laboratories, pharmaceutical manufacturers, and government agencies worldwide. Get involved in the power of genetic technology at some level. The FDA is the agency that's responsible for 80 percent of the United States food supply, according to a July 1st 2003 speech given by the Commissioner, Food and Drug Administration.

In that July 1, 2003 speech before Harvard School of Public Health, Mark B. McClellan, MD, PhD, Commissioner, Food and Drug Administration, said, "All of you—consumer advocates, representatives of the food industry, nutrition scientists, and other food experts—have a collective commitment to the issues we face at FDA that is integral to our ability us fulfill our mission. And your help is needed more than ever. Now more than ever, we all must work together to find better solutions."

It's important to read this speech and to look at materials on the Web at the Harvard School of Public Health. Click on the Harvard School of Public

Health's Web site at: http://www.hsph.harvard.edu/now/jul11/conference.html. Read the important facts there and check out the forums. The current headline at the Web site notes that "A July 2003 conference at the Harvard School of Public Health 'spurred' dialogue with the nation's food industry on the subject of "'Changing the American Diet' To Improve Health." Read the materials there and think for yourself about how the food you eat is processed and marketed.

In the speech, you'll find key words such as "consumer advocates" and "collective commitment." Your help indeed is needed to work together as a team, to share, with a purpose of finding better solutions. That's why it's important to listen and learn all you can about the future of nutritional genomics. The consumer's involvement is important. The theme emphasized "collaborating to improve the American diet." Collaborating with consumers also means working to decrease obesity and epidemics of diabetes in children.

You can look at the American diet from the point of view of those who work in DNA testing, from those who work with family history and ancestry research, or from those who work in food packaging and processing. It really hits home when you look from the point of view of the consumer or from marketing and product management. Everybody has to eat.

***

19

Consumer Surveillance

There's another branch of nutritional genomics that instead of only testing your DNA to find out which foods are healthiest for your genes, focuses on *manipulating plant micronutrients* to improve human health. See the article on the Internet, a PDF file on a Web site at: **http://www.ipef.br/melhoramento/genoma/pdfs/dellapenna99.pdf**.

The volume of imported food is growing each year. Consumers have a field cut out for them—surveillance. As FDA increases its examinations and sampling at borders, consumers can work together to research information about food imports and inspection.

A laboratory can only sample so many products. Consumers can take a role in food security, perhaps looking at industry to identify problems or threats. What the consumer's role entails is better information and collaboration. Everyone needs to keep costs down.

Plant biotechnology of food and feed is another area of consumer interest. If you buy food that comes from overseas, do you ever wonder who oversees the packaging and shipping of those products? Are there really enough inspectors to go around? Consumers worry about the widespread use of sugar in soft drinks. In addition to having your DNA tested, you need to understand how what you eat influences your health at all ages.

Another way consumers can oversee quality control is by forming public interest research groups funded by grant money, private donors, institutions, or the government. You can become a volunteer in nutritional genomics, an ombudsman, a lobbyist, or start your own consumer research interest group.

You can turn a hobby of nutritional genomics or DNA for ancestry and genealogy into a business by affiliating yourself with a university lab which you contract to do testing from your DNA testing clients. There are open doors for consumer involvement depending upon your skills and interests. Nutritional genomics needs public speakers and technical writers to relay to the public what innovations the experts are bringing to healthcare and food systems design.

From running a summer camp for teens interested in nutritional genomics internships or learning experiences to recruiting DNA donors to create a DNA bank or in researching and writing about genomics, there are a variety of doors. Consumers have power in numbers. You can even enter as a venture capitalist with a goal of raising funds even if you have no funds of your own and plenty of determination to learn to ropes.

Don't overlook nutritional genomics for the pet care industry from foods to medicine. Contact the veterinary schools about their research on how foods affect genetic signatures of pets or race horses. Check out the Web site for Research Diets at: http://www.researchdiets.com/.

Research Diets, a New Jersey company since 1984 has formulated more than 6,000 distinct *laboratory* animal diets for research in all areas of biology and related fields at hundreds of pharmaceutical, university, and government laboratories around the world. Nutritional genomics isn't only for humans or laboratory animals. Did you ever think about how your dog or cat could benefit by genetic testing to determine which foods are healthiest?

Talk to your veterinarian to see who is researching how nutraceuticals and better food can help your pet's health, especially when the pet is older. What about nutritional genomics for farm animals or pets? Find out who is doing what kind of nutrition research for better health.

Or you can tap the venture capitalist watering holes and open your own nutritional genomics business. Consumers can hire, outsource, or contract fee-for-service with licensed healthcare staff and geneticists specializing in nutritional genomics research. It's not difficult to raise funds. Billions already are invested in the gene technology industry. Sometimes parents of children with genetic conditions are the first to band together to form DNA banks for medical research. Don't overlook agri-nutrition. It's about what vegetables and fruits you eat that in turn impact your genes.

Medical schools and research institutions often seek federal grant money. Someone is investing billions for this research. Find out who are the investors. Who benefits from this knowledge first? Who has access? Is it the consumer who pays for testing, healthcare, nutraceuticals, and selected foods? It's up to the consumer to take a look at quality-control of the entire nutritional genomics arena.

Everyone is concerned about cell degeneration from a lack of critical nutrition. Unless you know how your genes are impacted by various foods and nutraceuticals, how can you can decide what's good or bad for your body other than looking at family history?

The consumer is concerned with how nutritional genomics as an area of research is applied to healthcare with a goal of disease prevention. For the consumer with no science background seeking a beginner's guide book to the power of gene technology, the first step is to realize that genomics has applications for many more areas of exploration than nutrition, healthcare, archaeology, genealogy, oral history, and population genetics.

So should you get screened? With the enormous variety of diet books on the market, healthy-eating advice abounds. How do you know what works to keep your arteries unclogged and your organs nourished for the combination of genes you inherited? People are different, but similar advice often is given to everyone.

What you need is a personalized report, time capsule, scrap book, or profile that has not only your genetic test results, but everything about your lifestyle, stress reactions, exercise. For example, if your body reacts to exercise by secreting way too much cortisol, should you be taking vitamin C? Will the vitamin restore to a healthier state or further damage your organs? If the vitamin is prescribed, what amounts of it will work best with your genetic expression or genotype? How do you find out whether this regimen is good for you or not? It is to these types of questions consumers want answers.

Information on food groups, vitamins, minerals, extracts, and nutraceuticals in an easy-to-understand format and binder are what consumers need. You also have to learn how your DNA test was interpreted. If someone interprets the test for you, how can you apply the test results to eating smarter foods for your genotype or selecting the right dosages of nutraceuticals? You'd have to keep paying for advice on every aspect of your nutrition as another branch of healthcare.

If you know the language of your genes or can learn it yourself, it makes your road to a healthier self clearer. Consumers need access to and knowledge of quality control. Who is performing the tests? What kind of medical supervision is officiating at the marriage of healthcare to nutritional genomics? Who is at the top overseeing the quality control—the consumer? It should be. You pay to fund the research in tax dollars. So form your consumer groups and look into quality control in the field of nutritional genomics, DNA testing, and related businesses that will open to serve the consumer and the licensed heath-care professional.

As the power of gene technology reaches the masses, all types of offerings will find a way to apply DNA test results to what you eat, how you exercise, what you wear, your prescribed medicines, therapies of various kinds, what music will change your physiological responses based on your genotype, and even what career, lifestyle, mate, childbearing plans, or hobby you choose.

DNA samples usually are obtained by rubbing a brush swab on the inside of your cheek, completing a questionnaire, and mailing the brush to a testing laboratory. Sometimes in the case of a rare genetic disease, blood samples are taken, but most DNA testing is done with a cotton or felt swab or mouthwash. After several weeks, your physician would get a report to interpret for you for nutritional genomics. In testing DNA for ancestry, the report would be sent directly to you.

How specifically is the report customized for you? Even tiny differences in your genes influence the way your body metabolizes foods and excretes toxins. These reports look for *variations*. Your feedback consists of dietary information. Some companies assess your present eating habits and foods with your genetic profile. Advice is offered. For example, some tests look at the type of meat you eat, whether your meat and fish are smoked.

The tests look at the types of vegetables. How many cruciferous vegetables do you eat—such as cabbage, cauliflower, Brussels sprouts, and broccoli? How many raw vegetables do you eat? How many sprouted legumes or whole grains not processed into flakes? Did you know that cauliflower is a low-carbohydrate vegetable, but broccoli is higher in carbohydrates? Did the sprouts you ate have bacteria on them that affected you?

This is important if you are insulin resistant or have too much insulin pouring outing each time you eat a high-carbohydrate vegetable. For example, pectin consumption results in an insulin release. What kind of antioxidants do you eat? How much folate are you taking in, and how does that folate impact your arteries—neutral, good, or worse?

What types of whole grains are you eating? Are you eating your whole oat groats or eating foods high on a high-glycemic index that rapidly turn to sugar in your blood and require more insulin to be secreted which can lead to other problems in excessive amounts.

Does your genetic profile reveal a need for a Syndrome X diet or other measures to halt the excessive insulin secretion each time you eat carbohydrates or proteins? What kind of weight problems or control do you have?

How much sugar do you eat? Do you drink liquid candy such as soda pop? How do diet drinks affect your genetic expression compared to sugared drinks? What kind of saturated fats do you eat and how does your body react to that kind of fat? Do you smoke? What damage has smoking caused your body? What kind of exercise do you get?

Walking moderately may be enough for your body, and too much exercise could bring about high cortisol levels—the stress hormone. Does your body react to exercise as stressful or relaxing? What allergies do you have? All these questions need to be compared against your genetic signature. What genes are assessed by the testing company? Do you have a genetic propensity to alcoholism? Which gene predisposes you to alcoholism or drug addiction when you are under stress? Did you inherit the anxiety gene? If so, what is the right career for you? Do you have the genes that predispose you to panic disorder? Find out how many genes are studied and which genes relate to what physical aspects of your health?

Look for a company with excellent quality control. Recommendations should be practical and proven. Your goal is to improve short and long-term health and prevent chronic diseases or at least delay them as long as possible. If the food you are prescribed don't make you feel better, find out why. The tests you take and the food or nutraceuticals prescribed should be scientifically proven in reputable studies. You can check out the studies in medical journal articles. Make sure they were not flawed studies or studies so old that new studies have proven opposite conclusions.

How can you tell for sure what risks you have and what needs your genes have for certain foods or nutraceuticals without testing *all* the genes in the human genome of an individual? If all the genes aren't tested, how can science really know everything there is to know about your individual gene expression? What's important to know now—going with what you have available? When will the research teams be ready to reach out to the average consumer of nutritional products and healthcare? Today there is a divide between what's going on in research and what's available to consumers. That's understandable because the research is still going on.

Without knowledge of all your genetic markers, how can you develop a plan for maximum health through nourishment? Why must this plan be in the hands of a licensed professional instead of an informed, self-taught consumer? If the consumer is armed with knowledge of what the individual's genes reveal and how it relates to certain foods and supplements, the consumer than can take back control and power over his or her own health.

Knowing your genes is only a beginning. How do you apply that knowledge to practical applications as in how certain foods affect your body chemistry and metabolism? If you have a plan A and a plan B, and you know the unexpected can always kick in, can you still have that feeling of a little more control over the way your body treats you? In a world so out of control, consumers of healthcare

are looking for a semblance more of control and power over their sense of well-being and mood.

Diet books written for the masses may not work well with your particular gene expression. Science no longer tells you that you have inherited some gene mutations or defects. So your gene expression causes certain chemicals in your body not to work normally. Instead, you are told that you have an individual gene expression requiring you to eat this instead of that.

If you eat this combination of food in the morning, that excessive insulin pouring out each time you eat high carbohydrate vegetables or fruits won't narrow your arteries so quickly. It sounds so positive. It's practical, and tells you specifically what you should eat and when and how the food affects particular organs or metabolic and chemical reactions in your body as a result of consuming a particular food or nutraceutical. At last, there's a positive solution.

You can use intelligent food to override your defective genes that put you at risk and nourish your gene expression by fulfilling your genetic signature's deepest nutritional needs. How fast are the professional dietitians and nutritionists tuning in to nutritional genomics?

The key for the average consumer without degrees in nutrition or genetics is to learn how to look at a printout of your genetic markers and be able to interpret what that means in terms of which foods and nutraceuticals to consume for your health based on what you are at risk or pre-disposed to come down with given the right interplay of environment, lifestyle, food, attitude, and perception of stress.

It's not as hard as you think to find this information in libraries and in online databases. Any subject that takes money away from professionals making a living giving advice in any field is going to require effort to learn. The key is to find sources willing to share information. You bet they are around, and a lot of scientists are willing to share. Some scientists are concerned about their careers and reputations. They should be. Some are cautious about with whom they share information.

On the other hand, some scientists do not share. It may not be the fault of the scientists, though. If you look in the archives of the Los Angeles Times, you'll perhaps find an article dated July 18, 2003 titled, "Whose DNA Is it, Anyway?" Read that article. It's about a person whose mother has Alzheimer's disease and who, according to the article, was "trying to coax Alzheimer's patients and their families to donate DNA" to a university. It's not only about one person because more than ten thousand people donated DNA for the cause.

How about you starting a nutritional genomics DNA bank, new support groups, non-profit organizations, online discussion groups, or Web sites such as

"Moms for Genome-Tailored Meals?" Think about how many consumers, not scientists, whose families have a condition or who are interested in a condition arrange events to encourage people to donate DNA to research.

How many consumers go out to assisted living complexes, senior centers, adult education classes, gerontology workshops, nursing homes and churches or family meetings to recruit people who have relatives with a certain condition such as Alzheimer's? How many consumers reach out to the actual Alzheimer's sufferers to recruit people to give DNA to science—to research, often located at or connected to various university laboratories? This is one way consumers get involved in science without necessarily having credentials in science.

All it takes is a deep interest to take action and learn more about the subject. Some research companies are independent, founded by scientists, but work closely with a variety of universities. Consumers are involved every day in science, usually on the end of recruiting people for donating DNA, planning events, fund-raising, writing about in the mass media, encouraging children to take an interest in science, acting as ombudsmen in nursing homes, running consumer awareness classes for seniors or parents, working with genealogy societies for various ethnic groups whose DNA is being studied by scientists for diversity, geography, or genetic diseases, and philanthropy are but a few ways to get involved in scientific research.

According to the article, you have a situation where DNA was used to search for Alzheimer's genes. Science itself has a great mission, to someday find a cure. The DNA is like a pointer. What the consumer is at risk for is that the research will stop. When the research stop for whatever reason, who controls the DNA you've donated? Read the Los Angeles Times article, "Whose DNA Is it, Anyway?" Think about your own DNA and thousands of others who have donated for testing. Who controls the DNA archive when the research is halted in midstream?

Who owns all the DNA samples—including yours? Who will inherit and use the samples and for what type of research? What if you have a group of family members suffering from a condition that inheritable and you don't know whether you'll get the disease or not?

What if you donate DNA along with others hoping it will point toward a cure, but the project ends? It's like the old question of who owns the living embryos in vitro when the project is over? Almost anyone with a nursing, social work, or other allied healthcare background can open a home-based or other business part time online or fulltime recruiting people for medical trials.

Around the world, people give their DNA for medical research. Why do people donate DNA? It's not always to find their ancestry or ethnic origins, because in many cases, results become statistical and no feedback is given to any individual or even to members of an entire group. What shows up are anonymous statistics in scientific journal articles or books. People often give DNA in the hope that science will find a cure for their or their family member's genetic disorder or inheritable disease.

People donate DNA to see whether a new test will work better for them, such as a test of racial percentages, or in the case of a disease, a test that will reveal risk that could be overcome with certain foods or medicines or even gene therapy. People hope that the medical field has a way of injecting normal genes into a child or person that will take hold and correct a defect in the existing gene expression. Part of gene therapy is about the introduction of new genes that could fix the old genetic problem, and often it works in certain cases.

What consumers can do for themselves is to take part in creating lending libraries for DNA. You need to have lending libraries open not only for scientists, but for the consumer to at least read about research. Of course, scientists who actually work in laboratories with DNA could enjoy the free and open access to the DNA managed by special DNA librarians. It helps research move forward. As a consumer you are a taxpayer.

Do you know your tax dollars are paying the cost of much of this scientific research, especially the research going on at state universities? If you are paying for this research, then the scientists are public servants to you, the consumer. You have the right to this information as much as any scientist, since you are paying the bills for the research.

The problem is that not all scientists freely share DNA all the time hoping some other scientist will find a breakthrough. It reminds me of competing journalists on different papers vying for the news scoop of the day. The DNA is collected and sometimes not shared at all.

The reason is that a scientist's career is at stake. In research, a scientist's reputation and job depends too often upon breakthroughs that help the career. Will sharing lift a scientist's career by his or her bootstraps? Sharing could mean the other guy finds the breakthrough first. Then in walk the patent attorneys. Scientists can file patents on genes.

What if your child is autistic? What scientists will share DNA with other scientists in research on autism or Alzheimer's or any other condition? What's your role as a consumer? It's to create a situation for cooperation. Consumers tend to form patient support groups and to build their own DNA banks. That's only one way of taking consumer action in a positive way to move research forward and point toward solutions and cures for the benefit of the health-care recipient—you, the consumer.

What can you do with nutritional genomics that is in your interest? You can continue to build DNA banks backed by consumers with an interest in nutritional genetics or DNA ancestry, or gene testing for disease markers. Share among yourselves. After all, you are not all scientists, and therefore have no career in science about which to be concerned. I speak as a consumer here rather than as a science communicator.

What you can start as a consumer is a movement to pool DNA samples into a managed library and DNA bank open to all scientists. Consumers could have access to reading about the research and learning to interpret their own DNA and other genetic tests of SNPs, since most consumers don't have access to laboratories to perform research. The time for the consumer's role in moving research forward is here, and consumers must learn all they can about how to interpret the entire human genome.

Your first step would be to make a list of the various institutes that study particular conditions of interest to you or written in your genes as your risk condition. Start with the ones funded by you as a taxpayer. Learn more about what your genes show you. When you talk to scientists, you'll find out how many will or will not share anything from information to DNA. Talk to the scientists. Get on emailing lists where scientists talk among themselves. If it really were true that scientists never share information, there wouldn't be any libraries or databases. Once anything is published and is not classified by the government as secret, it's shared through journals and libraries. The idea is to do the same with the raw material, the DNA through gene banks and DNA libraries as well as databases.

You already know some scientists don't like to share raw materials such as DNA with other scientists. Just ask most biology educators at universities to share their experiences with geneticists with consumers, let alone the media. If you think your DNA is a gift to the public like your donation dollars, think patents. What if you're a consumer who works with donors? What if your dream is to see children with certain diseases tested for little cost?

What if there are battles between scientists and hospitals for patents? What if a hospital takes out a patent on your DNA or those of a bunch of donors you as a consumer recruited from your efforts, support group, or fund-raising events? What if the hospital is awarded the patent and starts to charge fees?

What if that results in some testing programs closing? It happens over and over. What if the costs of developing certain types of DNA or other genetic tests are or remain too high? These are all the questions that consumers want answered by scientists, hospitals, research institutions, universities, and licensed healthcare professionals.

When a consumer so enthusiastic about learning all about genetics, DNA for ancestry, or nutritional genomics put effort into raising money and providing DNA donors, the consumer wants to be sure that what's used for research doesn't always go to pay for patents that result in fees and costs to the consumer or anyone else becoming unaffordable just to create or invent the new genetic tests.

Media has created videos romanticizing "gene hunters" poised on the cutting-edge of a new frontier—inner space. Gene hunting is entertainment. Yet there are little controls or rules—not even a law that requires scientists to share DNA with other scientists. If gene hunting is another Wild West frontier focused on individuality of gene expression, where are the public interest research groups here?

Back in 1975 I was a book author employed in a temporary job full time for ten months writing the consumer manuals and handling all incoming calls to the consumer complaint center of a public interest research group. My job was to write a book that ended up in the Attorney General's office. The book was a consumer manual of how to effectively complain.

Today, the consumer needs a guidebook or manual to learn how to effectively organize what should be in the public interest—research. That research could be anything from the frontiers of nutritional genomics to the study of food on a disease or the disease itself. Sometimes research is shared with the public long after everyone else knows about it. If it's archaeology, it's shared quickly. If it's gene research, that's another story. Sometimes patents get in the way, sometimes costs, and sometimes egos.

Databases are open to the public. The bioinformatics profession manages databases of bioscience information stored in computer software. You can get a certificate in bioinformatics at many community colleges or through extended study programs at various colleges and get your foot in the door of bioinformatics through an internship course.

On the other hand, you have scientists, universities, private companies, and research institutions vying for money and/or information being returned to them for their work in giving information to the databases that hold DNA information. It's all about your taxpayer dollars.

As consumers, you pay your taxes to the government. In return, the government takes your tax dollars and dangles that big carrot of federal grant money in front of the institutions employing the gene hunters. And if those scientists refuse to share DNA with other, competing scientists, well, they won't get the federal grant money that's really your tax dollars.

So who's at the root here who should be in control? It's you, the consumer because you supply the government with your tax money that makes up the federal grant money that goes to the many of the places that employ scientists who test DNA for research.

Write to the National Institutes of Health and ask who is sharing what with whom. How are your questions answered? You could get yourself a career as a grants writer and work your way up to become a grants administrator. That's a long career route where you never know whether or not you'll be chosen for the job. The highest you might get is as a serious student of grants administration research. Another route is to ask scientists why sharing DNA could destroy their careers. It's all a matter of who gets to publish first.

In my quest to find people to chat with, a few scientists turned me down not because I don't have a science degree, and not because I don't work for a major newspaper or magazine, but because I'm *the first journalist* to write a book on genomic nutrition before the scientists have had a crack at it. And the book isn't for scientists speaking to other scientists, but a consumer's guide to DNA-driven ancestry as well as a consumer's guide to nutritional genomics testing.

Consumers don't need to be slapped in the face with technical jargon unless the genetics terminology is defined in a glossary. The purpose of this book is to inspire consumers to look into the subject of DNA testing not only for deep ancestry, molecular genealogy, or archaeogenetics, but for tailoring intelligent foods to their genetic signatures. If the buzz words in the news are nutritional genomics, then now is the time to organize. The entire area of DNA testing needs direction, quality control, and applications to what consumers perceive as critical needs.

You see dozens of books published on population genetics, archaeogenetics, the peopling of various continents, mitochondrial Eve, ethnic DNA, diseases and DNA, ethnic genetic diversity, food intolerance, allergies, inherited illnesses, aging, and DNA for ancestry. How many general-reader type books are published on smart foods based on genetic signatures and SNPs rather than chemical, blood, or metabolic signatures? So positioning yourself first matters if you're a scientist who must publish and patent. If you're an educator with an academic institution, positioning yourself first in publishing is crucial to your career, even factoring in getting tenure.

A consumer's world is based on sharing. A scientist's world is based on fear of competition. It should not be that way, but it is. And few will admit it. What competition? There's no scarcity here. There's room for everyone? Then how come so many excellent PhDs in various sciences can't find jobs in their field?

If you want to check this out, talk to recruiters in the sciences and find out how many people are out of work with PhDs in a variety of sciences. Is it age discrimination? Ask the recruiters for their experiences. They will share with you in most cases. Try writing a book on careers in genetics. See what the recruiters have to say to you. Some scientists at some research institutions also will tell you that before they can share DNA, they have to get consent from everyone who donated DNA.

What do you do as a consumer? You don't work alone. Power is in numbers. You organize, get people together and pitch in as a growing group to start your own DNA bank, perhaps consisting of DNA donors. Then you'll have the problem of contacting universities and researchers at a wide variety of places to see whether they are interested.

For the average consumer who doesn't want to take any action, just find out what foods are healthier, the problem is simple. Just research your own genes yourself. All your genes interact as a team. And as all these genes work together, you usually can't point to one gene as being responsible for one action or disease. You can have a defect in one gene that causes a problem or only a risk.

Both consumers and geneticists are interested in finding out what mutation happens in a gene that leads to chronic illness at any age. When a gene oozes really bad protein or the wrong amount of protein—ah, oh.

The interaction between the rest of your genes and the chemical and metabolic systems go wild at the cellular level. You get sick. Gene hunters who study the mutations in your genes can study mutations for ancestry or for sickness. Different genes are studied for ancestry, and muations there have little to do with illness. So where does the consumer go to start? That depends on what you want from your DNA—ancestry, or a prescription for intelligent food?

If you're talking about illness, you start with patient support groups. For ancestry, it's the genealogy groups, ethnic groups, and DNA mailing lists online or start an archaeogenetics club or e-mailing list. You can start your own academic journal. For nutritional genomics, the field is wide open for you to start your own DNA collection banks. Who controls your DNA—you or the research 'industry'? You, of course own your DNA. You control the direction of any research performed on your DNA, and it's time to take DNA by the reigns and direct it—yes, you the consumer. Your tax dollars fund research. So take control of your own DNA.

In the field of nutritional genomics, you, the consumer must take charge of your DNA and begin to build collections in DNA banks where the DNA donors are in control of the direction of the research. Nutritional genomics is not only for parents of children with genetic defects that need the research so urgently. It's also for anyone who wants to know what foods to eat to stay healthier for longer periods of time.

You can start organizing family networks to review and compare the nutritional genomics industry. You can build collections of DNA in your own DNA banks and create databases and DNA libraries. Instead of re-inventing the wheel by duplicating the existing databases, fill the gaps where your efforts are needed and rewarded. Go to a variety of genetics, foods, and DNA-related conferences and let people know you are starting your own nutritional genomics DNA bank and/or intelligent food bank.

The skill you need is not a PhD in genetics, but the desire to persuade people to come together. You need to be a catalyst. If you're one of those unemployed people with science training and are good at public speaking, organizing people, and fund-raising, it could be the career you're waiting for.

If you are a parent with no particular skill other than public speaking and an interest to learn all you can about nutritional genomics from self-education, it's a way to take action. Be prepared to outwit Mr. or Ms. Medical Research Politics just as you have outwitted corporate politics and the games your mother never taught you. If you are lucky enough to be a homemaker, this is the perfect part time career—finding loopholes in medical research politics. It works best if you are not a doctor's wife. Being a doctor's mother is even better for this career. And if you're a patent attorney or work for one, this is another ball game.

The first loophole is to form your own group. Work as a group and put your power in the numbers. Work with scientists in a group also. If you have DNA to donate, give it to a group of scientists that you can organize as a group to work together. There's the problem of individual scientists sharing DNA with one another. If there's no group of scientists, ask them to form a group to work with your group. Organize for nutritional genomics.

Pool the scientists, not the DNA. Otherwise, you'll have a bunch of DNA on your hands and no place to store it. You've got to organize scientists together in a group to work with your organization. First, you need to start an organization. It doesn't take a lot of money. It's about people working together. Consumers aren't afraid of competition. Scientists are afraid of competition because positioning oneself first is the rule of the game if they are to keep their jobs or get ahead.

Consumers are parents, kids, or retirees, not scientists or medical researchers whose careers will be harmed by grouping together to share information or DNA. Most scientists will not share DNA with other scientists. You can go to scientists in other countries and talk to them. See if they will pool and share DNA for research. There's a far better solution. Start your own DNA bank. You organize parents—especially ones with a little money or prestige in their non-science careers. These elite parents have enough smarts to do fund-raising and hire people to collect the DNA to put in your own DNA bank.

What kind of people perform the best fund-raising and organizing? Public relations people—extroverts, people-people, public speakers, entertainers, and the media, for starters. Don't bother the introverted scientists who need to hold their jobs by positioning themselves first in publishing the fruits of their research. Contact public relations and marketing people such as fund-raisers and marketing communications managers.

Hire people not afraid to speak out and not afraid to learn about nutritional genomics. Start a DNA bank, a nutritional genomics resource center made up of many families. Collect a lot of DNA. Ask for private donations. Share your DNA bank's DNA with any scientist. Let the medical schools know about you. Again, you don't need any type of academic degree or credentials to organize a DNA bank. You just need to hire a staff to collect the DNA and make it available to medical schools and any other scientist for research. Who owns the DNA? The donors.

If you want to network with other parents and consumers who have done this, contact people who have organized their own DNA banks, such as the Autism Genetic Resource Exchange. They are on the Web at: http://www.agre.org/.

Talk to the medical schools that use the DNA banks. Read the Los Angeles Times article, "Whose DNA Is It, Anyway?" July 18th 2003. The article also mentions how the Autism Genetic Resource Exchange got started. You can do the same for researching nutritional genomics, a field that got started only back in 2000, and you don't have to have a background in any science to organize people, hire staff, and get started recruiting families to join you in creating your own DNA bank. Do you know how much you are needed by the medical schools?

When you speak to scientists, you may get cooperation, but you'll also come across someone who will tell you to scram because you don't have credentials, or in my case, because I make up stories for a living as a novelist, overlooking 35 of my non-fiction books. Don't let anyone distract your attention from your goal by trying to focus your attention on perceived shortcomings. Anyone who turns you away probably wants first position. True love is about making sure the loved one is positioned first. What nutritional genomics needs is a little tender loving care.

Who do you contact as your first customer? Try the *research-oriented* medical schools. They don't have anywhere near the resources to collect as much DNA as they need. Contact the heads of the molecular genetics departments at the medical schools.

Not everyone interested in nutritional genomics will want to start a DNA bank devoted exclusively to nutritional genomics. There's a place for all levels of participation. You can hire a DNA collector or learn how to collect DNA from people if you're a people-person. You can raise money from donations or ask your state for money. Have each donor sign a statement allowing the DNA to be shared by other research scientists.

Find someone who is associated with a reputable laboratory or with a laboratory in a medical college to store the DNA samples. Talk to biostatistics professionals. The more DNA you collect, the better it is for the numbers crunchers. Not much is learned from only a few DNA samples unless you're comparing Neanderthals to modern humans.

Nutritional genomics as an industry and area of science needs its own support groups. Most diseases have support groups. It's time support groups were formed for preventing or delaying genetic risks some people are pre-disposed to from occurring through learning more about smart foods tailored to individual genetic signatures.

What's your solution? Since scientific research is for the benefit of the consumer of healthcare and nutrition, start by going to other consumers and getting feedback by reviewing, comparing, and polling consumers on their experiences with testing. Are the consumers satisfied? What would they like to see improved? Today's consumers of genetic testing have enough money to spend on tests that help them make decisions about food, health, or ancestry research. You've seen Michael Moore's TV Nation. How about creating a video called Nutritional Genomics Nation? If not a video or a media presentation, how about a database and a Web site?

Still feeling a bit out of control? No need to. Just talk to the professionals in the field and make friends. Aside from professional competition that makes some scientists want to position themselves first in publishing and patenting, scientists really are friendly, welcoming people—especially when you are positioned to give them the kind of publicity in the media they want at this time. Nutritional genomics is a buzz word in the media. It's a hot topic this year. Scientists are finding this one-chance shot at getting positioned first in the media by publications of the highest repute.

Librarians at medical school libraries and publishers really like it when you read magazines that would otherwise be read by only a few. I used to spend months reading journals at the UCSD medical library to get ideas for topics that would result in books—either novels or nonfiction how-to series. So you, the consumer, also can learn the important connection between how to read the results of a genetic test and how to apply it to choosing the right food.

Right now, that key is in the hands of licensed healthcare professionals and scientists. It's time the consumer learned how to interpret the results of genetic testing and how to apply the results to a better way of eating. Self-education is the answer. And even if you put your life in your doctor's hands, how do you know whether he or she is qualified to make the connection between your genes and recommended foods or whether those foods will work with your specific gene expression?

Find out what shows up in your relatives and what you may or may not have inherited at the molecular level. A test of your biomarkers will at least give you that handle on your own healthcare, foods, and lifestyle. The point is to educate yourself about your individual genetic signature regarding risk and the implications genetic markers have on your health now and in your own future.

Keep your profile private, and use the information to choose the best possible combinations of foods to help prevent or delay any chronic illness for which you may be at risk. You don't have to give your genetic profile to anybody who could use it against you such as insurance companies and employers. What you can use your profile for is to choose the foods that your body needs. Let your gene expression show you what to eat, how much, and when.

When you look for a diagnostic lab, keep in mind that you want to deal with companies that provide their testing to physicians or similar licensed healthcare providers and not to anyone from the general public. The reason is that you need to know how to interpret these tests and how to apply what you learned.

For example, if genetic testing reveals your at risk for a certain chronic disease, how would you know that unless you can look at the genetic markers, see the risk, and be able to judge which foods and supplements would cut that risk. What I'd like to see in the future is that the consumer would directly be able to get that information from publicly available sources while maintaining privacy.

What I like about the company is that they have a Web site with clinician support where you can educate yourself about what to look out for in your own body so you can at least ask important questions when you see your doctor for testing.

And for physicians and clinicians, it's an excellent site to become more informed in specific areas.

The ideal situation would be to see the results of your testing and be told what risks you have. Then you'd be able to buy a book or look up enough medical journal articles. In the future, as a consumer I would like to see more consumer education so that your preventive care wouldn't always have to be solely in the hands of your managing physician.

You need to have more control over what you eat to lessen your disease risks. For you as a consumer to concentrate on focused prevention and treatment tailored to your individual genetic signature, you need knowledge. If knowledge is power, than those with the knowledge you need and don't have leaves you pretty powerless. So you have to educate yourself via books and medical journal articles, the Internet's Web sites and asking questions.

When you see your doctor, often you have only a few minutes to ask questions as doctors are pressed for time. Make a list of questions you want to ask. You might want to compare opinions and answers between alternative health care licensed professionals and your usual primary care physicians or HMO nurse practitioners.

Then check out with medical journal articles and news or even consumer feedback and reviews of what you're seeking. If you don't see enough Web sites that review genetic testing companies or research institutions and if you don't see reviews of biotech companies doing research, start your own Web site to review the many companies springing up.

Compare the nutritional genomics companies that deal with licensed healthcare professionals, and get answers, opinions, reviews, comparisons, feedback, polls, and consumer reports. You have consumer reports on cars, mattresses, and washing machines, why not a consumer report site or publication on genetic profiling for nutrition and health? Start a public interest research group on companies that look at genes or look at disease in various ways. You'll learn a lot from consumers just as teachers learn more from their students than from many of their books. Ask for feedback. It's your genes.

Learn the terminology. Find out how nutraceuticals affect your body before you decide what nutraceuticals to take—vitamins, minerals, antioxidants. What works for people with your risk, may not work with your specific genetic signature.

What foods can you as an individual eat to lessen the risk or delay the onset of the chronic ailment for which you are at risk? I'd like to see the entire genome tested, not only a few SNPs for the major chronic diseases, but all the genes.

Since cracking the human genetic code in its entirety is so new, science may not have the full impact of which genes react with what foods to build a healthier you. Does anybody really know today which genes are responsible for the way your body reacts to certain foods, medicine, or lifestyles?

Science can tell you the specific SNPs for certain chronic, degenerative diseases. Think about it, is that all there is—as far as all the genes responsible for those chronic diseases? What about the rest of your genetic signature? Are there hidden files at the cellular level?

Intelligent foods are attracting the attention of the big, international food design systems firms. Who else is interested in nutritional genomics? Rushing into the buzz about nutritional genomics include patent attorneys, genetics counselors, universities, pharmaceutical manufacturers, and physicians. Nutritional genomics also attracts the interest of naturopaths, the alternative health and health food industry, and merchants of nutraceuticals. Then there are the big, international food systems design companies. Anybody interested in health and healing or food and nutrition is paying attention to research in nutritional genomics. It's time for the consumer to do some research.

What is your stance on nutritional genomics? I visualize peering inside my genes every morning to know how my body responds to carbohydrates. I can feel too much insulin being released, hitting me like a bomb when I used to eat donuts and coffee the first thing in the morning. A few minutes later I was in tremors, caught between the sugar and caffeine. Needless to say, the whole process led to hardening of my arteries and more.

Only when I switched to a diet that calmed my body type, intelligent foods for my biomarkers, did I realize the connection between what I ate, how I exercised, and what changes were happening in my body in the past sixty-plus years. What I needed was foods tailored to my genotype. My relatives eating processed carbohydrates and trans-fatty acids didn't last very long. As the only survivor, I had to listen to what my body was communicating to me at the molecular level, to be aware. The best way to be aware is to listen to how your genes respond to what you eat.

Genes communicate beyond a chemical and metabolic level of consciousness through the language of biomarkers. They alert you to risk or no risk in a measurable, physiological way. And the language shows up not as words but as risks on genetic tests. Sometimes genes mutate or make mistakes in copying. There also are other reasons for defects in specific genes. Or you may not have any defects. Your biomarkers evolve over millenniums based on what your ancestors ate and the climate in which they lived for the past ten, twenty, or forty thousand years.

You can do something about risk such as lower your homocysteine levels with nutraceuticals and vitamins if the levels are too high. Look for clues in family history, your own blood tests and physical exams…and most important, your genetic risk profile.

That's why knowledge of healing foods tailored for your specific genetic profile should be your right. There are food and nutrient solutions to most risks that show up on tests. It's better to know how to overcome the obstacle than to remain in the dark about what foods will delay or prevent future or present chronic illness. Information about your genetic profile should always be private.

When should it not be private? What if you have a contagious disease or apply for a commercial pilot's license or drive trucks or busses, trains or boats? Then regular medical exams will reveal the expression of your genes as you age. What is your opinion on privacy? Have you looked at the polls on the subject of nutritional genomics? What do you think about genetic testing for nutrition and nutraceuticals? There are allied fields to explore as a consumer.

Research the publications and articles online or in print on pharmacogenetics and phenomics—the sciences of tailoring your medicine and healthcare to your genetic profile. Explore proteomics (drug discovery). Read about bioinformatics (managing bioscience information and statistics in computer databases and using computer science technologies with genetics research. You work with both computer programming and bioscience information.) The information is either on the Web or in medical and science journals you, the consumer can find on the shelves of university and medical school libraries.

You can purchase a library card and make use of medical and science libraries at universities. As a senior citizen in lifelong learning, I was able to purchase a library card through a retirement-age group at my nearest university. There are medical journals you can read in the library or subscribe to. Take notes. Use the Internet to read medical and scientific journals online. Use the library photocopy machine for articles you want to take home if you aren't online. Do your homework and teach yourself about the wonders that are out there in the burgeoning field of nutritional genomics.

It's true that a little knowledge can do a lot of harm, but what you're interested in is for the consumer to be able to self-educate at least to the level of the media, and to have the same access as the media to see what is evolving in fields that look at disease in new ways and in looking at health and preventive nutrition in new ways.

For example, when I used green tea extract to cure my gum problems, my dentist was happy for me. Right now nutritional genomics has buzz appeal. It's in the news. At the root is change. The way scientists look at your health today is that nutrients play your body like a violin. Your entire system is an orchestra, and you have to listen to the music.

How does an orchestra of musicians play together in sync? By being interactive. Your body is part a team. And the systems biology that nutritional genomics is all about works interactively. First science looked at the parts, and then the whole person. Now science looks at the orchestra playing together in sync, the systems biology at the genetic level. It all works together, and what you eat may be able to keep the symphony in sync, playing a healthier, more stable rhythm. Nutritional genomics is more about your individual genes that your ethnic group as a whole entity. What if you didn't inherit the gene to digest milk properly?

In the book **Archaeogenetics**, McDonald Research Institute Monographs, 2000, there's an excellent article based on a study of lactase diversity titled, "*Lactase Haplotype Diversity in the Old World,*" (chapter 36) page 305, by Edward J. Hollox, Mark Poulter and Dalls M. Swallow. Science knows that lactase persistence is a genetic trait. It shows up in various frequencies in different populations.

The point is you can't point to genetic markers to distinguish one ethnic group or race from another in order to prescribe a certain dosage of a drug or food because there is diversity in individuals. You have to look at the genes, the genetic profile first to see whether the individual inherited a specific genetic marker.

On the other hand, some members of some races react differently to some drugs and foods, and you can't prescribe for one person based on how the majority of his or her ethnic group or race reacts to the drug or food without seeing the genetic profile to see what the individual inherited. I say before you prescribe a dosage consider how the person's ethnic group reacts to that dosage....but know what genes the individual inherited first. That's at the root of nutritional genetics. Because it's such a hot buzz word in the news, there is lot of discussion about the changes going on.

Scientists are becoming mighty particular about the reputation of who they open up to because of the old adage that says your own scientific reputation in a relatively new field in the throes of change depends upon not only on who you talk to, with whom you're friends, but also on the reputation of the publication for whom the media person works that you open up to.

That's why it's nearly impossible for a freelance book author who works for no publication and is not under assignment by any editor or publisher to get an interview with a scientist willing to talk about nutritional genomics. I found *almost*, but not all doors closed as far as getting scientists to chat with me online about what's new and what's news in nutritional genomics. Yet, as you see here, some excellent scientists in the field did give me information at no cost to me for my book on their wonderful, new and changing field.

On the other hand, scientists working with DNA testing for ancestry or archaeogenetics, including those in Europe, greeted me warmly. They readily provided comments and quotes freely at no cost to me and with the friendliest of attitudes. So for those scientists who treated me equally well as any other media professional, thank you, I deserved that. I don't write for the tabloids. I don't put down anyone, ever. I write books for the general reader about choices.

I read scientific and medical journals and news and go to conventions and seminars with other media people writing about the sciences and belong to professional associations for journalists and authors who write about science. The scientists mentioned in this book from the nutritional genomics industry were most courteous and responded quickly to my email. They are friendly people who gave me the facts I needed for this book, responding immediately to my email. For busy people, that is awesome. Thank you again.

Scientists need to protect their careers and reputations, but they also need to remember that the consumer is the bread and butter of the nutritional genomics industry. What the consumer wants most from the nutritional genomics industry besides the profile and the prescribed foods and nutraceuticals are respect and privacy. Grandma knows best here. It's not a matter of preaching to the choir, of scientists selling to physicians, licensed health care professionals, or genetics counselors. The consumer's body as a recipient of healthcare is involved in the product, and the product is more than a genetic profile in a database. The key word is choice.

What's abuzz about nutritional genomics is that one size doesn't fill all ethnic groups when it comes to prescribing dosages of medicine or certain foods based on one's race or country of origin. For example, in the past, many people who lived in certain parts of northern or north central Europe for thousands of years mutated a gene to digest milk without symptoms such as gas, bloating, diarrhea, and cramps.

Many people who lived for thousands of years in Southeast Asia did not inherit a mutation to digest milk without symptoms. You can't assume the ethnic group reacts as a whole because there is diversity. Not all Northern Europeans can digest the lactose in milk, and not all Southeast Asians get the runs and gas from drinking milk because of lactase intolerance. That's why nutritional genomics (nutrigenomics) speaks out in the same way that archaeogenetics by DNA testing speaks about deep ancestry and population expansions. My all-vegetarian diet of soup and salad followed by a lot of muffins and frozen yogurt desserts eaten in my forties and fifties was making my insulin resistance/hyperinsulinism worse. The burgers and fries I ate up to the end of my thirties didn't help. Midway into my sixties decade I decided to eat smart foods, but which foods were right for me, now in my mid-sixties, with a history of parents and sibling dying young of hardened arteries, chronic anxiety, and genetic hypertension? Would my vegetarian diet be appropriate or would it be best to add salmon, and if so, how many times per week? What about green tea extract with polyphenols? Would the caffeine in it be harmful, or could I find a brand that was truly decaffeinated?

I've heard while chatting with people interested in DNA and nutrition, phrases such as *"eat 'bright' for your genotype."* It seems to be a word play on a NY Times best-selling book that I've read a few years ago titled *Eat Right 4 Your Type* by Dr. Peter D'Adamo. The Web site is at: http://www.dadamo.com/. There are lists of medical journal articles in the book, and I was able to read the medical articles referenced explaining the scientific basis of how lectins work. I followed the Blood Type Diet. I enjoyed the Web site designed to educate anyone about the scientific links between blood types and nutrition. Dr. Peter D'Adamo, is the author of *Eat Right 4 Your Type, Cook Right 4 Your Type,* and the *Complete Blood Type Diet Encyclopedia.*

The popularity of these diets recommended for the various blood types—O, A, AB, and B are backed by medical journal articles of scientific studies referenced. However, what if you are a fast burner, a fast metabolizer who needs some small amount of protein at every meal and some good oils to keep the insulin from pouring into your blood every time you eat a certain amount of carbohydrates? People are fast oxidizers, slow oxidizers, or mixed and balanced. So if you eat according to your blood type, what if you are a fast oxidizer and need that balance of carbs, proteins, and oils to normalize your hyperinsulinism or insulin resistance? Think about that before you research your diet, and see the need to customize the diet to your individual genetic signature and your metabolism. So do your research.

Eating according to your blood type is a theory, but backed by scientific studies. The entire concept of genetically individualized nutrition seems to have evolved starting with blood type. For example, there are scientific studies of how various lectins agglutinate the blood depending on one's blood type. You can see reviews of many diets at the Diet Reviews and Information Web site at: http://www.chasefreedom.com/eatrightforyourtype2.html.

It's as if the sciences of genetics and nutrition had its roots beginning with blood testing regarding how certain foods affect people with certain blood types. Before the entire human genome code was known, scientists used a metabolic and chemical approach to study how one's food intake influenced one's health. Blood was tested, physiological responses, metabolism, glucose, and other tests to measure anything from allergies to whether one had diabetes. Today we have the whole genome at our fingertips. Tests are costly, but the science is here now.

We don't have to wait another ten years to know how our genes react to food, exercise, environment, relationships, stress, or lifestyle. For example, if you look on the Web you may come across something like this: some people who prescribe nutraceuticals based on blood type might suggest that if you have blood type A, and your body is full of cortisol after heavy exercise, change the exercise or take some vitamin C. Is this medical education or medical advice? It's a thin line. And the person who puts this suggestion on the Web may or may not have an M.D. Naturopaths and nutritionists can make suggestions on the Web. How should you react?

You can ask your own physician when it comes to medical advice. On the other hand, is your physician trained in nutrition or nutritional genomics? Will you be referred to another specialist? Would that specialist be a nutritional genomics consultant, a naturopath, a genetics counselor, a nutritionist, a nurse practitioner, a nutraceuticals company, a testing lab, or who? That's why the consumer needs to take control of his own research about healthcare.

Scientists look at proteins in food and allergies, or whether a certain blood type became agglutinated when a certain food was consumed. Back in the era of World War One blood testing was linked to population genetics. Today, research on the entire human genome code has revealed a more comprehensive way of studying health and nutrition through nutritional genomics. Instead of only looking at blood type and the way food affects you—because your blood type is reflected in how every cell in your body reacts to a certain food—science looks at all your genes.

If you're a consumer with no science background, how can you make money in nutritional genomics? You can put up an informational Web site. Compare, review, and evaluate the genetic testing companies that have contact with consumers. You can compare, review, and evaluate the research companies and the university programs. Basically, you don't need a degree to develop a consumer-oriented information Web site and/or database or an annual book of facts published for consumers. What you can do is compare what's available to consumers in the field of genetic and genome testing. It can include nutrition, ancestry, pharmacogenetics, nutraceuticals, and related sciences such as phenomics.

You can provide services in a variety of categories even with no science background to start with by listing information on companies as a type of consumer report on the nutritional genomics industry, on the DNA testing for ancestry industry, and related health and nutrition or nutraceutical firms. If you have no money, begin by asking the reputable companies you list for funding.

Talk to experts who know which companies are reputable who can be asked to fund your online business. Ask the food industry for funding. Talk to the patent attorneys. If the universities have little money for research and can't help you, talk to those where the universities go for funding. What government agencies do the universities contact for funding their research in nutritional genomics? What large corporations? Which philanthropists? Start with the largest, international food companies and the healthcare industries.

Then work down to the nutrition companies and the people in alternative health care. If you only want to give out information, that's helpful to the consumer. Develop a database. You want the companies to sponsor your efforts. The first step is to write up a plan and keep knocking on doors. Do a bit of fund raising. If nothing happens, create an informational Web site anyway. I did at www.newswriting.net. My purpose was to give information on genetics, broadcast my talks, and promote my books with excerpts and articles.

Yours may be to compare and review companies involved in any aspect of the DNA testing or genomic profiling arena and possibly to review books and publications. You could include feedback from customers of DNA and genome profiling and testing firms, research organizations and institutions, and any company dealing with the public or the healthcare systems and food systems design corporations or the alternative health markets.

Suppose you weren't interested in any business but wanted to check out the reviews and comparisons of DNA testing companies at a Web site. So the possibilities are limited only by your creativity. If someone tells you that you haven't the credentials to do something, start a Web site reviewing and comparing the kind of companies in which you are interested, and let the consumers give you feedback as well as the professionals and experts in that field.

There are two voices to be heard—the consumer's and the expert from the most reputable companies in any field. That's what informational sites are about—selling facts to nontraditional markets, reviewing, and comparing. There's room for more than a few of these informational sites.

Certain people have different responses to nutrition than the majority of people. Twenty percent of people respond one way, twenty percent another way, and sixty percent still another. Do you respond to eating fruits by getting a bulging belly and too much insulin in your blood? Your DNA wants smarter food. Here's the big picture: Your DNA has cultural, biological, and nutritional core identities. Are you ready to look for the smile in your genes? Do mothers know best what foods their children will respond to according to the rules of nutritional genomics? Are the rules in place yet?

Your genes express their biological and cultural components in health or disease based upon the type of foods you eat. Your genes express their needs based on the type of exercise you do to stay fit as well as the way you perceive the stress in your environment. Your DNA is alive. It's conscious. It's a whole you in miniature. I'm a medical journalist interested in finding out my own response to nutrition. So I began by questioning scientists and other experts in genetics.

How about your response to what you eat? It's okay to want to eat smarter by learning to interpret the how-why-when-where-who-what-where of your entire genome and apply it to practical use in eating healing foods and choosing if needed, helpful nutraceuticals. The goal is to get consumers interested in looking at their own picture of health at the molecular level. It's a fantastic subject to learn. Question all authority and think for yourself. Eating for your genotype is here at last to free you from food cravings. Or is it available to all equally—to poor and rich alike? I say it is if you look deep enough into the research available to the public.

If you could get a printout of your entire DNA genome, what would it tell you about smart foods? How far would you go, how much would you pay to stay healthier and to prevent or delay chronic disease? What would you do to preserve your privacy and keep your employer, your HMO, or your primary care physician from looking at your entire genetic profile? The details are in the DNA.

20

Intelligent Nutrition or Smart Foods? Who Makes The Rules in Nutritional Genomics?

When a how-to book author who also writes mysteries, adventures, historicals, and romantic suspense novels turns to writing nonfiction books on DNA, from DNA-driven genealogy tips to nutritional genomics and archaeogenetics, most scientists with doctorates do a double-take. After all, archaeogenetics is an exhilarating hobby-turned-full time interest. What's encouraging is that some scientists in nutritional genomics, DNA-driven genealogy, and in other areas of genetics are speaking out to me, as you can see in the e-letter below, reprinted with permission:

From: David Crawford
To: Anne Hart
Sent: Thursday, July 17, 2003 9:00 AM
Dear Anne:

Kudos to you for standing up to the profit warlords, politics and professional jealousy!
I am also a "full-time volunteer who writes for the love of science."

Albert Swietzer said "a person who is truly happy is the one who has diligently searched and sought out how to serve others". That is our founding truth at Interface Medical Research, Inc.
We work as consultants to the nutragenomic and nutraceutical community. From basic research, product design and development and clinical studies, we are committed to serve the scientific community and public.

We would be tickled to be listed as a service provider in you book. Thank you!

Please use my letter as you see fit.

David S. Crawford, PhD
Research Director
Interface Medical Research, Inc.
545 Farr Avenue
Wadsworth, OH 44281

Thank you, scientists who were willing to talk to me about how foods, genes, and healthcare are linked. It's as if there's a pyramid with your genes at the top and nutrition—foods and nutraceuticals—at one end of the triangle linked to your healthcare at the other angle. It's a triumvirate of intelligent nutrition tailored to your genes.

What I like about the science of bioinformatics is that the field is about managing databases of biological data. The future of bioinformatics helps to ensure that the growing body of information from molecular biology and genome research is placed in the public domain.

Information should be available equally to self-taught freelance writers as well as staff journalists and to the potential consumer of genome testing and research. Information should be accessible freely to all facets of the scientific community in ways that promote scientific progress.

If you need to do some fundraising to establish a DNA bank or support group project, look to celebrities, retired celebrities, media and film producers, and entertainers. Each year movie stars frequently headline Hollywood fundraisers and similar events that bring in money, sometimes in the millions, to donate to causes, most often for various campaign committees or health causes.

You might make some phone calls to see whether celebrities might want to get involved in planning events that would bring people together to raise funds for nutritional genomics-related projects, perhaps on healing foods and individual genomes. You might develop a slogan such as "Don't let your *genomee* become your enemy." Think in terms of foods for the majority in the midst of individual genetic differences.

What foods are best for the majority, 60 percent of the population who usually follow food guidelines recommended by healthcare professionals, dieticians and nutritionists based on general physical exams? How do you respond individually to certain foods and diets? Does your immune system go down when you fast or consume sugary foods such as fruit juice, when you exercise or travel? What foods contribute to your well-being?

When you research a study, find out who has done the study and whether it has held up to the critics or was found to be flawed by someone credible. If the study cannot be shown to be flawed, consider it as evidence to explore further. For example, on July 22, 2003, the Bee News Services of the Sacramento Bee, a daily newspaper published an article titled, "Fish Diet May Help Seniors, Study Says." The sub-title read, "Weekly helpings may cut Alzheimer's risk, doctors argue." The article listed Chicago as the origin of the news article.

What's missing in the article is any mention of what study the news piece referred to. The subject of the news reads: "Older people who eat fish at least once a week may cut their risk of Alzheimer's disease by more than half, a study suggests." The only problem is that the study is not named.

The news article proceeds to mention that the study adds to the evidence that what you eat might influence your risk of developing the chronic illness. What study? It refers to "a growing body of scientific evidence." I would have liked to see the name of the study so I could read the details in a scientific journal that usually publishes abstracts of and articles connected with studies. Where can I find this "body of evidence?"

I want to read the research studies linking various foods to cutting risk of various illnesses for myself. How about you? Is the recommended food good for my genetic expression and everyone else's or only for the twenty percent of the population who can eat most diets and still remain healthy into old age? What about the other twenty percent who can't eat certain foods without damage to their arteries and organs?

What the article does discuss is that the evidence adds to accumulated information on the subject of reducing risk of developing several chronic illnesses such as cancer, Alzheimer's, or heart disease, if people eat a diet rich in fish, fruits, and vegetables and low in saturated fats from red meat.

There are statistics from the Alzheimer's Association in the article. Approximately four million people in the USA are afflicted with Alzheimer's now, and Alzheimer's cases are expected to rise to 14 million by 2050. The evidence presented in the news article was that 'researchers' found "that people 65 and older who had fish once a week had a 60 percent lower risk of Alzheimer's than those who never or rarely ate fish."

Who did the study? Who are the researchers? When was it done? No mention was given of the amount of mercury in the fish eaten. The article mentions that the meals included fish sticks and tuna sandwiches. No mention of whether the tuna was canned, canned with salt or no salt added, canned with oil or water packed, or whether it was fresh tuna or how it was prepared—grilled, fried, or baked, or boiled? Was the research done by the fish industry? What about the warning labels in supermarket's fish departments about the mercury in certain species of fish?

My neighborhood supermarket has a big sign posted at the fish counter for women of child-bearing age or pregnant women to avoid certain types of seafood with the names of the species mentioned due to the mercury levels. I see an article in the August 2003 issue of *Reader's Digest*, a cover story titled, "Hidden Dangers in Healthy Foods," that connects fish eating with individual reports of nervous system problems, illness, and hair loss due to mercury residues in fish.

Science writers should be accepted as media professionals by the science community without discrimination as to their credentials and should have equal access with all staff media of large newspapers, magazines, or broadcast networks to the growing body of information in genome research or any other area of molecular biology, especially areas related to preventive health and nutrition, foods, and vitamins and minerals.

The bioscience communicator whether self employed or staff employed acts as a go-between, a liaison between the consumer and the healthcare system or research laboratory, a type of educator and ombudsman. The communicator should be respected and allowed to access databases of biological data. Making the complex easier to understand for the consumer is news.

The scientific community should include journalists who specialize in writing about specific areas of science such as genomics without requiring writers to have science degrees. We attend enough genomics seminars, read enough journal articles, observe conventions, and read enough monographs and books to know what questions to ask the experts. In addition to looking toward the media as a liaison and ombudsman, the consumer of genetic testing is concerned about privacy issues.

For nutrition purposes only, results of tests for genetic risks should be private. The whole idea is to lessen the risk before a chronic disease develops. You can change how your genes express themselves based on food and nutraceuticals, but you can't swap your genes for some other genes.

Nutritional genomics is all about effects of intelligent nutrition—smart foods—on how my genes express themselves. Don't laugh at me. Smile. I need to know how specific dietary chemicals will nourish me at the molecular level. You also need to know about the effects of food on your own health.

Your genetic information—a printout of your entire genome—along with a profile interpreting it in terms of risk, disease, and recommendations for foods and/or supplements, should be kept private to be shared only by you and your healthcare professionals, and perhaps your heirs in a time capsule about DNA and medical history to be passed on to future generations. It's about what makes you tick at the molecular level, about DNA and health, behavior, and mood…If you are what you eat, why should you eat that way, and what will happen to your health at the molecular level when you stray?

Do you metabolize your meal fast or slow—burn your calories quick or over hours? Your genetic profile will give you clues, but you need to learn how to interpret it. And for the time being, you'll find it difficult to learn for yourself and take charge of your healthcare, because getting your entire genome tested is done in reputable places under the supervision of a physician managing your healthcare. You'll have to work hard to take control of your diet and your genes.

You have to work towards taking control of your genetic testing. The only way to do that for now is to learn how to interpret a genetic test of the entire genome and how to apply that to a diet that works for you. You need to get control of your genes. They are yours. The food you eat is yours. Why should putting diet and genes together cost a fortune now and be cheap in a decade? Knowledge is not only power. It saves you a bundle.

Okay, so the doctors need to eat, too. I should know. My son and son-in law are physicians. Just ask their wives about eating smart foods. The point is there are ways to learn how to look at a DNA test and look at a diet and understand how your genes respond to what you eat. Books are online and in the medical libraries. Read them, you autodidactic learners. Educate yourself. If a social worker can take a two-year masters degree program and become a genetics counselor, and if a nutritionist and a physician can learn to talk to people about what diets are healthiest for them, then you, too can learn to interpret a DNA test of your entire genome.

The trick is finding someone who comes highly recommended to give you your genomic testing. You do need to send your DNA to a lab. Where do you start? Talk to professionals in the field and find out what books and journals they read in order to understand how to interpret the DNA tests. Find out how your genes respond to what you eat. Very few people have the time to be interviewed, but keep asking around.

Find out whether anyone in the field or recently retired will teach an extended studies course for the public. Go to conventions and conferences where professionals in nutritional genomics congregate. Make friends. Read my romantic intrigue novel series on DNA. Or my books on tracing your ethnic DNA. Cook for your genes.

Have a nutritional genome feast/party. Invite people who can help you match your DNA to your diet or apply for a job in the field. Whatever you do, get a handle on, control of what you eat and how what you eat influences how your genes respond to that food.

Write about your dieting experiences. Go on the radio. Talk to people. Read professional publications. Sign up to receive the news of the industry. Ask yourself whether you want to get into an allied business from another angle where you'd rub elbows with people in the field. However you choose, get your entire genome tested, and find out how your genes respond to recommended diets compared to what you eat now. The whole industry is moving toward this becoming routine in about ten or more years. If you're as old as I, don't wait, start asking questions now, and find out what foods are healthiest for your genotype.

How does what you eat affect the expression of your genes? Nutritional genomics is causing a rhumba of rattlesnakes in the media. Do your research before you spend money on testing your DNA. If you don't know how to interpret your DNA test for nutrition, you'll need to find a physician who works with nutritional genomics professionals in the medical field and has had training in nutritional genomics.

For example, if the physician doesn't have a working contract with nutritional genomics professionals from reputable companies, how will the individual interpret a printout of your entire genome in relation to tailoring your foods? How many physicians today are trained in nutrition let alone nutritional genomics, a field that began around the year 2000?

So check out your healthcare professionals and the companies doing the research to make sure what they offer is what you want. Read the medical journals connected with nutritional genomics. You can find them in local university medical school libraries open to the public. Read some of the latest journals in the field. Go to conventions and attend meetings of associations of professionals who are connected with nutritional genomics and ask for reputable referrals. You can network with people in the field. Read about what research is being done by various companies.

When I was single and in my twenties, circa 1962, I joined clubs for single professionals where the scientists and technical people congregated so I could talk to them and enter the awe of their world in order to write about the aura of enthusiasm that engulfed them in such important work. I majored in professional writing/English and did get an M.A. degree, working my way through from the age of seventeen with a full-time day job, but since the age of eleven I had a deep interest in reading and daydreaming about science.

In 1962 my best female friend majored in chemistry and math, and I felt stifled by my day job typing important men's manuscripts and finishing my degree at night. Through chatting with the experts—the scientists, I found the inspiration to play with words and channel creative expression to novels featuring exciting, learned characters as scientists. I mean, they actually talked to me, a retired janitor's kid whose mom was a maid. (No—none of them proposed marriage, but I got to write romance and suspense novels about fictional scientists.) I needed something more—science news.

It was the high point of my life. Back in 1962 I ask each scientist who asked me to dance whether there was a new way to look at disease. I wanted to know then what my genes really wanted me to eat. What if genes could talk? What would they whisper to each of us? The DNA helix was big news with Watson and Crick visualizing the double helix. I ended up visualizing a double helix myself—writing the romance novel titled, *The Bride Wore a Double Helix*. The outcome was I wasn't left on my own. We talked about foods, as in how a quarter of a fried potato eaten each day for a year could cause someone with "specific genes" or "fat cells" to gain 'x' number of pounds.

Our conversations on the ballroom floor were about consumers getting their proteins analyzed. More than forty years later, science has spectrometers, and you can find bioinformatics databases that manage nutrition-related information. You can now have a genetic profile compared and cross-referenced to your blood chemistry, metabolic tests, or any other areas of your physical exam or genetic profile.

You can look at genetic risks and find the foods and nutraceuticals that will help you. If you know eating fruit all day will make your blood sugar problems worse, because too much insulin is released into your bloodstream due to a genetic situation, you have some more control. That's what it's all about—control. It's you armed with ways to override your genes as much as you can with healing foods and nutraceuticals, if needed.

Don't only get tested and then left on your own on the dancefloor. You need some kind of medical supervision so you can learn how your test was interpreted and why those foods were prescribed as well as the effects those foods will have on the expression of your genes. You need to look at yourself at the molecular level. When you've self-educated yourself, you can take more control over what you eat knowing why the food affects you the way it does—at the cellular levels. You'll know why you're salt sensitive or not, and what happens inside your cells when you eat foods that make react the way you do.

Everyone takes nourishment expecting to feel well after eating. Don't let the caricature of yourself hold you back from exploring how you react to dietary chemicals in food. Look at the nuances, the changes in concentration of the nutrients you take in. Look at how the chemicals in foods make you feel. Eat certain foods and take your pulse and blood pressure. What makes your physiological responses healthier?

How long does it take between eating and seeing the results in your health? You need to compare your genotype, that genetic information such as your genome, your DNA, and look at the way you respond to a particular food or meal. That's why a lot of diet books don't consider your individuality at the molecular level. Many stop at the metabolic or chemical level or just look at blood types. That's important too, but so is your entire genome, all your genes and markers and DNA. What nourishment do your genes require for you to become healthier, delay the onset of chronic diseases, and feel well?

Your genes shuffle each generation. They recombine. Food, like DNA moves beyond family history, ancestry, and molecular genealogy. You need to realize the cultural, biological, and nutritional components of your genome in order to eat smarter for your genotype.

Here's how to eat bright for your genotype. The new field, blooming since about 2000 is called nutritional genomics. Nutritional genomics link what you eat to your physician-supervised DNA test of your entire genome and your healthcare. Nutritional genomics is about food plus supplements when needed and any medicine and/or supplements termed nutraceuticals.

Your genes will give out the signals of what whole foods are needed to delay or prevent infirmity. Who is going to test that DNA, and who will teach you to interpret the test when you can no longer afford a supervising physician for healthcare and have to feed yourself on your own on what money or lack of it may come?

Tailoring your food to your genome reminds me of what I heard for the past thirty years in the holistic food fanfare world of eating brighter for your individual metabolic self. Only today it goes beyond the metabolism and moves to your entire genome—your genotype. That's why a lot of diets don't work for everyone, because food affects people in different ways. What's one person's nourishment is another's allergy.

I've been touting eating healthy since 1959. It's what the holistic food and health conferences and conventions were spouting for at least three decades. Now it's arrived: eating bright according to your genotype. You have to eat smart, that is intelligent diets tailored just for you if you want to be healthy. That is, unless you're lucky enough to be part of the 20 percent of the population who can eat any diet and stay healthy longer.

Ever since I avidly read, tried, and was successful at eating for my blood type after reading about it a few years ago, and later for my personality type. More than a decade ago I tried to eat right and take the right supplements for my menopause.

When my genes acted differently to hormone therapy, I stopped the therapy back in the early nineties when my healthcare professional insisted that going off of estrogen and progestin was tantamount to a diabetic going off insulin.

Nevertheless, after looking at my physiological responses to progestin, I dumped the pills and took up exercise and raw veggies. What works well for another's genome, didn't work for mine. After reading a book on body type and diet, I cut out the sugar, being a "thyroid body type," and my large, over-stimulated thyroid felt great again without the excess sugar. And since I began to eat whole grains and raw veggies, I tried to take charge of my own diet.

Before nutritional genomics, I had to read medical articles on blood type and diet or body type and diet, and figure out whether I was eating according to my genotype. Then I turned to books on eating for syndrome X, hyperinsulinism, and tried diets of 45% carbohydrates, 35–40% good oils, and 15% proteins to stop the excess insulin and insulin resistance after sixty from making my big tummy flatter on my thin, 124 pound frame. Without knowing my genotype, it was all based on metabolism, blood chemicals or blood type, ancestry, or body shape. Now with nutritional genomics, I can get to the molecular level, my entire genome…as soon as the price of testing becomes affordable to me…and available to consumers.

You can lobby for nutritional genomics to be available to everyone equally. How about the right to have your genome matched to a healthy diet? I want a good diet today that works for me. Why not match your genome to a diet that works to make you healthier? What's happening in this industry?

Consumer, be alert. I can't wait another ten years for nutritional genomics to be available to everyone on demand at little cost. I want it now. My solution is to write about it. My eight grandchildren will no doubt have it as their dads are physicians, and it will be a part of their children's lives. So what can you do now with your DNA as far as practical applications of DNA testing not only for ancestry or population genetics, but for nutritional genomics?

Scientists use the word 'interface' a lot as do business executives and people with day jobs. I have no day job. I'm a white-haired bag lady, so I don't have to interface. I interlace instead. What I do is greet you with a smile and hug. I don't interface with you. Only in nutritional genomics you have to study how to interface between your diet and your genes. Yes, that's you in there in the middle right between your diet and your genes. Now interface. Take a picture of your gene expression. Now smile. Snap. You've just developed a picture of your gene expression. It's profiled and digitally filed. You now have a picture of how you respond to what you eat. I'll say it again scientifically:

Nutritional genomics researches the interface between what you eat and your genetic processes. Scientists analyze your single nucleotide polymorphisms (SNPs)[1]. Your gene expression is profiled to get a picture of your response to your nutrition.

Individual nutrition is emphasized in the new field of nutritional genomics where you tailor what you eat, the nutritional supplements you take, and your entire health program of diet and exercise, work and lifestyles according to your own genetic profile—your genome. Forget about diets for large numbers of people.

Maybe those diets fit the sixty percent of us who can eat those things, but what about the twenty percent who need tailored menus according to our genes? Or what about the other twenty percent who can eat almost any food and still not develop the main diseases that could stem from eating the foods not right for our genotype—heart disease, hypertension, asthma, hardened arteries, cancer, loss of memory, and more? And are these degenerative diseases linked to changes in our genes? How do the different ethnic groups react to different foods?

All this research shines under the umbrella of nutritional genomics. This year it's tracing ancestry by looking at DNA test results, and next year the focus will be on nutritional genomics, how to tailor what you eat according to your genetic markers. That's where the focus of my next book is. Most people aren't aware of the research being done at universities or at companies targeting research on nutrition and your DNA.

There are new fields within nutritional genomics such as the study of how people's DNA change according to their diets. There's the study of pharmacodiagnostic drugs, and phenomics, tailoring your medicine, therapy, and healthcare or exercise and lifestyle to your genes.

Books abound on eating right for your chemical, blood type, and metabolic systems. At the molecular level, you'll see books popping up on eating bright for

---

1. SNPs see http://www.ornl.gov/TechResources/Human_Genome/faq/snps.html. SNPs are "DNA sequence variations that occur when a single nucleotide (A, T, C, or G) in the genome sequence is altered." The letters are pronounced as "snips."

your genotype, smart diets for your genetic markers. One diet book will not fit all. You have books on eating for Syndrome X, for diabetes, and other illnesses that appeal to groups, but what about eating bright for your genotype? Studies compare the immune systems of people from different ethnic groups or geographic areas.

This country faces a diabetes epidemic in children. Sugar is added to health foods such as soy milk or almond milk and other foods or beverages to bring people back by taste, even to 'addict' people to the sugar so they should buy more of the health food product in some cases.

Try finding an unsweetened almond, rice, or oat milk. You'll have to make your own from water and pureed grain. You'll have a hard time finding unsweetened soy milk, too but it's around if you look long enough at various shelves in health stores. Some health food products are drenched in salt so salt-sensitive seniors can't easily find a decent meat substitute that isn't loaded with salt, sugar, or fat. All this—in the midst of a revolution of research focusing on tailoring your food to your DNA to prevent or slow down premature degenerative diseases.

Make sure you're working with a reputable company that arranges to work with a managing physician. Beware of companies that test different aspects of your body to predict what to eat as the testing isn't of the complete genome and may not be done with a managing physician to interpret the results of the tests for prescribing your diet.

Don't waste money on incomplete or inaccurate testing. Watch out for testing companies that offer nutritional consulting based on genetic tests for a variety of complex diet-disease-or diet-health associations as they may not be reputable. Find out who is reputable before you spend any money. If you are testing for single gene defects such as PKU or hypolactasia, these can be tested for and diets arranged based on the test results.

Check out reputable companies that offer tests which are supervised. Some companies research the linkages between genes, diet and health. Don't go to a company that offers unsupervised tests. Learn the difference between a research company looking at how genes, diet, and health are linked, and a genetic testing service company. Beware of the unsupervised test or the company selling snake oil when it comes to finding out how your diet is linked to your genes and your health.

How do you know the difference? Talk to professionals in the research field before you think of any testing, and make sure everything done is supervised by a physician managing your personal healthcare who also is participating in research of how health, diet and genes are linked.

Think about all the Japanese who eat a dairy-filled American diet and land up with diseases not found in Japan when they kept the ethnic diet. What happens when other indigenous people take on a processed-food Western diet high in whatever disagrees with the nutritional expression of their genetic markers. It's a horizontal expression of a vertical desire. The remedy is tailoring your nutrition—foods and supplements—to your genotype as well as customizing your type of exercise and other lifestyle events.

Perhaps it's time to tailor our DNA to our multitude of ethnic diets and first find out which one we inherited or which diet works for our genetic expression. For me it's Omega 3 oils and low-carbohydrate vegetables, B complex vitamins, whole grains, and the Syndrome X diet, but how can we know for sure until our entire genome is passed through the test, and for now, the test is still expensive?

So we have to guess, perhaps at the mixture of ethnicities, family members and their lifelong genetic expression of what they ate. Look at what your families ate and what it did to them. Did you inherit any of those gene mutations? How has what you ate influenced your own health? The answer may lie in the SNPs, the genetic markers. Stay tuned for reporting on new research. These sciences include not only 'nutrigenomics,' but also tailored diets and understanding your genotype.

Related sciences that look at genetic variation to customize healthcare strategies include proteomics, metabolomics, bioinformatics, biocomputation, and phenomics. It's all about getting the details and the big picture of your nutritional status, requirements, and genotype. You can eat better according to your genotype without having to wait another ten or twenty years before testing your entire genome is affordable.

Start by observing your body's reaction to the nutritional and exercise environment. It's about personalized nutrition and medicine. Nutrients are not just nutrients. You have macronutrients, micronutrients, and antinutrients. The details come out in gene expression. Nutrients can alter your gene expression without changing your genes. Look at your metabolism, DNA, and gene expression the way you use your DNA test results to trace your ancestry. The door is open to personalize diets and customize your medical care. Under the umbrella of smarter nutrition, DNA plays a role that treats you as an individual rather than a member of a special group.

If you're an older individual, it's not possible to wait another decade to look for a customized smart diet to eat "bright" for your genotype. You can have your DNA tested now. Only be on the alert and research the company you're working with so you don't get scammed. A DNA test of your mtDNA or Y chromosome for deep ancestry is not enough of a test. Neither is a test of your racial percentages.

You need more clues—metabolic, chemical, and genetic clues. There are plenty of books on how to eat according to your metabolism or chemical clues, but you need more genetic information. What you need is your entire individual genetic makeup. If you can get a company to give your entire genome a pass-through and genetic printout, it could be helpful in preparing customized diets to help prevent and ease chronic disease.

The only obstacle is that you'd need either a professional to prepare the diet by interpreting your DNA test of your entire genome, and not all physicians can look at your genes and prescribe a diet. What you'd need would be a qualified, accredited, and experienced nutritional genomic specialist to consult with you, look at your genetic profile, and prescribe a diet. You'd need someone with enough experience in nutrition and genetics to know what foods would be best for you as a person. No diet fits all people, not even all ethnic groups. It must be individualized.

That's why it's called "intelligent nutrition." Smart menus are customized to your genomic profile. You eat at the molecular level. Eating smart for your genotype is science-driven nutrition. You can write a diet book for a specific individual, who's a member of a specific ethnic group of whom you can't prescribe a one-size-fits-all menu for that person within any group. There's too much individual diversity.

Where do you start? Research is the first step for any smart nutritional genomics consumer. You start by reading and going to conventions of food technologists and nutritional genomics professionals and/or students. Do your own homework. Start with reading about the metabolic diets based on an individual's chemistry. A decade ago, it was eating according to your blood type or metabolic type.

Now it's eating according to your entire genetic profile, your genome. The trend is becoming molecular, eating down to the atomic level in your molecules. That's because your genes express themselves at the molecular level, within and from the cells. The "eat according to your individual profile" movement began with alternative healing movements that always are keyed into scientific research at the genetic, molecular, and chemical/metabolic levels.

In most every alternative and holistic health magazine you find, the footnotes contain references to studies and medical journal articles. Check those out. It's a good use of time to learn how to read articles in medical journals and look up the terminology. If a study is flawed and is reviewed in another publication, read it. If a study holds up with time, keep a scrapbook of the study or articles for your reference.

Only now it's beyond looking at blood type or the lectins that agglutinate your blood from harsh reactions to foods, individual reactions. It's beyond chemical and metabolic, it's now genetic. You inherit and pass on recombination of genes that express themselves in various, individual ways, even within the same family. How you react to food is genetically determined and expressed not only in and through the genes, but also in behavior, mood, and sense of well-being.

Think of the potential for this science-driven food-for-health industry. You must explore the innovations and find professionals to network with to explore and evaluate the forthcoming innovations. Get on the mailing lists of the nutritional genomic research institutions. These research institutions may not do genetic testing of individuals for diets, but are engaged in the type of research you want to learn about before you put your health in the hands of a managing physician who must not only interpret your genome but prescribe diets and/or nutraceuticals.

You've heard of pharmaceuticals. Well, think nutrition and nutrition supplements. Research those nutraceuticals. Make sure everything is supervised, but that you still have an opportunity to learn about why and how your genes respond to what's prescribed, that is, you still have control over what goes into your body and knowledge of how it will affect your health. After you've finished being supervised for your own health, learn all you can about how genes and the diet work together to make you healthier. Then take control of your diet and celebrate your knowledge of a new language—a way to communicate with the expression coming from your genes.

Nutritional genomics is a science dedicated to researching smart diets customized for individuals with a purpose of creating a healthier population. Ask yourself as a consumer, how come it's so expensive right now to test the entire genome of a human to get a prescribed diet and so cheap to test the entire genome of a dog or race horse for breeding or diet? Lobby for costs to come down so diets can be prescribed according to one's genotype.

Besides the research scientists working or studying the field of nutritional genomics in academic or consumer consulting capacities, you have food industry leaders flocking to hear the scientists specializing in genetics and nutrition because the field is potentially a money-making enterprise for nutritional genomics-based food industries.

The whole idea of nutritional genomics got its wings around the year 2000 when studies revealed that a "dumb diet" as opposed to a "smart diet" (based on your gene expression) can fan the flames of your chronic disease risk. So to prevent or delay the chronic diseases for which your genes may be at risk, a diet prescribed only for your genes would help delay or prevent those chronic illnesses. You would need a printout of your genes based on DNA testing to find out for what diseases you may be at risk. Then the diet would be prescribed to prevent or delay those diseases.

Not only the diet, but the exercises and lifestyle and any other nutritional supplements would be recommended. The idea of an intelligent diet would be to nourish your genes at the molecular level based on what the genes needed to express themselves in the healthiest way possible for you as an individual.

So why wait a decade for science to come up with prescriptive diets? There's a lot you can do today with DNA testing and biotechnology before the impact on medical care, on the way foods are processed or not processed, and on your own health becomes influenced by the bottom line—profit.

Science has mapped the human genome. Back in 2001, you have the beginnings of the "marriage" between genomics—mapping your genes, and medicine. You have alternative medicine clawing at the door for decades demanding this knowledge and using it long before the medical fields ever thought of offering courses in nutrition beyond an introduction.

Now medicine and science is working to identify how genes work to change your health. In the meantime, the alternative health books have been emphasizing this all along—how to eat according to your body type, your blood type, your metabolic type, your chemistry, for years. Now medicine and the food industry is finally listening. The turning point came when the functions of various genes were identified with the aim of finding out how these functions affect your health, your risk of disease, and how the genes interplay with your lifestyle, environment, stress level, and even how you spend your day, let alone what you eat.

What you want to know now, especially if you're an older person who can't afford to eat the wrong foods for the next decade, is how and why your genes predispose you to sickness, premature aging, or obesity. While you're out there fighting for organic food, or worried about how genetically engineered crops will make you ill, here's one more concern: think how food is processed and how the process will express itself through your genes. Now customize your diet for your specific needs based on your genes—the expression of your genome based on the foods you eat and the lifestyle and exercise you do. In other words, eat "bright" for your genotype.

What is nutritional genomics? It's a market. It has legal and industrial implications for the food industry. We anthropology and bioscience communicators have been writing about alternative health for decades, exploring the whole oat grains and raw vegetables diets for years, looking at whether fish diets give us lower blood pressure or just a dose of mercury and writing about all sides of the whole foods spectrum, the details and the big picture.

Looking out toward the next decade, nutritional genomics professionals want to attract the baby boomers because of the large size of its population. However, the need is great right now among us parents of baby boomers, the senior silent generation. We are now over sixty and want to reverse our age-related conditions that pop up in the sixties and seventies decade of our lifestyles. We want nutritional genomics now, and we won't wait a decade. So how do we start, our little mtDNA or Y-chromosome DNA tests in hand that we sought to search for our deep maternal or paternal ancestry?

Don't you dare wait for boomers to demand smart diets prescribed for individuals according to their genome. We won't be scammed by diagnostics or counseling aimed at seniors by less than professional firms. We want the nutritional genomics professionals to be our entrepreneurs. To ensure this, we stick to the whole organic foods, but wonder which foods are best to nourish our genetic needs? Do we eat food cooked, or raw? What happens when too much of the raw food blocks certain vitamins or enzymes unless the particular food is cooked or frozen? What specific foods are good for us as individuals, and who will prescribe them? How much will it cost? How scientific is the information?

Could the genomics revolution actually be the alternative health movement and the whole foods movement joining together? Its purpose would be to look at the molecular expression of specific foods on the body. During the past 30 years, I've been attending holistic health conventions hearing lectures on the benefits of Omega 3 fatty acids, flax seed oils, fish oils, whole grains, low-glycemic diets for hyperinsulinism, syndrome X diets, raw vegetables, juicing with the pulp, the affects of too much fruit on triglycerides, how to flatten big bellies caused by eating sugar in people with hyperinsulinism, enzymes, food supplements, vitamins, and minerals, and what's needed to absorb calcium or magnesium, B-vitamins and TMG to reduce high homocysteine.

All of these are reactions expressed through specific genetic problems. And what I had to do 30 years ago was to take a specific, detailed look at how my body reacted physiologically, and change the food, customizing the food based on how that food influenced my physiology, that is the way my genes expressed themselves and reacted.

It always worked for me. Cut out the sugar, and the belly goes down. Exercise, cut the salt, and the symptoms change. Eat those whole grains that have a lower glycemic level, and the hyperinsulinism subsides until the next meal. Your genes are not my genes. What will work for you? Tailor eating to your own molecular level.

Know all about your genes and how food influences the way they express themselves. It is easy to guess at what the genes needed based on the symptoms and bodily response and also by looking at what the wrong diet did to family members who thrived on a Japanese, vegetarian, or Greek diet and succumbed on a high dairy, sugar, bread, coffee, canned tuna, hamburger, fries, too much yogurt, and chicken diet. The clue was to cut the calorie intake in half, and the disease retracted. So you got a handle on what the genes expressed.

A decade ago, I read books on the metabolic approach. I was a fast burner (metabolic), so I was supposed to look younger and thrived on eating protein and fat in the morning. Sugary cereals caused too much insulin and the shakes in the morning and weakness. Eating fish such as cooked salmon fried in olive oil mixed with egg whites in a patty with no salt, but celery seed, garlic, and onion powder felt healthy for breakfast. Oatmeal was too high glycemic and made me weak and shaky.

By working backwards from the symptoms, I could guess at my genes by what strengthened me after eating and gave me energy. Now I can work backgrounds from the general to the specific, from the effects of food on my feelings to the specific genes that put me at risk for the effects of too much insulin or insulin resistance and work backwards and forwards to prescribe myself low-glycemic foods that flatten my tummy and make me feel stronger and full of energy throughout the day.

I attended these holistic health fairs since 1970 when they arose out of the natural foods industry and the alternative medicine movement of naturopaths and nutritionists who were trying to get the medical industry to listen to them. Now they are listening as both sides—the alternative medicine and natural foods industries plug into the nutritional genomics research field from one end, and the medical and food industry from the other end. One group is visionary—looking toward the future. The other is benchmarking, based on what sold successfully in the past.

Prehistoric people ate berries in larger quantities than we eat today, when the berries were in season. Nutritional genomics researches diverse health effects for diverse genes. Flavinoid's effects on inflammation and on the nervous system are being researched. Ask yourself what can you eat that will protect your genes? Did prehistoric peoples have better nutrition then than the standard Western diets encouraged today? Did prehistoric peoples have similar genes to what we have today? Or did the slow-mutating regions of our DNA signal reactions to the change in diet and climate?

From the alternative food and health industries I learned about polyphenols. You find it in green tea, red wine extract, and other vegetable tannins. Anthocyanins are a type of polyphenol. According to Polyphenols Laboratories at: http://www.polyphenols.com/main.php?PHPSESSID=976bfb699f4c396cfb032a25fa7877fd on the Web, "Anthocyanins is a large water-soluble pigment group found in a large number of fruits, vegetables and flowers.

These are the pigments which give plants their brilliant colors ranging from pink through scarlet, purple and blue. Some pharmaceutical effects of anthocyanins have been suggested, for example in treatment of cardiovascular diseases and in ophtamology. The antioxidant potentials of anthocyanins are high."

According to Marilyn Sterling, R.D.'s article in the December 2001 Issue of *Nutrition Science News*, "Eaten in large amounts by primitive humans, anthocyanins are antioxidant flavonoids that protect many body systems. They have some of the strongest physiological effects of any plant compounds, and they are also things of beauty: anthocyanins provide pigment for pansies, petunias, and plums." (Anthocyanins are a separate class of flavonoids from proanthocyanidins, discussed in *NSN* 2000;5(6):231–4.)

Back in folk medicine of the 12th century, bilberry, a bioflavinoid, was fed to young women to induce menstruation. Bilberry (*Vaccinium myrtillus*) is one of several anthocyanins. British pilots during World War II took Bilberry to improve their night vision. So if your genomic profile calls for anthocyanins, at least you'll know what plant pigments in fruit juices or certain wines or in nutraceuticals such as bilberry extract will be of help.

The genes of the plants, such as the genes in the plant pigments feed your genes. The work goes on at the molecular level. In the past, bilberry was used to treat ulcers. If it's found that bilberry may increase the production of stomach mucus to protect the bacteria that causes ulcers from attacking as bilberry was a traditional treatment for ulcers in the past, then folklore herbal medicines may have come full circle to play a role in genomic profiling.

See the Web site at http://www.medpalett.no/index.php?lang=en.

Then there are the procyanidins, found in grape seed extract and other plant extracts. I take grape seed extract in a capsule. I heard about polyphenols and procyanidins from the alternative health market and sometimes on radio shows that most people would call the "woo-woo" industry.

I heard all about grape seed capsules on the Art Bell radio talk show, the show that showcases UFOs along with holistic health speakers and topics of past life and so on. Sometimes you hear about these products in mainstream media and sometimes in alternative health media. The benefits of nutritional genomics has made it to mainstream mass media with articles in daily newspapers and the NY Times news magazine.

What you as consumers need to do is to read the guidelines of the National Research Council. They are the group that offer guidelines of recommended dietary allowances (RDA) based on the results research. So with advent of nutritional genomics, research may open doors for those guidelines to shift.

Now biotechnology has joined up with talking business applications. The polyphenols found in berries is being researched to relieve arthritis suffering. For the last three decades the holistic food conventions have been touting the same blueberries, grapes, raspberries, blackberries, cherries, to help relieve the pain of arthritis. Grape seed extract which contain procyanidins have been said to help strengthen capillaries.

Now science finds out that eating compounds such as polyphenols found in dark berries inhibits the expression of a particular gene. You still have the arthritis, but the pain lessens or goes away. Does it work? Try it and see how the expression of your particular gene reacts to polyphenols.

The procyanidins in my grape seed extract and the polyphenols in the berries or green tea I consume daily not only strengthened my capillaries, it got rid of my 'rhoids. Then there's body shapes. In Ayurvedic medicine of ancient India, body shapes are divided into vatta, thin, pitta, athletic/muscular, and kapha, rounded, with a layer of fat under the skin. Vatta is less prone to arthritis and more prone to anxiety. I'm a vatta.

So just by looking at body shapes you gain a handle on something about the expression of your genes. In the health food stores there are recommended food and teas for vatta, pitta, and kapha types. Just read Deepak Chopra's books, and you'll learn.

The wisdom of the ancients around the world knew something about genetic expression. The body shapes are genetic and react differently to foods. A vatta sipping coffee will sometimes feel anxiety from the caffeine, whereas a kapha may need the jolt to the nervous system to get going. Even without a DNA test, you can look back at reactions to food from the various body shapes and see what foods hit you like a nerve-shattering bomb of hyperinsulinism after you eat, and what food calms you or makes you more alert.

Before you try anything, read the studies and find out which, if any, are flawed, and which stand up to science. Does it work? Only your individual gene can tell you. It speaks not in words, but in the expression of pain or no pain or in other symptoms of strength and energy or weakness and fatigue or chronic illness symptoms. Will your physician work with you? Who will? What about training physicians in nutrigenetics consulting? Will other professionals fill the gap?

What nutritional genomics needs are companies that will sell food that is good for our body organs. Right now I walk into a health food store and find the healthy alternatives to the milk that makes me sick filled with sugar, rice sweeteners, fructose, corn syrup, or other additions, including salt that makes me sicker than drinking milk. Soy milk marked 'plain' is not plain. It's sweetened. You have to look for unsweetened, and most supermarkets don't carry it because they answer when you ask, "Will it sell if it doesn't taste sweet?"

Why are food companies addicting us to sugar knowing we'll come back and buy more of the product because of sugar cravings not because we want the healthier, alternative products. So you have to make your own to leave out the sugar, salt, barley malt, rice flour, or other additives that sell the product by addicting you to the taste rather than selling the healthy alternative point.

A few physicians sometimes also produce and sell vitamins, minerals, or vegetarian food products. Are these products best for your particular genome? Does one size fit the masses or only a percentage of people? My last resort was to make my own soy, oat, rice, and almond milk by cooking the grain in water and pureeing it in a blender. That way I can drink some almond milk by blanching the almonds and removing the bitter skin, then putting it in the blender with water. I soak soy beans overnight, puree them in a blender with water, then cook for 20 minutes and strain into a bottle. Soy milk has to be cooked. Foods have to be researched. Some types of sprouts may contain salmonella or other bacteria on the seeds.

Food companies need to join forces with nutritional genomics specialists who need to join with physicians, HMO-employed nutritionists, and other health care professionals as well as with genetics counselors. If the food industry doesn't promote food good for specific parts of the body, the people who produce food supplements and other health and diet tools will step in and plug into the genomics field.

If all would work together, food could be produced to normalize cholesterol. However, there is a need to create foods for those who need special foods such as low-salt, low-sugar, and low-fat foods that don't take out one and put in the other. For example, low-fat foods often contain a lot more sugar and/or salt than regular food products. Again, taste is being sold, not health. Look at soy cheese.

Most of it contains 290 mg. of salt. Regular Swiss cheese often contains only 40 mg of salt. Or look at labels of some fat-free cookies with lots more sugar and salt added than you'd find in those unhealthier cookies full of trans-fatty acids, true, but less salt and sugar. My answer is to bake my own cookies.

I take a cup of oat bran mix it with a half cup of flax seed meal and some cooked whole oat groats, add a few almonds, a handful of millet soaked in carrot juice overnight, and sweeten with fruit juice or a banana. I add two or three egg whites and a quarter cup of lecithin granules and a little soy milk if needed. Then I roll out the cookies and bake until light brown at 350 degrees F.

Before you boy low-fat organic alternative food products such as soy milk, look at the label and see whether the sweeteners added will send your blood sugar up. If you are genetically at risk for diabetes II in your mature years, try making your own alternative milk products from soy or almonds or grains such as oat or rice. Or look for unsweetened products when you can find them. Talk to supermarkets and convince them that unsweetened products are not marked 'plain' in many cases, and that they will sell, especially to the senior citizens who often don't know where to look for unsweetened or no salt added products.

Know your genes. The old adage "know thyself" means know how your genes express yourself. There are not enough specialized physicians to go around to all the baby boomers who will be knocking down doors in a decade for healthcare. Don't wait for the boomers. Senior citizens must take nutrition into their own hands and do the research on their genes now, not in a decade.

If you're retired, it's time to make nutritional genomics your main hobby as far as researching the effect of food and exercise on your genes. Nutritional genomics is linked to many other scientific areas of study. It's the most important tool for self identity ahead of personality tests and DNA tests for ancestry.

Ally yourself with others researching all the disciplines and branches of the life sciences. Food industry people need to read up on nutritional genomics. You don't need a special degree to know that food changes the way your genes express themselves. The prescribed appropriate and specific food combinations won't change your genes, but it will allow what genes you have to express themselves in the healthiest way they can. Some people may need to change their food or nutritional supplements based on their gene expression.

You could be taking vitamins and minerals in the wrong amounts or emphasizing the wrong supplements. Find out which ones you need and how much according to your genetic needs. Again, you don't have to wait another ten years until this is available to the masses. Do your research homework.

There are ways to find out what works with your genes. Join organizations that focus on allying people in the life sciences. Join the various life sciences alliances. People in many disciplines need to join together. For example the food industry, genomic scientists, healthcare professionals, journalists in bioscience communications, and the alternative health care markets need to plug into one another's research and resources. What's missing from these alliances? It's the consumer. Take charge of your search for self-identity at the genetic level.

If you don't see a new development and you're a consumer, perhaps you can develop it yourself. You don't need a degree in nutrition to make your own soy milk in a blender if you see the food products on the market are adding too much sugar and salt or barley flour to the milk you see on the shelves. Write to the companies. If nothing happens, make your own nutritious food.

The future of nutritional genomics is based on whether it becomes profitable. If it doesn't become profitable, no it won't be shelved. It will move to the holistic health markets and the alternative medical markets. The food industry is quick to move in with its own strategies, but it will always be profit as long as there is profit in healthy food. The mission is to convince the buyers in the supermarkets that people will buy food that isn't loaded with sugar, salt, and flour when the fat is taken out.

Nutritional genomics can plug into the phenomics industry—the science of customizing and tailoring your medicine to your genome. The U.S. diagnostics industry is a $2.3 billion business. You want your genotype? Your key-words in your consumer search for companies are "science-based." Contact science-based companies. Most do research, not individual genetic testing, unless it's for single-gene problems. So do your homework and don't bother the research companies if you want a genetic test. What you want to do is search for a managing physician who works with the reputable, scientific-based companies to look at your entire genome before any diet is prescribed.

You want to talk with people who are supervised under an umbrella and find out what research and testing is being done by whom, why, and how, where and when. Don't get taken in by snake oil-type unsupervised testing. Know the company and get recommendations from scientists and attorneys in the genome research business. Talk or write to those that work with supervising physicians concerning your healthcare.

You can look at a printout of your genes, your genome and conference with health professionals in consultation with you and your nutritional genomics consultants to find out whether a diet alone will help or whether a 'nutraceutical' supplement, medicine, or drug will help you along with the diet. The first place to convince and check out is your HMO and your insurance company. Find out whether you'll be reimbursed for the expense of having your DNA tested and for any sessions with genetic counselors. Again, don't wait until this becomes common and genetic counselors are flooded with boomers demanding tailored diets and/or nutraceuticals.

What will happen is either the price of the testing will come down or the number of people demanding it will make the price of health care premiums go up, or testing may or may not be covered by insurance in the future when the demand becomes high. You can help yourself now by forming alliances with genetic counselors and research scientists in nutritional genomics. Or at least read their research and join the message boards in nutritional genomics on the Internet. If you don't ask questions now, the answers later could become expensive or scarce, if the price goes down.

Make business alliances with people in nutritional genomics fields. What you're doing is taking research off the shelves of medical school libraries and bringing science to yourself—as a consumer. Again, you don't need a special degree to research this field for your own use. You can become a collector of published research in the field. Clip articles in the news or save Web-based articles for your scrapbook on this subject. Contact people in the news to answer your questions. Get your DNA tested—the whole genome. It's worth the thousand bucks or so now, even if the price goes down to a hundred later. If you get the right diet, ask yourself would it be worth it to your health?

What you want to find out about yourself are the links between your genes and your HMO or other health care systems. If you understand yourself, identify yourself at the genetic, molecular level, you'll think differently about your own health. You'll think differently about the health of your children and grandchildren if you think in terms of nutritional genomics as preventative medicine. It could delay the onset of what your genes predispose you to get. If you're at risk, instead of worrying, the dietary change could delay the onset or prevent it.

Time is on your side when you know how your genes express themselves. It's just the molecular you expressing yourself creatively. And since the body is hardwired to try and heal itself, you could lend a helping hand with the specific diet tailored to your genes. A diet book can't fit a whole population any more than one shoe size will fit a whole group of people.

Think about your health at the molecular, cellular level, at the genetic level. It's all about using your DNA for ancestry in a new way, from genealogy to nutritional genomics. Now you can find out whether your vitamins and minerals or other supplements are actually being absorbed and whether you need most of them. If you do, then you'll know which ones are working and which are not.

Food industries, the agribusiness, genomics research, healthcare, and allied life sciences all are working together. The goal is to bring this to the consumer level now and not in the future. You have industries contributing millions to biotechnology centers. Many of these centers are university-based. It's time to bring the university ivory tower lab to the consumer. Food and drug industries are pouring millions into centers to further research. You have only a few scientists working on projects that bring the fields of genomics, nutrition, and health together.

You need more catalysts like this to bring people together to do more than 'interface.' You need consumer action to 'interact' with the scientists who are working in projects that 'interface' the genomics people with the nutrition researchers and the healthcare professionals and genetics counselors. You need consumer involvement and consumer-based research.

You need bioscience communicators to make the complex easy to understand by the consumer with little or no science background. You need the consumer to take charge of his or her own health, privacy, nutrition, and research. Extended studies courses would help bring the consumer into taking charge by learning what research is being done and comparing his own DNA genome test to various interpretations of the test by different professionals such as his or her physician, genetics counselor, nutritionist, and nutritional genomics scientist.

All these fields need to be brought together under an umbrella accessible to the consumer…not in the future, but now, and with the privacy issues intact for the consumer, and the research open to the public in publications or online to be read by consumers as well as by allied science professionals and the media.

Who will write the regulations? It should be up to the consumer to have privacy in individual genetic tests. Genetic profiles are the business of only the person taking the DNA test to be used for planning a diet or lifestyle change. And unless the tests are totally anonymous as to name and address, collections of genomes would benefit scientists studying ethnic groups, but only if there was total anonymity. Nobody wants to be labeled and have somebody else know the label can't be altered. It's too controlling. Privacy should come first. Privacy should be respected so people numbered not by name and address, but by anonymous numbers or letters would ensure the privacy of large groups of people.

Scientists would be eager to collect large numbers of genomes for databases. Your genes are private. They are your time capsule. Your genome test would be something you'd pass on in a family time capsule scrap book to your heirs or future generations to learn about what your genes expressed, what was in the family, and what may or may not have been inherited by heirs.

Each individual has a different genetic pattern due to recombination. Some genes are inherited, but not all. So a time capsule or book binder with your DNA test may be of interest for medical history to your grandchildren, but how much of this they have inherited wouldn't be known unless they had a test themselves and compared genes. The consumer could take an extension course or read about DNA and learn about what, how, and why, when and where the genes express themselves.

Check out the genetic testing products. Do your research so you won't get scammed by companies that could spring up anytime in the future anywhere and target certain customers such as senior citizens or Boomers. Learn which companies are reputable and check them out by satisfied customers and by the reputations of the people who work there. Make sure the companies have professionally trained and experienced people. Check with the universities and the genomic associations and centers to make sure you are being tested by a company they would use themselves.

Most people have no clue as to how to apply genetics to their personal life. Many fear privacy invasions. Few are told what risks and benefits surround eating bright according to your genotype. As regulations come into focus, ask yourself how to validate the testing. When you are able to validate the test, then you know you've learned enough to get tested. Always ask professionals how they validate the genetic test and learn from their answers. Then check out their answers with research and other opinions from various professionals in the field.

What can you teach yourself about the applications of nutritional genomics testing for your individual health concern? How do you apply the information you've read? How do you validate that information? If a company makes a health claim, find out how to prove it and test it before you pay for testing. Pass the word to parents and children about getting their children tested so genetic profiles will be used to prescribe better diets and exercise programs for children as well as parents. Go into the schools and speak at PTA groups on getting children tested.

With the epidemic of diabetes in children increasing, perhaps a genetic profile might help parents to plan healthier diets, beverages, and snacks for children. It might also help the fare in school cafeterias. The idea is to get the word out at the consumer level as to what research they can do as consumers regarding applications of testing and validation of the tests. Then what follows are the smart diets and diaries of changes in health for all ages. The time for tailored eating and lifestyle changes is now.

By the time we wait for cheaper DNA tests of our whole genome and the chance to apply the test results to a practical menu, in what shape will we be? Ask yourself, should what we eat need medical prescription? Learn for yourself. Powerful knowledge is out there for the public and the media. Take control of your genes, your food, and your nutraceuticals. Your destiny will be in the hands of nutritional genomic scientists until you learn the science of nutritional genomics and research how you are impacted at the molecular level by how and what you eat. Reactions to lifestyle, environment, relationships, and exercise, also are expressed through your genes and your body chemistry.

21

Your DNA Matches and Mine

Here are my DNA matches by geographic coordinates. The geographic center is 48.30N 4.65E, Bar sur Aube, France with a deviation of 669.62 miles as done by "Roots for Real." Now that I have this information, what practical applications can I have with this information? For starters, I can make a time capsule for future generations.

Information on geographic matches of mtDNA for Anne Hart from the database of Roots for Real:

Individual Co-ordinates	For further information contact:
Details	http://www.rootsforreal.com
Roots for Real 2003 Page 1.	address: PO Box 43708, London W14 8WG UK

First, let's take a look at the mtDNA sequences.

My low resolution (HVR1) Haplogroup and mutations relative to the Cambridge Reference Sequence (CRS) are given below. A value of CRS indicates no mutations. High resolution (HVR2) results are shown. This test was done by Family Tree DNA. For further information, their Web site is at http://www.familytreedna.com

HVR1 Haplogroup	H
HVR1 Mutations	16189C
	16356C
	16362C
	16519C
HVR2 Mutations	263G
	309.1C
	315.1C
	523-
	524-

Note: The geographic center on the map Roots for Real sent me is 48.30N 4.65E, Bar sur Aube, France with a deviation of 669.62 miles as done by "Roots for Real." This is where the origin could have been anytime in the last 10,000 years of my matrilineal line. The geographic center is the nearest location where my mtDNa sequences show up. Best matches are on the map as red dots and the yellow center arrow on the map Roots for Real sent me represents the geographic center.

The exact sequences are in the Roots for Real Database (and other mtDNA databases) for my markers of mtDNA haplogroup H with markers at 16189C, 16356C, 16362C, 16519C. Note that the database is constantly being updated and only represents the sample. This represents only the motherline, the matrilineal lines. It does not show any racial, linguistic, or religious background, just the deep matrilineal ancestry. It represents people currently living in certain areas of Europe in the database with the exact mtDNA sequences as I have.

Individuals	Co-ordinates	Details
1	42.67N 23.30E	Bulgarian
1	51.50N 0.17W	White Caucasian, England & Wales
1	52.25N 21.00E	Pole living in Germany
1	52.00N 7.50E	Munster area, Germany
1	41.17N 8.63W	Portugal north of Douro, 20
1	38.72N 9.13W	central Portugal between Douro and Tejo, 21
1	47.27N 11.40E	Innsbruck, Austria
1	53.15N 18.00E	Bydgoszcz (Bromberg) region between Pomerania and Kujawy, Poland

1	53.15N 18.00E	Bydgoszcz (Bromberg) region between Pomerania and Kujawy, Poland
1	53.15N 18.00E	Bydgoszcz (Bromberg) region between Pomerania and Kujawy, Poland
1	47.98N 7.85E	Freiburg, S Germany
1	57.90N 5.17W	Scotland (NW coast) EMBL:AY024788
1	55.83N 3.07W	mainland Scotland EMBL:AY025191
1	55.83N 3.07W	mainland Scotland EMBL:AY024734
1	55.83N 3.07W	mainland Scotland EMBL:AY024537
1	59.92N 10.75E	Oslo, Norway EMBL:AY025926
1	59.92N 10.75E	Oslo, Norway EMBL:AY025925
1	59.92N 10.75E	Oslo, Norway EMBL:AY025924
1	51.50N 0.17W	England EMBL:AY025564
1	64.15N 21.85W	Iceland, accno AF236959
1	40.40N 3.68W	Spain

Here's my racial percentages genotype from AncestryByDNA: 97% European, 3% East Asian. I'm supposed to take with a grain of salt anything 5% or less, but maybe I am 3% East Asian? Anyway, here is my genotype printout. The numbers represent catalogue numbers, and the letters are my DNA markers called genotypes.

SAMPLE NAME = Anne Hart	#976CC	#1044GG
SAMPLE ID = ANC0154	#977CC	#1047AA
MARKERS GENOTYPES	#978TT	#1048GG
#958CC	#979CC	#1049AA
#960GC	#980TT	#1050GA
#961GC	#993TT	#1051GG
#963CC	#1000TC	#1053GA
#964CC	#1015CC	#1055GA
#966CC	#1022CC	#1056GA
#969TT	#1029GA	#1057GA
#970CC	#1033GA	#1058GG
#971CC	#1036GA	#1060GA
#972CC	#1041AA	#1062AA
#973CC	#1043GA	#1064AA

This is the printout I received from Oxford Ancestors among other materials in August 2001 with three letters marked in red noting the "C" transitions at 16189C, 16356C, and 16362C:

**ATTCTAATTT AAACTATTCT CTGTTCTTTC ATGGGGAAGC
AGATTTGGGT ACCACCCAAG TATTGACTCA CCCATCAACA
ACCGCTATGT ATTTCGTACA TTACTGCCAG CCACCATGAA
TATTGTACGG TACCATAAAT ACTTGACCAC CTGTAGTACA
TAAAAACCCA ATCCACATCA AAACCCCCCC CCCATGCTTA
CAAGCAAGTA CAGCAATCAA CCCTCAACTA TCACACATCA
ACTGCAACTC CAAAGCCACC CCTCACCCAC TAGGATACCA
ACAAACCTAC CCACCCTTAA CAGTACATAG TACATAAAGC
CATTTACCGT ACATAGCACA TTACAGTCAA ATCCCTCTC
GCCCCCATGG ATGACCCCCC TCAGATAGGG GTCCCTTGAC**

My Matches and Yours

Note: The geographic center on the map Roots for Real sent me is 48.30N 4.65E, Bar sur Aube, France with a deviation of 669.62 miles as done by "Roots for Real." This is where the origin could have been anytime in the last 10,000 years of my matrilineal line. The geographic center is the nearest location where my mtDNa sequences show up. Best matches are on the map as red dots and the yellow center arrow on the map Roots for Real sent me represents the geographic center.

The exact sequences are in the Roots for Real Database (and other mtDNA databases) for my markers of mtDNA haplogroup H with markers at 16189C, 16356C, 16362C, the fast-evolving site, 16519C. About half the people in the world have 16519C and the other half do not have this sequence. It signifies some event. Usually when you have a sequence transition or mutation, it signifies some event. The slow-evolving event that caused my 16356C mutation shows up not only in my H mtDNA haplogroup, but also in the mtDNA haplogroup U4.

Interestingly H with a mutation of 16356C mtDNA also shows up in the similar geographic areas that U4 with the 16356C mutation or transition sequence. And that area is North Central and North Eastern Europe. It also is found in Armenia, Central Italy, and South Eastern Europe such as the Ukraine and Bulgaria. What event caused the 16356 mutation for both U4 and H?

Note that any DNA database is constantly being updated and only represents the sample done at a particular time. An mtDNA database sample represents only the motherline, also called the matrilineal lines. Your mtDNA sequence does not show any racial, linguistic, or religious background, just the deep matrilineal ancestry going back 10,000 or 20,000 years depending on the mutation sequences.

The more places your DNA is found, the more ancient your DNA is and the more time it had to travel longer distances. If your mtDNA or Y chromosome sequences are found only in a small area of the world, it is more recent and is either small or didn't have much time to expand to wider areas. The mtDNA listing below represents mtDNA matches of people currently living in certain areas of Europe in the database with the exact mtDNA sequences as I have which is H haplogroup with sequences at 16189, 16356, 16362, and if tested, 16519C.

***

22

What's The Oldest HomoSapien mtDNA in Europe?

The oldest mtDNA in Europe that's human, Homo Sapien and not Neanderthal or other archaic individual is U5. It had a common ancestor with its sister group, U6. The age of U5 is estimated at 50,000 but could be as old as 60,500 years. Where did U5 come from, as it's the first in Europe and evolved in Europe? The first place scientists find U5 in Europe is in Cyrenaica, and artifacts are found in Iberia. Syke's book says it shows up 45,000–50,000 years ago in Delphi, Greece.

It has a common ancestor with the Berber U6, found in a third of Moroccans, which is its ultimate starting point before it arrived in the Middle East and then went on into Europe. U6 in N. Africa is close to U5 in Europe, and U6 is close in age to U5. The female who was the ancestor of U5 and U6 lived in what today is Morocco and Algeria. U5 and U6 cluster with other Europeans and not with Sub-Saharan Africans. Today, U6 comprises a third of the Mozabite Berbers. There was gene flow between N. Africa and the Middle East. The ancestor of U5 and U6 lived in the Maghreb in N. Africa. U5 is found almost exclusively in Europe today.

U6 is found today in the Canary Islands, Iberia, N. Africa and Portugal. U5 is dominant in Scandinavia, particularly Finland, along with V and U4 there also. A large proportion of Canary Islanders are mtDNA haplogroup U6.

The medieval Guanches of the Canary Islands also had U6. There was a lot of interbreeding in paleolithic times between U5 and U6. The Berbers are high in U6 mtDNA today. Whereas U5 today is found all over Europe and is the oldest European mtDNA, and is found more in Scandinavia, particularly Finland.

Not all Berbers are U6. The largest cluster of Berbers from N. Africa is H, especially in the city of Mzab. But when you test the Berbers with H, you find a sequence 16213 that has been found so far only in Europe, possibly suggesting a European origin for the H sequences in this N. African indigenous population of

Berbers. They have numerous people with red hair and freckles, speaking Berber languages. The Kabyl peoples of Algeria also often show this trait. Did back migration to Africa also take place in paleolithic times?

Yes. Another common haplotype 16148–16343 belongs to the Berbers who have U3, a common haplogroup in the Middle East (Iraq) and also in Europe. And mtDNA U is found there, unrelated to U6, native to N. Africa. J is there, but J comes from Syria and Turkey. Some gene flow did come from further south in Africa, because a few Berbers are L3b, L2 and L3a, but sub-Saharan gene flow is only 14% among the Berbers of Morocco and the people of the Canary Islands in modern times.

So what this study shows is that European and Middle Eastern sequences in the Berbers came from Europe within the last 10,000 years. They came from geographic areas that today are known as Sicily, Malta, and Spain. People from the Nile Valley and the Levant migrated to Morocco in ancient times adding more mtDNA diversity. About an eighth of Berbers in North Africa came from Sub-Saharan Africa. The rest back-migrated from Europe, especially from Spain and mostly from the Levant and Mesopotamia. Peoples who originally lived in the Nile Valley also went east to Morocco. So what do all these expansions and migrations show in modern times? That U5 is the first echo out of Africa into Europe, but that it shows up as the first Europeans in two places, Delphi and Spain around 50,000 years ago. Europeans didn't migrate from North Africa to Spain. They migrated from Eastern Europe, Central Asia, the Caucasus, Anatolia, and the Levant to Europe.

For more information on this subject read, "The Emerging Tree of West Eurasian mtDNAs: A Synthesis of Control Region Sequences and RFLPs," American Journal of Human Genetics: 64:232–249, 1999, V. Macaulay, et al.

So U5 turns out to be the most ancient mtDNA in Europe (50,000 years to 60,500) and U6 in N. Africa. What's interesting is that U5 and U6 are "sister mtDNA groups" with a common ancestor in N. Africa. Each mtDNA group has a sister group.

For example H and V are sister groups, with a common ancestor. And J and T are sister groups. U and K are sister groups. Each sister group has a common ancestor that had in its signature both J and T or H and V or U and K.

Why the Deep Genetic Split Between Two European Groups of mtDNAs?

There's also a deep genetic split between some of the European groups. For example X is split deeply from H by certain transitions such as 16223T instead of 16223C in some, but not all X mtDNAs. There's a deep genetic split between (H, I, J, and K) and (T, U, V, W, and X).

What kind of event took place in Paleolithic times to cause this huge split between these European groups? Did the split take place before or after arrival in Europe? Was it the isolation of the Ice age that caused it?

Were the two groups separated, for example, in different parts of the world? It doesn't seem so, because H, I, J and K are in one group, and H lived in Paleolithic times in France and Spain, whereas "I" mtDNA haplogroup lived in the Middle East or Central Asia and J and K lived in Syria and also later, K lived in the Alps (from 17,000 years ago)…but also is found in the Middle East and all over Europe. So what split the two groups?

Look at the other group (T, U V, W and X). Note that X usually is grouped with I and W, and is found rarely in Europe and heavily in the Middle East and Caucasus, especially in Georgia. T is all over the British Isles, but also in the Arabian Peninsula and the Middle East.

U is found all over N. Africa, Europe, and the Middle East and is the oldest in Europe. and W and X, some in Europe, but most in the Middle East, N. India, and Caucasus, except for the X that went to the new World via the Central Asia through Siberia route, and is found among certain Native American tribes like the Ojibwa and Sioux, Lakota, and a few other tribes. What do you think caused the split between (T U V, W and X) and (H I J and K)?

Cro Magnons

If you like studies about Cro-Magnon fossils, read the article, "Morphological Evolution in Prehistoric Skeletal Remains," in the book, Archeogenetics, (Mc Donald Institute Monographs).

So, who were the Cro-Magnons? Their common ancestors were U6 from N. Africa and U5 from Europe. They had broad faces, were tall, slender, and long boned. Skeletal remains from caves in Spain such as Longar show they are most closely related to today's Swedes. The mtDNA studies on the Cro-Magnon fossils from the prehistoric Basques show they are slightly different from today's Basques, but today's Basques are similar to medieval Basques.

The Paleolithic samples showed they were closer to modern Swedes than to modern Basques. The Pico Ramos caves and other prehistoric Basque area samples showed the Paleolithic peoples were closer to one another than to anyone modern, but Basque population's ancient and modern did group together. mtDNA J was absent from the Longar site cave of Paleolithic Cro-Magnon samples, but the predominant prehistoric mtDNA was H, a high amount of H as if that's the dominant population there 22,000 years ago.

Other mtDNAs were identified—U, T and X. Interestingly, some other mtDNA haplogroups showed up that didn't fit in anything modern. Those just disappeared 22,000 years ago or so. Either they didn't survive to reproduce or they had only sons.

Fascinating….Even 20,000 years ago, H was still the dominant type in Europe as it is today—47% of Europeans are H. What was it about that group that had so many daughters survive to modern times, and what was it that made the other mtDNA groups smaller in size, at least in Europe?

H mtDNA haplogroup has been found to exist so far in only 6 percent of the Middle East today, but is the dominant type in the Caucasus at a smaller number than in Europe. Was it something in the food supply that didn't exist elsewhere in Europe at the time of the last maximum Ice Age? Or did more female infants survive? What happened so that today 47 percent or more of Europeans have H mtDNA haplogroup, including me? What has your research shown? Isn't reading about archaeogenetics fascinating and awesome?

The following is the mtDNA test result from the individual's personal DNA test with ancestral mutations that evolved over thousands of years as compared to the Cambridge Reference Sequence (CRS) shown below (with permission). The HVS-1 and HVS-2 mtDNA test was done by Family Tree Genetics.

Let's take a look at how online research can give some clues, but not a definite decision on where the matrilineal line originated. First we'll look at HVS-1, the low-resolution haplogroup mutations in the mtDNA, that is the matrilineal lines that go back about 21,000 years in Europe, and perhaps are even older before the ancestor migrated to Europe. Somewhere mtDNA haplogroup H hand a previous ancestor, possibly in the Middle East who was pre-HV. That ancestor gave rise to another ancestor HV, which in time gave rise to H.

Finally, H migrated from the Middle East or Western Asia to Europe and settled in the Ice Age "refugium" in southern France or Spain, near the Mediterranean in the foothills of the Pyrenees. Then about 13,000 years ago, V mtDNA (split off from H and migrated to Scandinavia and to the Berbers in N. Africa. H mtDNA became a "clan mother" during the Ice Age by successfully reproducing enough daughters to pass on the haplogroup. After the Ice Age, H expanded all over Europe from Iceland to the Urals and is a "pan European" type today. H mtDNA also is found in the Middle East. About 46% of Europeans and about 25% of Middle Eastern peoples currently have H mtDNA. V mtDNA possibly arose in Europe.

Searching the Mitomap Table at: http://shelob.bioanth.cam.ac.uk/mtDNA/hvr1n.html
we find the slowest mutating sequence, 356 shows up in individuals from the following geographic locations: UK, Havik, Karelina, Portuguese, Turkish and Volga-Finnic.

Then we look at the next sequence which would be 356C and 362C and find it shows up in a Bulgar and a Karelian. When I get to the exact sequences 189C, 356C, 362C, it shows up in a person from the UK, Spain, Austria, and Bulgaria.

We see that the main mutation is 356 or actually 16356C in the individual with mtDNA H haplogroup, mutations at 16189C, 16356C, 16362C, 16519C. The 189 mutation or 16189C is a fast mutating site. So is the mutation 362 or to be exact, 16362C as it's usually written. So what we focus on for ethnicity is the slowest mutating site, which is 356 or 16356C.

In the tables at http://shelob.bioanth.cam.ac.uk/mtDNA/hvr1n.html on the Web, we see that 16356C is found in the Volga Finnic people of North Eastern Europe, but also in Turks, Bulgarians, and Portuguese. So does this very frum (religiously obedient) orthodox Ashkenazi Jewish person with ancestors from Bialystock, Poland and Bessarabia Rumania have a female ancestor who in ancient times—very ancient times came from Karelia (nation within and just east of Finland)? Or would the individual's Yiddish speaking ancestors have come from Bulgaria, Turkey, or Portugal? People of Southeastern Europe are closely related genetically to people of Northeastern Europe.

The time frame here could be tens of thousands of years or as recent as a common ancestor perhaps 250 years ago. Let's look a bit deeper into the ancestry and the sequences. Keep in mind that H haplogroup is pan-European and found all over the Middle East. The sequences place the individual in various possible geographic locations as far as ancestry.

These particular sequences also show up in one to four individuals recorded in other databases in Siena, Italy, in Crete, in Norway, Germany, Poland, Scotland, Bulgaria, Spain, England, Iceland, Croatia, Albania, Portugal, Austria, Bashkortostan in the Urals, the Ukraine, and among the Komi of Finland.

How can you tell where (geographically) the person's female ancestral line originated? That's part of the history and geography of human genes.

You can see where the mtDNA haplogroup started out, going from the Middle East to Europe, probably to the north of Spain or south of France 20,000–21,000 years ago. You see H in the Middle East about 28,000 years ago in the Fertile Crescent and in Anatolia and what was Armenia. You see it in the refugium in the Pyrenees 21,000 years ago, and in England 12,000 years ago where a young male skeleton was found in Gough's Cave in Somerset having H mtDNA haplogroup.

At the end of the Ice Age 12,500 years ago, H mtDNA traveled from the Cromagnon caves of southern France and Spain to what was to become England. All people in with European ancestors with H mtDNA descend from one single female with H mtDNA haplogroup who lived about 20,000 years ago in the Pyrenees area where there was a "refugium" from the Ice Age where the fishing was abundant and the campsites offered good shelter.

Yet, we still have to look further to get a handle on what happened to these particular sequences of H mtDNA. We know by looking at databases of Eastern Europe that those particular sequences 189C, 356C, 362C, 519C are found in North Central and North Eastern Europe primarily. They show up in Europe at a very low level, but where they show up most frequently is in Germany, Poland, Scotland, Norway, and Sweden.

Today, matches for these sequences are found frequently in England, Belgium, the Netherlands, the US, and Germany. The sequences are found all around the Adriatic Sea. So it's pan-European. On a low level, the sequences also show up in Turkey, Bulgaria, England, Spain, Portugal, Crete, and Italy. On a slightly higher level the sequences show up in Germany, Scotland, Poland, the Ukraine, and among the Komi in Finland.

We know Finland has one of the least diverse mtDNA pool. So all we can surmise from looking at the mtDNA databases is that the mtDNA haplogroup with those particular sequences show up at a low level in Europe and show up at a slightly higher level in North Central and North Eastern Europe. The difference in levels would be four individuals from Scotland and Germany have this sequence verses three individuals from Poland, Norway, Sweden for example. And the source would be a database with about 10,000 sequences in it.

We also know that without examining any particular sequence, H mtDNA in general, appears 25% of the time in the Middle East and more than 46% of the time all over Europe—from Iceland to the Urals. Those particular sequences of the individual do appear in Iceland, Norway, Sweden, Scotland, England, Germany, the Ukraine, (Komi group) of Finland, and Austria....but also in Bulgaria, Turkey, and as far east as Bashkortostan in the Urals. So we have to look at where it appears most—North Central Europe and North Eastern Europe.

The problem in looking at ancestry is that the point of origin is not necessarily where the mtDNA sequences appear in the largest numbers today, but where it is found most diverse. The more diverse the sequences the longer it has been in that place. This gives a clue to the origin or coalescence point.

H mtDNA in Turkey (ancient Anatolia) has a coalescence point of about 28,000 years or more compared to 20,000 years in Europe. However, the sequences of H in the Middle East would be different from the sequences of H in Europe in many cases, but not all.

You have this little movement call back-migration, where from ancient times people traveled from Europe back to the Middle East and stayed there. And you have waves of migration moving from southwest to northeast. About 15% of the Middle East is comprised of Europeans moving back there in prehistoric times, and about 20% to 26% of Middle Easterners, mostly Neolithic farmers after the end of the Ice Age moved into Europe. It took them much longer to reach the northern countries, but some did arrive and stay, introducing by trade the idea of farming to the Paleolithic hunters.

Where else can you check the history and geography of your mtDNA or Y chromosome? There are several excellent databases on the Web where you can check your sequences to see how many individuals from which countries have the same sequences as your own. Keep in mind that this information does not necessarily tell you where your particular ancestors came from, but the databases give you an idea where sequences like yours are found.

The question is, when you find someone with your sequences, does it mean that at one time you and that person shared the same common ancestor perhaps 300 generations ago or more? If you share the same single female mtDNA H haplogroup ancestor who lived 20–21,000 year ago in Europe, yes, you share at least the same first H mtDNA clan mother. And perhaps if your sequences match today, you share a slightly more recent ancestor. However, those mutations could pop up anywhere in anyone's line, related or not with or without the same mtDNA H. It all depends on what the other mutations are along with your sequences. So you'd have to check out that person's lineage to see how it might have crossed the path of your ancestors to get a handle on your search.

Let's look further at HVS-1 sequences. First I searched the Internet at Macaulay's Founder Tables to see where my exact mtDNA sequences are showing up at: http://www.stats.ox.ac.uk/~macaulay/founder2000/tableA.html

Be aware that this table from the United Kingdom may not necessarily contain the same sequences from the same countries as some of the databases in Eastern Europe. So check out as many databases online or in scientific journals as you are able to access. For example, I found the exact sequences in Macaulay's Table listing the following sequences found in individuals claiming ancestry from the following geographic areas 189 356 and 362 for England, Spain, Austria, and Bulgaria. The tables at the Web site list abbreviated names for the countries.

How you search in what country currently your mtDNA HVS-1 haplogroup appears is by realizing that the letters stand for names of countries. For example:

Nu means Nubia. Eg stands for Egypt. Be stands for Bedouin. Ye is Yemen. Iq is Iraq. In is India. Sy stands for Syria. Pl stands for Palestine (Arab Palestinian). Dz is Druze, a tribe of people living in Lebanon. Tk is Turkey. Ku is Kurd. Ar is Armenia. Az is Azerbaijan (in Central Asia), Cauc stands for people living in the areas of the Caucasus mountains such as Georgia, Armenia, Ossetia, Chechnya, Kabarda, Adyge, the Circassians, for example, the peoples of the north and south Caucasus mountains. MdE stands for Eastern Mediterranean.

SE is Southeastern Europe. MdC is central Mediterranean—Italy. Alp means the area of the Alps, such as Austria. NC stands for North Central Europe—Germany. MdW stands for Western Mediterranean, such as Spain and Portugal. Bs stands for Basque. NW stands for Northwestern Europe, namely, England/Britain. Sca means Scandinavia—Sweden, Iceland, Norway, Finland, and Denmark. NE is North Eastern Europe.

Look at the sequences of H mtDNA 189, 356, 362. The mutation (519C) wasn't included as some mtDNA research testing only test 400 or so base pairs. You'll see where the exact sequences appear in the table in only four countries: Spain, England, Austria, and Bulgaria. If you look at the sequences that contain the 356C mutation without the 362 mutation, you'll see it appears in more places and that more people have that sequence. The conclusion is that H mtDNA 189C, 356C, 362C, 519C appears at low levels in Europe. Perhaps it went through a bottleneck. Perhaps it went through extinctions, even though it's the most frequent mtDNA haplogroup in Europe—which is H.

However, it does appear in Europe in those four areas in that particular database. In an Estonian unpublished database, it appears in many more countries, including the one's in Macaulay's database. The Estonian database has those exact sequences appearing among the Komi of Finland, in the Ukraine, in Croatia, Albania, Bulgaria, Turkey, Crete, but more frequently in Germany,

Poland and Scotland, and also in England, Iceland, Sweden, Norway, and in Bashkortostan, a country in the Urals.

According to the Macaulay database, those exact sequences aren't found east of Bulgaria. In the unpublished Estonian database Turkey and Bashkortostan are added, but also Crete. Since Albania and Crete show these sequences in the Estonian unpublished database, history notes there was a migration in medieval times from Albania to Greece and to Crete, and the population was absorbed.

So what are the origins of this person—Volga Finnic? Germanic? Polish? Bulgarian? Can we ever know for sure? The exact sequences are found in England, Netherlands, and Belgium. Where did the most recent ancestor of this individual originate? Will we know more in the future about tracing our ancestors from a great grandparent back in time several hundred or several thousand years?

We have the first ancestor—H mtDNA and the present, with the particular sequences living in the US. We know for generations the ancestors lived in Poland and then in Rumania. When did the ancestors arrive in Poland, and did they come from Germany or Lithuania? If Lithuania, when did they arrive there from what other geographic place? It's a fascinating migration as the sequences reveal.

So if you want to match your own sequences to a table that shows what nations your sequences appear in today, click on Macaulay's statistical tables from Oxford University in England and see whether your particular sequences are found in that database. The Web site again, is: http://www.stats.ox.ac.uk/~macaulay/founder2000/tableA.html.

Note that to the left of the table all the countries of the Middle East are listed starting with Nubia (in Africa) at the extreme left with Egypt next and then Bedouins, Yemen, Iraq, India, Syria, Palestine, Druze, Turkish, Kurdish, Armenian, Azerbaijan, Caucasus, Mediterranean East. Then you have to the right half of the table all the European countries going from the central Mediterranean—Italy north to the Alps and Austria followed by Germany, then down to Spain and Portugal and then to the Basques in Western Europe, then up again to England in NW Europe, then northeast to Scandinavia and finally to North East Europe.

The table more or less follows several of the routes of migrations of ancient and prehistoric peoples out of Africa (Nubia) into Egypt an then into the Middle East, to India, and migrating west to the Middle East and then north to the Caucasus, then back to Europe and finally, the last migration route from southwest Europe to Northeastern Europe passing through central Europe, England, Scandinavia, and finally reaching Northeastern Europe. Where are you on any of the databases? Also check out the mtDNA Concordance Tables at: http://shelob.bioanth.cam.ac.uk/mtDNA/hvr1o.html.

The individual from a long line of orthodox Jewish families was concerned that the mtDNA sequences didn't show up on the left side of the Macaulay Database statistical tables. Why didn't they show up with an ancestry in Iraq, Palestine, or Syria? Did the Jews flee oppression from those areas and move into Western and then Eastern Europe?

Why is the ancestral mother matching with all those Central and Eastern Europeans and also with individuals in Spain and Portugal and England? Note that only one person from Spain, one person from Austria, one person from England, and one person from Bulgaria shows up for that exact mtDNA sequences of HVS-1. So we dig a little further and look at the high resolution mtDNA mutations for HVS-2. And what do we find? This table below is from the individual's personal DNA reading for the high-resolution HVR2 mutations regarding deep maternal ancestry.

HVR2 Mutations 263G	309.1C
	315.1C
	523-
	524-

We find that the HVR2 Mutations (or HVS-2) are found in most people all over Europe and all over Western Asia and the Middle East, or are they? Look where they do appear. The HVR2 mutations also are very common sequences. Remember that the HVS-1 appears only at a low level in Europe. But look at the high-resolution HVR2 mutations. They appear in France and in the Orkney Islands, in Scotland, Bulgaria, and Turkey.

HVR2 can distinguish mtDNA haplogroup H from mtDNA haplogroup U4. If you see the letter "A" at position 00073 in HVR2 (or HVS2) the mtDNA haplogroup is H. If you see a "G" at position 00073 in HVR2, the mtDNA haplogroup is U4. Both H and U4 contain the mutation sequence 356 or 16356C. However, U4 and H show up in different countries but they also show up in the same geographic ancestral areas. For example, U4 doesn't show up among the Ossetians of the Caucasus Mountains, but H shows up among this ethnic group. Sequences 189, 356, 362 haplogroup H doesn't show up in the Caucasus, at least in the databases, but again, you have to think about how many people are in the sample. Some databases contain at least 10,000 individuals from a variety of countries.

Check out some of the founders at: http://www.stats.ox.ac.uk/~macaulay/founder2000/tableB.html.

Note that the tables of founders at V. Macaulay's site of Oxford University, UK is excellent for checking your sequences on the Internet. I highly recommend looking at all the tables on these sites. So, is this individual with mtDNA haplogroup H, sequences at 16189C, 16356C, 16362C, 16519C Germanic, Northeastern European, Southeastern European, Middle Eastern or what do you think? What we are looking at here is the person's ancient female ancestor with the latest mutations since H mtDNA evolved thousands of years ago and is found all over West Eurasia.

What we have to narrow down is the current mutation sequences. That's the mutations or transitions in the sequences that evolved away from the Cambridge Reference Sequence (CRS) and see where in geographic location the sequences as they appear today turn up. Most helpful are tables online you can check out and the various statistical databases that match your mtDNA sequences to the actual parts of the world in which they appear. However, it won't tell you for sure whether you're related to anyone in that location today. Check out Vincent Macaulay's Web site at:

http://www.stats.ox.ac.uk/~macaulay/ and look over his tables. Perhaps you'll find your mtDNA sequences there. Also for males, interested also in Y chromosome haplogroups, check out the various Y chromosome tables at: http://freepages.genealogy.rootsweb.com/~dgarvey/DNA/markers.htm

If you want to see how genetically close Ashkenazim are to other nationalities, look at the Human Races Calculator on the Web at:

http://racearchives.com/calc/jewsychroms.asp?popid1=1&popid2=9&dbname=jewsychroms
Compare 6 Jewish and 23 non Jewish Y-Chromosome Analysis. You'll see that Ashkenazic male Y chromosomes match 95% with Syrian male Y-chromosomes, 94% with Greek Y-chromosomes, and 95% with Roman Jews.

Also see at that site: http://racearchives.com/calc/haplo_profiles.asp?dbname=europeanmtdnabyregion&po

You can select a population and view its profile as far as how distant or close the population is from another population. You can also view Y-chromosome haplotype frequency tables from the Hammer study at http://www.rasch.org/rmt/rmt161g.htm. The article is titled, "DNA and the Origins of Jewish Ethnic Groups. In the article at this Web site, the Ashkenazi male Y chromosomes are positioned on a table between the Jewish non-Ashkenazi and the Turks. Other studies report Ashkenazi male Y chromosomes close to the Greeks and still other studies close to the Kurds. So you have to look at a large number of studies to see whether there is agreement over time where the Ashkenazi stand.

The article, "In DNA, New Clues to Jewish Roots," by Nicholas Wade, at: http://www.racesci.org/in_media/raceanddna/dna_jewish_nyt_May2002.htm reports that the Ashkenazim show "less diversity in its mitochondrial DNA, perhaps reflecting the maternal definition of Jewishness. But unlike the other Jewish populations, it does not sow signs of having had very few female founders."

What the pattern in the Ashkenazic population does suggest is that it's a mosaic of separate populations "formed the same way as the others." According to the article, it reports that Dr. Harry Ostrer, a medical geneticist at New York University said, "the 26 specific genetic diseases found among Ashkenazim usually attributed to 'founder effects' could be explained by the idea of small populations."

So what you get from reading more and more research articles on Ashkenazic genetic origins is that the whole idea of Jewish communities founded by Jewish men and local women is against the Jewish tradition. What do the genes say? Tradition says the first Jewish communities in Eastern Europe were formed by families who fled oppression from one land and were invited by princes to populate other lands, usually to the East.

What was the primary catalyst that brought the Jewish communities together in medieval times through the 18th century? How do you trace Jewish DNA for ancestry and family history through 3,000 years by looking at a merging mosaic of communities? How much trade, traffic, intermarriage, migration, and rapport occurred between the Ashkenazic and non-Ashkenazic Jewish communities of Europe, the Middle East, Africa, Central Asia, India, and China in historic times? And when the records can't be readily accessible, how much detail can we read in the DNA?

***

23

From Which Families Do You Descend?

Are you sure you're connecting with the correct family? Thousands of families could have the exact same name as the vital link in your recorded ancestors. DNA tests of the Y-chromosome for males can reveal which person of the **same surname** could be your relative. Or mtDNA tests might point to maternal lineages from the same female founding mother with different and changing surnames as well as ancient, prehistoric founders of your lineages. What steps do you take if you want to search family history records?

There are basically seven steps in searching those ancestors from whom you descend, regardless of what country. Whether you descend from Scandinavians, Germans, Greeks, Italians, Syrians, Hungarians, Asians, Latin Americans, Africans, or Asians, you begin looking for written records with civil registration of populations. These records may be in schools, churches, synagogues, pagodas, temples, family Bibles or other religious written records, hospitals, midwives' files, census, military service records, pension and medical records, or with libraries of family history. When the written records stop, you can turn to DNA testing to see whether a group of males are related if they have the same surname, or females, with different surnames, but descended from a founding mother.

From whom do you descend? It's easy to look at old books on early New England genealogy, but what if your family lived under the Ottoman Empire for the past five hundred years? The Ottoman Empire lasted from 1300 until 1922. In 1924 Kemal Ataturk abolished the Muslim caliphate and founded the Republic of Turkey.

So regardless of the language spoken by your ancestors—Slavic, Arabic, Greek, Judezmo (Ladino), Uralic, Yiddish, Romanian, or Turkish, the Ottoman Empire controlled and kept careful census records in Turkish using Arabic script in the following countries of Europe and the Middle East known today, but not necessarily before 1924 as the following names:

Hungary, Yugoslavia, Croatia, Bosnia, Albania, Macedonia, Greece, Romania, Moldova, Bulgaria, southern Ukraine, Turkey, Georgia, Armenia, Iraq, Kuwait, Cyprus, Syria, Lebanon, Israel/Palestine, Jordan, Eastern and Western Saudi Arabia, Oman, Bahrain, eastern Yemen, Egypt, northern Libya, Tunisia, and northern Algeria.

In addition, more recent records in Arabic or in the language of the country emphasized are kept in the national archives and in the courts dealing with property-related issues, assets left behind, divorce decrees, and other legal documents. Often when a population is forced out, the individual country, such as present-day Egypt, may then determine that assets, property, and religious items taken away or left behind such as Judaica, Hellenica, and Armenica are declared antiquities of that country that cannot be removed.

Records in the former Ottoman Empire were archived at various locations depending upon your religious ethnic affiliation because only Moslem males were tallied for conscription into the army until 1881. So their records are kept in one archive. Jews and Christians were tallied separately in the census so they could be taxed separately. Records are found in the archives pertaining to tax collection during the Ottoman rule. Moslems weren't taxed. So your first step is to search records according to the religion and ethnic group. Start your search with the list of countries under the Ottomans and pick your ancestor's homeland. Then focus on the religion and the ethnic group.

Here are the basic seven steps applied to searching your Middle Eastern and European genealogy in the former Ottoman Empire records—census, population registers, and beyond. What do you check first before looking in localities and jurisdictions of records kept in other countries?

First check the Family History Library catalog (Salt Lake City, UT) for books, microfilms, civil registration, localities and jurisdictions. Then search the **Ottoman Census and Population Registers** by country named in Turkish the *Nüfus Defter*. Check the online library catalog at **Bogaziçi University (a.k.a. the University of the Bosphorus).** See http://www.geog.port.ac.uk/hist-bound/country_rep/ottoman.htm. Also check out the flyer: **http://www.hanover.edu/haq_center/Bogazici-Flyer.html**

Records are found in the Ottoman Census archives in Turkey pertaining to tax collection during the Ottoman rule. Moslems weren't taxed. Before 1881, each census focused on tabulating male names to find Moslem men to conscript into the military service and non-Moslem men to pay personal tax. Search records according to the religion and ethnic group. Start your search with the list of countries under the Ottomans and pick your ancestor's homeland.

Step 1: Translating Names

Don't skip generations. Each generation is a vital link in countries where thousands have the same name. Check out the Middle EastGenWeb Project at: http://www.rootsweb.com/~mdeastgw/index.html.

Some religious names are used by Moslems and Christians. Christian European names are translated into Arabic in Arabic-speaking countries. For example, in Lebanon, Peter becomes Boutros and George becomes Girgis or Abdul Messikh, meaning servant of the Messiah (Christ). Shammout (strong), Deeb (wolf), or Dib (bear). Nissim (miracles) is used by Jewish Levantines, and Adam is used by Jewish, Christian, and Moslem families.

Women had the choice of taking their husband's surnames or keeping their maiden names. Neutral names, used by Moslems, Jews, and Christians such as Ibrahim (Abraham) or Yusef (Joseph) came from the Old Testament. In Lebanon, Christians often used neutral names such as Tewfik (fortunate).

Arabic-speaking and Turkic-speaking countries didn't use surnames until after the end of the Ottoman Empire. Then in Lebanon and Syria many Christians took as their surnames European or Biblical first male names such as the Arabic versions of George, Jacob, Thomas, and Peter which were in Arabic: Girgis, Yacoub, Toumas, and Boutros. Others took popular surnames describing their occupations such as Haddad meaning 'smith.'

After 1928 in Turkey, but not in any of the other Middle Eastern nations, a modified Latin alphabet replaced Arabic script. Four years later (1932) the Turkish Linguistic Society simplified the language to unify the people. Surnames were required in 1934 and, old titles indicating professions and classes were dropped. (See "Turkey" Web site at: http://www.kusadasitravel.com/turkey.html.)

In Middle Eastern countries under the former Ottoman Empire, such as Lebanon/Syria, each child was given a first name but most people in the Middle East had no surname until 1932. Also the father's given name was given as a middle name such as Yusef Girgis, meaning Yusef (Joseph), son of George. It came in handy in the days before surnames were required. Now it's used as a middle name.

Many Syrian and Lebanese families, particularly Christians, after 1932 took similar names such as Peter Jacobs or George Thomas. The name 'Thomas' in

Lebanon is spelled in translation as either Touma or Toumas. Many Assyrian males in Northern Iraq took the popular name Sargon, an ancient king.

When surnames in Lebanon became a requirement, you have very popular names such as Peter George Khoury in America being Boutros Girgis Khouri in Lebanon or Syria when translated into Arabic. In the Levant, daughters have a first name and their father's given name meaning "daughter of Yusef" or Ayah Yusef. Translated into English at Ellis Island, it could have become Aya Joseph.

When surnames became a requirement, many included professions or place names, especially Halaby (from Aleppo) or Antaky (from Antioch). The largest Lebanese community in America is in Dearborn, Michigan.

In Lebanon, most names were Christian prior to 1870, and the Christian names could also be European, especially Greek names like Petros (Peter) which later becomes Boutros in Arabic. If you're searching Assyrians, check out the Assyrian Nation Communities Web site at: http://www.assyriannation.com/communities/index.php.

After 1870, in Lebanon and Syria names in Christian families became Arabic rather than European due to increasing pressure by the Ottoman Empire on Christians to use Arabic instead of Greek names. After the demise of the Ottoman Empire at the close of World War I, Hellenistic names such as Kostaki (Constantine) became popular in Beirut.

The distinctly Christian Lebanese surnames Khoury (priest) or Kourban sprang up again when Lebanon became a French protectorate. Neutral, Greek, and Old Testament names also return. You see many French first names in Christian families between 1914 and 1950.

After the 1950s, Christian and French first names dwindle, and Arabic names appear. If your ancestors were Moslem, instead of a surname prior to 1932, you were known as "son of" (Ibn) as in Ibn Omar, for a male, and for a married woman with children called, "mother of" (om) as in Om Kolthum, (mother of Kolthum).

You'd be called mother of your first born son, (Om___Name of first born son) (Om Ahmed). If you had no sons, you'd be called mother of your first born daughter (Om Rania) (Om___Name of first born daughter). Single women often were called "daughter of" as in Bint Ahmed (daughter of Ahmed).

Arabic women's first names were plentiful, popular, and used at mainly at home. Examples include Samara, Zobaida, Rayana, Rania, Anissa, Dayala, Azma, Aya, or Salwa. Children had first names.

If your ancestors were Armenian living in the Levant you might have the name Ter or Der before a surname designating descent from an Armenian Apostolic priest followed by a name ending in ian or yan meaning "son of" such as Manvelian or a place name such as Halebian (from Aleppo) when translated into English. If you're Armenian searching Turkish census records, the pre-1920 border of Armenian habitation usually was south of Lake Van, near Mush (in Armenia), and Bairt and Dersim (in Turkey). Each religion had a different status under the former Ottoman Empire—Moslems first class and conscripted into the military; all other religions, not conscripted, but taxed.

Step 2: Narrow the Categories

Categorize the religion—not only Catholic, but Melkite Catholic or Maronite Catholic. Antiochian Syrian Orthodox or Roman Catholic? Byzantine Catholic (Byzantic) or Greek Orthodox? Lebanese immigrant to Cairo, Egypt and Coptic Orthodox? Moslem? Jewish? Druze? Armenian Apostolic? Sephardim? Ashkenazim? Protestant? Greek Orthodox? Greek Catholic? Bulgarian or Romanian Orthodox? Serbian? Croatian?

Color-code cards or files noting the date, religion, ethnic group, and town. When did the immigrant arrive in the US from a Middle Eastern country? Was it before or after the end of the former Ottoman Empire? For example, Antioch, now in Turkey used to be in Syria before World War II. And before 1918, Syria and Lebanon was one province under the Ottoman Empire. So use old and new **maps** to see what country to emphasize at which dates.

Step 3: National Archives in the Country of Origin

Maps of old neighborhoods show locations of houses. Start with the national archives in the country of origin. For Syria that would be the Syrian National Archives in Damascus, Aleppo, Homs, or Hama where court records are archived for the years 1517 to 1919. If the relatives lived before the end of the Ottoman Empire or before World War I, also search the census records of the former Ottoman Empire in Turkey rather than the archives in the country of origin that may not have existed before the end of the Ottoman Empire.

Records stand alone rather than in groups of catalogs. Check separate Jewish genealogy sources and synagogue documents for the Jewish records of Mizrahi and Sephardim, such as marriage ketubim, bar mitzvah records, births, deaths, rabbinical documents such as a *'Get'* for a divorce or a pedigree called a *Yiccus*.

If you're checking Sephardic (Jewish) records of the former Ottoman Empire, there's an excellent article on Jewish genealogy published in *Los Muestros* magazine, a publication of Sephardic and Middle Eastern Jewish genealogy titled *Resources for Sephardic Genealogy* at: http://www.sefarad.org/publication/lm/010/cardoza.html. Also see the magazine, Los Muestros at: http://www.sefarad.org/publication/lm/010/som10.html for archived Sephardic genealogy articles.

Another excellent publication of Jewish genealogy, Avotaynu maintains a Web site at: http://www.avotaynu.com/. If you're looking for Jewish records in the Middle East, also check the Sephardic associations, for example, Sephardim.com at http://www.sephardim.com/. Look for memorabilia, diaries, house keys, and maps of neighborhoods.

For Sephardic genealogy in the former Ottoman Empire, contact the Foundation for the Advancement of Sephardic Studies and Culture Web site at: http://www.sephardicstudies.org/cal2.html to learn how to interpret calendars and how to read birth certificates. You'll learn how to decipher the handwritten entries using Arabic script. Regardless of the religion of the individual, this site shows you how to read the certificates written with certain types of scripts.

The site also shows the dialects spoken in the various areas of the Ottoman Empire. Also there is information on how to read the Arabic script but Turkish language writing on gravestones, especially in Turkish cemeteries. The site shows you how to read the alphabet encountered in genealogical research in the former Ottoman Empire. Emphasis is on interpreting Sephardic birth certificates.

Step 4: How to Translate and Locate without Surnames

What's in the census? Ottoman census records for the period 1831–1872 were compilations of male names and addresses for fiscal and military purposes. Instead of population counts, the Ottoman records contain the name of the head of household, male family members, ages, occupation, and property.

You won't find surnames in old records. Most Middle Eastern countries didn't require surnames until after the fall of the Ottoman Empire. If you're searching Middle Eastern genealogy before 1924, begin by familiarizing yourself with the record keeping and social history of the Ottoman Empire.

Turkish language written in Arabic script is the key to searching genealogy records in European and Middle Eastern areas formerly ruled by the Ottomans. You'll need an Arabic-English dictionary or instruction guide that at least gives you the basic Arabic script alphabet.

You'll also need the same type of phrase book with alphabet translation for modern Turkish written using Latin letters. You can put the both together to figure out phrases.

Find in your town a graduate student or teacher from abroad who reads Arabic script and modern Turkish. Hire the student or teacher to copy the records you want when overseas. Barter services. Or contact the Middle East history and area studies, archaeology, or languages departments of numerous colleges. Who teaches courses in both Turkish and Arabic? Contact private language schools such as Language School International, Inc. at: http://www.languageschoolsguide.com/listingsp3.cfm/listing/4092.

Step 5: What Religious Group Will You Search?

Social history is the key to genealogy. Records that existed under the Ottoman Empire listed names of the head of household and parents, residence, dates and places of birth and baptism, marriage, death and burial. Records also have entries for ages for marriage and death.

Baptisms included names of the godparents. Deaths sometimes included the cause of death. For Christians, entries sometimes identified residence for those not of the parish. Check the state archives in the country of your ancestors and also in Turkey. Then check the court, notary, and property records.

Contact the parish churches to look at parish registers and synagogues to look at the Jewish registers. If you're checking Bulgaria, Macedonia, or Greece, numerous pre-1872 registers are located in Greece. "The Bulgarian Orthodox Church was subordinate to the Patriarchate in Greece before 1872," notes researcher, Khalile Mehr, formerly of the Family History Library, Salt Lake City, UT. Mehr is the author of numerous articles on searching genealogy records in Macedonia, Bulgaria, and Croatia, copyrighted by Family History Library. Find out whether a country had its state church subordinate to another country's church with records archived in a different language.

Step 6: Check Business, School Alumni, Medical, Military, Marriage, and Property Records

Research wills and marriage records in order to track down property records. Search medical and dental records, hospitals, orphanages, prisons, asylums, midwives' records, marriage certificates, business licenses, work permits, migration papers, passports, military pensions, notaries, sales records of homes or businesses, or any other court, military, or official transaction that might have occurred.

If you're Armenian, check out the Turkish Armenian Reconciliation Commission (TARC) at: http://www.asbarez.com/TARC/Tarc.html. Or for the Balkans, look at the Center for Democracy and Reconciliation in Southeast Europe's Web site at: http://www.cdsee.org/teaching_packs_belgrade_bio.html.

Step 7: Search the 'Annual' Census and the Population Registers

Check recorded births and deaths in the first Ottoman census of 1831. Each census focused on tabulating male names to find Moslem men to conscript into the military service known as "The Army." Before 1881, the annual census registered only the male population. Search the names of committee members. Committees were set up each year to register the males in order to keep tabs on migrations in and out of each district. When the census wasn't taken, the Population Register of Moslem males kept careful records of migrations.

Find out in which local district or *'kaza'* your ancestor lived. Ottomans called their annual census the *sicil-i nüfus* after 1881 or the *nüfus* between 1831–1850. You can research the **Ottoman Census and Population Registers** named in Turkish the **Nüfus Defter**. Ottoman population demographics and statistics adjusted to satisfy tax desires, since the non-Moslem population was taxed but not conscripted into military service. The annual census didn't cover every year.

Check the Ottoman census for the years1881–1883, and 1903–1906. Family historians can search each census as well as separate registers to view supplemental registration of births, marriages, divorces, and deaths. After 1881, the census takers counted all individuals (not only Moslem males) in the census and in the population registers. Sometimes people who thought all genealogy records were destroyed in fires in their native country are surprised to learn that census records may be archived far away inTurkey.

If you need text or a Web site translated into numerous languages, check out the Systran Web site at: http://www.systranbox.com/systran/box. You can translate free an entire Web site or 150 words of text.

◆ ◆ ◆

Web Resources on Former Ottoman Empire Genealogy Web Sites

Albanian Research List: http://feefhs.org/al/alrl.html

Armenian Genealogical Society: http://feefhs.org/am/frg-amgs.html

Historical Society of Jews from Egypt http://www.hsje.org/homepage.htm

Jewish Genealogy: http://www.jewishgen.org/infofiles/

Lebanon Genealogy http://genforum.genealogy.com/lebanon

http://www.mit.edu:8001/activities/lebanon/map.html

Lebanese descendants of the Bourjaily Family (Abou R'Jaily) http://www.abourjeily.com/Family/index.htm
Descendants of Atallah Abou Rjeily, born about 1712

Lebanese Club of New York City:
http://nyc.lebaneseclub.org/
http://www.rootsweb.com/~lbnwgw/lebclubnyc/index.htm

Lebanese Genealogy: http://www.rootsweb.com/~lbnwgw/

Middle East Genealogy: http://www.rootsweb.com/~mdeastgw/index.html

Middle East Genealogy by country:
http://www.rootsweb.com/~mdeastgw/index.html#country

Sephardim.com: http://www.sephardim.com/

Syrian and Lebanese Genealogy: http://www.genealogytoday.com/family/syrian/

Syria Genealogy: http://www.rootsweb.com/~syrwgw/

Syrian/Lebanese/Jewish/Farhi Genealogy Site (Flowers of the Orient) http://www.farhi.org

Turkish Genealogy Discussion Group: http://www.turkey.com/forums/forumdisplay.php3?forumid=18

Turkish Telephone Directories Information: Türk Telekomünikasyon (Telecommunication) http://ttrehber.gov.tr/rehber_webtech/index.asp

Croatia Genealogy Cross Index: http://feefhs.org/cro/indexcro.html

Eastern Europe: http://www.cyndislist.com/easteuro.htm

Eastern European Genealogical Society, Inc.: http://feefhs.org/ca/frg-eegs.html

Eastern Europe Index: http://feefhs.org/ethnic.html

India Royalty: http://freepages.genealogy.rootsweb.com/~royalty/india/persons.html

Romanian American Heritage Center: http://feefhs.org/ro/frg-rahc.html

Slavs, South: Cultural Society: http://feefhs.org/frg-csss.html

Ukrainian Genealogical and Historical Society of Canada: http://feefhs.org/ca/frgughsc.html

Rom (Gypsies): http://www.cyndislist.com/peoples.htm#Gypsies

See: McGowan, Bruce William, 1933- **Defter-i mufassal-i liva-i Sirem : an Ottoman revenue survey dating from the reign of Selim II./ Bruce William McGowan.**
Ann Arbor, Mich.: University Microfilms, 1967.

See: **Bogaziçi University** Library Web sites:
http://seyhan.library.boun.edu.tr/search/wN{232}ufus+Defter/wN{232}ufus+Defter/1,29,29,B/frameset&FF=wN{232}ufus+Defter&9,9,

or http://seyhan.library.boun.edu.tr/search/dTaxation+--+Turkey./dtaxation+turkey/-5,-1,0,B/exact&FF=dtaxation+turkey&1,57,

Jurisdictions and localities in Bulgaria:
Michev N. and P. Koledarov. ***Rechnik na selishchata i selishchnite imena v Bulgariia, 1878–1987*** (Dictionary of villages and village names in Bulgaria, 1878–1987), Sofia: Nauka i izkustvo, 1989 (FHL book 949.77 E5m).

***

24

Merging a Mosaic of Jewish Communities by DNA

Say you're Jewish with ancestors from Eastern Europe, an Ashkenazim whose ancestors spoke Yiddish in the old country. How do you interpret your own DNA test results to look at not only your own family history, but your ancestors as far back as you can go genetically? You know the records may stop at one point back in time, but the DNA will not stop. It will mutate over thousands of years, but it will show you an open road on where you came from, even in a general sense.

I highly recommend the book, *The Beginnings of Jewishness* by Dr. Shaye Choen, professor of Jewish literature and philosophy at Harvard. When you begin to trace your DNA think about when the Jewish communities were formed. Were they formed in Southern Italy in the third and fourth century? Or were the communities formed soon after the Islamic conquests from the seventh to the ninth centuries CE?

If you consider Jewish tradition that says families practiced matrilineal descent to determine Jewishness, and your DNA search takes you to the seventh to ninth centuries CE, you can see how small communities were formed in places such as Alsace in France or in the Rhine Valley. Look at articles written by Dr. Hammer, Dr. Goldstein, and the population geneticists who are researching Ashkenazic mtDNA, the founding mothers.

Look at DNA to see founding events. Read the article by M.F. Hammer, A.J. Redd, et al. (2000) "Jewish and Middle Eastern non-Jewish populations Share a Common Pool of Y-Chromosome Biallelic Haplotypes." Proc. National Academy of Science, USA, Vol 97, Issue 12, 6769–6774, June 6, 2000. Then compare the study with later articles such as "High-resolution Y chromosome haplotypes of Israeli and Palestinian Arabs reveal geographic substructure and substantial overlap with haplotypes of Jews." Oppenheim, Ariella, et al. *Human Genetics* 107(6) (December 2000): 630–641.

Check out "The Y Chromosome Pool of Jews as Part of the Genetic Landscape of the Middle East." Oppenheim, Ariella, et al. *The American Journal of Human Genetics* 69:5 (November 2001): 1095–1112. I also recommend reading the following articles: "Study finds close genetic connection between Jews, Kurds." Traubman, Tamara. *Ha'aretz* (November 21, 2001).

According to the article, it's the Sephardic Jewish males who closely resemble the Kurdish Jewish males. The Ashkenazim were found to differ from the Sephardic and Kurdish males, even though in most previous studies, the Y chromosome of Ashkenazim clusters closely with Sephardic Jews. Yet in this study the Kurdish and Sephardic Jewish populations differed to some degree from Ashkenazic Jews. Studies are still suggesting Ashkenazic males mixed with European peoples during their diaspora in varying degrees, except for the Ashkenazic males with the Cohen Modal Haplotype on their Y chromosomes.

The researchers suggested that the approximately 12.7 percent of Ashkenazic Jews carrying the Eu 19 chromosomes—signature of Eastern Europeans, descend paternally from eastern Europeans. The Eu 19 chromosome shows up in 56% of Polish men and 60% of Hungarian men. Not all Ashkenazic males carry the Eu19 chromosome.

Read the article, "Are today's Jewish priests descended from the old ones?" Zoossmann-Diskin, Avshalom. *HOMO: Journal of Comparative Human Biology* 51:2–3 (2000): 156–162. See the article at: http://www.ariga.com/genes.shtml titled, "Doctor finds fault in the contentions that the "Cohen modal haplotype" designates Israelites and that most Jewish priests have a common ancestor." 27 February 2001.

According to the article at the Web site, "Zoossmann disputes point that the Cohen Modal Haplotype is the signature for the Hebrew nation. * The Cohen modal haplotype is the most common haplotype among Southern and Central Italians*1, Hungarians*2, and Iraqi Kurds*3, and is also found among many Armenians*4 and South African Lembas*5. This calls into question the notion that the haplotype was a marker for the ancient Hebrew population."

You'd have to compare Dr. Zoossmann's article with the articles that discuss the Cohen Modal Haplotype such as "*Y Chromosomes of Jewish Priests,*" in Nature volume 385 by Michael F. Hammer, Karl Skorecki, and their colleagues, January 2, 1997, and the article, "*Origins of Old Testament Priests*" by Karl Skorecki, David Goldstein, et al. in Nature volume 394. Before you begin your research, look at the Cohen Modal Haplotype Web site at: http://www.geocities.com/hrhdavid/cmhindex.html.

The point is that whenever you have a research study and a resulting article, another scientist will find some flaw in it. That's the way the scientific process moves forward. My highest recommendation on how to understand all this is to look at articles and research done by scientists with enough genetics background to consider the possibilities. There are many sites on the Web discussing Jewish genetics, many written by historians and many written by scientists and physicians. Then there are articles summarizing secondary sources. You have to do your own homework and be the judge. New studies come out each year with new information. The door is open for new results and findings.

The majority of Ashkenazic Jewish men carrying the Eu 9 chromosomes and similar markers descend paternally from Judeans who lived in Israel two thousand years ago. So the point is you can't make a judgment about the ancestry of any particular Jewish male—or female—until you look at all the genetic markers and match them to the geographic location where they are or were.

There's always the Cohen Modal Haplotype found in some Ashkenazic males to research. I recommend the article, "The Priests' Chromosome? DNA analysis supports the biblical story of the Jewish priesthood." Travis, J. *Science News* 154:14 (October 3, 1998): 218. Jewish women don't have a Cohen Modal Haplotype because it's a male Y-chromosome passed from father to son for thousands of years. The new puzzle to solve is the Ashkenazic females. Since Jewish tradition passes the Jewish heritage as a religion and ethnic community from mother to children, what geographic areas did the women come from?

I suggest following studies on mtDNA and other markers to see where the mothers came from or at least who the mothers' genes are closest to in current times. So much has been studied and written about male Y-chromosomes and ancestry and so little on women. The door is just opening and currently much of the data is yet unpublished.

So patience will reveal new information on women as soon as the newer studies are published. In the meantime, keep informed by looking at the latest articles in the various human genetics journals. Contact your local Jewish genealogy society and see what speaker is available who is researching mtDNA and other female ancestry markers. Universities around the world have population genetics departments doing research. Keep in contact with them and their publications and/or conferences.

Once you found your DNA trail—mtDNA for women, Y-chromosomes for men, and even if you took a racial percentages DNA test, what do you do with the information? You create a time capsule, a scrap book, a type of DNA and oral history journal. You use photography, text, sound, music, speech, video, and you put your ancestral record archive in an oral history—transcribed and put to audio or video, to sound, text and imagery. You even may want to put your information on the Web or in an archive to be viewed by future descendants. The whole tapestry links you back to that first ancestor and to the one in the future. It's a rich experience in history and time.

How do you write, tape, and transcribe an oral history of your DNA along with your genealogy and family history records and photos? Assuming you're a beginner in genealogy with no science background and interested in family history, where do you begin your search? What's the cultural component behind a trait as biological as your genes?

If you're a family historian, an oral history researcher, or a person fascinated with ancestry, here's how to understand the results of DNA tests. Different people have different, sometimes opposite opinions on whether DNA testing is a useful tool in the hands of family historians. If you are a carrier of a genetic disorder, DNA testing is useful in researching your family history to find out who was the first carrier in your ancestry back in time.

Here the debate unfolds as scientists, authors, physicians, media people, owners of DNA testing companies, genealogists, historians and researchers comment, write, and opine on DNA testing and genealogy.

What is the distribution of mtDNA haplogroup K? We know that haplogroup K is a clade of haplogroup U. Previous studies revealed that 27% to 32% of Ashkenazim have haplogroup K and that 9.0 of Ashkenazim have the mtDNA haplogroup H that follows the CRS. The rest have mutations of haplogroup H that does not follow the sequences of the CRS without mutations. What is the distribution of mtDNA haplogroup K in the rest of the populations sampled?

Take for example one sample appearing in the chapter, *Anatolian and Trans-Caucasus Populations*, Archaeogenetics: DNA and the Population Pehistory of Europe, McDonald Institute Monographs, 2002, page 223. The highest figure is 31.7% in Georgians sampled have haplogroup K mtDNA. Next are Armenians sampled with 27.8% haplogroup K mtDNA. Following are Turks with 24.7%, and Siena, Italy samples, Tuscans, Saradinians, French, and Albanian samples with 35.8% haplogroup K mtDNA. Haplogroup K mtDNA also appears at 25% in the Nile Valley, and at 20% in Ethiopians, and 1.2% of Indians. It also appears at 11% in Estonians, Finns, and Karelians, and at 18.6% in Russians, Poles, Czechs, and Slovaks.

Where did it originate? In the Middle East, probably, but about 17,000 years ago it developed in Europe in the area where Venice, Italy is today and then migrated into the Alps.

***

25

How Do You Use DNA Testing To Interpret Family History Records?

How eager are people to take a DNA test for family history research? Most DNA tests require only that someone swish mouthwash around in his or her mouth and send it for testing to a laboratory. So what can a DNA test really tell you about your own ancestry—distant or not so distant? And most of all, how do you interpret and use the results?

Here's a letter from Dr. Mark Humphrys:
Lecturer
School of Computer Applications,
Dublin City University,
Glasnevin, Dublin 9, Ireland.

"Dear Anne:

Here's a summary of the position as I see it: Why everybody in the west is descended from Charlemagne: We all know that all humans are related. So a good question is: When was our Most Recent Common Ancestor (MRCA)?

Surprisingly, the answer to that question is a lot more recent than DNA studies would suggest, since we are searching all lines of descent, rather than just the lines genes traveled on. You do not inherit all your ancestor's DNA, but only a small part of it. And yet, even if you inherit NONE of their DNA (which is not only possible, but probable, as you go back far enough), they are still your ancestor.

To find the answer to the MRCA, we need to look beyond DNA studies. Mathematical models suggest that, if humans picked mates randomly, the MRCA is in historical times, perhaps c. 1200 AD! This is an amazing result, suggesting that we do not have to go back into prehistory to find an ancestor of every single human! But obviously humans do not pick mates randomly—they tend to mate with people in their local geographic area.

Computer simulations that take this into account suggest that even with a high degree of local mating, the MRCA is still in historical times, perhaps c. 300 AD. If we consider just the West, the MRCA may be as recent as c. 1000 AD.

How realistic are these models? Well, there has been a growing collection of REAL, proven descents from medieval figures in genealogy. For instance, my own children have proven descents—through many different lines—from Charlemagne, who lived around 800 AD. He is the ancestor of most of the royal houses of Europe and so is a natural focal point for the genealogies of the West.

My web page, Royal Descents of Famous People, is a large and growing list of famous people in the west who are all proven descendants of Charlemagne. And if all these people have proven descents, every step of the way, how many more people must have descents in reality that cannot be proved because of the scarcity of records? It must be a much greater number. My work is a strong indicator that everyone in the West descends from medieval royalty.

In short, work by a number of people—my genealogical study, other people's computer simulations and mathematical models—all confirm each other's findings that the MRCA for a large, interbreeding area such as the West, is within recent recorded history. Another finding of those models is that not long before the MRCA, if someone is the ancestor of anyone alive today, they are the ancestor of all people alive today. Since Charlemagne is probably around that early date for the West, and since he is a proven ancestor of some people, it is likely that he is the ancestor of all people in the West.

Everybody in the West is descended from Charlemagne: In conclusion, if you have west European ancestry at all, it seems virtually impossible for you not to be descended from Charlemagne, who lived around 800 AD. 90 percent of the world (including all the West) is descended from Confucius.

For the MRCA of the whole world, we need to consider extremely isolated aboriginal populations. If they were truly isolated, we may have to go back thousands of years to get a common ancestor with them. For people who did not live in isolated enclaves though—the West, Middle East, more or less all of Asia, most of Africa—the MRCA is highly likely to be in recent historical times (late BC, possibly even AD). Anyone with ancestry from these areas is, for example, almost certainly a descendant of Confucius, who lived around 500 BC and who is a proven ancestor of some people alive today in China, hence probably ancestor of all people in the world except the extremely isolated.

This exciting consensus is fairly new, and is supported by three independent fieldsof (a) genealogy, (b) mathematical models, and (c) computer simulations.

The findings are robust with respect to barriers such as religion, class difference, etc. All one needs is a tiny amount of crossing of such barriers in the population in the past in order to get everyone today (of different religions etc.) with a recent common ancestor.

The only thing that can push back the MRCA before historical times is total geographic isolation of populations from each other, which we know did not happen for most of the world.

People are inspired (rightly so) by DNA studies of ancient common human ancestors tens of thousands of years ago. And yet the fact is that we are almost certainly all descended from any historical figure in classical times that left descendants."

---

end quote

Web pages on MRCAs:

http://computing.dcu.ie/~humphrys/FamTree/Royal/ca.html

http://computing.dcu.ie/~humphrys/FamTree/Royal/ca.genetic.html

http://computing.dcu.ie/~humphrys/FamTree/Royal/ca.math.html

http://computing.dcu.ie/~humphrys/FamTree/Royal/famous.descents.html

The Web pages include a fantastic computer simulation by a gentleman named Rohde at MIT to work out the MRCA for a non-random mating model. He confirms much of Chang's work, and in general it is another strong indicator that the MRCA for the world (or 99 percent of it) is post-1000 BC—maybe even AD.

Regards

—Mark
Dr.Mark Humphrys

Lecturer
School of Computer Applications,
Dublin City University,
Glasnevin, Dublin 9, Ireland.
http://computing.dcu.ie/~humphrys/

Steve Olson, author of the book, Mapping Human History in a telephone interview with me on December 3, 2002 at 1:50 PM EST answered my question: What do you say about using DNA as a tool for genealogy—to extend family history research?

"The most valuable use of DNA testing is to demonstrate how closely related we are to each other, both as individuals and as members of human groups," Steve Olson says. "I'm a skeptic about the Seven Daughters of Eve (book) because I believe that everyone in the world is descended from all seven of those women. As I point out in my book, Mapping Human History.

"As I point out in the book, I believe that everyone living today is descended from most people who lived a few millennia ago," Olson explains. "So genetic tests need to be interpreted very carefully or you draw false conclusions about being descended from a relatively small number of people."

Does Steve Olson think DNA testing as a tool is useful to genealogists? "No, I don't feel DNA testing can tell you things that can't be discovered in other ways. I should probably say here, though, that I'm fairly skeptical about DNA testing for genealogical purposes, and I'm particularly critical of the Seven Daughters of Eve idea."

The Seven Daughters of Eve, is a book written by Bryan Sykes, from Oxford Ancestors. Bryan Sykes, MA PhD DSc, is Professor of Human Genetics, University of Oxford, and with Oxford Ancestors, who comments in this book. Sykes has a different opinion about DNA testing and genealogy/family history research.

Oxford Ancestors is the world's first company to harness the power and precision of modern DNA-based genetics for use in genealogy. The motto on the Oxford Ancestors Web site at: <http://www.oxfordancestors.com/> reads: "Putting the genes in genealogy."

Oxford ancestors is based on more than a decade of research into human populations and their origins carried out by Professor Bryan Sykes, Professor of Human Genetics at the University of Oxford and his team in the world-renowned Institute of Molecular Medicine in Oxford, England.

So as you read in this book, and any other books you research, there are many different experts in genetics, medicine, and the media with different opinions.

Science is supposed to be skeptical by nature as well as open-minded for change. Facts provide an open door to further inquiry, and facts need to be checked as new information comes in. So what stance will you take? Will you take the skeptical side on DNA testing? Or do you reason that DNA testing is one more tool that adds genes to genealogy? Where will you stand in this debate—observer, skeptic, or DNA researcher working with genealogy and genetics?

Perhaps you're a genealogist or oral historian doing family history research and you have no science background in genetics. Now that the molecular revolution has introduced DNA testing for ancestry research, here's how to use your curiosity.

You may want to find information about tests of racial percentages or DNA testing and ancestry. Make up your own mind. Now you can combine DNA research and oral history research into one archive. How do you interpret DNA tests and how do you plan, record, and transcribe an oral history for your family or others?

DNA testing for ancestry is offered by companies that combine DNA testing (done by commercial laboratories and/or university laboratories) with surname genealogy. Various surname groups on the Web also offer discounts with DNA testing companies and laboratories. Ask for the facts.

Whom do you go to first, assuming you have no background in either science or genealogy? Scientists? The media? DNA testing companies? University laboratories? Oral historians? Genealogists? Biological anthropologists? DNA mailing lists on the Web? Whom can you trust?

And where does the truth lie—in technology? It's like twelve blind men asked to describe an elephant. Each touches a different part of its body and replies, "it's a tail," or "it's a trunk." Everyone has a different opinion of the value of DNA testing in researching family history or ancestry.

What do the changing scientific facts say, for now? Question authority, and do your research. Then decide what tools are the best for you in your own family history quest.

Go online to Pub Med, an NCBI research engine on the Internet for medical and scientific articles in professional journals. Read the article titled, "The common, Near-Eastern Origin of Ashkenazi and Sephardi Jews Supported by Y-Chromosome Similarity," Santachiara Benerecette, AS, Semino O, Passarino G, Torroni A, Brdicka R, Fellous M., Modiano G, Diparrtimento de Biologia Cellulare, University della Calabria, Cosenza, Italy. Annals of Human Genetics Jan; 57 (Pt 1): 55–64, 1993. Compare the study to studies done on Jewish Y-chromosomes a decade later. What new information did you find?

In the 1993 study about 80 Sephardim, 80 Ashkenazim and 100 Czechosolvaks had a DNA test of their Y-chromosomes. These males were examined for the Y-specific RFLPs. The aim of the study was to look at the origin of the Ashkenazi gene pool by looking at specific genetic markers useful to estimate the paternal gene contribution to these men's origins. To find out ancestry, scientists examine specific genetic markers in the DNA.

The genetic markers of the Ashkenazim and the Sephardim were compared with each other and with the Czechoslovakians. The Czechoslovakians turned out to be very different from the two Jewish groups. Who are the Ashkenazim closest to? The Sephardim first, and then the Lebanese. Both the Ashkenazim and the Sephardim appeared to be closely related to the Lebanese. The conclusion of the study was that Ashkenazi and Sephardi are both close genetically to the Lebanese as far as their Y-chromosome-specific RFLP markers. According to the Pub Med abstract, "A preliminary evaluation suggests that the contribution of foreign males to the Ashkenazi gene pool has been very low (1% or less per generation.)

Also check out, Santachiara Benerecetti AS, Semino O, Passarino G, Morpurgo GP, Fellous M, Modiano G (1992) "*Y-chromosome DNA polymorphisms in Ashkenazi and Sepharadi Jews.*" In: Bonné-Tamir B, Adam A (eds) Genetic diversity among Jews: diseases and markers at the DNA level. Oxford University Press, New York, pp 45–50

In the nineties decade, most of the studies of Ashkenazic DNA ancestry markers focused on looking at the paternal lineages, the Y-chromosomes, unless the studies were specific for genetic diseases research. Population genetics that looks at ancestry focused widely on the male line of descent, and articles or research tracing the maternal lineages were difficult to find in public searches of scientific journals or even in the media.

With the new decade, more studies of the maternal lineages of Ashkenazim and non-Ashkenazic Jews are more readily available. Still, there is research yet unpublished that continuously adds new information to the research.

The question remains, who were our founding mothers and from where did they come—from ancient Israel, from Western Asia, from Europe or are we what we appear to be—a mosaic of Jewish communities that merged and intermarried over the centuries? Is there yet to be found an unbroken line yet found that links us to any one ancient civilization, or is there? Or are we mixed with the memories of expansions and migrations over thousands of years?

The mtNDA reveals the haplogroups found among non-Jews with similar origins predating any organized religion. We females are part of the K mtDNA haplogroup clan and the H mtDNA haplogroup clan that gave rise to millions of Europeans, West Asians, and Middle Eastern peoples. What new information. Are Jews a race?

The answer is they are part of the human race. The debate will go on regarding how many genes tie Jews to their ancestral palm-latitudes home in the Levant and before that in the Fertile Crescent. Is there unity in diversity? And why such a debate on origins at all? It's more political than religious. Anyone who takes Judaism seriously can become a Jew, have a child and rear it as Jewish, and so on for thousands of years since Abraham smashed the idols in his dad's sculpture shop in ancient Sumeria.

If Jews weren't linked to the Middle East, why would caricature stereotyped images portray them in cartoons with typical Semite features and Assyroid skulls? And yet if European Jews didn't look Northern European as well, why would they be forced to wear armbands with six-pointed stars in Germany and North Eastern Europe during World War II, unless they looked so much like Germans you couldn't point them out in public as appearing different in coloring or features?

Seeking a core identity is the reason why Jews want to explore ancestry by DNA testing. The feeling you belong to a certain group of people who wrote certain books of wisdom in ancient times is powerful. You want to belong to a family, a family within the family of humankind. You want to celebrate holidays with others celebrating the same holidays. Jews come in all colors and in all types of DNA. And yet, there is this paternal lineage link to the Lebanese for the Ashkenazim and the Sephardim. Now what about our founding mothers? Like my former teacher once remarked, "How could a young, Jewish merchant from Baghdad or Persia traveling to Poland or to the Jewish communities in the Rhine Valley in the ninth century resist marrying a redheaded or blonde 'shiksa' goddess on the road to Speyer?" The answer could be, more genes outflowed from the Jewish males into the general local population than came in and remained.

So the origin of Ashkenazic communities will always be up for debate and theory. The details remain in the genes that reveal ancestral lines. How would historians account for the Ashkenazic Levites some, but not all, of which have genes close to Sorb males? Did Sorb males convert to Judaism to escape oppression from the Germans who invaded their Western Slavic territories to convert them to Christianity and Germanize them at a time when they were pagans resisting Christianity and the German language?

Did they team up with the Jewish communities living in East Germany in those early medieval centuries, perhaps around the year 800 CE? All scientists know is that the many Jewish communities founded independently eventually merged as more and more people flowed east into Poland upon invitation from the Polish princes. The Ashkenazic families migrated from Germany, France, and Hungary eastward. By 1000 CE, most Jewish communities were already founded and scholarship began to flourish, such as the writings of Rashi.

Who were these Ashkenazic Jews of Germany and France? They were the Roman Jews, the Jews who lived in ancient Rome who migrated northward from Southern Italy and Rome after the fall of Rome. They went were the trade flourished—trade that didn't require owning land. They minted coins and changed money, cut diamonds, or worked with gold. They sewed clothing and peddled wares. They traveled and wrote books.

They were musicians and healers, rabbis and scholars. And they went were they were welcome. Some followed the northward movements of the Roman army. Sephardic Jews traveled the same routes, but instead of going north, went to Spain and N. Africa. Other Jews fleeing oppression branched into Iraq, Syria, Greece, and Persia. Some went out to Bukhara and Central Asia from villages in Iraq. Some went to India and some to Ethiopia.

Sometimes they found local women and married them, and other times, the women came from other Jewish communities. So you can't at least at this time in research make a sweeping statement of where all the women came from. You can only say they came mostly from other merging Jewish communities. Further scientific research again, will let the details in the genes speak their ancestral history. So keep updated and read the emerging articles in the human genetics journals. You can access them online through most of the medical research engines on the Internet. Or you can visit a university library, especially a medical school library, and read the newest articles at the library as well as the latest books on Jewish genetic diversity or any other topic in genetics. The particular field looks at population genetics and evolutionary genetics. Read the article that will take you back before there were borders or organized religions titled, "*Tracing European Founder Lineages in the Near Eastern mtDNA Pool.*" Martin Richards, Vincent Macaulay, et al. American Journal of Human Genetics, 67: 1251–1276, 2000. See where our mothers really came from. (The term "our mothers" refers to the human race.) You'll notice in the articles you read that the British often use the term "Near Eastern" and that the Americans use the term "Middle Eastern" for the same area.

Perhaps you want to find out the percentage of various races in your ancestry. How do know where to begin your journey into the past and future? What if you're a foundling, an orphan, or have no knowledge of your own ethnicity? Can a DNA test at least tell you how many races are in your recent or ancient past? What facts do genetic markers really tell you about ancestry?

If you want to start your ancestry search with DNA testing, first you take the DNA tests along with tests of racial percentages if you desire. Even your DNA has a cultural component to its molecular biology. Then you interpret the results making the complex easy to understand for yourself or your clients. Your DNA testing service can help you find answers. So can many Web sites as well as this book and other books recommended here.

Next in your family history search, you collect letters, diaries, oral history transcriptions, home sources, artifacts, memorabilia, Census research, wills-and-probate records, medical histories, land records, slave ownership records, if it applies to your or your client. Pay particular attention to social histories to fill some gaps left by lack of women's records.

Search through church, synagogue, mosque, pagoda, or temple records, vital records from the US government such as military records, social security information, and government pensions for retired government employees, employment and tax records, if any exist and are available. Check school records from elementary through college, if any, social histories, ethnic histories, and religious school records.

Go to the family history Web sites, the ships' passenger lists. I highly recommend a book for searching women's ancestry, titled, Discovering Your Female Ancestors, by Sharon DeBartolo Carmack, Betterway Books, Ohio 1998, ISBN # 1-55870-472-8. The book's subtitle emphasizes "Special strategies for uncovering hard-to-find information about your female lineage."

Marriage records often were in different languages representing the former country or languages of the ethnic group. You may need to translate a different alphabet to find a maiden name on a marriage certificate never registered, but obtained from clergy.

Then you review and analyze the records. Study the social history of the times and location of this individual.Add family history and migrations to social history, and you have the beginnings of an outline to write a biography of the ancestor as a family history.

Learn to interpret the results of your own DNA test and expand your historical research ability to trace your ancestry. "An interesting idea was expressed by a colleague from Canada, Dr. Charles Scriver," explains geneticist, Dr. Batsheva Bonné-Temir. "At a meeting which I organized here in Israel on Genetic Diversity Among Jews in 1990, Dr. Scriver gave a paper on 'What Are Genes Like that Doing in a Place Like This? Human History and Molecular Prosopography.' He claimed that a biological trait has two histories, a biological component and a cultural component." Dr. Charles Scriver is founder of the DeBelle Laboratory of Biochemical Genetics in Canada. He also established screening programs in Montreal for thalassaemia and Tay Sachs Disease.

According to Bonné-Tamir, at the 1990 meeting in Israel on Genetic Diversity Among Jews, Dr. Charles Scriver stated, "When the event clusters and an important cause of it is biological, the cultural history also is likely to be important because it may explain why the persons carrying the gene are in the particular place at the time."

The term, "when the event clusters" refers to an event when genes cluster together in a DNA test because the genes are similar in origin, that is, they have a common ancestral origin in a particular area, a common ancestor.

"When I look at my own papers throughout the years," says Bonné-Tamir. "I find that I have been quite a pioneer in realizing the significance of combining the history of individuals or of populations with their biological attributes. This is now a leading undertaking in many studies which use, for example, mutations to estimate time to the most recent ancestors and alike."

What lines of inquiry are used in genetics? Dr. Charles R. Scriver wrote a chapter in Batsheva Bonné-Temir's book, titled What are genes like that doing in a place like this? Human History and Molecular Prosopography. The book title is: Genetic Diversity Among Jews: Diseases and Markers at the DNA Level. Bonné-Tamir, B. and Adam, A. Oxford University Press. 1992.With permission, an excerpt is reprinted below from page 319:

"When a disease clusters in a particular community, two lines of inquiry follow:

1. Is the clustering caused by shared environmental exposure? Or is it explained by host susceptibility accountable to biological and/or cultural inheritance?

2. If the explanation is biological, how are the determinants inherited? These lines of inquiry imply that a disease has two different histories, one biological, the other cultural. One involves genes (heredity), pathways of development (ontogeny), and constitutional factors; the other, demography, migration and cultural practice.

Neither history is mutually exclusive. Such thinking shifts the focus of inquiry from sick populations and incidence of disease to sick individuals and the cause of their particular disease. The person with the disease becomes the object of concern which is not the same as the disease the person has." (Page. 319).

After hearing from Dr. Scriver by email, I then emailed Stanley M. Diamond. He contacted writer, Barbara Khait, and got permission for me to reprint in this book some of what she wrote about Diamond's project. It's the chapter, "Genetics Study Identifies At-risk relatives" from Celebrating the Family published by Ancestry.com Publishing.

Check out the Web site at: http://shops.ancestry.com/product.asp?productid=2625&shopid=128.
Here's the reprinted article. Persons interested may go to the Web site for more information. I found out about Stanley M. Diamond from Dr. Scriver, since he mentioned Stanley M. Diamond's project in the book chapter Scriver wrote for Batsheva Bonné-Temir's book on Genetic Diversity Among Jews: Diseases and Markers at the DNA Level. Barbara Khait's chapter follows.

◆ ◆ ◆

"In 1977, Stanley Diamond of Montreal learned he carried the betathalassemia genetic trait. Though common among people of Mediterranean, Middle Eastern, Southeast Asian and African descent, the trait is rare among descendants of eastern European Jews like Stan. His doctor made a full study of the family and identified Stanley's father as the source.

"Stan was spurred to action by a letter his brother received in 1991 from a previously unknown first cousin. Stan asked the cousin, "Do you carry the beta-thalassemia trait?" Though the answer was no, Stan began his journey to find out what other members of his family might be unsuspecting carriers.

"Later that year, Stan found a relative from his paternal grandmother's family, the Widelitz family. Again he asked, "Is there any incidence of anemia in your family?" His newfound cousin answered, "Oh, you mean beta-thalassemia? It's all over the family!"

"There was no question now that the trait could now be traced to Stan's grandmother, Masha Widelitz Diamond and that Masha's older brother Aaron also had to have been a carrier. Stan's next question: who passed the trait onto Masha and Aaron? Was it their mother, Sura Nowes, or their father, Jankiel Widelec?

"At the 1992 annual summer seminar on Jewish genealogy in New York City, Stan conferred with Or. Robert Desnick, who suggested that Stan's first step should be to determine whether the trait was related to a known mutation or a gene unique to his family. He advised Stan to seek out another Montrealer, Dr. Charles Scriver of McGill University-Montreal Children's Hospital. With the help of a grant, Dr. Scriver undertook the necessary DNA screening with the goal of determining the beta-thalassemia mutation.

"During this time, Stan began to research his family's history in earnest and identified their nineteenth century home town of Ostrow Mazowiecka in Poland. With the help of birth, marriage, and death records for the Jewish population of Ostrow Mazowiecka filmed by The Church of Jesus Christ of Latter-day Saints (LOS), Stan was able to construct his family tree.

"Late in 1993, Dr. Scriver faxed the news that the mutation had been identified and that it was, in fact, a novel mutation. Independently, Dr. Ariella Oppenheim at Jerusalem's Hebrew University-Hadassah Hospital mad e a similar discovery about a woman who had recently emigrated from the former Soviet Union.

"The likelihood that we were witnessing a DNA region 'identical by descent' in the two families was impressive. We had apparently discovered a familial relationship between Stanley and the woman in Jerusalem, previously unknown to either family," says Dr. Scriver.

"It wasn't very long ago when children born with thalassemia major seldom made it past the age of ten. Recent advances have increased life span but, to stay alive, these children must undergo blood transfusions every two to four weeks. And every night, they must receive painful transfusions of a special drug for up to twelve hours.

"The repeated blood transfusions lead to a buildup of iron in the body that can damage the heart, liver, and other organs. That's why, when the disease is misdiagnosed as mild chronic anemia, the prescription of additional iron is even more harmful. Right now, no cure exists for the disease, though medical experts say experimental bone-marrow transplants and gene-therapy procedures may one day lead to one.

"Stan's primary concern is that carriers of thalassemia trait may marry, often unaware that their mild chronic anemia may be something else. To aid in his search for carriers of his family's gene mutation of the beta-thalassemia trait, he founded and coordinates an initiative known as Jewish Records Indexing-Poland, an award-winning Internet-based index of Jewish vital records in Poland, with more than one million references. This database is helping Jewish families, particularly those at increased risk for hereditary conditions and diseases, trace their medical histories, as well as geneticists."

Says Dr. Robert Burk, professor of epidemiology at the Albert Einstein College of Medicine at Yeshiva University, and principal investigator for the Cancer Longevity, Ancestry and Lifestyle (CLAL) study in the Jewish population (currently focusing on prostate cancer), "Through the establishment of a searchable database from Poland, careful analysis of the relationship between individuals will be possible at both the familial and the molecular level.

"This will afford us the opportunity to learn not only more about the Creator's great work, but will also allow (us) researchers new opportunities to dissect the cause of many diseases in large established pedigrees."

Several other medical institutions, including Yale University's Cancer Genetics Program, the Epidemiology-Genetics Program at the Johns Hopkins School of Medicine, and Mount Sinai Hospital's School of Medicine have recognized Diamond's work as an outstanding application of knowing one's family history and as a guide to others who may be trying to trace their medical histories, particularly those at increased risk for hereditary conditions and diseases.

In February 1998, in a breakthrough effort, Stanley discovered another member of his family who carried the trait. He found the descendants of Jankiel's niece and nephew—first cousins who married—David Lustig and his wife, Fanny Bengelsdorf. This was no ordinary find—he located the graves by using a map of the Ostrow Mazowiecka section of Chicago's Waldheim Cemetery and contacted the person listed as the one paying for perpetual care, David and Fanny's grandson, Alex.

"It turned out Alex, too, had been diagnosed as a beta-thalassemia carrier by his personal physician fifteen years earlier. The discovery that David and Fanny's descendants were carriers of the beta-thalassemia trait convinced Stan, Dr. Scriver, and Dr. Oppenheim that Hersz Widelec, born in 1785, must be the source of the family's novel mutation.

"'This groundbreaking work helps geneticists all over the world understand the trait and its effects on one family,' says Dr. Oppenheim.

"A most important contribution of Stanley Diamond's work is increasing the awareness among his relatives and others to the possibility that they carry a genetic trait which with proper measures, can be prevented in future generations. In addition, the work has demonstrated the power of modern genetics in identifying distant relatives, and helps to clarify how genetic diseases are being spread throughout the world."

For more information about thalassemia, contact Cooley's Anemia Foundation (129-09 26th Avenue. Flushing, New York, 11354; by phone 800-522-7222; or online at www.cooleysanemia.org). For more about Stanley Diamond's research. visit his Web site (www.diamondgen.org).

Thalassemia is not only carried by people living today in Mediterranean lands. The first Polish (not Jewish) carrier of Beta-Thal was discovered in the last few years in Bialystok, Poland. Stanley Diamond met with the Director of the Hematology Institute in Warsaw in November 2002, and the Director of the Hematology Institute in Warsaw indicated that they now have identified 52 carriers. Check out these Web sites listed below if the subject intrigues you.

"Genealogy with an extra reason"...Beta-Thalassemia Research Project.
http://www.diamondgen.org
JTA genetic disorder and Polish Jewish history
www.jta.org/page_view_story.asp?intarticleid=11608&intcategoryid=5
IAJGS Lifetime Achievement Award
http://www.jewishgen.org/ajgs/awards.html
Jewish Records Indexing—Poland
http://www.jri-poland.org

***

26

Molecular Genealogy Revolution

Geneticists today are making inroads in new areas such as phenomics and ancestral genetics. Batsheva Bonné-Tamir, PhD, http://www.tau.ac.il/medicine/USR/bonnétamirb.htm or http://www.tau.ac.il/medicine/ at Tel-Aviv University, Israel, is Head of the National Laboratory for the Genetics of Israeli Populations (with Mia Horowitz) and Director of the Shalom and Varda Yoran Institute for Genome Research Tel-Aviv. She is also on the faculty of the Department of Human Genetics and Molecular Medicine, Sackler School of Medicine.

Dr. Bonné-Tamir states that "One of my most impressive conclusions from the advancement in the last few years and the accumulation of knowledge in the fields of genetics and medicine, is the molecular revolution based on immense sophistication of lab techniques. This is really responsible for the recent increased emphasis on the human-socialanthropological aspects that affect biological diversity."

Bonné-Tamir explains, "At a meeting in 1973, in my paper on Merits and Difficulties in Studies of Middle Eastern Isolates, I said that 'The Middle Eastern isolates have emphasized again the fertile and necessary interrelationship between history and genetics.'"

Do historical events influence genes? "Comparative studies in population genetics are often undertaken in order to attempt reconstruction of historical and migratory movements based on gene frequencies," says Bonné-Tamir. "The Samaritans and Karaites offer opportunities in the opposite direction, for example, to learn the influence of historical events on gene frequencies."

In another paper in 1979 on Analysis of Genetic Data on Jewish Populations, Dr. Bonné-Temir wrote that "Our purpose in studying the differences and similarities between various Jewish populations was not to determine whether a Jewish race exists. Nor was it to discover the original genes of 'ancient Hebrews,' or to retrieve genetic characteristics in the historical development of the Jews.

"Rather, it was to evaluate the extent of 'heterogeneity' in the separate populations, to construct a profile of each population as shaped by the genetic data, and to draw inferences about the possible influences of dispersion, migration, and admixture processes on the genetic composition of these populations."

In 1999, Dr. Bonné-Temir organized an international symposium on Genomic Views of Jewish History. "And unfortunately, the many papers presented were never published," says Bonné-Temir.

Molecular Genealogy Research Projects

Certain mtDNA haplogroups and mutations or markers within the haplogroups turn up in research studies of Ashkenazim when scientists look at the maternal lineages. For example, 9.0 of Ashkenazic (Jewish) women have mtDNA haplogroups that follow the Cambride Reference Sequence (CRS) (Anderson et al. 1981).

That means their matrilineal ancestry lines follow the reference sequence that all other mtDNA haplogroup markers are compared with. The CRS shows a specific sequence of mtDNA haplogroup H found in more than 46% of all Europeans and 6% of Middle Eastern peoples. You can view a table of mtDNA sequences titled "Frequently Encountered mtDNA Hapotypes" at Table 3 in the article, "*Founding Mothers of Jewish Communities: Geographically Separated Jewish Groups Were Independently Founded by Very Few Female Ancestors,*" Mark G. Thomas et al, American Journal of Human Genetics, 70:1411–1420, 2002.

The only exception is that the Ashkenazic mtDNA haplogroup maternal ancestral lines, show up at only 2.6% with one mutation away from the CRS 343 creating U3 mtDNA haplogroup instead of the H mtDNA haplogroup. The CRS is H haplogroup. Where does U3 mtDNA show up at the higher rate of 17%? In Iraqi Jews. Ashkenazi mtDNA shows up at 9.0 percent following the CRS with H mtDNA haplogroup. Yet 27.0% of Moroccan Jews have mtDNA following the CRS. So does more than 46% of all Europeans and 25% of all Middle Eastern people have MtDNA following the CRS. That's haplogroup H of the Cambridge Reference Sequence.

The table in the article mentioned above has many sequences of mtDNA listed. Ashkenazi mtDNA was compared to MtDNA in other Jewish groups—Moroccan, Iraqi, Iranian, Georgian, Bukharan, Yemenite, Ethopian, and Indian. These were compared to non-Jewish Germans, Berbers, Syrians, Georgian non-Jews, Uzbeks, Yemenites, Ethiopian non-Jews, Hindus, and Israeli

Arabs. The percentage of frequencies of HSV-1 mtDNA sequences were listed in the samples from sites 16090–16365.

Looking only at Ashkenazi mtDNA, the mutations 184 and 265T show up in 2.6% of the Ashkenazi mtDNA. Yet in Jews from Bukhara, these same mutations show up in 15.2% of mtDNA.

And 129 and 223 mutations show up in 2.6% of the Ashkenazi mtDNA. Yet in Bukharian Jewish mtDNA, this mutation shows up at 12.1%. The rest of the Ashkenazi mtDNA stands at 9.0% for matching the CRS. That's H haplogroup, the most frequent mtDNA haplogroup found in Europe. Only 1.3% of Ashkenazic mtDNA has one mutation at 274. Yet 20% of Yemenite Jewish mtDNA shows this same mutation at 274.

Be aware that not every Jewish person has been tested for mtDNA or Y-chromosomes. You first have to research how significant samples are in speaking for the majority of any population. They do have scientific credibility, but you must always look at the sample size.

Men carry their mother's mtDNA but pass on to their sons their Y-chromosome. Women pass on their mtDNA haplogroup only to their daughters.

Genomic views of any ethnic group's history are important for further study. Whether you are taking the skeptic's position or the genomic view of your cultural history, biology does have a cultural component that needs to be analyzed scientifically. Finding flaws or benefits in research studies of any kind is the way to find inroads to truths. How else can facts change and knowledge progress?

Molecular genealogy has joined efforts with molecular genetics. How can this information help you in family history research? Ugo A. Perego, MS. Senior Project Administrator, Molecular Genealogy Research Project, Brigham Young University, http://molecular-genealogy.byu.edu, says, "I believe that DNA is the next thing in genealogy—the tool for the 21st century family historians. In the past 20 years, the genealogical world has been revolutionized by the introduction of the Internet.

"An increasing number of people are becoming interested in searching for their ancestors because through emails and websites a large world of family history information is now available to them. The greatest contribution of molecular methods to family history is the fact that in some instances family relationships and blocked genealogies can be extended even in the absence of written records.

"Adoptions, illegitimacies, names that have been changes, migrations, wars, fire, flood, etc. are all situations in which a record may become unavailable. However, no one can change our genetic composition, which we have received by those that came before us.

"Currently, DNA testing is an effective approach to help with strict paternal and maternal lines thanks to the analysis and comparison of the Y chromosome (male line) and mitochondrial DNA (female line) in individuals that have reason to believe the existence of a common paternal or maternal ancestor.

"A large database of genetic and genealogical data is currently been built by the BYU Center for Molecular Genealogy and the Sorenson Molecular Genealogy Foundation. This database will contain thousands of pedigree charts and DNA from people from all over the world.

Currently it has already over 35,000 participants in it.

"The purpose of this database is to provide additional knowledge in reconstructing family lines other than the paternal and maternal, by using a large number of autosomal DNA (the DNA found in the non-sex chromosomes).

This research, known as the Molecular Genealogy Research Project is destined to take DNA for genealogists to the next level." For additional reading, please visit the BYU's Molecular Genealogy Research Project's two Web sites. Another good source of information is at www.relativegenetics.com, a company specialized in Y chromosome analysis for family studies.

What Are Your Genes Doing In That Temporary Container?

Have you ever wondered what your genes are doing in your "temporary container" before they move on and change and where they have traveled during the past 40,000 years or more? When you have your DNA tested, work with the lab and DNA testing company, and ask them to explain to you how the 25 Y-chromosome markers you had tested as a male or mtDNA as a female help you determine your ancestry or find any matches similar to your DNA in a database. Read the frequently asked questions files of DNA testing companies online. Ask questions by email.

Sons inherit their mtDNA from their own mothers. Then mothers pass mtDNA on to their daughters. Sons also inherit the Y-chromosome from their fathers, but do not pass it to their daughters. Women don't have Y chromosomes.

The mtDNA is passed down from mother to son and mother to daughter through the cytoplasm (the cell contents surrounding the nucleus) in the egg. Only the daughters pass on their mtDNA to their daughters. And sons pass their Y-chromosomes to their sons, but those sons carry the mtDNA of their own mothers.

Even though both sons and daughters are formed from the union of a sperm and an egg, only the daughters will pass their mtDNA (which is the same as their great grandmother's on their maternal side and still further back to the first founder of their mtDNA group. Only daughters will pass the mtDNA on to the next generation.

Each lab has different methods of reporting their results, but you can use DNA test results as a tool for learning family history. Join DNA mailing lists and research or ask questions of the DNA testing firms specializing in genealogy by genetics about how many mutations occur in how many generations.

Here's how you can do your own research independently of any laboratory that tested your own DNA if you're curious about Y-chromosome DNA tests. Males take Y-chromosome DNA tests to find out paternal ancestry lineages. Males also can have their mtDNA checked, but women don't have a Y chromosome. So women only can check their maternal lineages with mitochondrial (mtDNA) tests. According to Alastair Greenshields who runs Ybase at http://www.ybase.org, "Ybase is a free and open database which allows people to enter their Y-chromosome haplotype details independently of the laboratory they were tested at. Anyone can contribute to it and anyone can explore it."

The database can accept results for any of the 36 Y-chromosome markers currently in use along with other genealogical information.

"Ybase is searchable for exact haplotype matches and/or near misses,"says Greenshields. "Surnames, variant spellings and other relevant names can also be searched for, which is especially useful for the genealogist wishing to locate and contact others that share their surname and have had their DNA tested."

Most researchers that recognize their Y-chromosome cannot identify an actual individual and are happy to share their results online in an effort to find their DNA cousins and genetic roots. "Genealogical research coupled with DNA testing is already proving a very powerful method of substantiating ancestry," Greenshields explains. "And Ybase is sure to grow in line with this upward trend, benefiting the genealogical community as a whole."

Are male and female genetic lineages studied for different purposes?

"Ybase is solely intended for Y-DNA and not mtDNA. A database of the latter would be so general and broad, given the nature of how mtDNA is passed on, as to be of little value to a researcher,"Greenshields says.

Below are a couple of explanations on Y-DNA Alastair Greenshields wrote for two different people with entirely different backgrounds who needed the whole thing explained to them. "They explain essentially the same terms but have been 'dumbed down' to varying extents. Please feel entirely free to copy them verbatim or adjust as you deem necessary," says Greenshields.

To answer my question to Greenshields on how to interpret the results of DNA tests for Y-chromosome analysis for ancestry, he's explained it well in terms that most people can understand.

"Imagine a very long rope, some of which is lying across your desk. This is the DNA strand," Greenshields says. "It just happens to be the length of rope called 'Your Y-chromosome'.

"Now look at the bit that lies across your desk and grasp the rope with both hands, about half a meter apart. This is a 'marker' or 'locus' (Latin for 'place'). We'll call it DYS19.

"In between your hands, imagine that bit of rope is divided into 14 equally-spaced segments. If you look very closely at the segments, you can see that each one has a bit of writing on it, which reads TAGA.

"This is simply the DNA code for each repeat. Therefore the marker DYS19=14 repeats. Or if I ask you, 'What allele have you got for DYS19?' You can tell me '14'. (Allele effectively means the number of repeats.) For example, DYS19 has about nine possibilities (between 11 and 19).

"If you do the same at lots of different markers or loci (plural of locus), you'll get a whole series of numbers (DYS19=14, DYS388=15, DYS461=11 etc.). This is your 'haplotype'. It doesn't matter whether it is 20 or 200 numbers in length. This series of numbers is still called your haplotype.

"Now you are going to make some rope for your new son. You are pretty good at making rope and usually you can copy your own precisely, but this time you made it slightly too short. There are now 13 repeats. (Technically this is called a 'mutation' which can occur when an enzyme mis-types the DNA code). It still works perfectly well so your son keeps it and is very happy! There, you have it—DNA in a nutshell! (Be aware that your repeats, like the stock-market, may go up as well as down, but for entirely different reasons.)

Genealogy and the Y Chromosome

"DNA for the use of genealogy usually requires an analysis of your Ychromosome," Greenshields explains. "Only males have this particular chromosome and the DNA code held within it is passed down from father to son (virtually) unchanged. Provided there is an unbroken paternal line between two males, that is both share a g-g-g-g-g-grandfather, their Y-chromosome DNA will be the same."

When the DNA is analyzed, many small sections are looked at. Presently, the testing companies look for anywhere between 10 and 26 sections or 'markers.' "At any one of the markers, the code will repeat itself, for example, 15 times," Greenshields explains. "If the marker is called DYS19, we can give the result DYS19=15."

If you analyze several of these markers, you end up with a 'haplotype.' Thus you can compare haplotypes to see if you are related. "I say 'virtually' unchanged, as the DNA can change slightly over time due to 'mutations'—small errors formed when the DNA is copied," says Greenshields. "When comparing the haplotypes from two people, this will show up as a 'mismatch'—where, for example, DYS 19=14."

These mutations are useful by themselves and occur at a fairly steady rate over time. "It also gives us the great variability that we observe over populations," says Greenshields. "If the mutations did not occur, every male would have identical Y-chromosomes. Also, within DNA/genealogy studies, if many mismatches occur when comparing two male haplotypes, we can say that they are not related."

So you can see DNA can be a very useful tool when comparing 'suspected' relatives. "But there is one caveat, however," Greenshields emphasizes. "You must share a surname, or have a very good reason to believe you are related. DNA alone will not identify your relatives from any other random person."

For example, you will probably share the same haplotype to at least someone in your home-town, but having a similar or same surname will raise the probabilities significantly. "DNA is a tool that should be overlaid on the existing genealogical records," notes Greenshield. "It can be an excellent way of deciding on further avenues of research, or indeed defining that there are several distinct lines within your family name."

Understanding HLA Genes (White Blood Cells)

HLA genes are white blood cells. Anthropologists look at white blood cells called by the scientific name of HLA genes to study genetic drift. If you have a research need to learn about tissue typing, you might want to read about understanding the HLA genes. Tissue typing is usually done on white blood cells, or leukocytes. The markers are referred to as human leukocyte antigens (HLA). A good starting point is to read the definitions and excerpts about tissue typing testing on the 'About' Health and Fitness Web site at: http://thyroid.about.com/library/immune/blimm25.htm?terms=Hum an+Leukocyte+Antigen.

"My personal view of the HLA genes and genetic genealogy is that the two will almost certainly not be mixed in the near future on a commercial basis," says Greenshield. "My reasoning is two-fold. First, the L in HLA stands for leukocyte/leucocyte (Gk leuko=white). The test involves looking at 'white' blood cells." This would involve taking a blood sample (or possibly sampling a site of infection) and a trained phlebotomist being on hand. The Brigham Young University (BYU) study used to have participants give blood (of which HLA typing is possibly involved)."

Ugo A. Perego, MS, Senior Project Administrator, Molecular Genealogy Research Project, BYU emphasizes, "We stopped using blood last summer. All our collections are now based on a simple 45 seconds rinse using mouthwash."

"Your average genealogists are only willing to swab the inside of their mouths," says Greenshields. "Even a pin prick of blood would, possibly, be too much to ask of one's suspected relatives."DNA testing for ancestry only requires a swab of felt or cotton on the inner cheek or rinsing your mouth with mouthwash.

"Secondly, the HLA system can provide information or guides to genetic disposition to disease. Given the apprehension about giving a cheek swab to be 'junk-DNA tested' for the Y-chromosome, having HLA alleles typed, compared, and possibly posted to the net would discourage all but the hardy," says Greenshields. "I believe HLA testing for genealogical studies will remain under the domain of research into genetic drift and anthropological studies."

If you ever need a tissue donor or have to get your tissue typed for medical reasons, that's when the HLA genes play a major role in tissue typing. Interestingly, when small communities are isolated for long periods of time, and bottlenecks pare down the population to only a few founders, genetic drift may occur. That's when the anthropologists and evolutionary biologists look at the HLA genes.

With some DNA testing companies offering racial percentages tests, Y-chromosome tests, and mtDNA tests, what is being done with mtDNA testing for maternal lineages? How can we trace female relatives or ancestors who leave no written records of their name or existence?

Will studies of HLA genes be used by others in various fields as research now is used by anthropologists studying genetic drift, scientists studying tissue differentiation, or physicians looking at how white cells fight infection?

Dr. Peter Reed has a PhD in Human Genetics from the University of Oxford and was a pioneer in the use of STR genetic markers in medical research.He explains here that HLA genes primarily determine how our blood cells recognize and react to other cells present in our bodies. In particular, this makes HLA genes important in how our body responds to 'foreign bodies.'

For example, the HLA genes as white blood cells, fight infection.

When bacteria or viruses enter our bodies, the HLA genes are there to do battle. When organs or tissue are transplanted, HLA genes have to be considered.

They would attack the foreign tissue placed in the body. "When people talk about blood typing or tissue matching, on the whole, they are referring to determining some aspect of the set of HLA genes," says Reed. "HLA genes are perhaps the most variable of all human genes.

Across the population, some HLA genes have dozens of different forms of genes or 'alleles.'"

The result is that two randomly chosen people are unlikely to share identical HLA genes. "Even within families, there is a good chance that each family member has a different set of HLA genes," says Reed.

"That's why finding a 'suitable match' for a transplant can be difficult." It's also the reason why we all react differently to infections. On the plus side, HLA genes (white blood cells) can deal with all the different infections we get during our lives, usually without being aware of them.

Apart from the obvious medical importance of a role in responding to infection and transplantation, there is another role. It's perhaps one of the primary reasons that HLA genes are some of the most intensely studied of all human genes.

"This relates to the role HLA genes play in determining how our blood cells respond to the other cells of the body," says Reed. "In certain circumstances, some of a person's own cells are mistaken as 'foreign bodies.' These cells are responded to as if they were an infection." The technical jargon for this is called auto-immunity.

"This can result in disease. This sort of problem is believed to be one of the underlying causes many fairly common diseases, more appropriately termed 'conditions'. Such conditions include Rheumatoid Arthritis and Juvenile Diabetes," says Reed. "A connection between particular types of HLA genes and certain conditions was first recognized more than thirty years ago." Since then many connections between HLA and human conditions have been identified.

These genes are obviously very important in human health, and are often suspected as being the major genetic causes of numerous conditions. Consequently, there are a number of clinical programs where HLA genes are screened (particularly in children) to research and even determine the risk of later disease.

"One aspect of the high variability of HLA genes, is that certain types (alleles) of certain HLA genes have been found to be geographically/ethnically distributed," says Reed. "For example, some alleles of some HLA genes may be more frequent in Japan than in England. Therefore there is some possible utility in the use of HLA genes in determining ancestry from different geographical locations. However, because the HLA genes are only a small fraction of all our genes, examining HLA genes alone is not likely to be very informative."

"Because HLA genes, like almost all our other genes, are shuffled and mixed as they are passed on from parents to children, it's difficult to determine the exact set of HLA genes of even one or two generations previous," Reed says. "So they have little utility in determining recent ancestry."

There could be some utility of HLA for genealogy. "This could be so in certain circumstances," Reed explains, "but the hurdles mentioned above will need to be overcome. I'm exploring this further." According to Ann Turner, Genealogy-DNA List Administrator, at: http://lists.rootsweb.com/index/other/Miscellaneous/GENEALOGYDNA.html, an excellent Web site for explaining HLA is located at the Web site: http://www.med.umich.edu/trans/public/hla/hla_&_you.html.

"This HLA Web site diagrams the inheritance patterns. It says HLA is on gene 6, but it means chromosome 6," reports Turner. "You also can learn about linkage disequilibrium at this site. Some genes in the HLA system are close to one another. That makes the alleles, which are a form of a gene, also linked together closely and inherited as one unit, or haplotype."

That's the original context for the word 'haplotype.' Also look at: http://www.hokkaido.bc.jrc.or.jp/laboratory/laboratory500_eng.htm, Turner notes.

Understanding Your Maternal Lineages—mtDNA

MtDNA shows ancestry passed from mother to daughter from a single common ancestor or founder. Every human owes his or her ancestry to the ultimate "Mitochondrial Eve" the first woman to walk out of Africa and head towards Yemen around 154,000 years ago, give or take several thousand years.

People who have their DNA tested for ancestry research want their results kept private and not used against them, by employers or insurers looking for genetic risks, or by groups that use the results of DNA tests to begrudge them of anything from their rights to religious and personal choices to their core identity. DNA analysis for ancestry does not report on disease risk.

It's okay to have databases with your DNA matches so you can contact what might turn out to be someone who shared a common ancestor with you 250 years ago or thousands of years ago or more recently. People who were reared in adoptive families or foster homes would cherish the idea that they had ancestors who lived in a particular geographic area that they could visit or look to for a core identity.

Sometimes the country you live in provides the core identity and other times some people want something more personal or religious as a core identity within a core identity. For example, someone may be Jewish, but knowing they are Sephardic, Ashkenazic, Mizrahi or another group provides an open door to sampling travel, identity, food sampling, religious and ethnic customs, history and more, perhaps a search for ancestors or relatives.

If we study mitochondrial "Eve" then we have to study Y-chromosome "Adam." We have to ask whether mtDNA diversity is higher than Y-chromosome diversity because mtDNA developed and mutated at a different rate than Y chromosomes if we look to prehistoric ancestry lines.

Usually, studies of mtDNA show either the female population had a more diverse genetic history or some communities were founded by very few female founders. For example, H haplogroup of mtDNA found in a large percentage of Europeans may have begun in the Dordogne valley of what is now France and/or in northern Spain about 21,000 years ago, but before that, H haplogroup may have had an ancestor somewhere else.

That common ancestor was one woman who had at least two daughters who survived to have more daughters and who lived somewhere in the Middle East. At some point back in time, H haplogroup arose from a still more ancient common ancestor, another woman, who lived outside of Europe.

What we see now are the mutations that occurred over thousands of years since haplogroup H mtDNA reached Europe and expanded to cover today all of Europe from Iceland to the Urals. H haplogroup mtDNA today is found in places as far apart as Bashkortostan in the Urals and Iceland, Scotland, Spain, Norway, Austria, Turkey, Crete, Ukraine, Italy, and Bulgaria.

MtDNA haplogroups are classified as A, B, C, D, E, F, G H through J, K, M, N, O, P, R, and T through Z. Then some are given little sub classifications such as U1, U2, U3, U4, U5, U6, U7, and various types of U found in mostly in India. New mtDNA haplogroups are still be uncovered. M is a super haplogroup divided into various groups of M such as M1 and M11. As ancient burials are uncovered, different mtDNA haplogroups turn up that are not here today because they are very ancient and did not survive because some women had only sons and some daughters didn't survive to reproduce.

Using Your Own DNA Test Results as a Genealogy Tool

It's good to have a mentor to answer questions about your test results until you are able to do your own research on the Web. If you're a lay person, where can you learn enough molecular genetics to get a handle on DNA test results and untangle ancestral roots? If you've ever wondered why your genes are not where you thought they were supposed to be (in geographic location on a map), that topic of research is called molecular prosopography. See the Web site: www.linacre.ox.ac.uk/research/prosop/prosopo.stm.

Prosopography is an independent science of social history embracing genealogy, onomastics and demography. Prosopography is all about human history and genes that travel because your genes have both a cultural and a biological component. The cultural component includes onomastics which is the study of the origin of a name and its geographical and historical utilization.

Onomastics includes the study of how and when place-names were originated and used. Then there's toponymics. Toponymics is the study of names related to a place or region. See http://libraryweb.utep.edu/onomastics.html or http://www.kami.demon.co.uk/gesithas/biblio/bib08.html. And you probably know demography, is the interdisciplinary study of human populations. Demography deals with social characteristics of the population and their development. So you'd find more information on demography by researching population studies. Phenomics is the science of customizing, tailoring, and individualizing medicines and other health treatments to the total human genome of one person.

The age of one medicine or hormone fits all is gone. As a tool, phenomics also can be applied to herbal remedies, food supplements, vitamins and minerals, hormones, and other formulas adjusted to an individual's total genome. If you have a genetic risk for a certain disease, perhaps you can find out what way there is to prevent it by using phenomics as a tool for customizing your treatment or working on prevention strategies of lifestyle, diet, or medicine.

Family history DNA testing is a new way to approach biological research. Genealogy and genetics are forms of hunting and gathering that persist. First you start with transcribed oral history. We are foragers in molecular family history.

Molecular genealogy uses DNA testing (human genetics) as a tool for untangling ancestral and recent family roots. Here's an introduction to family DNA testing to be used with oral history gathering and genealogy.

Start your family history time capsule, gift basket, scrapbook, genetic genealogy, or begin a small business publicizing DNA testing for genealogy. The place for genetic genealogy is in an archives, library, museum, or good storage place.

Future generations need a DNA history of as many ancestors as they can find willing to participate and to create oral histories. Genetics is the most mathematical/statistical of the biological sciences. We have fields such as bioinformatics that combine computers and biological information. Family historians need a bridge to fill the gap between such a mathematical science as genetics and genealogy, often based on records and oral histories.

The oral history would be transcribed on acid-free paper in hard, bound copy. Photos and other memorabilia could be added. Then the basic archive would be copied onto disks such as a CD, DVD, or other, stored in a computer and on video and audio tape.

Another copy would be saved as a multimedia presentation with text, sound, voice, photos, illustrations, and video/audio and saved on a disk to be played on screen with a home entertainment player or in a computer.

You could put a smaller file online on a Web site. This molecular biography would represent not only the life of a person, but a history of the person's DNA test results, racial percentages, ethnicity, if known, and anything else about the DNA sequence as far as geographic location or even medical history, if desired, in a more private file for relatives.

This is where genetics joins with genealogy.

We not only have a family history to archive, but now a genome, or at least a record of the matrilineal and patrilineal ancestry by DNA. We have the markers and the sequences. The idea is to learn enough about DNA testing and genealogy to understand what those sequences and markers mean.

What can we learn about ancestry through the mitochondrial DNA (for women and men) the Y chromosome only for men, and other markers on the genome? What should we look at to view the percentages of races such as Native American, African, East Asian, or Indo-European (Europe, Middle East, and India)?

What do these sequences tell us about our ancestry? If there's no such thing as race, what geographic locations of our ancestors are we viewing back in time when we look at the genetic markers?

What dates are we looking at—a few generations ago or 21,000 years? What do our transitions and mutations mean over a long span of time? What foods, medicines, therapies, and climates are best for our customized, individual molecular profiles?

How do we read and interpret those genetic markers? Where does genealogy and oral history fit in? Family history—genealogy—now has joined up with molecular genetics and evolutionary anthropology. And included with genealogy is the tradition of transcribing and recording oral history, diary journaling and restoration, time capsules, biography, scrap booking, videography, and photography.

The genome has reached the genealogist. Family history today is multimedia and molecular, historical and futuristic. "Progress in our knowledge of the genome and of its function has been extremely rapid since the development, in the mid-eighties, of the Polymerase Chain Reaction," says Professor of Genetics, Guido Barbujani, (Department of Biology, University of Ferrara, Italy.)

Dipartimento di Biologia, Universita' di Ferrara via L. Borsari 46, I-44100 Ferrara, Italia. See his Web site at: http://www.unife.it/genetica/Guido/Guido.html.

Dr. Barbujani's fields of interest include human population and molecular genetics and evolution, and I've read many of his articles in the various journals of genetics and research books, such as Archaeogenetics: DNA and the population prehistory of Europe published by the McDonald Institute Monographs.

"By that method, minimal quantities of DNA can be studied, which has opened the field for a number of previously hard-to-imagine applications, ranging from gene therapy to the prediction of interactions among genes, from the sequencing of entire genomes to the retrieval of DNA sequences from extinct organisms," Barbujani explains.

"DNA technologies proved so powerful that people tend to forget about their limitations. Still, limitations exist, especially in the field of genealogical reconstructions, and future technical advancements are unlikely to be of great help.

"Consider this: Each of us has two parents, four grandparents, eight grand-grandparents, and so on. In principle, only ten generations ago (around 1750 AD) we had 1024 different ancestors. In fact, chances are our ancestors were less than 1024, because consanguineous marriages likely occurred at various stages. But even if we had only 200 independent ancestors ten generations ago, each of them contributed to our 30,000 or so genes.

"On the other hand, only one of them transmitted to us her mitochondrial DNA and, if we are males, from only one of them did we inherit our Y chromosome," Barbujani reveals. "The other 198 or 199 ancestors' contributions to our genotype are of course equally important, but there is no easy way to figure them out."

"Indeed, at every generation recombination created new associations of genes along our chromosomes, except for the mitochondrial DNA and for part of the Y chromosome, which do not recombine. In this way, traits of DNA coming from different ancestors have been assembled in a mosaic that cannot be disentangled a posteriori, in which each piece has a different, and possibly very different, origin. In short, it is an illusion to think that our mitochondrial DNA (or our Y chromosome) may allow us to understand our family history.

"These are small parts of our genome, and hence contain information on but a small bit of our biological history," says Barbujani. "Other ancestors have transmitted to us many more genes than the ancestors from whom we inherited our mitochondrial DNA, and they may have come from different parts of the world."

"That may sound frustrating to some, but population genetics has something important to tell us in this regard. Population histories are much easier to reconstruct than individual histories, because chance phenomena have a much greater impact on the latter.

"When a large number of individuals are jointly analyzed, rather robust evolutionary inferences may be drawn, even if some members of the sample have had an unusual family history. By combining measures of genetic diversity, among populations and among individuals, with the evidence coming from mitochondrial and Y-chromosome genealogies, population geneticists have shown very clearly that each population contains a large proportion of all humankind's alleles, around 85 percent, on average.

"This finding has several implications. One is: should most humans disappear because of some global catastrophe, and should only one community survive, the loss of genetic diversity would be very limited, around 15 percent. That might or might not be reassuring, but is true.

"Secondly, although many tend to think that humans come in clear racial clusters, that is not true; if, on average, populations contain 85 percent of the global human diversity, two individuals from very distant localities can be just 15 percent more different genetically than members of the same population (unless the latter are relatives, of course). Third, if genetic diversity is so high among members of the same population, the only possible explanation is that those populations incorporated, through time, contributions from other populations at a rather high rate.

"In other words, our ancestors spent most of their evolutionary time in communities connected by extensive migratory exchanges, and not in isolated groups. Through migration, alleles of African, Asian and European origin ended up all over the world, and no biologically recognizable race evolved in our species.

Therefore, it is impossible to define our origin by studying our DNA, but if it were possible, we would probably find that our roots are spread over much of the world.

"As Jonathan Marks remarked, today convincing people that there is no such thing as a human race is probably as difficult as, in the 17th century, to convince people that the earth rotates around the sun and not vice versa. However, this is a scientific fact, and perhaps the single most significant result of human evolutionary studies. Everybody can tell a Nigerian from a Japanese person, but if we move from Nigeria to Japan we shall never find a sharp boundary separating two well-distinct groups.

"Rather, we shall notice that the genetic features of people change continuously, in a gradient, and that each community harbors substantial biological differences among its members. The best way to summarize these concepts, I think, is by a slogan invented by the French anthropologist André Langaney: Tous parents, tous différents. We are all relatives, and we are all different."

The Cohen Modal Haplotype and the Armenian Modal Haplotype

If you want to look for signatures of Middle Eastern origin in Ashkenazic males you first look for repeat patterns that support the hypothesis that haplotypes with high DYS388 repeat. If they repeat, then you can hypothesize that there is an origin in the Middle East. The DYS388 repeat becomes one of the signatures of ancestry of Middle Eastern or Southeast Asian origin (Bradman et al. 2000).

For example, in Armenians, the DYS388 marker of the Y-chromosome repeats 15 times. In Georgians the repeat is 12 times. Does it repeat in the majority of Ashkenazic males? The Armenian modal hg2 haplotype is shared with other Middle Eastern countries, but the Armenian modal hg1 haplotype is a one-step relative of the English modal haplotype 3 and the Frisan modal haplotype 50 of the hg1 haplotypes. The Armenian hg1 haplotype also is the Turkish modal hg1 haplotype.

Scientists know that the Cohen Modal Haplotype shows up in non-Jews in Hungary, Italy, and Armenia, but there's also the modal hg2 haplotype in Turkey and Azerbaijan. Armenians share haplotypes that occur frequently in Jews and in other populations. The Cohen Modal Haplotype is strikingly similar in Sephardic and Ashkenazic Y-chromosomes—the paternal lines. There are regional DNA differences in Armenia. And Armenia had a vibrant Jewish community in medieval times that disappeared several hundred years ago.

The community existed contemporaneously with Jewish communities at the tail end of the era of the Khazars (a few were still around in 1200 after Khazaria was destroyed in 965 CE). Jewish communities flourished in the 13th century in neighboring Georgia and also in the Caucasus, in Iran, Azerbaijan, Daghestan (original homeland of the Khazars), the Crimea, (the refugium for Byzantine Jews and people fleeing Germany and Hungary).

The Armenian Jewish community was a neighbor to the Jewish communities of the Ukraine, where Jews from Germany and Central Europe joined existing communities of Jews from Anatolia and Greece fleeing to the Ukraine from both West and South. Byzantine Jews lived in the Ukraine and the Crimea, as well as Khazars who may have joined the nearby Jewish communities. What is known is that Jewish men traveled far and wide within existing Jewish communities. Medieval Armenian Jews had Hebrew names. The city of Eghegis contains 62 medieval Armenian Jewish gravestones dated around the year 1266 CE with Aramaic and Hebrew inscriptions.

The medieval Armenian Jewish community was small—perhaps consisting of only about 150 people, according to Frank Brown, writing in the Jerusalem Report ("Stones from the River"). The stone inscriptions contain dates ranging from the middle of the 13th century to 1337.

See the Web site on Jewish Armenian communities in medieval times at: http://www.sefarad.org/publication/lm/045/4.html.

DNA of the Levites

The Levites are different. Levites, unlike Cohanims, have some Y-chromosomes in three different groups showing a heterogeneous, a diverse, variable origin. Some contemporary Levites may not be direct patrilineal descendants of a paternally related group. There is a term, "the Ashkenazic Levite Modal Haplotype." See the Scientific Correspondence section of Nature, Vol. 394, 9 July 1998.

As noted in the first chapter, population geneticists estimate that Cohanim and Levites each make up ~4% of the Jewish people (Bradman et al. 1999.) See the article, *"Multiple Origins of Ashkenazi Levites: Y Chromosome Evidence for Both Near Eastern and European Ancestries,"* Doron M. Behar, Mark G. Thomas, Karl Skorecki, Michael F. Hamer, et. al. American Journal of Human Genetics 73:768–779, 2003.

Also, see the article titled, *"Geneticists Report Finding Central Asian Link to Levites,"* September 27, 2003, by Nicholas Wade, New York Times. According to that article which refers to a report published in the "current issue" of the

American Journal of Jewish Genetics, a study was prepared by population geneticists in Israel, the US, and England, based on a six-year study. The conclusion was that "52 percent of Levites of Ashkenazi origin have a particular genetic signature that originated in Central Asia, although it is also found less frequently in the Middle East," according to the NY Times article. That signature is the R1a1 Y-chromosome haplotype. Interestingly, that R1a1 signature is found in 60% of Hungarian men and 56% of Polish men. It's also found all over Europe, including in some of the Basques. A much rarer signature, R1a* with the M17 mutation sequence in that particular study was found in one Armenian male and in on man from Belarus. Currently research is ongoing in Chuvashia to see if there is a genetic link with Ashkenazic males. The Chuvash language is closest to the medieval Khazar language. However, the R1a1 Y-chromosome haplotype is also found in Sorbs of Lusatia in East Germany and Western Poland. The Sorbs are the westernmost Slavs who are said to have contributed linguistically to the Yiddish language in medieval times. Interestingly, about 52% of Ashkenazic Levites have Y-chromosomes close to the Sorbs. Scientists are researching whether the Ashkenazic community received the R1a1 signature from the Sorbs, from the Khazars, from Central or West Asia, or from anyone else. The R1a1 marker came into the Ashkenazic population about 1,000 years ago in medieval times from either one male or a very few males. However, you can't exclude the possibility that it came from a Syrian male rather than a European because ten percent of Syrian men also carry the same R1a1 Y-chromosome haplotype.

Scientists say it isn't likely since it entered 2,000 years later than the Cohen Modal Haplotype found in the Middle East and Mediterranean areas. Yet the Cohen Modal Haplotype also shows up in some non-Jewish Hungarians as does the R1a1. Also, the Cohen Modal Haplotype occurs in Armenians, Italians, and many peoples who have had contact with the Mediterranean or who live around the Mediterranean. Due to so many migrations, you can exclude possibilities, only estimate probabilities based on the idea that one male 1,000 years ago introduced R1a1 to the Ashkenazim. The Cohen Modal Haplotype has a common ancestor in the Middle East in the Jewish population that goes back 3,400 years, and in the Middle East non-Jewish and Jewish population in general that occurs around 7,800 years ago as a common ancestor.

When the question of whether a Khazar contributed the R1a1 to Ashkenazim, it's possible, but it's also possible that during the expansion of Islam 1,000 years ago a Syrian could have been the contributor as well or a Sorb from Eastern Germany/Western Poland around the year 800. It's all a matter of looking at the time frame the genetic marker entered the population. It's also about proximity, probability, and the estimate that in Europe Ashkenazim may resemble the host country more than non-Ashkenazic Jews. So the research is continuing, especially with the Chuvash of Chuvashia. Research is updated periodically in the various journals of genetics.

Middle East male markers such as the DYS 388 have longer repeat patterns than European similar markers. Studies look at the marker DYS 388 in the Y-chromosome to see its repeat patterns. Scientists also use a model as a method to estimate the coalescence time of Cohanim chromosomes and in certain studies have dropped DYS388 in analyzing Cohanim chromosomes to estimate coaslescent time. Ashkenazic and Sephardic Cohanim chromosomes have been dated to an estimate of 2,684 to 3,221 years before present.

The earlier date goes back to the Exodus and the latter date goes back to the destruction of the first Temple in 586 BCE. Studies conclude that the origin of "priestly Cohanim Y-chromosomes" originated sometime during or right before the Temple period in the history of Judaism. There's always uncertainty because the mutation rate varies.

In some of the older anthropology textbooks, Jews and Armenians have been classed together as "Armenoid or Assyrian" in characteristics such as skull shape and features.

What actually happens is when a broad-skulled person from any country marries a long and narrow-skulled person, the child sometimes is born with an "Armenoid" skull, that is a flattened occiput, narrow skull, long or oval face, and a convex nose with complexions varying from fair to swarthy and hair color from blonde or red to brown or black. These characteristics for decades had been assigned to the "Armenoid" peoples where Ashkenazim and Armenians had been grouped together before DNA testing was used by physical anthropologists, population geneticists, or archaeogeneticists.

For more information on DNA studies of the Levites see the article, "The Origins Of Ashkenazic Levites: Many Ashkenazic Levites Probably Have A Paternal Descent From East Europeans Or West Asians," Bradman, N1, Rosengarten, D and Skorecki, K21, The Centre for Genetic Anthropology, Departments of Biology and Anthropology, University College London, London, UK. And 2 Bruce Rappaport Faculty of Medicine and Research Institute, Technion, Haifa 31096, Israel and Rambam Medical Center, Haifa 31096, Israel. (See the Bradman Index at http://dna6.com/abstracts/bradman.htm which is online as part of an index to a list of abstracts and posters at http://dna6.com/abstracts/index.htm

Here are some definitions you might want to peruse before we go into the next chapter on personalizing family history records with the results of DNA tests. If you are a historian or genealogist, it would be useful to be able to discuss possible DNA testing with your clients.

Molecular tools to family history research open doors to new subjects.

Useful DNA Definitions for Historians and Genealogists Interested in Molecular Anthropology/Archaeology

* Genome. A person's genome is one set of his (or her) | genes. The human genes, which control a cell's structure, | operation, and division, are located in the cell's nucleus. The | full human genome (estimated at 50,000 to 100,000 genes) is present in every cell-nucleus, even though many genes are| inactive in cells which have some specialized functions (the| "differentiated" cells).

* Genes and Chromosomes. Genes are composed of segments of DNA. In normal cell-nuclei, the DNA is distributed among 46 chromosomes (23 inherited at conception from a person's father, and 23 from the mother). Each chromosome consists of one very long strand of DNA and numerous proteins, which are required for successful management of the long DNA molecule. The longest chromosomes each "carry" thousands of genes. Every time a cell divides, the cell must duplicate the 46 chromosomes and must distribute one copy of each to the two resulting cells.

* The Code. The DNA of each chromosome is composed of units—nucleotides" of four different types (A, T, G, C). These nucleotides are linked to each other in linear fashion. The sequence of the four types of nucleotides is critical, because the sequence produces the "code" which (a) determines the function of each particular gene, (b) identifies the gene's start-point and stop-point along the DNA strand, and (c) permits certain regulatory functions. The code of the human genome consists of more than a billion nucleotides.

* The Mitochondrial DNA (mtDNA). Outside the nucleus, human cells also have some "foreign" DNA located in structures called the mitochondria. This small and separate set of DNA does not participate in the 46 human chromosomes, and is not part of "the genomic DNA." The mitochondria are inherited from the mother.

These genetic term definitions are from the book titled: Confirmation that Ionizing Radiation Can Induce Genomic Instability: What is Genomic Instability, and Why Is It So Important? John W.
Gofman, M.D., Ph.D., and Egan O'Connor, Executive Director, CNR. Spring, 1998. The excerpt of definitions from Dr. Gofman's essays, such as what is quoted above is reprinted with permission. See excerpts from the book at the Web site at: http://www.ratical.org/radiation/CNR/GenomicInst.html.

All the definitions from Dr. Gofman's essays are available for reproduction in other publications. Please do cite the title and above URL so people who wish to study the complete work can do so. For more information see the Web site at: CNR page (http://www.ratical.org/radiation/CNR/). For further information, contact the publisher, David Ratcliffe, "rat haus reality press" at: http://www.ratical.org/rhrPress.html.

***

27

Personalizing Ethnic Family History Records with DNA Testing

You can choose a DNA testing company that provides a database for you to find DNA matches to your mtDNA or Y-chromosome or let you take racial percentages tests. Then you can find your match and interview the individual or chat and create DNA Match time capsules of your own, a journal, audio tape, video, or other archive where your family and the other family shares ancestral history events in an oral or transcribed archive or record.

Or you can work with famiy diaries of events uniting ancestors. Diaries and DNA testing personalize family history records. DNA family databases and scrap booking are more valuable when linked together. It's time to compile a written and illustrated family tree time capsule in any or all of various media—print, pictures, video, audio, for the Web, in a scrapbook form of stories, anecdotes, experiences, photos, journal writings, and memorabilia that also includes DNA testing.

DNA test results and autobiographies also may highlight the important events that you want remembered. In the future, families may be able to archive the sequences of their entire genome—all their genes—into a database to be kept along with photos, video, audio, crafts, and memorabilia marking the life experiences, events, rites of passage, and highlights focused around a central issue.

You can piece together records of women's clubs, diaries, and DNA tests and look at your maternal lineages. String together military pension or service records, village societies, or Census records and city directories, and link your paternal lineage from voter lists and court records to Y-chromosome test results.

The file or database could be passed onto other family members and the genome given to health care professionals to customize therapy or treatment, tailor foods and vitamins or supplements, or create a living video biography. For now, whole genome DNA testing is expensive, and what are affordable include the matrilineal and patrilineal lineages and the percentages of races.

For family historians, there are also the surname databases and message boards on the Web. And for tracing female ancestors, there are marriage certificates, birth records, court records, church and synagogue records, records of teachers, factory workers, census records, social and immigrant records, women's publications, clubs, insurance records, deeds and wills, and other records that reveal more than the words, "and wife."

The 1850 US Census was the first census to name all members of a household, their birthplaces, and whether married within the year. By 1870, the US Census asked whether one's parents were foreign born. And a decade later, the 1880 US Census named relationships to the head of household.

By 1900, the US Census included the number of years married, number of children born and who of the children were living. They also added immigration and naturalization data. Native Americans had special censuses. So you can also look at school censuses and state and local censuses as well as city directories published until 1976.

Years ago when people had no phones, the city directory was one way to locate families. In creating a "family history memorabilia time capsule" or database that includes DNA testing, you can include small crafts such as braided hair embroidery as art, craft, needlework, preserved clothing or wedding gowns restored and wrapped.

You'd index where the craft work or clothing is located, and the index or list of memorabilia would go into your scrapbook, database, or time capsule. What new item that you might add, would be the family's DNA, the male and female mtDNA and the male Y chromosome, plus a racial percentages DNA test.

Records may be copies and stored in many ways—as Web sites, printed books, diaries, on CDs and DVDs, as video and audio tapes, as oral history transcripts printed out in text, or as a photo scrapbook with captions. Or you can create a multimedia presentation combining text, voice, video, photographs, music, and commentary.

You would ask as many relatives as you can find to lightly rub a felt tip or pad, cytobrush, or a swish of mouthwash in their mouths and send the samples packaged and labeled separately to a DNA for genealogy testing company. With all those entries in your time capsule, your descendants (or your clients' if you do family history research) will have a better idea of who any particular family was as people (rather than some anonymous photos).

You can even search antique stores and flea markets and the people listed on the Web. Some genealogists rescue old photos and are listed on the Web. Check to see whether any photos found in certain locations might be your family members or those of your clients if you are a genealogist or family history researcher.

So you wipe a felt swab or small brush across the inside of your cheek and mail it to a DNA testing company. Or you swish a type of mouth wash and expectorate it into a container and mail it back to a DNA testing company emphasizing testing and/or researching genealogy by DNA analysis. Some of these companies may also have a division that tests DNA for forensic purposes, and other firms may give the DNA to a laboratory for actual testing and then send you the results with information on how to further research your ancient lineages or genealogy for more recent DNA matches.

What comes back to you in the mail a few weeks later are a print out, perhaps a CD, or a mailing on paper and/or email table of some of your DNA sequences. Now it's your job to find out what the sequences mean. Most companies have frequently answered questions message boards and some firms will email you answers to your questions. Other companies may offer to store your DNA or a certain length of time. So check with the company on what it offers regarding DNA testing and genealogy questions answered.

"The state of identity testing is such that people should have a specific hypothesis that they want to test," says Harry Ostrer, M.D, Professor of Pediatrics, Pathology, and Medicine Director, Human Genetics Program New York University School of Medicine.

"Do Susan Smith and I share a common matrilineal ancestor? "Do Jeffrey Jones and I (if male) share a common patrilineal ancestor?" Dr. Ostrer asks. "Hoping to discover something unanticipated is unrealistic. It is very unlikely that amateur genealogists will discover that they had Amerindian ancestry unless they had a strong reason to expect so.

"The problem of course is that many people are searching for roots and hoping that genetic testing will fill the gap for absence of familial oral histories," Dr. Ostrer explains. "Unfortunately, there are no shortcuts to the work of the genealogist. With the Internet, email and a heightened awareness of genealogy, the tools—word-of-mouth and access to vital records—are more accessible."

According to Dr. Harry Ostrer's article, "A Genetic Profile of Contemporary Jewish Populations," in Nature Reviews/Genetics, Vol. 2, November 2001 (Science and Society), p. 895, Macmillan Magazines Ltd, "The Ashkenazi Jewish population in Eastern Europe expanded rapidly, growing from an estimated

10,000–15,000 people in 1500 to 2 million in 1800 and 8 million in 1939 (REF.34)."

Compare that profile to the group of people you're researching. What's the individual's genetic profile? How does it compare to the oral history profile or the written record profile, either social or medical, ethnic, or industrial?

Consider the group of people you may belong to on one or both sides of your family. Then find out about the genetic history of your people in the same way as you research the genealogy or family history—through reading articles on the molecular genetics history of your ethnic group. You can also talk to relatives and even trace each ancestor's medical histories as far back as oral history or written records take you.

How can you use DNA testing information together with oral histories, diaries, military and court records? You can do online genealogy searches. You can explore by reading and interviewing scientists or listening to tapes or videos. Research your own genetic origins or any ethnic group you want to study. If you're not Jewish, the techniques are universal. Learn about how DNA techniques and molecular genealogy meet. Apply the methods to your own family history research.

Dr. Ostrer of New York University School of Medicine, Human Genetics Program conducted genetic analysis of Jewish origins. See the Web site at: http://www.med.nyu.edu/genetics/jewishorigins.html

According to the Web site, "The next step in Jewish genetic demography will be to understand the patterns of Jewish migration that formed the historical communities. Clearly most of these communities no longer exist, but their genetic structure can be discerned by studying the DNA of their descendants."

Consider looking at any other family memorabilia such as photographs or paintings, even crafts made in the past. Oral history, DNA testing, and genetic history work together with written, medical, and oral family history. Consider the times and background of the dates.

Search the industrial revolution era or before up to the present. What conditions did the person you're searching live under—agricultural or industrial? Was the family confined to a tiny apartment in an urban setting or on a farm?

Once you receive the results of DNA testing, you'll have a collection of sequences. Now is your chance to learn how to interpret those equences by asking questions on the DNA mailing lists and reading up on the subject of what the sequences mean in plain words.

To find out where to get your questions answered about interpreting the sequences, first check with the company that tested your DNA as various labs use different markings. Then contact the message board at Roots Web.com and subscribe to the genealogy digest known as GENEALOGY-DNA-D.

Subscribe to the mailing list and receive frequent email, or just the digest, or read the messages at the Web site. There, you may ask your questions about how to interpret your own sequences or others. To subscribe to GENEALOGY-DNA-D, send a message to GENEALOGY-DNA-D-request@rootsweb.com that contains in the body of the message the command subscribe and no other text.

No subject line is necessary, but if your software requires one, just use subscribe in the subject, too. To contact the GENEALOGY-DNA-D list administrator, send mail to: GENEALOGY-DNA-admin@rootsweb.com. Your first step is to ask the company who tested your DNA to tell you how to interpret your sequences. Most companies have a frequently asked questions section on the Web site, and others ask to be emailed questions.

Are you interested in researching, collecting information, scrap booking, genealogy, or writing about family histories or your genetic history? How can you create a time capsule for future generations of printouts of part of your DNA or mtDNA or Y chromosome sequences?

What will future generations do with this information? Can it help unite people who are distant or close relatives or those with the same common ancestor in the very distant past? Family historians and genealogists now have a new branch of genealogy to learn—molecular family history.

If you're interested in DNA and Jewish genealogy, write to: JewishGen, Inc. 2951 Marina Bay Dr., Suite 130–472 League City, Texas 77573. On the Internet, the Web site is at: http://www.jewishgen.org/. There is a special interest group on Jewish DNA research and genealogy. It's called Genealogy by Genetics, and the Web site is at: http://www.jewishgen.org/dna/. You can subscribe to the mailing list.

The Genealogy by Genetics special interest group of the Jewish Genealogy Web mailing lists on the Web have a partner, Family Tree DNA (FTDNA) that tests your DNA and if you select, racial percentages. The DNA testing partner maintains databases. They will be integrating the FTDNA database library with existing JewishGen databases to provide users with the ability to connect with lost branches of their families.

There are several excellent books written on how to interpret DNA tests for people without a science background. See the bibliography at the back of this book. I particularly found Alan Savin's book very informative in bringing together DNA testing knowledge to genealogists.

Alan Savin of Maidenhead, England, is author of the 32-page book, DNA for Family Historians (ISBN 0-9539171-0-X). See the Web site: http://www.savin.org/dna/dna-book.html.

This excellent book that I highly recommend explains and explores in layman's language how family historians-genealogists can use DNA research and test results for family history research. The book also has case studies and makes genetic theory easier to understand by those without a background in genetics. It discusses the practicality of DNA testing for family historians as genetics joins genealogy. And it includes discussion of some of the problems of using DNA testing as a tool for family history research.

What I like about this book is that it's written at a reading level that is clear to understand without a science background. And the reader will find a good introduction, historical background, explanation of DNA fingerprinting, mitochondrial DNA testing, Y chromosome DNA testing (for males), collecting and analyzing DNA, future developments, and an excellent bibliography that includes Web sites, magazine articles, and books. So when I contacted Alan Savin by email, he related to me his story of how he introduced genetics into genealogy.

"I believe I was the first family historian in the world to use DNA for genealogical research back in 1997," says Alan Savin. "I originated the phrase 'genetic genealogy'. Realizing its potential, I wished to share this with everybody, hence the writing of the book. It is still selling well, especially in the USA, with orders being received worldwide. I have been approached recently for the book to be translated into German. It has been well received and recommended by a spectrum of reviewers from many genealogical publications, DNA testing companies themselves, e.g. Family Tree DNA and the media, e.g. the BBC."

"As stated in its introduction 'one of my primary aims is to explain this area of genetics in a language easily understood by a genealogist or any lay person'. Further books are planned in the series to develop the themes."

Savin says, "I could be said to be the father of genetic genealogy and I have seen my idea grow with the help of others. I keep a close watch on its development, behind the scenes, and look forward to seeing the science reach its maturity."

Aside from reading books on DNA, what else can a family historian do when there are no records to be found? Hobbyists and professional genealogists who wish to extend their family trees by confirming a link where no conventional source records exist would be interesting in having their DNA tested. Sometimes DNA tests may be used to determine whether a person is part of a larger group of people: for example, Jews will be able to confirm they are of Cohanim lineage.

DNA tests are excellent for individuals who want to perform surname-based family tree reconstruction projects.

An excellent article containing another version of the quote below (used with permission) titled, "Tangled Roots? Genetics Meets Genealogy" by Kathryn Brown appeared in the publication, Science, 1 Mar 2002.

Commenting on the role of DNA testing companies, Peter Underhill, a molecular anthropologist at Stanford University admits, "My concern is that people comprehend the relatively low level of resolution offered by these tests. Because the tests analyze relatively few markers along Y DNA or mtDNA, millions of people may share a given molecular profile. I think these companies have a role to play, as long as the science is done well."

Terry Melton, PhD is President/CEO/Laboratory Director of Mitotyping Technologies, LLC, 1981 Pine Hall Drive, State College, PA 16801. "The most important contribution of this science to genealogy is the ability of mtDNA to trace the maternal line long distances throughout a family tree," says Dr.Melton."My favorite pedigree is one from a paper by Sigurdardottir (American Journal of Human Genetics 66:1599) showing fifteen generations of an Icelandic family where living individuals typed from extreme tips of the family (whose ancestral female dates back to 1560) have the same mtDNA profile.

"In addition, mtDNA can sometimes be used to illuminate ethnic ancestry (in a very general way). Mitochondrial DNA types are correlated with the region of the world where the ancestral lineages originate. There have been dozens of scientific papers written on this subject.

"Unlike Y chromosome typing, which should follow the patriline and family name), mitochondrial DNA is more difficult to correlate with recorded genealogy, since female names are lost in marriages," Melton explains. "However, the possibility remains that if a family can locate two (even very distant) maternal relatives in their tree, the mtDNA typing can confirm the matrilineal connection."

In addition to checking DNA test results with databases and tables on the World Wide Web or in other records, the family historian can compare results and read further about DNA to learn more.

Family members who have their DNA tested also can also track lineages and more recent genealogy by looking at the tell-tale clues that old, antique photos offer as well as use old and new city directories that list people who may not even have had a telephone.

Marjorie Rice rescues old family photos from antique stores and flea markets using the skills and sharp eye of a genealogy researcher to get them back into the hands of family members. An article about her work is on the Web at: http://www.ancestry.com/library/view/news/articles/6590.asp.

She looks on the back of the photos to see whether there are family names and/or photographer imprints on the front. She posts the names and locations on surname message boards on the Internet. To date, she has restored 409 photos to family members. See article about her work at: http://www.ancestry.com/library/view/news/articles/6590.asp.

Besides putting dozens of family photos from the early 20th century in my own computer database by scanning and saving, I wrote to various genetic scientists, physicians, and researchers in the field of evolutionary biology and genetics for their opinions regarding the application and use of DNA testing for family historians and genealogists, even for people who want to track and record their own lineages, family trees not only by surname, but by DNA to find out what they can.

Some people are puzzled when "and wife" is listed instead of a female and her maiden name on documents. And with women for hundreds of years taking their father's or husband's surname, doors can open to researching female lines when mtDNA is tested.

Ancient ancestry in female lineages may be traced somewhat by mtDNA. It is inherited by women and passed on to their daughters. Y chromosomes are inherited by men and passed on to their sons. Both show us clues to ancient ancestry or ethnicity even in some small ways that show expansions and migrations of people across geographic distances for thousands of years. Mutation rates and genetic drift due to isolation of small communities show the researcher where the people had sought refuge and how they expanded in clines or gradients of genes.

Where the genes are most diversified shows researchers a clue to where the genes originated rather than where they might be today. Where the genes look alike or very close, shows the people have migrated to an area only recently in the eons of time. What are various geneticists' and genome scientists' opinions of DNA testing for genealogy research?

Richard Villems, MD, Dr. Sci, head of the Department and Professor of Evolutionary Biology at the Institute of Molecular and Cell Biology, Tartu University, Tartu, Estonia, replied to my question by email, noting that, "The answer is straightforward and short: yes, DNA 'testing' is a very powerful method for genealogy research—specifically so as far as maternally inherited mitochondrial BSA and paternally inherited Y chromosome, are concerned.

"Although the current practical use is, technically speaking, far from a possible state-of-art level in case of mtDNA, the latter is, if a full sequence of mtDNA is analyzed, a very precise tool to resolve genealogy in a phylogenetically correct way already. As far as Y chromosome is involved, it would be even more so, because Y chromosome is huge compared to mtDNA—some 60 millions of nucleotides compared to about 16,500 in mtDNA. However, realistically speaking, it would take a huge technological effort to reach a stage where a phylogenetic resolution would be "final"; we are just at the beginning of a long way.

"In theory, even a son will differ from his father, in average, in a few Y-chromosomal mutations—therefore the ideal resolution would indeed allow reconstructing the biological history of this chromosome in minute details. At present, this time is still far away because of an enormous cost of such a work. Nevertheless, the fact that we do know what is possible, one may predict that any man can calculate how exactly he is related to, say, to his contemporary PM of the country—or, say, to the Secretary General of the Chinese Communist Party (if that exists at that future time anyway).

"As far as autosomal genes are involved, I am pessimistic—Mendelian segregation and recombination are probably too powerful in creating noise that such a clear-cut resolution cannot be expected—never.

"Hobbyists and professional genealogists who wish to extend their family trees by confirming a link where no conventional source records exist would be interesting in having their DNA tested. Sometimes DNA tests may be used to determine whether a person is part of a larger group of people: for example, Jews will be able to confirm they are of Cohanim lineage. DNA tests are excellent for individuals who want to perform surname-based family tree reconstruction projects."

Dr. Richard Villems also wrote me this reply, when I inquired about what ethnicities my own mtDNA might reveal, "Your motif in HVS-1 is beyond any reasonable doubt within haplogroup H. Every even slightly experienced in mtDNA researcher knows that although transition in 16356 is a good guess that a particular mtDNA belongs to U4, there are enough 16356 mutations also within H. And what does it mean that 'research showed recently' that 16356C is U4—this is a very well known fact already at least for 4–5 years! But there are exceptions one ought to know as well.

Moreover, this combination you have is well present all over Europe—plenty in Scandinavia, in Estonia as well—but also in Germany and, to make it really a pan-European—also in the Adriatic area as well as in Eastern Europe down to the slopes of the southern Urals, among Turkic-speaking Bashkirs."

I also wrote to another scientist who works with human genetics, Dr. Vincent A. Macaulay, Dept of Statistics, University of Oxford, UK, who replied similarly, "Your sequence (16189–16356–16362) is almost certainly in haplogroup H. I have several exact matches to your sequence in my database which are confirmed as H using other markers in the mtDNA molecule.

"Position 00073 is in HVSII, which is not in the part of the molecule that Oxford Ancestors sequenced. I think they have confused 00073 and 16073 (which is in HVSI) in their reply to you. If you had HVSII sequenced, I would be confident that 00073 would display an "A". The 16356 mutation has happened more than once, so it does not always imply haplogroup U4.

"For your information this sequence has not been observed east of Bulgaria. In my database, there are sequence matches in UK, Spain, Portugal, Germany, Austria and Bulgaria. I hope this helps: I would suggest that you seek further clarification from Oxford Ancestors."

Another scientist in genetics, Dr. Antonio Torroni, Institute of Biochemistry, University of Urbino, Italy, also wrote to me that he found one person from Crete in his mtDNA database with my mtDNA sequences. So where did my own founding female lineage come from and which country represented my direct ancestor—or did all of those countries?

Since only a sample from each country was tested and put in the various database, I wondered whether an ancestor might have been not yet tested, not in the database, and in some geographic area not yet mentioned. The journeys for the founder types have only just begun. What geographic part on Earth, what ethnicity could I ultimately call my own down to the bones? Would I ever find out?

Dr Peter Reed has a PhD in Human Genetics from the University of Oxford and was a pioneer in the use of STR genetic markers in medical research. "To non-enthusiasts genealogy is often considered an obsession with the past," says Reed. "Yet the combination of genetics and genealogy enforces how family history is integral to what we are today."

All our ancestors have contributed to our personal genetic makeup, and by examining our own genes we are viewing the DNA of our ancestors.

"Current uses of genetics in genealogy only examine a fraction of our entire genetic makeup and have largely developed from anthropological research," Reed explains. "However, the driving force of the current 'genetic revolution' has been research into human physiology and psychology, and it is from this work that new applications for genealogy will be developed."

"In five to ten years knowledge of our entire personal genetic code will be feasible, and with this knowledge will come an ability to much better understand how our ancestors contributed to our genetic makeup.

"In a few generations from now not only will knowledge of an ancestor's full genetic code be as indispensable for family history as birth and death dates are today, but family history (both genetic and social) will be a vital instrument in personal health care."

* * *

28

The Phenomics Revolution: My Positive Experiences with DNA Testing

"DNA testing is an exquisitely precise tool for answering certain types of genealogical questions, but it is clear that this technique is, and will continue to be, a disappointment to many who see it as a way of leaping over the 'brick walls' in their conventional research," says genealogy researcher, John F. Chandler. "DNA testing is at its best in demonstrating that two people or two lineages are not related within a genealogical time frame."

"When used for the purpose of proving that two people are related, it is notoriously often misconstrued. By itself, DNA testing can only show a general relationship, not a specific one," Chandler admits. "For the future, if the field continues to grow, there is some hope that DNA will offer a realistic chance to hunt for relations by looking for exact matches, but the growth will have to be at least three orders of magnitude before this comes to pass."

I wanted to find out more about Ancestry Informative Markers (AIMs) and my own ancestry than only what my mitochondrial DNA (mtDNA) could tell. I wanted to find out the racial percentages in my genealogy search and how this information could help me write my ancient or recent past journal.

The DNA of the mitochondria is the energy generators transmitted through the egg cells, according to the New York University School of Medicine's Human Genetics Program that offers genetic analysis in various studies. See the Web site at: http://www.med.nyu.edu/genetics/jewishorigins.html. The DNA of the Y chromosome told about male ancestry for males.

Women don't carry the Y chromosome and so are tested for DNA by either their mtDNA for ancestry, or their nuclear DNA. Most DNA testing companies offer mtDNA testing for women and Y chromosome testing for men, and men also have mtDNA, and may have that tested also to learn about their female lineages as well as their male-Y-chromosome lineages.

Some DNA testing companies also offer the racial percentages DNA testing. The racial percentages such as East Asian, Indo European, African, and Native American are DNA tests of Ancestry Informative Markers (AIMs).

I wanted to find out what I could about my ancient and recent ancestry that extended beyond records in city directories and other cross-reference files. I took the racial percentages test to find out I was 97% Indo-European and 3% East Asian. (I was told to take anything 5% or under with a grain of salt.) Was the test only revealing back to my great great grandmother and no further back? Or did it reveal the "real me" as I was 21,000 years ago? Or was I looking at a printout of the maternal lineages that contributed to my genes 500 years ago or one generation back from my birthdate? The search for a core identity and ancestral DNA records of the deep past was on.

Of all the three DNA tests that I took to find out what I could about my unknown ancestry, the DNA test that I found most helpful was the one that looked at the percentages of the various races in my ancestral history.

I took The DNAPrint ANCESTRYbyDNA test in order to look at my personal panel of Ancestry Informative Markers (AIMs). "Our test can only indicate to what percentage a person is Native American, African, East Asian and Indo European," says Carrie Castillo, Corporate Communications, DNAPrint genomics. For further information, contact, DNAPrint genomics, Inc. 900 Cocoanut Ave, Sarasota, FL 34236.

The test uses markers that have been characterized in a large number of well-defined population samples. These markers are selected on the bases of showing substantial differences in frequency between population groups and, as such, can tell about the origins of a particular person whose ancestry is unknown.

After the analysis of these Ancestry Informative Markers (AIMs), in a sample of a person's DNA, the probability that a person is derived from any of the parental populations and any of the possible mixes of parental populations is calculated.

The population (or combination of populations) where the likelihood is the highest is then taken to be the best estimate of the ancestral proportions of the person. Confidence intervals on these point estimates of ancestral proportions are also being calculated.

For example, the Duffy Null allele (FY*0) is very common (approaching fixation or an allele frequency of 100%) in all sub-Saharan African populations and is not found outside of Africa. So a person with this allele is very likely to have some level of African ancestry.

Knowing the percentage of your races may be one consideration when planning for personalized medicine. How many know the entire history of one's own ancestry? If your parents and grandparents had genetic-related degenerative diseases that are high in certain populations, would you want to know whether you carried genes from a particular ethnic group? I wanted answers to questions such as these because I'm a freelance writer and write novels about DNA.

Ten generations is roughly 250 years and within the time frame of genealogical interest, especially when we are considering the settlement of North America, because they only look at two (2) chromosomes. Ychromosomal analysis and mtDNA analysis each could only provide information on a very small proportion of a person's ancestors.

I went for my third and last DNA test to the AncestryByDNA test because it relies on sequences throughout one's genome. So the results I received from AncestryByDNA said more about a greater number of ancestors. Also, I received a print-out from Roots for Real's mtDNA database of the geographic area in latitude and longitude where the possible origin of my particular mtDNA sequences may have occurred in the past 10,000 years.

Roots for Real, in London, England, noted that my mtDNA had a "reach" of 669 miles from the possible center in what today is the town of Bar Sur Aube, France. If I drew a circle in all directions extending about 669 miles, I would be able to connect the dots on the map where my mtDNA currently is found. Yet the matches of people with my own mtDNA sequences occurred in places as far apart as Spain and England, Iceland and Bulgaria, and in an Eastern European database, as far east as the southern Urals in Bashkortostan. Most of the matches were in north central Europe such as Norway, Iceland, Sweden, Italy, Scotland, England, Orkney Islands, Germany, Austria, and Poland, but just as many matches occurred in Portugal and Spain, Bulgaria, Crete, the Adriatic, Turkey and among the Komi of Finland.

I can conclude the mtDNA is very ancient and pan-European. In the Eastern European database, there were matches in Turkey, Bulgaria, Italy, Crete, and the Ukraine and among the Komi of Finland as well as among the Bashkirs of Bashkortostan in the Urals. Another British database, had only Spain, England, Bulgaria, and Austria listed where my mtDNA sequences appeared. So the matches are dependent upon what samples go into which country's database. In a U.S. database from one of the companies that tested my mtDNA, it included matches in Ireland, Germany, and Latvia so far. Databases are updated regularly and again, depend upon what samples go into the database.

Who knows where my mtDNA will appear next? You can see how wide a geographic region your mtDNA will expand to from generation to generation. The same goes for the Y-chromosome in its travels for males.

FamilyTree DNA also offers AncestryBy DNA's racial percentages test along with their mtDNA and Y-chromosome tests. My goal was to find out what I could for what I could afford, about my ancient ancestry through DNA testing. Perhaps it could tell me something more when oral and written records weren't within reach.

Could these tests tell me anything other than that I resemble millions of people with the same mtDNA sequences, or if I were male, Y-chromosome markers? Could it provide a DNA match with someone who possibly shared a common ancestor with me hundreds of years ago? Who was my most recent common ancestor, and would that person show up in a database? The answer is only if that person took the test and asked for his or her email to be made public for contact with another DNA match.

After all, my entire genome wasn't being tested. That's not affordable on my budget. And my entire genome would reveal more personal information that the mtDNA, even the high and low resolution HVS-1 and HVS-2 sites. I still matched with millions of people. Only in my case, my mtDNA sequences are found only at a low frequency in Europe. What caused their extinctions? Or did they have more sons than daughters?

The places where my recent relatives lived were mostly unknown to me, and all I could go by was looks as perceived by others based on stereotypes based on cartoons and caricatures. Online database tables told me someone with my matrilineal DNA (mtDNA) lived in Crete. It was as good a starting point as any. So I went out and bought feta cheese for lunch melted on pizza dough.

Then one to four people per thousand with my exact sequences of mtDNA also lived in Scotland and Norway and Siena, Italy (Tuscany) and Turkey and Bulgaria, and Iceland and Austria. What part of the world did my ancestors belong to for any length of time? What would it mean to me other than an ethnic costume for Halloween or a trip to an ethnic restaurant or listening to music of that geographic place?

The first table I researched contained my HVS-1, low resolution sequences of mtDNA. Yet when I had my mtDNA tested for HVS-2, high resolution, different geographical places turned up—the Orkney Islands off the coast of Scotland, France, England, Bulgaria, and Turkey.

Well which is the real me—Bulgaria and Turkey or Scotland and Orkney Islands? Grandma had red hair and blue eyes, but that was the paternal side. And red hair is found in all those countries.

What could DNA testing tell me? I saw an announcement on the Web about Oxford Ancestors from http://www.OxfordAncestors.com and had my mtDNA tested. That's the matrilineal line of ancestry for women. A little cytobrush arrived in the mail. I rubbed the inside of my cheek, and mailed it across the Atlantic to Oxford, England, and two months later, a pretty chart came back with a page on how the mtDNA (mitochondrial DNA) testing was done, a pretty chart linking my mtDNA clan of Helena (or H haplogroup) to everyone else's' in Europe. (A year later the chart changed to how everyone in the world is linked by mtDNA). I also received a printout of the letters of my mtDNA.

The letters were a printout of CGAT, the letters of everyone's DNA, and the chart didn't have the letter of my Haplogroup, but instead, a name given by Oxford Ancestors to the first female ancestor of the H haplogroup clan in Europe, Helena. There was no mention on the chart that Helena was a moniker for the H haplogroup which is one of the most common in Europe, making up about 47% or more of Europeans and 6% of Middle Easterners.

The only clue to my sequences were three little red letters—C showing how I differed in three places from the Cambridge Reference Sequence (CRS) by transitions or mutations of nucleotides, which I was compared against. I still wanted more—to find what ethnic groups had larger numbers of the same sequences of mtDNA as me. Also, I wanted to know where the most variation occurred, in what part of the world.

Perhaps that would give me a clue to the origins. It's not so much how many people today live in an area with your sequences, but how much variation there is that tells you how old your mtDNA is. So I was looking for a founder female, a single person or a coalescence point. For H haplogroup, the coalescence point in Europe had been about 21,000 years.

Now I asked whether H mtDNA haplogroup arose from a common ancestor? If so, where did the ancestor come from and was she also an H, or an HV—since H and V split off from HV and HV evolved from pre-HV—somewhere and sometime before 28,000 years ago. H mtDNA had arose in the Pyrenees 20,000 years ago, but H had a common ancestor elsewhere.

Even HV had a common ancestor, pre-HV, somewhere in the Middle East, further back it time. Or did H arise by itself as a mutation in Europe during the height of the Ice Age 21,000 years ago between the coast of SW France and N. Spain's Pyrenees?

The printout of letters didn't mean much to me then. I needed sequences for mutations written in numbers. I found the sequences numbers by asking at one of the genealogy and DNA mailing lists on the Web whether anyone could tell the sequences in numbers from my printed-out letters. All I had were the three mutations in red—three "C" letters.

So to answer in part these questions, I wrote to an acquaintance at the Whitehead Institute at MIT (genomics research division) who told me that the three "C" letters (mutations) could be put on another table that I never received or saw before. This table now had numbered sequences.

The table was available to anyone on the Web. I emailed the acquaintance my low-resolution sequence numbers of the HVS-1. Now I had at last sequences to compare so I could look up what countries of the world these appeared in presently, even though there was no way to tell whether my particular ancestors lived in any particular location at any time.

I looked on the Web at Victor Macaulay's HVS-1 tables at: http://www.stats.ox.ac.uk/~macaulay/founder2000/tableA.html and found that my sequences H haplogroup mtDNA of 16189C, 16356C, and 16362C all showed a mutation from T to C and were found with that transition on the table to appear in Spain, England, Austria, and Bulgaria.

Any place else? I emailed an mtDNA research team member, Kristiina Tambets, Estonian Biocentre and Department of Evolutionary Biology, Tartu University, Tartu, Estonia, who took part in a research study appearing in an article that I read in *Archaeogenetics*, a book published by The McDonald Institute Monographs, edited by Colin Renfrew and Katie Boyle, and Tambets emailed me several sequences, including mine, from her unpublished database.

The few sequences she sent me from her unpublished database classified my sequences as H2b, but she emphasized that I should be cautious as the material was not published. And my sequences in her database of thousands also included other countries listed under H2b, which included my sequences, (a division of haplogroup H).

The sequences listed as well countries that were found to contain at least one or more persons with my HVS-1 sequences, such as: Bashkortostan, Turkey, Crete, Croatia, Albania, Hungary, Portugal, Germany, Iceland, Spain, Norway, Sweden, Komi (Finland group) including the other countries—Spain, Portugal, England (UK), Germany, Austria, and Bulgaria as well as the four areas listed on the Web on Macaulay's tables—England, Spain, Austria, and Bulgaria for H mtDNA haplogroup, HVS-1 region, with transitions at 16189C, 16356C, 16362C. Family Tree DNA also found another transition—16519C as they tested more than 400 base pairs. Oxford Ancestors tested 400 base pairs. I eventually also had my HVS-2 high resolution mtDNA area tested.

The number of samples obtained from each country also was listed in each of the database sequences or tables that I studied. Victor Macaulay's mtDNA (mitochondrial DNA) tables also are on the Web at: http://www.stats.ox.ac.uk/~macaulay/founder2000/tableA.html.

My acquaintance from MIT whom I met from Internet correspondence on a mailing list of interest to people working with DNA, wrote that I was probably close to Bulgar/Turk or Karelian or a mixture of all three, when he looked at my sequences back in 2001. He didn't say how he came to that conclusion. However, he asked me to look at all my mtDNA sequences when searching the tables online.

I assume that the sequences might have shown up among these ethnic groups. So I didn't really learn the process of how he found it out.

My research showed England, Spain, Austria, and Bulgaria on Macaulay's tables online. So I needed to learn much more about how to look up sequences on tables that are on the Web for all to access. If I removed the 189, a fast-mutating site, and only looked at 356 and 362, then it fit well into Karelia, Bulgaria, or Turkey. The 356 mutation, fit with Armenia and central Italy. What if I looked at all my sequences? Then other countries were on the tables.

In 2001, I received a certificate from Oxford Ancestors saying I was a Helena (haplogroup H). I thought this was awesome, because I was reared thinking I was in ancient times, at least in part Middle Eastern or from the Caucasus mountains, where U haplogroup is common and H is found at lesser frequencies than in Europe.

In 2002 I received a certificate from Oxford Ancestors saying I was an Ulrike (haplogroup U4). I wrote Vincent Macaulay asking how can one tell which is which with the same mutations?

From his tables and email, he said to find out whether I had a G at position 00073 at HVS-2. He emailed me a note saying that chances are almost certain that I had an A at that position 00073 on HVS-2, if I'd look. Certainly the three mutations I had were in his online database tables. An A at position 00073 of HVS-2 would make me H haplogroup. A G at position 00073 would make me a U4, but the only way to be sure is to do a high resolution test of the mtDNA for HVS-2.

So I wrote to Dr. Villems from Estonia, where Kristiina Tambets database sequences came from and to Bryan Sykes, from Oxford Ancestors. Bryan Sykes, MA PhD DSc, is Professor of Human Genetics, University of Oxford, and with

Oxford Ancestors. He has written several books on the history and geography of human genes, including The Seven Daughters of Eve and a book on the sons of Adam, genetically speaking. Anyway, Villems also agreed I'm probably an H, not a U4 (what Oxford Ancestors named Ulrike).

The question came up because one of my sequences, 16356C also is found in mtDNA haplogroup U4. Dr. Sykes and I corresponded by email several times, and Dr. Sykes is most helpful and emailed me on January 15, 2002. He wrote, "David Ashworth from Oxford Ancestors has shown me you message and the replies you received from Drs. Macaulay and Villems about whether your DNA sequence places you in the clade of U4 Ulrike or H Helena.

"David tells me that on your original certificate, issued in August 2000, you were placed in the clan of Helena but that when you were sent a replacement you had become a daughter of Ulrike instead. Of course your actual DNA sequence hadn't changed, but the assignment of you clan had.

"It may help if I explain how that is done. Clans are defined by a mixture of two sorts of genetic markers, the variants in the control region sequence and the variants at a number of other sites around the mtDNA molecule now generally called SNPs (short for Single Nucleotide Polymorphisms). These are usually designated as +4643Rsa1 or +11329Alu1 etc as you have pointed out in your messages.

"Vincent's classification on the website and in the papers you refer to contain a complete list of these variants. What they mean is that a restriction enzyme recognises the variation and either cuts or does not cut the DNA at that point. Since your hobby is reading about DNA, I am sure I don't need to explain what a restriction enzyme is. So, take +4643Rsa1. That means the enzyme Rsa1 cuts at base number 4643. The variant-4643Rsa1 means that the enzyme does not cut the DNA.

"Vincent and I, with Martin Richards, spent a great deal of time correlating the control region variants with the SNPs by analysing both on several hundred (it may even have been thousands) of mtDNAs and this was an important part of distinguishing the different clans. On the whole these types of variant are more stable than some of the control region sequence variants but not always and ideally every DNA should be tested for both. However, that would put the price to customers up hugely because each one of the SNPs had to be done separately—although I know that Oxford Ancestors are looking into offering this service.

"But even that would not guarantee completely accurate assignment in every single case. Only sequencing the entire mtDNA circle of sixteen and a half thousand bases at astronomical cost would do that—though even that would not be any good unless you had at your fingertips a database of thousands of other complete sequences with which to compare it and only a handful have been completely sequenced to date.

Also, as more work is reported, the evolutionary networks will change. "What I am getting at is that no system is foolproof. The Oxford Ancestors service, to keep it affordable, only sequences the control region. Then the sequence is compared to a database which holds other sequences which have been examined for SNP variants. If the customers control sequence matches up with one of these then it is assigned to the same clan.

In other cases, where there is not an exact match, the database is searched for close matches or sites which are characteristic of particular clans. In the case of your sequence which has variants from the reference sequence at 189, 356, 362 (we delete the 16 prefix for HVS1) two of the three variants are quite unstable—that means they can mutate back and forth. Position 189 is one of the least stable of all and position 362 is not very far behind. Position 356 is far more stable and is also characteristic of clade U4, whose clan mother is Ulrike. However, it is not completely stable and does crop up in other clans—one of which is Helena.

"So the sequence 189, 356, 362 could be in the clan of Ulrike mutating at the unstable positions 189 and 362 away from the U4 root sequence of 356. Or it could be in the clan of Helena with a rare variant at 356 taking it away from 189, 362.

One way of telling the two apart is to look at the variant at 073. This is actually in HVSII and not HVSI and that was a source of confusion in some of the email exchanges I have read.

"Oxford Ancestors doesn't do the 073 test, as you know, so the sequence was assigned on the balance of probabilities to Ulrike. I have now had a chance to compare the sequence to some new research data of my own from Britain in which we did do the 073 test and found five exact matches which carry A at 073, indicative of clade H. So I think you are probably correct and are indeed a daughter of Helena rather than Ulrike. This means that the mutation at 356 would have occurred on a Helena background rather than the 189 and 362 variants occurring on an Ulrike background.

"That might explain why you were originally issued with a Helena certificate in August 2000. At that time, the service was being sent out from my laboratory before Oxford Ancestors acquired its own premises. That means that whoever did that first assignment, and it may well have been me, did recognize the ambiguous nature of the 356 mutation in that particular sequence but that piece of information was not properly transferred to the new set-up—and that is my fault.

"I must thank you for clarifying the assignment of this particular sequence. It is a changing field and your observation has helped it move on one more stage further. I am sure Dr Ashworth will want to issue a new Helena certificate. And of course, I hope you are pleased to have moved back to your original clan."

◆   ◆   ◆

Belonging to Helena's clan sounds more personal that what it means—a member of haplogroup H, presumably one of the Seven Daughters of Eve, the founding "clans" or haplogroups found in Europe in prehistoric times, according to Bryan Sykes book titled *The Seven Daughters of Eve*. Even H haplogroup had to have a common ancestor before heading for Europe in the middle of the Ice Age 21,000 years ago.

That ancestor had to come from presumably somewhere in the Middle East. Only in a book titled, *The Real Eve* by Stephen Oppenheimer, (Carroll & Graff, NY 2003), Oppenheimer theorizes that H mtDNA came from the Caucasus to Europe and before that from Central Asia and before that near India or in Northwest India. H mtDNA in India has the most diversity.

Then I read Spencer Wells' book, *The Journey of Man: A Genetic Oddysey*. I highly recommend this book if you want to follow the origins of the male Y-chromosome around the world as men journeyed out of Africa to populate the rest of the world through the southern route to Australia.

Wells emphasizes that Europe was populated from Central Asia and not from the Middle East originally. It was only after the agricultural revolution, the Neolithic area, that a smaller percentage of Middle Eastern migrations into Europe took place. Europeans spent thousands of years on the cold steppes of what today is Central Asia, namely Kazakhstan before turning west and moving into Europe to populate both Eastern and Western Europe. He shows it in the M173 Y-chromosome markers found all over Europe today and traces European markers from their roots in Central Asia. I also recommend his video or DVD, *The Journey of Man*. Show it to your class or group if you're a facilitator or educator trying to inform people about the joys of DNA-driven anthropology or genealogy.

Where you have the most diversity, you have a place of origin for a particular mtDNA haplogroup. H mtDNA had an ancestor from which it split off, and that ancestor was HV. Today, in India, H and HV mtDNA has the most diversity.

It's the area of northwest India around Kashmir and the Indus and in the Pujab and Pakistan that theoretically is the original homeland of H and HV.

Haplogroup H mtDNA, denoting female lineage, is the most frequent cluster in the Middle East and the Caucasus. It's present at a frequency 25 percent. In Europe, it's found much more frequently at around 46 percent. There are more haplogroup H people in Europe than there are in any other part of the world.

The mtDNA haplogroup H represents a female ancestor who lived in the Middle East up to 28,400 years ago. Some surviving daughters then went to Europe, and some stayed in the Middle East. However, the European daughters' mtDNA with Haplogroup H have an "age estimate" of only up to 21,400 years. So the arrival of or mutation to H in Europe began more recently in Europe than it did in the Middle East. However, mtDNA H came into Europe from the East, paralleling Y-chromosome "I" found mostly in Europe. Yet today there are more Europeans with haplogroup H than there are in the Middle East with the same matrilineal lines of mtDNA haplogroup H.

There also was a back-migration from Europe to the Middle East in prehistoric times. About 15% of people in the Middle East today represent this European back-migration. Keep in mind that until 44,000 years ago, the Levant was an extreme desert. The route into Europe more likely came from the Caucasus, West Asia or Central Asia into the Ukraine and the Balkans and from there, West into Europe. Places of refuge during the ice age were in northern Spain and southwest France, in the Balkans, and in the Ukraine.

North Africa was populated by people migrating from the Middle East and from Europe, and only later, from a small number of people migrating from sub-Saharan Africa during the grassland stage when the Sahara was grassland with heavy rains right after the end of the last ice age. Hunters followed the herds of savannah-type African animals and left drawings in caves.

Other mtDNA haplogroups of female lineages are much older, if we look at mtDNA coalescence times and founder effects. For example, mtDNA haplogroups T and J both date to around 50,000 years before the present in the Middle East, but more recent dates in Europe.

Haplogroup J is found today at its highest frequency in Arabia, making up 25 percent of the Bedouin and Yemeni population.

It's also found in Europe, also a more recent arrival. Perhaps mtDNA haplogroup J entered Europe around 10,000 years ago or less, probably with the agricultural revolution of the Neolithic age, after the last Ice Age ended around 12,500 years ago.

The oldest mtDNA in Europe is the female lineage of U mtDNA haplogroup. It's around 50,000 years old in the Middle East, perhaps around 45,000 years ago in Europe, and from U comes many sub divisions such as European-specific U5, North African U6, and Indian U2i, all showing an origin of around 50,000 years ago. The first female lineage out of East Africa into the Middle East was L3, but mtDNA haplogroup U also followed out of Northern Africa and went to Europe as well as most of the Middle East, where it's found today.

An interesting theory is that 50,000 years ago, the route from Northern Africa was closed until 44,000 years ago, so people coming from Asia or Africa into Europe through the Middle East had to come from India across the southern route until the route from the Middle East opened up to get to Europe, and that was around 44,000 years ago.

At that time the Fertile Crescent around what is today Iraq opened up as well as the Levant. By then, people could use a northern route across the Bosphorus from Turkey to Greece and enter Europe. Those in what is today Greece then would have gotten trapped in the Balkans by 25,000 years ago, unless they made it to the refugiums in southwestern France and northwestern Spain near the Mediterranean and at the foot of the Pyrenees. Some made it.

That's why we have the wonderful cave paintings of large animals in southwestern France, particularly the Dordogne valley. Art bloomed when the latest ice age was raging at its maximum. The open tundra or grasslands became a steppe for animals and hunters.

As soon as the last ice age ended and forests took over most of Europe, the wonderful art subsided into a dark age that began around 12,000 years ago. Then the early farmers came up from the Middle East. They now make up about 20 percent of Europe's population. The idea of farming grain born out of gathering seeds and roots to eat caught on during the next few thousand years. There was an agricultural frontier that separated hunters from farmers, and it lasted thousands of years.

What puzzled me is why today H haplogroup is nearly missing from the Arabian Peninsula. What made it leave, the changing climate or low population frequency? For more information I turned to an excellent article titled "Tracing European Founder mtDNAs" by Richards, et al published in the American Journal of Human Genetics, 67:1251–1276, 2000. I read about mtDNA haplogroup V, which appears to have expanded in Europe around 13,000 years ago. Then I re-read Bryan Sykes's book, *The Seven Daughters of Eve*. It was time to take the next step.

I went to Family Tree DNA for more information. I had my HVS-2 (high resolution) mtDNA tested to find out more about what my ancestry might be and whether I was a U4 or an H haplogroup. The two single female ancestors who founded H and U4 in prehistoric times came from different parts of Eurasia.

Family Tree DNA sent two small felt-tipped serrated pads with a push stick that pushes the felt pad from the holder into a small vial of preservative solution. I brushed the inside of my cheek several times, put the two pads into the two tiny vials of solution, and sent it back. I had a little difficulty poking the stick to drop the pads, and in the end had to pull the pads firmly from the sticks and drop them into the vials with my fingers. (I washed my hands first.)

Family Tree DNA tested more base pairs. I found that indeed there was an A at position 00073 in HVS-2, just like Victor Macaulay told me I'd probably find if I looked. That made me a member of H mtDNA haplogroup. Family Tree DNA also found a new transition to the letter C at position 16519. Family Tree DNA tested 540 base pairs as compared to the 400 tested at Oxford Ancestors, which is close to the minimum that needs to be tested to find out one's mtDNA haplogroup results.

Bennett Greenspan of Family Tree DNA emailed me an answer to my inquiry of what question he would like to see answered in a book on genealogy through DNA testing. "It would be to explain the direct line of descent...male to male to male, with no mixing of the genders. That's what people seem to have a hard time with." I enjoyed the attention to answering email quickly, excellent DNA testing by Arizona University and information about Dr. Hammer's DNA research. There are informative online newsletters, databases for online searching and mtDNA or Y-chromosome matches to search online at the Web site. I received fine service from Family Tree DNA and would highly recommend them to anyone interested in DNA-driven genealogy.

What I learned from email correspondence from Family Tree DNA is that my mtDNA won't tell me when the last mutation on a string happened, but my mtDNA in its form as H haplogroup appears on the CRS (Cambridge Reference Sequence) without all the mutations that I had occurred during thousands of years of evolution, is a little over 20,000 years old. (I visualized H mtDNA as the sequence of the Cro-Magnons of the Mesolithic caves of SW France and Northern Spain.)

Could my ancestors have painted that fine art work on caves at Lascaux or Altamira during the height of the Ice Age? Since then my mtDNA has mutated away from the Cambridge Reference Sequence (CRS), and perhaps the last mutation that occurred in my HVR-sequence happened 2–10,000 years ago. Should I look to Europe or the Levant for my founding matrilineal line?

And I wish there was a way to also see what ethnicities my patrilineal line also contributed. I don't have the Y chromosome to trace the Adams in my family, but can a test of percentages of races give me any clue? What tests were within my budget?

Therefore every person I find that matches me exactly on HVR-1 is a nice clue because I am closer to them then to any one else on the direct female side. I like the way Family Tree signs "email me anytime" at the end of each email address. Response time was fast in answering my questions.

So I'm thankful for all this information I found useful and practical.

They have databases, search engines, and surname projects. Yet I still needed to find out more about what races are in my ancestry. I moved on and found another DNA testing company that even has a chat room.

There are about 3 billion DNA letters (technical name: "base pairs") in human DNA. You can picture these letters as beads on a keychain, each one labeled with an A, C, G, or T. The letters stand for adenine, cytidine, guanine, and thymidine, always referred to as A,C,G, and T.

These nucleotide bases are joined together, one after the other like a molecular string of pearls called DNA (deoxyribonucleic acid).

Oxford Ancestors tested only 400 of my HVS-1 sequences. Family Tree DNA tested 540 and found a new mutation or transition16519C in HVS-1. Then when I had HVS-2 tested at Family Tree DNA, I found the transitions also were very common on the CRS and were 309.1C, 315.1C, and deletions at 523-and 524-. Any mutations I had that showed up in my HVS-2 results were more common in the population, and the rare sequence was on the CRS.

I looked up HVS-2 sequences on Victor Macaulay's tables on the Web and found additional countries such as Orkney Islands off the coast of Scotland, Norway, Turkey, Bulgaria, Scotland, and England matching my HVS-2 results. So far, I had no clue as to whether my maternal line was European or Middle Eastern or both. You have to look at HVS-1 and HVS-2 together and not look at them separately when you are trying to find out your nearest common ancestor.

Since my family is so multicultural, I wanted to find out what the DNA said. Family Tree DNA had a database and a name search, but I wasn't able to get into the database or understand what I had to do to check for matches with similar names, but I was very happy with the results.

Now I had low and high resolution mtDNA. That still wasn't enough. There was no way I could afford to have the entire genome checked. Some Web sites say they will check the whole genome for ancestry or other for a fee that I couldn't afford. Some people want their whole genome done for medical reasons or to make sure medicine is matched to their personal genome.

In a few years, perhaps anyone can have his or her nuclear DNA and entire genome checked for ancestry or to match the right diets or medicines to individual genetic situations. Right now, a whole genome check is priced in the thousands at some sites and at varying prices too high for me.

Also, what I learned from Family Tree DNA, that I never found out from Oxford Ancestors, is that when I had the HVS-2 tested, Family Tree had included some very fast moving, potentially unstable markers to attempt to break down the time when I find a match on HVS-1.

This new information helped me move forward in my research for my own ancestry through DNA, something I heard couldn't be done yet, (other than in the distant past.) Yet I was getting closer. Now it was time to move on to the percentages of races that are found in my DNA.

That's when I turned to a BioGeographical Ancestry test from AncestryByDNA. It's a company that recently presented a complex genetics classifier for personalized medicine. AncestryByDNA's CEO is Dr. Tony Frudakis, of DNAPrint genomics, Inc. 900 Cocoanut Ave, Sarasota, FL 34236941-366-3400941-952-9770 fax tfrudakis@dnaprint.com. The Web site is at http://www.dnaprint.com/. This racial percentage test also is given by Family Tree DNA and is updated frequently.

The Web site for testing ancestry such as percentages of various races is located at: www.ancestrybydna.com. They even have a chat room. DNA testing is about personalized medicine as well as finding out about ancestry and other genetic-related questions. It's also about an informatics platform for genotype pattern recognition.

Here's your gateway to understanding the genetic basis for complex trait determination. DNA testing is about looking at pioneers in the emerging field of post-human genome phenomics. By putting the genomics puzzle together, you become part of the phenomics revolution with ancestry testing or other genotyping services. It's about research on individual genetic responses to certain medications as much as about finding your ancient or racial percentage ancestry.

From the DNA kit I was sent with printed information, I mailed in a DNA sample obtained from rubbing a soft piece of felt inside my cheek on a felt-tipped, serrated swab, which I air-dried and placed in an addressed return envelope and mailed back to DNA Print genomics and asked to be tested for percentages of various racial groups such as African, East Asian, European, and Native American.

Some post offices don't want you to mail biological material, such as DNA. Most post offices, though allow you to put the envelope in a regular mail box, since it only required two first-class stamps to return, weighing around two ounces or less. The envelope contained a signed form with my address so my printed out results could be returned. You can also get your results by email. To be safe, I placed the return envelope with the two swabs inside of another mailing envelope in case a mailing machine cut a hole in the self-addressed return envelope containing my two pieces of felt with my DNA on it from inside my cheek.

If you air-dry the DNA on the small felt tip before you put it in the envelope, you won't get saliva wetting inside of the mailing envelope. When checking for racial percentages from each parent, the DNA testing goes beyond only testing the mtDNA for women and Y chromo-some and/or mtDNA for men.

Those results are limited to only a very small number of your ancestors in your ancient past, such as your female founders for mtDNA (women) or Y chromo-some and mtDNA for my ancestors who lived perhaps in Ice Age refugiums thousands of years ago.

The test for racial percentages relies on sequences throughout your genome, so a DNA testing company can say more about a greater number of ones ancestors. That's why I went to the BioGeographical Ancestry (BGA) test. More companies are offering it, but I saw it first at ancestrybydna.com. BioGeographical Ancestry (BGA) is the term given to the biological or genetic component of race.

I was looking for what the BGA offered—a simple and objective description of my ancestral origins in terms of the major population groups such as: Native American, East Asian, Indo-European, and sub-Saharan African. BGA estimates represent the mixed nature of all of us in current populations.

Most of us have racial mixing in the distant past that we don't know about yet. I wanted to find out because my ancestors lived both in Europe and along the Silk Road in ancient and/or medieval times. I'm interested in reading about molecular anthropology. I write novels about the work of fictional molecular anthropologists and similar researchers.

My experiences reading in anthropology taught me that anywhere in the world there have been mixing, even among groups isolated for thousands of years. I wanted to find out the percentages of different races I had in my background.

The phenomics revolution is new and fascinating. The reasons why you would look at more of your genome is useful, not frightening. And all the new ways to look at your genome are becoming more personalized and relevant. With phenomics, the future moves towards personalized medicine. And DNA testing offers a more personalized genealogy for us history buffs and novelists.

***

29

Finding Female Ancestors by Searching for Maiden Names

Let's say you tested your mtDNA to find your female lineage, but you want to look up female relatives and don't know their maiden names. My own great grandmother's death certificate didn't list a maiden name. The next step would be to look at her marriage certificate.

I could also look at the City Directory for a particular city in 1889 to see where she lived, looking first under her father's name. I could try the US Census records for 1890 also in that city.

My grandparents were small children at that time living in a large house and farm in upstate New York, and all I know to begin my search is my great grandmother's first name and her father's last name, and I have my mtDNA, female lineage high and low resolution sequences.

Those would be the same in my great grandmother as in my present grand daughter. The mtDNA stretches back to a single female founder 21,000 years ago in Europe and further back in time, the usual cereal belt. I know my sequences are quite popular in Crete but also in Norway and Iceland. I could ask my mtDNA "Quo Vadis?" Where are you going? That's what DNA-driven genealogy is all about—asking where is the DNA going geographically as well as by mutation of sequences every 10,000 to 20,000 years.

Population genetics follows the 'clines' or direction of migrations from southeast to northwest or from northeast to southwest by what the genes presently reveal. In archaeogenetics, the routes of migration found genetically mostly agree with the shards of pottery, tools, and other relics left behind that archaeologists using DNA testing are researching.

The genealogist, family historian, personal historian, oral historian, and ancestry searcher looking through archives and other records will find that at a point records stop. For example, how do you find maiden names of ancestors? In modern times, you have city directories. For ancient times, you have genetic testing and looking at tools and utensils.

City Directories often list maiden names and the names of all residents living in a home, particular before homes had telephones. Where else can you look on the Web to start besides the usual beginning genealogy Web sites? You might order marriage certificates or death certificates as well as birth certificates to find maiden names.

The National Center for Health Statistics in Hyattsville, Maryland is also on the World Wide Web at: http://www.cdc.gov/nchs/howto/w2w/newyork.htm. For example, to view an address where to write to in order to purchase a copy of a marriage certificate for New York State, the Web site will help provide information as to where to write to in order to purchase copies of marriage licenses for New York, for example, from the year 1880 forward.

For other states, check the Web site at The National Center for Health Statistics under each state. You can also search foreign countries for records. You could also look at the Ellis Islands records, if a relative came to Ellis Island, and even view a picture of the ship at the Ellis Island Online Web site at: http://www.ellisisland.org/.

One good place to start is http://www.cdc.gov/nchs/products.htm for publications and products. Click on http://www.cdc.gov/nchs/nvss.htm for Vital Statistics. Click here http://www.cdc.gov/nchs/releases/96facts/mardiv.htm for marriage and divorce statistics.

***

30

The DNA Testing Companies of Interest to Family Historians

The Power of DNA Technology in Every Home. This is the slogan of the GeneTree DNA Testing Center that supplies DNA testing applications directly to the consumer. "We would like to demonstrate how we provide the science of DNA analysis applications directly to the consumer, allowing them to conduct their own research projects in the comfort of their own home," says Terrence C. Carmichael, MS, founder of GeneTree DNA Testing Center.

"By examining the Autosomal STRs, Y-chromosome STRs and mtDNA sequence analysis and RFLP, GeneTree is helping people (such as genealogists, anthropologists, and just those generally interested) uncover their deep ancestral migration patterns, establish biological relationships with relatives 1–50 generations apart, and uncover the mysteries of past and present relationships. These services are wonderful, and prove to be of great value to the consumer, whether it is for immigration purposes, assistance with genealogy or anthropology research, or for answering the simplest of questions, such as 'are you my father?'"

After receiving his MS degree, Carmichael went on to receive a Professional Designation in Marketing and Sales from UCLA. Terry has worked at the DNA laboratory bench for 4 years and spent 9 years providing Product Development, Technical Consulting, and Marketing for the DNA purification industry, working for companies such as Bio-Rad and QIAGEN. In 2000, Carmichael co-authored a book titled, "How to DNA Test your Family Relationships".

Having started 2 successful businesses, Terry is a visionary. He has applications submitted for 2 separate patents; one for applying DNA profiles to identification cards and the other for a new high-throughput DNA purification product held by Bio-Rad Laboratories. Below are the Web sites for some of the products offered by GeneTree DNA Testing Center.

GeneTree Products:
http://www.genetree.com/servlet/moonshine/goto?page_url=/products/productgroups.jsp
Y-Chromosome Information:
http://www.genetree.com/servlet/moonshine/goto?page_url=/products/product.jsp&id=7
mtDNA Information:
http://www.genetree.com/servlet/moonshine/goto?page_url=/products/product.jsp&id=8
Native American Assessment:
http://www.genetree.com/servlet/moonshine/goto?page_url=/products/product.jsp&id=20
Biography: Terrence C. Carmichael, MS
Terrence Carmichael, MS
(888) 404-GENE (ext. 207)
****

GeneTree DNA Testing Center
3150 Almaden Expressway, #203
San Jose, CA 95118-1253
Phone: (888) 404-GENE/(408) 723-2670
Fax: (408) 723-2671 http://www.genetree.com/
****

The Power of DNA Technology in Every Home FamilyTreeDNA

Here's how genealogy and DNA testing interacted together for the Craycraft Surname Project from FamilyTreeDNA. The Craycraft DNA project involves the surnames: Cracraft, Craycraft, Cracroft, Craycroft, and Craecraft. For purposes of this article, sent to me by FamilyTreeDNA, the use of the surname Craycraft refers to all the possible spellings noted above, unless otherwise indicated.

The surname first appears as Cracroft, in Hogsthorpe, England in the early 1200's with one Walter de Cracroft son of Humphrey Fitz Walter. It is later found in Hackthorn, England in the early 1600's, and today a Cracroft family still resides in Hackthorn Hall. Currently there are about 10,000 people, with this surname, and the surname is found in the following countries: England, USA, Canada, and New Zealand.

The first known emigrants to the Americas with this surname were John Cracroft and his wife Ann who emigrated from Lincolnshire, England in about 1665. The passenger list has John and Ann recorded as Creacroft or Creacraft. Today the descendents of John and Ann mostly spell their surname as Craycroft. Later, Joseph Cracraft/Cracroft came to the America's circa 1702 from Lincolnshire, England. He had 5 sons.

The objectives of the DNA project for phase I were the following:

1. Verify the documented genealogy research for the Joseph line in the US.
2. Verify the documented genealogy research for the John line in the US.
3. Determine whether the US Craycrafts are related to any England Cracrofts.

To meet the objective of verifying documented genealogy research, multiple participants were required due to the many branches between now and the identified most distant ancestor. The Lines tested are shown below, labeled by the most distance ancestor's name.

Joseph Cracraft, born Lincolnshire, England: Descendents of 3 of his 5 sons. John Cracroft, emigrated from Lincolnshire, England in about 1665 Cracroft family residing in Lincolnshire England, descendents of Walter de Cracroft.

The results from Phase I:

1. The documented genealogy of the descendents of 3 sons of Joseph was verified. A search continues for descendents of the 2 other sons.
2. The documented genealogy of John's descendents was verified.
3. The DNA results of Joseph and John's descendents matched as well as these DNA results matched those of the present day Lincolnshire family of Cracroft.

The surprises uncovered:

1. It was rumored that a female ancestor had a male child out of wedlock, and the child assumed the family surname. This was confirmed by the descendents in this branch having different DNA.

2. An adoption or non paternity event was discovered when a descendant's DNA results did not match. This event was confirmed by testing descendents from a branch earlier in the line, which did match the Craycraft DNA results. A Phase II of the project is planned.

====================================
Spot Light: Austin Research Validation
====================================

Validating research with DNA testing: Descendents of John Austin, JR (1726–1795)

Records in the US in the 1700's are scarce, and researching in this time period is very difficult. After 26 years of research, there was a preponderance of circumstantial evidence that showed that William Austin (b. Bet. 1750–1760, Hallifax County, Virginia or Surry County, North Carolina) was a son of John Austin, Jr. (born 14 Sep 1726, James City County, Cornwall Parish, Virginia (now Halifax & Pittsylvania).

After 26 years of exhaustive research, the researchers turned to DNA testing to find the answer. The objective of the testing was to determine whether the circumstantial evidence from research showing that William was a son of John Austin, Jr. could be proven or disproved.

To accomplish this objective, two participants were selected. One participant was a documented descendent of William Austin. The other participant was a documented descendent of one of the sons of John Austin, Jr., Isaiah. The results of the DNA test confirmed the research.

====================================
Spot Light: Roper Surname Project
====================================

Objective: Determine if any of the Roper Lines are related. There are many Roper Lines in the US and England that can not be connected by documents. For Ropers with ancestors who resided in new Kent County Virginia, the court-

house fires destroyed all records before 1864, severely limiting the ability to trace a family tree past 1864 and connect with any other Roper Line. To date, 25 participants have been tested. The geographic representation of these participants include:
18 US
2 Canada
1 Australia/England
4 UK

The surnames tested are Roper and Rooper. The results to date have exceeded expectations and been a significant contribution to Roper genealogy research. The major results include:

1. The majority of Roper Lines in the US match 24/25 or 25/25. One of these lines can trace their ancestor back to John Roper, born 1611.

2. County Norfolk in England has been identified as the ancestral homeland.

3. Most of the branches of the US Ropers have a different mutation, enabling a branch to be identified by the mutation.

In addition, research was confirmed when a participant who descended from a female Roper did not match, as would be expected. The next phase of the Project involves more testing in England, concentrating on the Counties of Norfolk and Kent, where Ropers from Norfolk migrated.

From CL of Savannah, GA:

"I used FamilyTreeDNA for our guys. They do an extra strand, I believe. Our Rosenblaths go back to the 1600's. However when they came to America their name was misspelled many different ways. It was hard to tell who was related to whom. Recently we had a confirmed "Rosenplot" relation in TN, originating in Canada as with my Grandmother who is descended from Rosenplot, submitted his DNA test to see if he was related to an unknown Rosenblath in Shrevesport LA.

"It comes to find out we have several relations, a Robertson who thought they were Scottish because of their name, and several more differently spelled Rosenplots/Rosenblaths. It has brought all of us who were strangers together in a very meaningful way. We share news about our families. Some have been able to 'paper' prove the connection while others are still looking for a few generations old grandfather in the tree we have. It has been great! A lot of new contacts!"

***

31

What is DNA?

"The human genome is about 3 x 109 base pairs long, which would weigh about 40 pg picograms: 1 pg=10–12 grams) per genome," reports Michael Onken, and this description appears on the science Web site of Ricky J. Sethi, MadSci.ORG Administrator at MadSci.ORG (http://www.madsci.org) at the Washington University School of Medicine, http://www.madsci.org/. "Human cells are diploid, i.e. each contains two copies of the genome, so the nuclear DNA from a human cell would weigh about 80 pg. If we want total cellular DNA, then we need to include mitochondrial DNA (mtDNA).

"The human mitochondrial genome is about 16,000 base pairs long. There are about 10 copies of the genome per mitochondrion, and there are on the order of 1,000 mitochondria per cell. This gives us about 0.2 pg of mtDNA per human cell.

"There are on the order of 1014 cells per adult human, many of which are without nuclei, like skin cells and red blood cells. This would give us just under a kilogram of chromosomal DNA and on the order of a few grams of mitochondrial DNA in the average human body."
(This excerpt is reprinted with permission of Ricky J. Sethi, MadSci.ORG Administrator, at the Washington University School of Medicine. See the Web site at http://www.madsci.org/.)

Knowing how many genes a human has in the future will help not only genealogists and other family and oral historians trace ancestors and keep records of lineages, but physicians will be able to tailor medicines to help people based on how their individual genes react to different elixirs, drugs, natural supplements, herbs, foods, and medicines.

Combined with the knowledge of rainforest tropical plants and their cures, the human genome is headed towards individualization and customization, with an appropriate mixture of food, medicine, or therapy based on one's individual genetic makeup.

To the person without a science background, knowing one's genes also is a way to connect people to their common ancestors in the past and to those descendants. Family history can be researched not only for medical reasons, but for historical reasons, and to show how people are related to one another down through the ages.

To understand how DNA testing relates to history and family records, let's look at some basics of genomics such as what are cells and what is DNA. Then we can think about ways we can use the results of DNA tests in the realm of family history.

Credit for the following Primer below and Dictionary at the back of this book is acknowledged to the U.S. Department of Energy Human Genome Program as the source for both and included also here is the U.S. Department of Energy Human Genome Program's Web site for more information on the Human Genome Project and its applications: www.ornl.gov/hgmis.

This document may be cited in the following style: Human Genome Program, U.S. Department of Energy, Genomics and Its Impact on Medicine and Society: A 2001 Primer, 2001. For printed copies, please contact Laura Yust at Oak Ridge National Laboratory. Send questions or comments to the author, Denise K. Casey. Site on the Web designed by Marissa Mills. This primer was prepared by Denise Casey, Human Genome Management Information System, Oak Ridge National Laboratory. You can find this Primer on the Web at:

http://www.ornl.gov/hgmis/publicat/primer2001/1.html and the index to additional publications at:

http://www.ornl.gov/hgmis/publicat/primer2001/index.html

Genomics and Its Impact on Medicine and Society: A 2001 Primer
(Courtesy of the U.S. Department of Energy Human Genome
Program:
http://www.ornl.gov/hgmis

The Basics

Cells are the fundamental working units of every living system. All the instructions needed to direct their activities are contained within the chemical DNA (deoxyribonucleic acid).

DNA from all organisms is made up of the same chemical and physical components. The DNA sequence is the particular side-by-side arrangement of bases along the DNA strand (e.g., ATTCCGGA). This order spells out the exact instructions required to create a particular organism with its own unique traits.

The genome is an organism's complete set of DNA. Genomes vary widely in size: the smallest known genome for a free-living organism (a bacterium) contains about 600,000 DNA base pairs, while human and mouse genomes have some 3 billion. Except for mature red blood cells, all human cells contain a complete genome.

DNA in the human genome is arranged into 24 distinct chromosomes—physically separate molecules that range in length from about 50 million to 250 million base pairs. A few types of major chromosomal abnormalities, including missing or extra copies or gross breaks and rejoinings (translocations), can be detected by microscopic examination. Most changes in DNA, however, are more subtle and require a closer analysis of the DNA molecule to find perhaps single-base differences.

Each chromosome contains many genes, the basic physical and functional units of heredity. Genes are specific sequences of bases that encode instructions on how to make proteins. Genes comprise only about 2% of the human genome; the remainder consists of noncoding regions, whose functions may include providing chromosomal structural integrity and regulating where, when, and in what quantity proteins are made. The human genome is estimated to contain 30,000 to 40,000 genes.

Although genes get a lot of attention, it's the proteins that perform most life functions and even make up the majority of cellular structures. Proteins are large, complex molecules made up of smaller subunits called amino acids.

Chemical properties that distinguish the 20 different amino acids cause the protein chains to fold up into specific three-dimensional structures that define their particular functions in the cell.

The constellation of all proteins in a cell is called its proteome. Unlike the relatively unchanging genome, the dynamic proteome changes from minute to minute in response to tens of thousands of intra-and extra cellular environmental signals.

A protein's chemistry and behavior are specified by the gene sequence and by the number and identities of other proteins made in the same cell at the same time and with which it associates and reacts. Studies to explore protein structure and activities, known as proteomics, will be the focus of much research for decades to come and will help elucidate the molecular basis of health and disease.

* * *

32

Human Genome Project

Reprinted with permission of the US Dept. of Energy, Human Genome Program, http://www.ornl.gov.hgmis

HUMAN GENOME PROJECT
Genomics and Its Impact on Medicine and Society: A 2001 Primer

A Little Bit of History

Though surprising to many, the Human Genome Project (HGP) traces its roots to an initiative in the U.S. Department of Energy (DOE). Since 1945, DOE and its predecessor agencies have been charged by Congress to develop new energy resources and technologies and to pursue a deeper understanding of potential health and environmental risks posed by their production and use. Such studies have since provided the scientific basis for individual risk assessments of nuclear medicine technologies, for example.

In 1986, DOE took a bold step in announcing its Human Genome Initiative, convinced that DOE's missions would be well served by a reference human genome sequence. Shortly there-after, DOE and the National Institutes of Health developed a plan for a joint HGP that officially began in 1990.

Ambitious Goals...

From the outset, the HGP's ultimate goal has been to generate a high-quality reference sequence for the entire human genome and to identify all human genes. Other important goals are to sequence the genomes of model organisms to help interpret human DNA, enhance computational resources to support future research and commercial applications, and explore gene function through mouse-human comparisons. Potential applications are numerous and include customized medicines, improved agricultural products, new energy resources, and tools for environmental cleanup. The HGP also aims to train future scientists, study human variation, and address critical societal issues arising from the increased availability of personal human genome data and related analytical technologies.

...and Exciting Progress

Although the HGP originally was planned to last 15 years, rapid technological advances and worldwide participation have accelerated the expected completion date to 2003. In June 2000, scientists announced biology's most stunning achievement: the generation of a working draft sequence of the entire human genome. In addition to serving as a scaffold for the finished version, the draft provides a road map to an estimated 90% of genes on every chromosome and already has enabled gene hunters to pinpoint genes associated with more than 30 disorders.

HGP resources have spurred a boom in spin-off sequencing programs on the human and other genomes in both the private and public sectors. To stimulate further research, all data generated in the public sector are made available rapidly and free of charge via the Web.

HGP Spinoff Projects

- **Microbial Genome Project**
  - www.sc.doe.gov/ober/microbial.html
  - www.ornl.gov/microbialgenomes/
- **Microbial Cell Project**

  microbialcellproject.org
- **Genomes to Life**

  doegenomestolife.org
- **Environmental Genome Project**

  www.niehs.nih.gov/envgenom/home.htm
- **Cancer Genome Anatomy Project**

  www.ncbi.nlm.nih.gov/ncicgap/
- **SNP Consortium**

  snp.cshl.org

Human Genome Project Goals 1998–2003

Human DNA Sequencing

The HGP's continued emphasis is on obtaining by 2003 a complete and highly accurate reference sequence (1 error in 10,000 bases) that is largely continuous across all human chromosomes. Scientists believe that knowing this sequence is critically important for understanding human biology and for applications to other fields.

A "working draft" of the sequence was completed 18 months ahead of schedule, in June 2000. The achievement has provided scientists worldwide with a road map to an estimated 90% of genes on every chromosome. Although the draft contains gaps and errors and does not yet meet the standard of accuracy outlined above, it provides a valuable scaffold for generating a high-quality reference genome sequence. HGP scientists make human DNA sequence available broadly, rapidly, and free of charge via the Web.

Sequencing Technology

Although current sequencing capacity is far greater than at the inception of the HGP, further incremental progress in sequencing technologies, efficiency, and cost-reduction are needed. For future sequencing applications, planners emphasize the importance of supporting novel technologies that may be 5 to 10 years in development.

Sequence Variation

Although more than 99% of human DNA sequences are the same across the population, variations in DNA sequence can have a major impact on how humans respond to disease; to such environmental insults as bacteria, viruses, toxins, and chemicals; and to drugs and other therapies.

Methods are being developed to detect different types of variation, particularly the most common type called single-nucleotide polymorphisms (SNPs), which occur about once every 100 to 300 bases. Scientists believe SNP maps will help them identify the multiple genes associated with such complex diseases as cancer, diabetes, vascular disease, and some forms of mental illness. These associations are difficult to establish with conventional gene-hunting methods because a single altered gene may make only a small contribution to disease risk.

Functional Genomics
Efficient interpretation of the functions of human genes and other DNA sequences requires that strategies be developed to enable large-scale investigations across whole genomes. A first priority is to generate complete sets of full-length cDNA clones and sequences for human and model-organism genes. Other functional-genomics goals include studies into gene expression and control and the development of experimental and computational methods for understanding gene function.

Comparative Genomics
The functions of human genes and other DNA regions often are revealed by studying their parallels in nonhumans. HGP researchers have obtained complete genomic sequences for the bacterium Escherichia coli, the yeast Saccharomyces cerevisiae, the fruit fly Drosophila melanogaster, and the roundworm Caenorhabditis elegans. Sequencing continues on the laboratory mouse. The availability of complete genome sequences generated both inside and outside the HGP is driving a major breakthrough in fundamental biology as scientists compare entire genomes to gain new insights into evolutionary, biochemical, genetic, metabolic, and physiological pathways.

Ethical, Legal, and Social Implications (ELSI)
Rapid advances in the science of genetics and its applications present new and complex ethical and policy issues for individuals and society. ELSI programs that identify and address these implications have been an integral part of the U.S. HGP since its inception. These programs have resulted in a body of work that promotes education and helps guide the conduct of genetic research and the development of related medical and public policies.

Bioinformatics and Computational Biology
Continued investment in current and new databases and analytical tools is critical to the success of the HGP and to the future usefulness of the data it produces. Databases must adapt to the evolving needs of the scientific community and must allow queries to be answered easily. Planners suggest developing a human genome database, analogous to model organism databases that will link to phenotypic information. Also needed are databases and analytical tools for studying the expanding body of gene-expression and functional data, for modeling complex biological networks and interactions, and for collecting and analyzing sequence-variation data.

Training

Future genome scientists will require training in interdisciplinary areas including biology, computer science, engineering, mathematics, physics, and chemistry. Also, scientists with management skills will be needed for leading large data-production efforts.

See Previous Goals

1998–2003 Third Five Year Plan

1993–1998 Second Five Year Plan

1991–1995 Original HGP Goals

This document may be cited in the following style:
Human Genome Program, U.S. Department of Energy, *Genomics and Its Impact on Medicine and Society: A 2001 Primer*, 2001.

For printed copies, please contact Laura Yust at Oak Ridge National Laboratory. Send questions or comments to the author, Denise K. Casey. Site designed by Marissa Mills.

***

33

What We've Learned So Far

❖

Genomics and Its Impact on Medicine and Society: A 2001 Primer

Achievement of a Draft Human Genome Sequence

Reprinted with permission of the US Dept. of Energy, Human Genome Program, http://www.ornl.gov.hgmis

In February 2001, HGP and Celera Genomics scientists published the long-awaited details of the working-draft DNA sequence. Although the draft is filled with mysteries, the first panoramic view of the human genetic landscape has revealed a wealth of information and some early surprises. Papers describing research observations in the journals *Nature* (Feb. 15, 2001) and *Science* (Feb. 16, 2001) are freely accessible.

Although clearly not a Holy Grail or Rosetta Stone for deciphering all of biology—two early metaphors commonly used to describe the coveted prize—the sequence is a magnificent and unprecedented resource that will serve as a basis for research and discovery throughout this century and beyond. It will have diverse practical applications and a profound impact upon how we view ourselves and our place in the tapestry of life.

One insight already gleaned from the sequence is that, even on the molecular level, we are more than the sum of our 35,000 or so genes. Surprisingly, this newly estimated number of genes is only one-third as great as previously thought and only twice as many as those of a tiny transparent worm, although the numbers may be revised as more computational and experimental analyses are performed.

At once humbled and intrigued by this finding, scientists suggest that the genetic key to human complexity lies not in the number of genes but in how gene parts are used to build different products in a process called alternative splicing. Other sources of added complexity are the thousands of chemical modifications made to proteins and the repertoire of regulatory mechanisms controlling these processes.

The draft encompasses 90% of the human genome's euchromatic portion, which contains the most genes. In constructing the working draft, the 16 genome sequencing centers produced over 22.1 billion bases of raw sequence data, comprising overlapping fragments totaling 3.9 billion bases and providing sevenfold coverage (sequenced seven times) of the human genome. Over 30% is high-quality, finished sequence, with eight-to tenfold coverage, 99.99% accuracy, and few gaps.

High-Quality Version Expected in 2003

The entire working draft will be finished to high quality by 2003. Coincidentally, that year also will be the 50th anniversary of Watson and Crick's publication of DNA structure that launched the era of molecular genetics. Much will remain to be deciphered even then. Some highlights follow from Nature, Science, and The Wellcome Trust philanthropy (an HGP funder).

By the Numbers

- The human genome contains 3164.7 million chemical nucleotide bases (A, C, T, and G).

- The average gene consists of 3000 bases, but sizes vary greatly, with the largest known human gene being dystrophin at 2.4 million bases.

- The total number of genes is estimated at 30,000 to 40,000, much lower than previous estimates of 80,000 to 140,000 that had been based on extrapolations from gene-rich areas as opposed to a composite of gene-rich and gene-poor areas.

- The order of almost all (99.9%) nucleotide bases is exactly the same in all people.

- The functions are unknown for more than 50% of discovered genes.

The Wheat from the Chaff

- About 2% of the genome encodes instructions for the synthesis of proteins.

- Repeated sequences that do not code for proteins (Òjunk DNAÓ) make up at least 50% of the human genome.

- Repetitive sequences are thought to have no direct functions, but they shed light on chromosome structure and dynamics. Over time, these repeats reshape the genome by rearranging it, thereby creating entirely new genes or modifying and reshuffling existing genes.

- During the past 50 million years, a dramatic decrease seems to have occurred in the rate of accumulation of these repeats.

How It's Arranged

- The human genome's gene-dense "urban centers" are composed predominantly of the DNA building blocks G and C.

- In contrast, the gene-poor "deserts" are rich in the DNA building blocks A and T. GC-and AT-rich regions usually can be seen through a microscope as light and dark bands on the chromosomes.

- Genes appear to be concentrated in random areas along the genome, with vast expanses of noncoding DNA between.

- Stretches of up to 30,000 C and G bases repeating over and over often occur adjacent to gene-rich areas, forming a barrier between the genes and the "junk DNA." These CpG islands are believed to help regulate gene activity.

- Chromosome 1 has the most genes (2968), and the Y chromosome has the fewest (231).

How the Human Genome Compares with Those of Other Organisms

- Unlike the human's seemingly random distribution of gene-rich areas, many other organisms' genomes are more uniform, with genes evenly spaced throughout.

- Humans have on average three times as many kinds of proteins as the fly or worm because of mRNA transcript "alternative splicing" and chemical modifications to the proteins. This process can yield different protein products from the same gene.

- Humans share most of the same protein families with worms, flies, and plants, but the number of gene family members has expanded in humans, especially in proteins involved in development and immunity.

- The human genome has a much greater portion (50%) of repeat sequences than the mustard weed (11%), the worm (7%), and the fly (3%).

- Although humans appear to have stopped accumulating repetitive DNA over 50 million years ago, there seems to be no such decline in rodents. This may account for some of the fundamental differences between hominids and rodents, although estimates of gene numbers are similar in both species. Scientists have proposed many theories to explain evolutionary contrasts between humans and other organisms, including life span, litter sizes, inbreeding, and genetic drift.

Variations and Mutations

- Scientists have identified about 1.4 million locations where single-base DNA differences (SNPs, see Goals Box: Sequence Variation) occur in humans. This information promises to revolutionize the processes of finding chromosomal locations for disease-associated sequences and tracing human history.

- The ratio of germline (sperm or egg cell) mutations is 2:1 in males vs females. Researchers point to several reasons for the higher mutation rate in the male germline, including the greater number of cell divisions required for sperm formation than for eggs.

Applications, Future Challenges

Deriving meaningful knowledge from the DNA sequence will define research through the coming decades to inform our understanding of biological systems. This enormous task will require the expertise and creativity of tens of thousands of scientists from varied disciplines in both the public and private sectors worldwide.

The draft sequence already is having an impact on finding genes associated with disease. Genes have been pinpointed and associated with numerous diseases and disorders including breast cancer, muscle disease, deafness, and blindness. Additionally, finding the DNA sequences underlying such common diseases as cardiovascular disease, diabetes, arthritis, and cancers is being aided by the human SNP maps generated in the HGP in cooperation with the private sector. These genes and SNPs provide focused targets for the development of effective new therapies.

One of the greatest impacts of having the sequence may well be in enabling an entirely new approach to biological research. In the past, researchers studied one or a few genes at a time. With whole-genome sequences and new automated, high-throughput technologies, they can approach questions systematically and on a grand scale. They can study all the genes in a genome, for example, or all the gene products in a particular tissue or organ or tumor, or how tens of thousands of genes and proteins work together in interconnected networks to orchestrate the chemistry of life.

***

34

After the Human Genome Project (HGP), the Next Steps...

❖

Genomics and Its Impact on Medicine and Society: A 2001 Primer

The words of Winston Churchill, spoken in 1942 after 3 years of war, capture well the HGP era: "Now this is not the end. It is not even the beginning of the end. But it is, perhaps, the end of the beginning."

The avalanche of genome data grows daily. The new challenge will be to use this vast reservoir of data to explore how DNA and proteins work with each other and the environment to create complex, dynamic living systems.

Systematic studies of function on a grand scale—functional genomics—will be the focus of biological explorations in this century and beyond. These explorations will encompass studies in transcriptosmics, proteomics, structural genomics, new experimental methodologies, and comparative genomics.

- **Transcriptomics** involves large-scale analysis of messenger RNAs (molecules that are transcribed from active genes) to determine when, where, and under what conditions genes are expressed.

- **Proteomics**—the study of protein expression and function—can bring researchers closer than gene-expression studies to what's actually happening in the cell.

- **Structural genomics** initiatives are being launched worldwide to generate the 3-D structures of one or more proteins from each protein family, thus offering clues to their function and providing biological targets for drug design.

- **Knockout studies** are one experimental method for understanding the function of DNA sequences and the proteins they encode. Researchers inactivate genes in living organisms and monitor any changes that could reveal the function of specific genes.

- **Comparative genomics**—analyzing DNA sequence patterns of humans and well-studied model organisms side by side—has become one of the most powerful strategies for identifying human genes and interpreting their function.

MEDICINE AND THE NEW GENETICS: GENE TESTING, PHARMACOGENOMICS, AND GENE THERAPY

DNA underlies every aspect of human health, both in function and dysfunction. Obtaining a detailed picture of how genes and other DNA sequences function together and interact with environmental factors ultimately will lead to the discovery of pathways involved in normal processes and in disease pathogenesis. Such knowledge will have a profound impact on the way disorders are diagnosed, treated, and prevented and will bring about revolutionary changes in clinical and public health practice. Some of these transformative developments are described below.

Gene Tests

DNA-based tests are among the first commercial medical applications of the new genetic discoveries. Gene tests can be used to diagnose disease, confirm a diagnosis, provide prognostic information about the course of disease, confirm the existence of a disease in asymptomatic individuals, and, with varying degrees of accuracy, predict the risk of future disease in healthy individuals or their progeny.

Currently, several hundred genetic tests are in clinical use, with many more under development, and their numbers and varieties are expected to increase rapidly over the next decade. Most current tests detect mutations associated with rare genetic disorders that follow Mendelian inheritance patterns. These include myotonic and Duchenne muscular dystrophies, cystic fibrosis, neurofibromatosis type 1, sickle cell anemia, and Huntington's disease.

Recently, tests have been developed to detect mutations for a handful of more complex conditions such as breast, ovarian, and colon cancers. Although they have limitations, these tests sometimes are used to make risk estimates in pre-symptomatic individuals with a family history of the disorder.

One potential benefit to using these gene tests is that they could provide information that helps physicians and patients manage the disease or condition more effectively. Regular colonoscopies for those having mutations associated with colon cancer, for instance, could prevent thousands of deaths each year.

Some scientific limitations are that the tests may not detect every mutation associated with a particular condition (many are as yet undiscovered), and the ones they do detect may present different risks to different people and populations. Another important consideration in gene testing is the lack of effective treatments or preventive measures for many diseases and conditions now being diagnosed or predicted.

Revealing information about risk of future disease can have significant emotional and psychological effects as well. Moreover, the absence of privacy and legal protections can lead to discrimination in employment or insurance or other misuse of personal genetic information. Additionally, because genetic tests reveal information about individuals and their families, test results can affect family dynamics. Results also can pose risks for population groups if they lead to group stigmatization.

Other issues related to gene tests include their effective introduction into clinical practice, the regulation of laboratory quality assurance, the availability of testing for rare diseases, and the education of healthcare providers and patients about correct interpretation and attendant risks.

Pharmacogenomics: Moving Away from "One-Size-Fits-All" Therapeutics

Within the next decade, researchers will begin to correlate DNA variants with individual responses to medical treatments, identify particular subgroups of patients, and develop drugs customized for those populations. The discipline that blends pharmacology with genomic capabilities is called pharmacogenomics.

More than 100,000 people die each year from adverse responses to medications that are beneficial to others. Another 2.2 million experience serious reactions, while others fail to respond at all. DNA variants in genes involved in drug metabolism, particularly the cytochrome P450 multigene family, are the focus of much current research in this area.

Enzymes encoded by these genes are responsible for metabolizing most drugs used today, including many for treating psychiatric, neurological, and cardiovascular diseases. Enzyme function affects patients' responses to both the drug and the dose. Future advances will enable rapid testing to determine the patient's genotype and drastically reduce hospitalization resulting from adverse reactions.

Genomic data and technologies also are expected to make drug development faster, cheaper, and more effective. Most drugs today are based on about 500 molecular targets; genomic knowledge of the genes involved in diseases, disease pathways, and drug-response sites will lead to the discovery of thousands of new targets.

New drugs, aimed at specific sites in the body and at particular biochemical events leading to disease, probably will cause fewer side effects than many current medicines. Ideally, the new genomic drugs could be given earlier in the disease process. As knowledge becomes available to select patients most likely to benefit from a potential drug, pharmacogenomics will speed the design of clinical trials to bring the drugs to market sooner.

Gene Therapy, Enhancement

The potential for using genes themselves to treat disease or enhance particular traits has captured the imagination of the public and the biomedical community. This largely experimental field—gene transfer or gene therapy—holds potential for treating or even curing such genetic and acquired diseases as cancers and AIDS by using normal genes to supplement or replace defective genes or bolster a normal function such as immunity.

More than 500 clinical gene-therapy trials involving about 3500 patients have been identified worldwide (June 2001). The vast majority (78%) take place in the United States, followed by Europe (18%). Although most trials focus on various types of cancer, studies also involve other multigenic and monogenic, infectious, and vascular diseases. Protocols generally are aimed at establishing the safety of gene-delivery procedures rather than effectiveness, and no cures as yet can be attributed to these trials.

Gene transfer still faces many scientific obstacles before it can become a practical approach for treating disease. According to the American Society of Human Genetics' Statement on Gene Therapy, effective progress will be achieved only through continued rigorous research on the most fundamental mechanisms underlying gene delivery and gene expression in animals.

♦ ♦ ♦

SOCIETAL CONCERNS ARISING FROM THE NEW GENETICS

Genomics and Its Impact on Medicine and Society: A 2001 Primer

Since its inception, the Human Genome Project has dedicated funds toward studying the ethical, legal, and social issues surrounding the availability of the new data and capabilities. Examples of such issues follow.

- **Privacy and confidentiality of genetic information.** *Who owns and controls genetic information? Is genetic privacy different from medical privacy?*

- **Fairness in the use of genetic information by insurers, employers, courts, schools, adoption agencies, and the military, among others.** *Who should have access to personal genetic information, and how will it be used?*

- **Psychological impact, stigmatization, and discrimination due to an individual's genetic differences.** *How does personal genetic information affect self-identity and society's perceptions?*

- **Reproductive issues including adequate and informed consent and the use of genetic information in reproductive decision making.** *Do healthcare personnel properly counsel parents about risks and limitations? What are the larger societal issues raised by new reproductive technologies?*

- **Clinical issues including the education of doctors and other health-service providers, people identified with genetic conditions, and the general public; and the implementation of standards and quality-control measures.** *How should health professionals be prepared for the new genetics? How can the public be educated to make informed choices? How will genetic tests be evaluated and regulated for accuracy, reliability, and usefulness?* (Currently, there is little regulation at the federal level.) *How does society balance current scientific limitations and social risk with long-term benefits?*

- **Fairness in access to advanced genomic technologies.** *Who will benefit? Will there be major worldwide inequities?*

- **Uncertainties associated with gene tests for susceptibilities and complex conditions (e.g., heart disease, diabetes, and Alzheimer's disease).** *Should testing be performed when no treatment is available or when interpretation is unsure? Should children be tested for susceptibility to adult-onset diseases?*

- **Conceptual and philosophical implications regarding human responsibility, free will vs. genetic determinism, and concepts of health and disease.** *Do our genes influence our behavior, and can we control it? What is considered acceptable diversity? Where is the line drawn between medical treatment and enhancement?*

- **Health and environmental issues concerning genetically modified (GM) foods and microbes.** *Are GM foods and other products safe for humans and the environment? How will these technologies affect developing nations' dependence on industrialized nations?*

Commercialization of products including property rights (patents, copyrights, and trade secrets) and accessibility of data and materials. *Will patenting DNA sequences limit their accessibility and development into useful products?*

***

35

Interviewing for Personal Histories: How to Interview Older Adults. Intergenerational Writing for Genealogy and Life Stories.

STEP 1: Send someone enthusiastic about personal and oral history to senior community centers, lifelong learning programs at universities, nursing homes, or senior apartment complexes activity rooms. You can reach out to a wide variety of older adults in many settings, including at libraries, church groups, hobby and professional or trade associations, unions, retirement resorts, public transportation centers, malls, museums, art galleries, genealogy clubs, and intergenerational social centers.

STEP 2: Have each personal historian or volunteer bring a tape recorder with tape and a note pad. Bring camcorders for recording video to turn into time capsules and CDs or DVDs with life stories, personal history experiences, memoirs, and events highlighting turning points or special times in people's lives.

STEP 3: Assign each personal historian one or two older persons to interview with the following questions.

1. What were the most significant turning points or events in your life?

2. How did you survive the Wars?

3. What were the highlights, turning points, or significant events that you experienced during the economic downturn of 1929–1939? How did you cope or solve your problems?

4. What did you do to solve your problems during the significant stages of your life at age 10, 20, 30, 40, 50, 60 and 70-plus? Or pick a year that you want to talk about.

5. What changes in your life do you want to remember and pass on to future generations?

6. What was the highlight of your life?

7. How is it best to live your life after 70?

8. What years do you remember most?

9. What was your favorite stage of life?

10. What would you like people to remember about you and the times you lived through?

STEP 3:

Have the student record the older person's answers. Select the most significant events, experiences, or turning points the person chooses to emphasize. Then write the story of that significant event in ten pages or less.

STEP 4: Ask the older person to supply the younger student photos, art work, audio tapes, or video clips. Usually photos, pressed flowers, or art work will be supplied. Have the student or teacher scan the photos onto a disk and return the original photos or art work or music to the owner.

STEP 5: The personal historian, volunteer, student and/or teacher scans the photos and puts them onto a Web site on the Internet at one of the free communities that give away Web site to the public at no cost....some include http://www.tripod.com, http://www.fortunecity.com, http://www.angelfire.com, http://www.geocities.com, and others. Most search engines will give a list of communities at offering free Web sites to the public. Microsoft also offers free family Web sites for family photos and newsletters or information. Ask your Internet service provider whether it offers free Web site space to subscribers.

1. Create a Web site with text from the older person's significant life events
2. Add photos.
3. Add sound or .wav files with the voice of the older person speaking in small clips or sound bites.
4. Intersperse text and photos or art work with sound, if available.
5. Add video clips, if available and won't take too much bandwidth.
6. Put Web site on line as TIME CAPSULE of (insert name of person) interviewed and edited by, insert name of student who interviewed older person.

STEP 6: Label each Web site Time Capsule and collect them in a history archives on the lives of older adults at the turn of the millennium. Make sure the older person and all relatives and friends are emailed the Web site link. You have now created a time capsule for future generations.

This can be used as a classroom exercise in elementary and high schools to teach the following:

1. Making friends with older adults.
2. Learning to write on intergenerational topics.
3. Bringing community together of all generations.
4. Learning about foster grandparents.
5. History lessons from those who lived through history.
6. Learning about diversity and how people of diverse origins lived through the 20th century.
7. Preserving the significant events in the lives of people as time capsules for future generations to know what it was like to live between 1900 and 2000 at any age.

8. Learning to write skits and plays from the life stories of older adults taken down by young students.

9. Teaching older adults skills in creative writing at senior centers.

10. Learning what grandma did during World War 2 or the stock market crash of 1929 followed by the economic downturn of 1930–1938.

What to Ask People about Their Lives

Step 1
When you interview, ask for facts and concrete details. Look for statistics, and research whether statistics are deceptive in your case.

Step 2
To write a plan, write one sentence for each topic that moves the story or piece forward. Then summarize for each topic in a paragraph. Use dialogue at least in every third paragraph.

Step 3

Look for the following facts or headings to organize your plan for a biography or life story.

1. SLOGAN. Ask the people you interview what would be their slogan if they had to create/invent a slogan that fit themselves or their aspirations: One slogan might be something like the seventies ad for cigarettes, "We've come a long way, baby," to signify ambition. Only look for an original slogan.

2. CRUSADE. Ask the people you interview or a biography, for what purpose is or was their crusade? Is or was it equality in the workplace or something personal and different such as dealing with change—downsizing, working after retirement, or anything else?

3. IMPACT. Ask what makes an impact on people's lives and what impact the people you're interviewing want to make on others?

4. STATISTICS: How deceptive are they? How can you use them to focus on reality?

5. How have the people that you're interviewing influenced changes in the way people or corporations function?

6. To what is the person aspiring?

7. What kind of communication skills does the person have and how are these skills received? Are the communication skills male or female, thinking or feeling, yin or yang, soft or steeled, and are people around these people negative or positive about those communication skills?

8. What new styles is the person using? What kind of motivational methods, structure, or leadership? Is the person a follower or leader? How does the person match his or her personality to the character of a corporation or interest?

9. How does the person handle change?

10. How is the person reinforced?

Once you have titles and summarized paragraphs for each segment of your story, you can more easily flesh out the story by adding dialogue and description to your factual information. Look for differences in style between the people you interview? How does the person want to be remembered?

Is the person a risk taker or cautious for survival? Does the person identify with her job or the people involved in the process of doing the work most creatively or originally? Does creative expression take precedence over processes of getting work out to the right place at the right time? Does the person want his ashes to spell the word "love" where the sea meets the shore? This is a popular concept appearing in various media.

Search the Records in the Family History Library of Salt Lake City, Utah

Make use of the database online at the Family History Library of Salt Lake City, Utah. Or visit the branches in many locations. The Family History Library (FHL) is known worldwide as the focal point of family history records preservation.

The FHL collection contains more than 2.2 million rolls of microfilmed genealogical records, 742,000 microfiche, 300,000 books, and 4,500 periodicals that represent data collected from over 105 countries. You don't have to be a member of any particular church or faith to use the library or to go online and search the records.

Family history records owe a lot to the invention of writing. And then there is oral history, but someone needs to transcribe oral history to record and archive them for the future.

Interestingly, isn't it a coincidence that writing is 6,000 years old and DNA that existed 6,000 years ago first reached such crowded conditions in the very cities that had first used writing extensively to measure accounting and trade had very little recourse but to move on to new areas where there were far less people and less use of writing?

A lot of major turning points occurred 6,000 years ago—the switch to a grain-based diet from a meat and root diet, the use of bread and fermented grain beverages, making of oil from plants, and the rise of religions based on building "god houses" in the centers of town in areas known as the "cereal belt" around the world.

Six thousand years ago in India we have the start of the Sanksrit writings, the cultivation of grain. In China, we have the recording of acupuncture points for medicine built on energy meridians that also show up in the blue tattoos of the Ice Man fossil "Otsi" in the Alps—along the same meridians as the Chinese acupuncture points.

At 6,000 years ago the Indo European languages spread out across Europe. Mass migrations expanded by the Danube leaving pottery along the trade routes that correspond to the clines and gradients of gene frequency coming out of the cereal belts.

Then something happened. There was an agricultural frontier cutting off the agriculturists from the hunters. Isn't it a coincidence that the agricultural frontiers or barriers also are genetic barriers at least to some degree?

***

36

Oral History

Here's how to systematically collect, record, and preserve living peoples' testimonies about their own experiences. After you record in audio and/or video the highlights of anyone's experiences, try to verify your findings. See whether you can check any facts in order to find out whether the person being recorded is making up the story or whether it really did happen.

This is going to be difficult unless you have witnesses or other historical records. Once you have verified your findings to the best of your ability, note whether the findings have been verified. Then analyze what you found. Put the oral history recordings in an accurate historical context.

Mark the recordings with the dates and places. Watch where you store your findings so scholars in the future will be able to access the transcript or recording and convert the recording to another, newer technology. For instance, if you have a transcript on paper, have it saved digitally on a disk and somewhere else on tape and perhaps a written transcript on acid-free good paper in case technology moves ahead before the transcript or recording is converted to the new technology.

For example, if you only put your recording on a phonograph record, within a generation or two, there may not be any phonographs around to play the record. The same goes for CDs, DVDs and audio or video tapes.

So make sure you have a readable paper copy to be transcribed or scanned into the new technology as well as the recordings on disk and tape. For example, if you record someone's experiences in a live interview with your video camera, use a cable to save the video in the hard disk of a computer and then burn the file to a CD or DVD. Keep a copy of audio tape and a copy of regular video tape—all in a safe place such as a time capsule, and make a copy for various archives in libraries and university oral history preservation centers. Be sure scholars in the future can find a way to enjoy the experiences in your time capsule, scrapbook, or other storage device for oral histories.

Use your DNA testing results to add more information to a historical record. As an interviewer with a video camera and/or audio tape recorder, your task is to record as a historical record what the person who you are interviewing recollects.

The events move from the person being interviewed to you, the interviewer, and then into various historical records. In this way you can combine results of DNA testing with actual memories of events. If it's possible, also take notes or have someone take notes in case the tape doesn't pick up sounds clearly.

I had the experience of having a video camera battery go out in spite of all precautions when I was interviewing someone, and only the audio worked. So keep a backup battery on hand whether you use a tape recorder or a video camera. If at all possible, have a partner bring a spare camera and newly recharged battery. A fully charged battery left overnight has a good chance of going out when you need it.

What Should Go Into an Oral History?

Emphasize the commitment to family and faith. To create readers' and media attention to an oral history, it should have some redemptive value to a universal audience. That's the most important point. Make your oral history simple and earthy. Write about real people who have values, morals, and a faith in something greater than themselves that is equally valuable to readers or viewers.

Publishers who buy an oral history written as a book on its buzz value are buying simplicity. It is simplicity that sells and nothing else but simplicity. This is true for oral histories, instructional materials, and fiction. It's good storytelling to say it simply.

Simplicity means the oral history or memoirs book or story gives you all the answers you were looking for in your life in exotic places, but found it close by. What's the great proverb that your oral history is telling the world?

Is it to stand on your own two feet and put bread on your own table for your family? That's the moral point, to pull your own weight, and pulling your own weight is a buzz word that sells oral histories and fiction that won't preach, but instead teach and reach through simplicity.

That's the backbone of the oral historian's new media. Buzz means the story is simple to understand. You make the complex easier to grasp. And buzz means you can sell your story or book, script or narrative by focusing on the values of simplicity, morals, faith, and universal values that hold true for everyone. Doing the best to take care of your family sells and is buzz appeal, hot stuff in the publishing market of today and in the oral history archives. This is true, regardless of genre. Publishers go through fads every two years—angel books, managing techniques books, computer home-based business books, novels about ancient historical characters or tribes, science fiction, children's programming, biography, and oral history transcribed into a book or play.

The genres shift emphasis, but values are consistent in the bestselling books. Perhaps your oral history will be simple enough to become a bestselling book or script. In the new media, simplicity is buzz along with values. Oral history, like best-selling novels and true stories is built on simplicity, values, morals, and commitment. Include how one person dealt with about trends. Focus your own oral history about life in the lane of your choice.

Steps to Take in Gathering Oral Histories

Use the following sequence when gathering oral/aural histories:

1. Develop one central issue and divide that issue into a few important questions that highlight or focus on that one central issue.

2. Write out a plan just like a business plan for your oral history project. You may have to use that plan later to ask for a grant for funding, if required. Make a list of all your products that will result from the oral history when it's done.

3. Write out a plan for publicity or public relations and media relations. How are you going to get the message to the public or special audiences?

4. Develop a budget. This is important if you want a grant or to see how much you'll have to spend on creating an oral history project.

5. List the cost of video taping and editing, packaging, publicity, and help with audio or special effects and stock shot photos of required.

6. What kind of equipment will you need? List that and the time slots you give to each part of the project. How much time is available? What are your deadlines?

7. What's your plan for a research? How are you going to approach the people to get the interviews? What questions will you ask?

8. Do the interviews. Arrive prepared with a list of questions. It's okay to ask the people the kind of questions they would like to be asked. Know what dates the interviews will cover in terms of time. Are you covering the economic depression of the thirties? World Wars? Fifties? Sixties? Pick the time parameters.

9. Edit the interviews so you get the highlights of experiences and events, the important parts. Make sure what's important to you also is important to the person you interviewed.

10. Find out what the interviewee wants to emphasize perhaps to highlight events in a life story. Create a video-biography of the highlights of one person's life or an oral history of an event or series of events.

11. Process audio as well as video, and make sure you have written transcripts of anything on audio and/or video in case the technology changes or the tapes go bad.

12. Save the tapes to compact disks, DVDs, a computer hard disk and several other ways to preserve your oral history time capsule. Donate any tapes or CDs to appropriate archives, museums, relatives of the interviewee, and one or more oral history libraries. They are usually found at universities that have an oral history department and library such as UC Berkeley and others.

13. Check the Web for oral history libraries at universities in various states and abroad.

14. Evaluate what you have edited. Make sure the central issue and central questions have been covered in the interview. Find out whether newspapers or magazines want summarized transcripts of the audio and/or video with photos.

15. Contact libraries, archives, university oral history departments and relevant associations and various ethnic genealogy societies that focus on the subject matter of your central topic.

16. Keep organizing what you have until you have long and short versions of your oral history for various archives and publications. Contact magazines and newspapers to see whether editors would assign reporters to do a story on the oral history project.

17. Create a scrapbook with photos and summarized oral histories. Write a synopsis of each oral history on a central topic or issue. Have speakers give public presentations of what you have for each person interviewed and/or for the entire project using highlights of several interviews with the media for publicity. Be sure your project is archived properly and stored in a place devoted to oral history archives and available to researchers and authors.

Aural History Techniques

1. Begin with easy to answer questions that don't require you explore and probe deeply in your first question. Focus on one central issue when asking questions.

2. First research written or visual resources before you begin to seek an oral history of a central issue, experience, or event.

3. Who is your intended audience?

4. What kind of population niche or sample will you target?

5. What means will you select to choose who you will interview? What group of people will be central to your interview?

6. Write down how you'll explain your project. Have a script ready so you don't digress or forget what to say on your feet.

7. Consult oral history professionals if you need more information. Make sure what you write in your script will be clear to understand by your intended audience.

8. Have all the equipment you need ready and keep a list of what you'll use and the cost. Work up your budget.

9. Choose what kind of recording device is best—video, audio, multimedia, photos, and text transcript. Make sure your video is broadcast quality. I use a Sony Digital eight (high eight) camera.

10. Make sure from cable TV stations or news stations that what type of video and audio you choose ahead of time is broadcast quality.

11. Make sure you have an external microphone and also a second microphone as a second person also tapes the interview in case the quality of your camera breaks down. You can also keep a tape recorder going to capture the audio in case your battery dies.

12. Make sure your battery is fully charged right before the interview. Many batteries die down after a day or two of nonuse.

13. Test all equipment before the interview and before you leave your office or home. I've had batteries go down unexpectedly and happy there was another person ready with another video camera waiting and also an audio tape version going.

14. Make sure the equipment works if it's raining, hot, cold, or other weather variations. Test it before the interview. Practice interviewing someone on your equipment several times to get the hang of it before you show up at the interview.

15. Make up your mind how long the interview will go before a break and use tape of that length, so you have one tape for each segment of the interview. Make several copies of your interview questions.

16. Be sure the interviewee has a copy of the questions long before the interview so the person can practice answering the questions and think of what to say or even take notes. Keep checking your list of what you need to do.

17. Let the interviewee make up his own questions if he wants. Perhaps your questions miss the point. Present your questions first. Then let him embellish the questions or change them as he wants to fit the central issue with his own experiences.

18. Call the person two days and then one day before the interview to make sure the individual will be there on time and understands how to travel to the location. Or if you are going to the person's home, make sure you understand how to get there.

19. Allow yourself one extra hour in case of traffic jams.

20. Choose a quiet place. Turn off cell phones and any ringing noises. Make sure you are away from barking dogs, street noise, and other distractions.

21. Before you interview make sure the person knows he or she is going to be video and audio-taped.

22. If you don't want anyone swearing, make that clear it's for public archives and perhaps broadcast to families.

23. Your interview questions should follow the journalist's information-seeking format of asking, who, what, where, where, how, and why. Oral history is a branch of journalistic research.

24. Let the person talk and don't interrupt. You be the listener and think of oral history as aural history from your perspective.

25. Make sure only one person speaks without being interrupted before someone else takes his turn to speak.

26. Understand silent pauses are for thinking of what to say.

27. Ask one question and let the person gather his thoughts.

28. Finish all your research on one question before jumping to the next question. Keep it organized by not jumping back to the first question after the second is done. Stay in a linear format.

29. Follow up what you can about any one question, finish with it, and move on to the next question without circling back. Focus on listening instead of asking rapid fire questions as they would confuse the speaker.

30. Ask questions that allow the speaker to begin to give a story, anecdote, life experience, or opinion along with facts. Don't ask questions that can be answered only be yes or no. This is not a courtroom. Let the speaker elaborate with facts and feelings or thoughts.

31. Late in the interview, start to ask questions that explore and probe for deeper answers.

32. Wrap up with how the person solved the problem, achieved results, reached a conclusion, or developed an attitude, or found the answer. Keep the wrap-up on a light, uplifting note.

33. Don't leave the individual hanging in emotion after any intensity of. Respect the feelings and opinions of the person. He or she may see the situation from a different point of view than someone else. So respect the person's right to feel as he does. Respect his need to recollect his own experiences.

34. Interview for only one hour at a time. If you have only one chance, interview for an hour. Take a few minutes break. Then interview for the second hour. Don't interview more than two hours at any one meeting.

35. Use prompts such as paintings, photos, music, video, diaries, vintage clothing, crafts, antiques, or memorabilia when appropriate. Carry the photos in labeled files or envelopes to show at appropriate times in order to prime the memory of the interviewee. For example, you may show a childhood photo and ask "What was it like in that orphanage where these pictures were taken?" Or travel photos might suggest a trip to America as a child, or whatever the photo suggests. For example, "Do you remember when this ice cream parlor inside the ABC movie house stood at the corner of X and Y Street? Did you go there as a teenager? What was your funniest memory of this movie theater or the ice cream store inside back in the fifties?"

36. As soon as the interview is over, label all the tapes and put the numbers in order.

37. A signed release form is required before you can broadcast anything. So have the interviewee sign a release form before the interview.

38. Make sure the interviewee gets a copy of the tape and a transcript of what he or she said on tape. If the person insists on making corrections, send the paper transcript of the tape for correction to the interviewee. Edit the tape as best you can or have it edited professionally.

39. Make sure you comply with all the corrections the interviewee wants changed. He or she may have given inaccurate facts that need to be corrected on the paper transcript.

40. Have the tape edited with the corrections, even if you have to make a tape at the end of the interviewee putting in the corrections that couldn't be edited out or changed.

41. As a last resort, have the interviewee redo the part of the tape that needs correction and have it edited in the tape at the correct place marked on the tape. Keep the paper transcript accurate and up to date, signed with a release form by the interviewee.

42. Oral historians write a journal of field notes about each interview. Make sure these get saved and archived so they can be read with the transcript.

43. Have the field notes go into a computer where someone can read them along with the transcript of the oral history tape or CD.

44. Thank the interviewee in writing for taking the time to do an interview for broadcast and transcript.

45. Put a label on everything you do from the interview to the field notes. Make a file and sub file folders and have everything stored in a computer, in archived storage, and in paper transcript.

46. Make copies and digital copies of all photos and put into the records in a computer. Return originals to owners.

47. Make sure you keep your fingerprints off the photos by wearing white cotton gloves. Use cardboard when sending the photos back and pack securely. Also photocopy the photos and scan the photos into your computer. Treat photos as antique art history in preservation.

48. Make copies for yourself of all photos, tapes, and transcripts. Use your duplicates, and store the original as the master tape in a place that won't be used often, such as a time capsule or safe, or return to a library or museum where the original belongs.

49. Return all original photos to the owners. An oral history archive library or museum also is suitable for original tapes. Use copies only to work from, copy, or distribute.

50. Index your tapes and transcripts. To use oral history library and museum terminology, recordings and transcripts are given "accession numbers."

51. Phone a librarian in an oral history library of a university for directions on how to assign accession numbers to your tapes and transcripts if the materi-

als are going to be stored at that particular library. Store copies in separate places in case of loss or damage.

52. If you don't know where the materials will be stored, use generic accession numbers to label your tapes and transcripts. Always keep copies available for yourself in case you have to duplicate the tapes to send to an institution, museum, or library, or to a broadcast company.

53. Make synopses available to public broadcasting radio and TV stations.

54. Check your facts.

55. Are you missing anything you want to include?

56. Is there some place you want to send these tapes and transcripts such as an ethnic museum, radio show, or TV satellite station specializing in the topics on the tapes, such as public TV stations? Would it be suitable for a world music station? A documentary station?

57. If you need more interviews, arrange them if possible.

58. Give the interviewee a copy of the finished product with the corrections. Make sure the interviewee signs a release form that he or she is satisfied with the corrections and is releasing the tape to you and your project.

59. Store the tapes and transcripts in a library or museum or at a university or other public place where it will be maintained and preserved for many generations and restored when necessary.

60. You can also send copies to a film repository or film library that takes video tapes, an archive for radio or audio tapes for radio broadcast or cable TV.

61. Copies may be sent to various archives for storage that lasts for many generations. Always ask whether there are facilities for restoring the tape. A museum would most likely have these provisions as would a large library that has an oral history library project or section.

62. Make sure the master copy is well protected and set up for long-term storage in a place where it will be protected and preserved.

63. If the oral history is about events in history, various network news TV stations might be interested. Film stock companies may be interested in copies of old photos.

64. Find out from the subject matter what type of archives, repository, or storage museums and libraries would be interested in receiving copies of the oral history tapes and transcripts.

65. Print media libraries would be interested in the hard paper copy transcripts and photos as would various ethnic associations and historical preservation societies. Find out whether the materials will go to microfiche, film, or be digitized and put on CDs and DVDs, or on the World Wide Web. If you want to create a time capsule for the Web, you can ask the interviewee whether he or she wants the materials or selected materials to be put online or on CD as multimedia or other. Then you would get a signed release from the interviewee authorizing you to put the materials or excerpts online. Also find out in whose name the materials are copyrighted and whether you have print and electronic rights to the material or do the owners-authors-interviewees—or you, the videographer-producer? Get it all in writing, signed by those who have given you any interviews, even if you have to call your local intellectual property rights attorney.

How Accurate Are Oral/Aural Histories?

Cameras give fragments, points of view, and bits and pieces. Viewers will see what the videographer or photographer intends to be seen. The interviewee will also be trying to put his point of view across and tell the story from his perspective. Will the photographer or videographer be in agreement with the interviewee? Or if you are recording for print transcript, will your point of view agree with the interviewee's perspective and experience if your basic 'premise,' where you two are coming from, are not in agreement? Think this over as you write your list of questions. Do both of you agree on your central issue on which you'll focus for the interview?

How are you going to turn spoken words into text for your paper hard copy transcript? Will you transcribe verbatim, correct the grammar, or quote as you hear the spoken words? Oral historians really need to transcribe the exact spoken word. You can leave out the 'ahs' and 'oms' or loud pauses, as the interviewee thinks what to say next. You don't want to sound like a court reporter, but you do want to have an accurate record transcribed of what was spoken.

You're also not editing for a movie, unless you have permission to turn the oral history into a TV broadcast, where a lot gets cut out of the interview for time constraints. For that, you'd need written permission so words won't be taken out of context and strung together in the editing room to say something different from what the interviewee intended to say.

Someone talking could put in wrong names, forget what they wanted to say, or repeat themselves. They could mumble, ramble, or do almost anything. So you would have to sit down and weed out redundancy when you can or decide on presenting exactly what you've heard as transcript. When someone reads the transcript in text, they won't have what you had in front of you, and they didn't see and hear the live presentation or the videotape. It's possible to misinterpret gestures or how something is spoken, the mood or tone, when reading a text transcript. Examine all your sources. Use an ice-breaker to get someone talking.

If a woman is talking about female-interest issues, she may feel more comfortable talking to another woman. Find out whether the interviewee is more comfortable speaking to someone of his or her own age. Some older persons feel they can relate better to someone close to their own age than someone in high school, but it varies. Sometimes older people can speak more freely to a teenager.

The interviewee must be able to feel comfortable with the interviewer and know he or she will not be judged. Sometimes it helps if the interviewer is the same ethnic group or there is someone present of the same group or if new to the language, a translator is present.

Read some books on oral history field techniques. Read the National Genealogical Society Quarterly (NGSQ). Also look at The American Genealogist (TAG), The Genealogist, and The New England Historical and Genealogical Register (The Register). If you don't know the maiden name of say, your grandmother's mother, and no relative knows either because it wasn't on her death certificate, try to reconstruct the lives of the males who had ever met the woman whose maiden name is unknown.

Maybe she did business with someone before marriage or went to school or court. Someone may have recorded the person's maiden name before her marriage. Try medical records if any were kept. There was no way to find my mother's grandmother's maiden name until I started searching to see whether she had any brothers in this country. She had to have come as a passenger on a ship around 1880 as she bought a farm. Did her husband come with her?

Was the farm in his name? How many brothers did she have in this country with her maiden surname? If the brothers were not in this country, what countries did they come from and what cities did they live in before they bought the farm in Albany? If I could find out what my great grandmother's maiden name was through any brothers living at the time, I could contact their descendants perhaps and see whether any male or female lines are still in this country or where else on the globe.

Perhaps a list of midwives in the village at the time is recorded in a church or training school for midwives. Fix the person in time and place. Find out whom she might have done business with and whether any records of that business exist. What businesses did she patronize? Look for divorce or court records, change of name records, and other legal documents.

Look at local sources. Did anyone save records from bills of sale for weddings, purchases of homes, furniture, debutante parties, infant supplies, or even medical records? Look at nurses' licenses, midwives' registers, employment contracts, and teachers' contracts, alumni associations for various schools, passports, passenger lists, alien registration cards, naturalization records, immigrant aid societies, city directories, and cross-references. Try religious and women's clubs, lineage and village societies, girl scouts and similar groups, orphanages, sanatoriums, hospitals, police records. Years ago there was even a Eugenics Record Office. What about the women's prisons? The first one opened in 1839—Mount Pleasant Female Prison, NY.

Try voters' lists. If your relative is from another country, try records in those villages or cities abroad. Who kept the person's diaries? Have you checked the Orphan Train records? Try ethnic and religious societies and genealogy associations for that country. Most ethnic genealogy societies have a special interest group for even the smallest villages in various countries.

You can start one and put up a Web site for people who also come from there in past centuries. Check alimony, divorce, and court records, widow's pensions of veterans, adoptions, orphanages, foster homes, medical records, birth, marriage, and death certificates, social security, immigration, pet license owners' files, prisons, alumni groups from schools, passenger lists, military, and other legal records.

When all historical records are being tied together, you can add the DNA testing to link all those cousins. Check military pensions on microfilms in the National Archives. See the bibliography section of this book for further resources on highly recommended books and articles on oral history field techniques and similar historical subjects.

Does Writing Your Life Story As A Genealogy and/or Novel Affect Your Memory?

Oral history depends on memory and the ability to speak. I also think of oral history as aural history, based on the ability to hear someone's experiences and remember them to pass on to the next generation or the world.

To find out the effects of oral history on memory and on creative writing on memory, we'd have to ask the people who write their life story and/or genealogy in their older years what it did for them, their memory, and their ability to think and feel. Make use of introverted feeling in writing a commercial or salable life story for the new media. Think in three dimensions for older adults is a different highway. How did DNA testing influence a genealogy search for family history facts?

Did the individual create a time capsule? How was the time capsule saved—on the Web? On a CD, DVD, video, or audio tape? In a scrapbook of photos, with various memorabilia? Did anyone rescue old photos from antique stores and flea markets by searching for photographer's prints on the front or back of the photo or names on the back of the photo and dates or locations?

1. When turning your salable life story, corporate history, or biography into an adventure action romance novel, don't set up your main characters in the first chapter to be in transit traveling on board a plane, train, or ship going somewhere. The action actually starts or hits them after they have already arrived at their destination.

Start your first chapter when your characters already get to their destination place or point in time. A first chapter that opens when your main character is on a plane or train is the kiss of death from many editors point of view and the main reason why a good novel often is rejected. So cut out the traveling scene from your first chapter and begin where the action starts for real, at the destination point. Does anyone visit antique stores, malls, or flea markets to search for family history memorabilia? What about attic, basement, or garage sales?

2. Use a lot of dialogue when turning a biography or your life story into a salable novel, especially in a romance, adventure action, or suspense novel or in one where you combine romance with adventure and suspense.

   Use no more than three pages of narrative without dialogue. Let characters speak through the dialogue and tell the reader what is happening. Get characters to speak as normally as possible. If the times and place dictate they speak in proverbs, so be it. Proverbs make the best novels as you turn your proverb into a story and play it out as a novel. Otherwise, have normal speech so you can be the catalyst and bring people together who understand clearly what one another means.

3. Put your characters on the stage and have them talking to one another. If you have introspection in your book, don't use introspection for your action line. Action adventure books keep characters on stage talking to the audience.

4. Use magazines and clothing catalogues to make a collage of what your character might look like. This inspiration may go up on a board in front of you or on the wall to see as you work. Get a picture in your mind of what your characters look like. If they don't exist in art history, draw them yourself or make a mixed media collage of what they look like, speak like, and stand for. Some ideas include the models in "cigar" magazines, catalogues, and fashion publications as well as multi-ethnic and historical illustrations and photos.

5. Research history and keep a loose-leaf notebook with tabs on the history of places you want to research. The history itself is great for ideas on what plot to write. Look at or visit old forts and similar places. Plug in characters to your research. Look at forts of foreign settlements in the country of your choice, U.S. or any other place. Record the dates in your files. Create a spreadsheet in Excel or any other type of spread sheet with your dates from historical research as these will relate to your characters and help you develop a real plot.

6. Keep a notebook for each novel or biography you write. Put everything related to each book in a notebook. Have one notebook for historical research and one for the novel you're writing or true storybook.

7. When sending out your book manuscript make a media kit for yourself with your resume, photo, list of works in development if you are not yet published, and any other material about your own experience in any other field. Your own biography and photo presented to the press also can be used to let an editor know when you send out your manuscript of what's in development and what you've done.

8. Write down the point of view before your book is begun. Whose point of view is it anyway? Who tells the story? If you're writing a romance novel from your life story or a military romantic suspense novel, true story, radio script, or other genre, agree on the point of view before you start.

9. Who's telling the story and how does she or he know how the other characters know what to say?

10. It is not necessary to continue ethnic stereotypes in your book. If one of your characters is a music agent, for example, and a lot of music agents are of one ethnicity or speak with a certain accent, it's not necessary to continue the stereotyping roles. Pick something new for a change.

11. Otherwise it becomes cliché. Research numerous ways of telling the same story. The readers needs to learn facts or experiences, anecdotes, oral histories transcribed, and stories that have not been generalized. Use a series of incidents, action and relationship tension to balance your plot with your dialogue.

12. If you're turning a biography into a romance novel, you need to balance the relationship tension with the mystery, action, or other plot. You must have some event occur on both sides, on the sexual tension side and on the mystery or action side to balance out the book.

Record and Transcribe Life Stories as Highlights, Turning Points, Skits, or Significant Events:

Recording and transcribing personal histories in their original form and/or as skits is a team project of the person whose life is being turned into a time capsule and for the personal historian with a video camera, tape recorder, and notepad. For every action in a life story, there's an equal and opposite reaction that's primarily character-driven and secondarily plot-driven. And in an autobiography or anyone's life story, relationship tension occurs. Then the plot moves on.

If it's a romantic suspense or mystery within a life story, such as true confession, true crime, or biography, usually twenty-four short chapters makes a book-length story. Note how the memories are brought up by associations with various words, places, or questions.

A diary is written in first person as a journal or log, but a biography can be of you or your client. Even in a memoirs book or diary, you have to balance action with interaction between the heroine and the hero.

You can be the only person in your diary, but the action and interaction needs to be balanced with something out there in the external world—either forces of nature or another person—or the competition.

If you keep the competition out of your diary, put in the memories, actions, and warmth of the friends, including pets. If there are no other people, put in some force of spirit, some other push and pull, or tension, for balance with something outside yourself. This can be a job, school, a hobby, or what you choose as the force that pulls in an opposite direction existing with the force or person that pulls in your direction.

Try putting the relationship tension between the hero and heroine in the even-numbered chapters, and the mystery, historical events, or action plot events in the odd-numbered chapters.

In a 24-chapter-historical romance, this alternating action chapter, romantic tension chapter balances the plot smoothly. Most historical romance novels have 22–24 chapters. If you analyze the best-selling ones, you'll see that chapter one has an opening scene on the action side so you see what's happening.

The first action-oriented introductory chapter that shows us what's happening is followed by the second chapter on the romantic tension side showing us when and how the heroine meets the hero or has a re-union with the hero. In the second chapter, the writer takes the heroine somewhere in place or time. The heroine in the second chapter is defined. Either she's a 90's woman, or she's in her place in history or rebelling against it. You tell the story. If you're male, you'd use a hero.

Romantic life stories featuring genealogy combined with biography usually are 10–12 chapters long. Historical romances are twice that size at 22–24 chapters. The writer decides whether it's best to turn a biography into a historical romance or a life story into a mystery, suspense, action adventure, young adult novel, romance, or other genre.

If you are not fictionalizing genealogy or biography into a story, keep your time capsule book, database, or other media true to facts and historical records only. You might want to add your DNA testing records of relatives along with a family tree or other database or time capsule.

For those who want to turn factual biography into a novel, in turning a biography into a romance, the romantic tension side is about girl meeting hero in the first chapter. In the second chapter, the hero takes her somewhere in place, space, time, or state of mind. An oral history may be written as true life story in the form of a novel or play, skit, or anecdote of experience.

The oral history highlights a life experience within a time frame set in one or more locations with all the nuances of that place. It's basically a life story, but it can be transcribed with that certain something, including—charisma, liveliness, action, forward movement, drama, tension, and unique experiences, problems solved, and goals.

***

37

Diaries Plus DNA Equal Time Capsules

What do diaries and DNA tests have in common? Both are handled as evidence.

Evidence and diaries: Diaries, like deleted files hidden in the cache pits of computers, and DNA tests can be used as evidence. Diaries also hold the seeds of a story. You could write a novel or a screenplay from a diary. A diary also is a history. So preserve a diary as you would restore and preserve a valuable work of art from the past. Diaries are meant to be passed to future generations for a glimpse into a world that can be experienced by generations far into the future. Keep a file of dates listed in the diary and any objects that surrounded the diary from the same era.

A story with a central issue needed little explanation when my mom wrote in her diary in the style of a telegram: "October 25, 1926: First day of honeymoon. On train to Miami. Today I died." What central issues and themes tell a story in the diaries that cross your path?

Keep the dates and topics organized if you are working with restoring and preserving diaries. There should be a central issue or theme. How old was the person writing the diary? How many years did the individual keep the diary? What kind of objects were near the diary, packed together?

What kind of dust or other stains were on the diary—sawdust? Farm materials and plants? The first corsage from the senior prom? How about recipes, household hints, or how-to tips for hobbies? Was the diary or journal personal and inner-reflected, or geared toward outer events in the world? Were anecdotes about people and/or pets included, or was it about the feelings of the author of the diary?

Find out what other clues the mystery of the diary unfolds, from the lipstick or nail polish stain to the sawdust and coffee stains, or that faint smell of tobacco, industrial lint, or is it lavender, jasmine or farm dust and straw? Look inside the box in which the diary was packed. It's all evidence and clues waiting to be examined just like a mystery novel. A diary is a story, and everyone life deserves a novel, story, or biography and eventually, a place in a time capsule.

How to Restore a Diary

Make a book jacket for your diary to preserve and restore it. Use acid-free paper. Call a library or museum and ask for a brand or type of long-lasting acid-free paper and where you can buy some. Put a title and label on the dust jacket with the name of the diary's author and any dates, city, state, or country.

You can also speak to the art history department of most universities and find out what kind of paper is best to use for a book jacket to restore and preserve a diary. Treat it like a work of art. The same can be done for photo scrap books.

If torn, mend the diary. Your goal is to improve its condition. Apply a protective plastic wrapper to your valuable dust jacket. Give the diary a dust jacket in good condition. It should start to look more like a valuable book in good condition. If the diary is dingy and dirty, bleach it white on the edges. Put a plastic cover on the diary. The white pages of a diary without ink can be bleached with regular household bleach, but don't let the vapors of the bleach soak through to reach the ink because it will bleach out the writing.

Repair old diaries and turn them into heirlooms for families and valuable collectibles. The current price for repair of handwritten diaries and books is about $35 and up per book or bound diary, if you like to specialize in mending old dairies and family or personal books for a fee.

Some old diaries contain recipes and also served as personal, handwritten cookbooks containing recipes created by a particular family or family cook. These were valuable books preserved as if they were family scrapbooks, unlike the recipe databases in computers we have today. They are works of art, like an old tapestry embroidered with the story of a family's major turning points and events.

For more repair tips on bound diaries and books, I recommend *How to Wrap a Book*, Fannie Merit Farmer, Boston Cooking School.

How do you repair an old diary to make it more valuable to the heirs? You'll often find a bound diary that's torn in the seams. According to Barbara Gelink, of the Collector's Old Cookbooks Club, San Diego, to repair a book, you take a bottle of Book Saver Glue (or any other book-repairing or wood glue), and spread the glue along the binder.

Run the glue along the seam and edges. Use wax paper to keep the glue from getting where it shouldn't. Put a heavy glass bottle on the inside page to hold it down while the glue dries.

To remove tape, tags, or stains from a glossy cover, use lighter fluid or cleaning fluid (away from sparks, flames, or heat lamps). Dampen a cloth with nail polish remover if lighter fluid is too smelly and flammable for you.

Another way to remove something pasted on a plastic book cover is to use the finest grade sandpaper. Many books you'll find at goodwill will have adhesive price tags on the book. It's not usual to find diaries, even bound diaries in old book stores, but they show up in garage sales and in some antique stores and flea markets along with old photos.

To bleach the "discarded book stamp" that libraries often use, or any other rubber stamp mark, price, date, or seals on the pages of a book or on the edges, use regular bleach, like Clorox. It turns the rubber stamp mark white. The household bleach also turns the edges and pages of the book white as new.

To preserve a valuable dust jacket, a tattered jacket with tears along the edges needs extra firmness. A protective plastic wrapper can improve the condition of a book if it has a jacket cover.

To look for old diaries, or old family photos, look in garage sales, flea markets, and antique shops. Attend auctions and book fairs. Two recommended auction houses for rare cookbooks include Pacific Book Auction Galleries, 139 Townsend, #305, San Francisco, CA 94107, or Sotheby's, New York, 1334 York Ave., New York, NY 10021. Pacific Book Auction Galleries recently put a large cookbook collection up for an auction.

Hunt for diaries, old photos, and other old clues to family history in thrift shops and antique stores. Some diaries also combine cookbooks with personal histories and transcribed oral histories, but these are very rare. Genealogy also can have a person's collection of favorite recipes or anything else the person collected organized and archived along with family history and genetics records.

If you're into scrap booking with family history, photos, and recipes, for nostalgia, look for cookbooks printed by high school parent-teacher associations. Some old ones may be valuable, but even the one put out by the depression era San Diego High School Parent Teacher Association for the class of 1933–34 is only worth $10.

To find out-of-print and mail-order cookbooks, contact Charlotte F. Safir, 1349 Lexington Ave #9-B, New York, NY 10128-1513, (212) 534-7933. She specializes in hard-to-find cookbooks and children's books. Astor House Books, PO Box 1701 Williamsburg, VA 23187 (804) 220-0116, specializes in cookery and gastronomy. Amber Unicorn Books, specializes in rare cookbooks. They're at 2202 W. Charleston Blvd, #2, Las Vegas, NV 89102 (702) 384-5838.

Little Treasures (Joyce Klein) Cookery Books has British and American cookbooks and general stock. They're at 7517 W. Madison, Forest Park, IL 60130 (708) 488-1212. Send your wants because they have no catalog.

Cornucopia, run by Carol A. Greenberg, has cooking and food literature and domestic history, household management, herbs and kitchen gardens, hotels and restaurants, etiquette and manners, pastimes and amusements, and needlework old and rare books. They search for out-of-print books, and are interested in material from the 19th century through 1940. Write to: Little Treasures at PO Box 742, Woodbury, NY 11797, (516) 692-7024. Greenberg is always grateful for quotations on old, rare, and unusual materials in fine condition.

The Collector's Old Cookbooks Club has half their members in other states and half in San Diego County. They send a newsletter to each member after the monthly meeting.

You could specialize in being a diary restoration specialist and book finder for genealogy groups. Perhaps you want to deal in collectors' valuable diaries, largely first editions. Mostly diaries are only editions. Some people had them bound like a blank notebook, and wrote in them. So they look like first editions of books.

Are diaries worth as much as rare cookbooks? How much are the thousands of rare cookbooks worth today? A helpful guide is the Price Guide to Cookbooks & Recipe Leaflets, 1990, by Linda J. Dickinson, published by Collector Books, at PO Box 3009, Paducah, KY 42002-3009.

See Bibliography of American Cookery Books, 1742-1860. It's based on Waldo Lincoln's American Cookery Books 1742-1860, by Eleanor Lowenstein. Over 800 books and pamphlets are listed. Order from Oak Knoll Books & Press, 414 Delaware St., Newcastle, Delaware 19720 (302) 328-7232.

Louis & Clark Booksellers specialize in rare and out-of-print cookery, gastronomy, wine and beverages, baking, restaurants, domestic history, etiquette, and travel books. They're at 2402 Van Hise Avenue, Madison, WI 53705 (608) 231-6850.

Cookbooks and diaries are not that much distant from each other, although diaries of a famous person would have monetary value as would old cookbooks that are rare. Think of the events in the life story and history value of what real people's lives were like many decades or centuries ago. Any restored diaries would be valuable to descendants of anyone, and you can't put a price tag on these people's lives as expressed in diaries.

Get acid-free storage envelopes or boxes for diaries. Handle old diaries and books with gloves, and get rid of mildew safely without destroying the pages or fading the ink. It's here that a library can be helpful. Ask questions about storage from historical societies and libraries. Make copies of the diaries. Work with the photocopies when you decipher the writing.

Store old diaries in a dry, cool place where there are no bugs. Lining the storage place with plastic that's sealed will keep out vermin and bugs. Without moisture, you can keep out the mildew and mold. Store duplicates away from originals.

Call archivists and historians in your area and ask for their advice. Was something placed on a certain page, such as a dried rose, letter, or a special book mark? What meaning did it have? Date the diary. List the date it was begun and when it was ended, or look for clues for a time frame. List the geographic location of the events in the diary or the writer.

Of what kind of materials is the diary made? Is it improvised, created at low cost by the author? Or is it fancy and belonging to someone of wealth? What is the layout like? Does it show the education of the writer or anything personal? Was it a masculine almanac or calendar or a feminine expression of memorabilia and sentiment? Or was it written by a man writing poetry or letters and being romantic?

Can you tell the personality traits of the writer of that diary? What was the writing tool, a quill or a pencil? What's the handwriting like? Would a handwriting analyst know what to say about it? Does the handwriting and words express anger, joy, sadness, or what? Is it full of detail, or is the reader given the big picture? Why was the diary written? What is its central message? Do you see patterns, concepts or facts?

Transcribe the diary with your computer. What historical events influenced the writing of the diary? What language is it in or dialect? Study the historical meaning of the diary so you can get to know the writer across the chasms of decades or centuries.

Are there vital records such as deeds to real estate mentioned in the diary? What photographs are in the diary? Any artifacts mentioned or pressed between the pages? See whether you can relate to the diary author and find some type of context of the story from entry to entry. Are there any genealogical records or mention of family names, such as a great, great, great grandmother's maiden name? What about recorded events of inherited diseases or medical histories?

If the grandfather's dad went blind with glaucoma, it's has a genetic element that heirs should know about whether anyone inherited it or not. See what the diary unfolds that can be read as family history, world history, or used in the phenomics mode, to customize treatment or therapy based on genetics.

Treat the diary as a precious work of art, including the photos in there, if any. Don't touch the old right side up photos with bare fingers because the emulsion would quickly come off. Some genealogists specialize in working with old diaries, and novelists or screenwriters would probably be interested in a unique story. Restore a diary, and if a building such as a restaurant wants to hang old photos in the dining hall, let the photos be copies rather than the originals.

If you want to record the memories of relatives or friends, list 100–200 inspiring questions. Give the questions to relatives and friends. Create the questions with the goal of triggering recall of memories and experiences, events, and highlights of their life stories. Give them a diary and make a dust cover that's fancy. Put a plastic cover over the dust jacket.

Use a hardbound book with a question on each page. Or give several pages per question if the person might like to write for personal expression. You could ask the person if he or she wanted more or less pages. About 150 questions or pages are fine to last a lifetime of a minimum of one-page summarized life events or answers to the questions you ask.

The questions should be important to both you and the person who's going to write in the diary or journal. This is one way to create a genealogy diary gift that develops into a biography from a personal journal or diary.

Make sure there's enough pages and room to express what's required. Print the questions large and clear so you can elicit recorded responses to the questions. Leave room to attach photographs on acid-free, archival-quality paper.

If the person really won't write, create a tape recorded or video diary where the person can record his voice on tape, in a computer, or with a video camera on a tripod poised next to a desk to record the person talking without requiring anyone else in the room when recording.

Give the person a remote control device for the video camera to click on and off without walking toward the camera or tape recorder, or computer microphone. If the person will write at all, stick with the personal ambiance of a diary gift that someone can hold in any location or take traveling.

What questions to write? List thought-starters. Write questions about their childhood, values, dreams, and goals. Create a section for recording time as a mother or father, grandparent, how a child's name was chosen, how the marriages were arranged, or what things each child did to make a parent proud.

Put in blank pages and a place for each relative to put in the results of their DNA testing, fingerprints, print of hand or palm, or any other personal information, even medical notes or other memories in the memorabilia. Leave room for children's photos, small drawings, and meaningful relics. Use the kind of binding where someone can add pages.

File and archive the DNA test results, racial percentages tests, along with any other information from memory. Add an oral history, perhaps a pocket for a tape, or room for a transcript of an oral history to be added, such as a loose-leaf binding or similar binding so pages may be added to folders or plastic envelopes holding mementos.

Have a space for the first job and for a father's memories.

The genealogy thought-starter questions can come bound in a fancy, hard-bound loose-leaf that will last as pages are added. Or better yet to give each his or her own individual expression without the judgment of another relative, give a separate book to each family member so he or she can keep memories or traditions private until passed to heirs. You might want to create a separate book for each grandparent, parent, and other relative.

***

38

Mapping Your Personal Anthropology with Genetic Genealogy

Source: Facts & Genes" <http://www.familytreeDNA.com/facts_genes.asp>.
Reprinted with Permission from:
Facts & Genes, from Family Tree DNA
Copyright 2002, Family Tree DNA),
November 21, 2002 Volume 1, Issue 5
=====================================

Family Tree DNA enjoys hearing about the results of your DNA testing. One customer wrote: "Just a word of thanks to you for the work you are doing on this project. The information I received today has already helped direct my genealogical work into a more focused and well-researched area, and has saved innumerable hours of work! Thanks for making this testing available, and for providing it in a financially accessible form. It's appreciated!"

Send comments, suggestions, tips, questions, and tell Family Tree DNA about your Surname Project to: editor@familytreedna.com.

Family Tree DNA is pleased to announce that the ANCESTRYbyDNA test is now available. The ANCESTRYbyDNA test was developed by DNAPrint Genomics, Inc., and is available through Family Tree DNA.

The ANCESTRYbyDNA test will measure a person's Personal Anthropology and their corresponding ancestral ethnic proportions. The result of the test is a report showing your percentages of each ethnic ancestry or major human population group. For example, your result could be 18% Native American, 70% European, and 12% African.

Perhaps you have wondered whether you have any Native American ancestry, or maybe you are just curious to find out more about yourself. The ANCESTRYbyDNA test will unlock the secrets to your ancestors contained in your DNA.

The ANCESTRYbyDNA test analyzes your DNA to determine which of the major human populations your ancestors belonged to, and what percentage you have inherited of these groups. These four geographical areas and the corresponding major human population groups are: Native American, East Asian, European, and sub-Saharan African.

This test, developed by DNAPrint Genomics, utilizes SNP's that are diagnostic of a person's continent of origin. SNP's are deep ancestral locations along the human genome, and have a different result when tested with different peoples.

To order the ANCESTRYbyDNA test, click on this link: http://www.familytreedna.com/products.html#dnaprintorder

Surnames

Are you wondering why the XYZ surname project has over 50 participants, and you only have 6 participants so far? Do you look at your Web site and correspondence, and wonder what is their secret to recruiting participants? The answer may be that they have a larger population of their surname from which to recruit participants. Your Surname Project may actually have a higher percentage of the surname participating than the project with over 50 participants.

It is common knowledge that Smith is the most frequent surname in the US. The chart below shows the 10 most frequent surnames in the US in the 1990 census. For each surname, the percentage represents the percentage of persons in the US with this surname, and the Rank is the ranking of the surname with 1 being the most frequent. For example, in the chart below, eight surnames are more frequent than Moore.

Surname	%	Rank
SMITH	1.006	1
JOHNSON	0.810	2
WILLIAMS	0.699	3
JONES	0.621	4
BROWN	0.621	5

DAVIS	0.480	6
MILLER	0.424	7
WILSON	0.339	8
MOORE	0.312	9
TAYLOR	0.311	10

Assume that a person started a Smith Surname project. There are over 2 million Smith's in the US, of which over 1 million would be males. This is quite a few people. If they signed up 50 people, they have only signed up a very small percentage of the Smith surname.

Compare this to the surname Mumma, which is .001 % of the population, and its Rank is 15,109. There is a much smaller pool of Mumma potential participants. If you look at the surname, Norin, its number is so small in the US 1990 census, that it does not even get a result when the 1990 US census Surname Frequency is searched.

You can find out what percentage of the US population holds your surname by going to the US Government census site at: http://www.census.gov/genealogy/www/freqnames.html

The site also covers the methodology that the Census Bureau used to come up with the percentages and rank for the surnames.

The US population on April 1, 2000 was 281,421,906 people. If you would like a rough idea of the males with your surname in the US, first search the site <http://www.census.gov/genealogy/www/freqnames.html> to get the percentage for your surname. Multiply that percentage times the population of the 2000 census. In their rough calculation, they will assume that 50% are males, so now divide by 2. This is an estimate of the number of males with your surname. To estimate the number of adult males, multiply by .7. The formula is:

Percentage * 281,421,906/ 2 * .7 = adult males with surname

You can also find out how common your surname is in the UK at the site: http://www.taliesin-arlein.net/names/search.php. There are 269,353 surnames in the UK database, representing 54,412,638 people. This database is provided by the Office of National Statistics of the UK, and gives an actual count of the number of persons for each surname. Their database is an extract of an Office of National Statistics database, and provides a list of surnames in use in England, Wales and the Isle of Mann in September 2002.

The US Census population database and the Office of National Statistics of the UK database used different methodologies to come up with their results. Rare surnames will not get a search result in the US census site, whereas they will in the UK site, even if there are only a few persons with the same surname. Names shared by fewer than five people have been excluded from the UK list.

Now that you have an idea of the size of your potential prospect pool, lets assume that only 1/3 are interested in genealogy, so you now divide by 3. The end result is a very rough approximation of the number of potential participants available. If you are only using the Internet to find your participants, cut this number in half for the US. Other countries have a smaller percentage of persons on the Internet than the US.

As your first step, you have probably posted your project to as many sites and mailing lists that are applicable and allow such postings. You have probably also put up a web site, even if it is only one page. Most likely you have contacted all those persons whom you had contact with in the past regarding genealogy.

Here are some suggestions to consider for making more people aware of your project:

1. Consult the Directory of Family Associations. If there is a Family Association for your surname, contact them and offer to write an article for their publication about your project.

2. Register your web site with familysearch.org. Everyone searching on your surname at Familysearch.org will find your web site. You must first register yourself with familysearch.org to be able to submit your website for consideration.

3. Visit your local Family History Center, and offer to show the Genealogy by Genetics video to the staff and patrons. This might not find you any participants, but if every Group Administrator takes an hour to do this, then all the Surname Projects might find participants.

4. Review your web site. It needs to be easy to understand for those not familiar with DNA testing, and clearly present the benefits to the participant. What will they gain from participating? How will it help them in their research? What might the results tell them?

5. Find out if there are any genealogy clubs or organizations in your area, and volunteer to show the video, and answer questions.

DNA testing for genealogy is a new field, and we are all pioneers. Most likely you have learned a lot about the field as a result of your testing. Those of us who have learned about DNA testing and how to interpret the results are aware of the benefits and how the testing can assist us with our genealogy research. Convergence explains why a haplotype will match others with a different surname. DNA testing for genealogy is not a substitute for genealogy research, but is instead a companion.

The majority of those interested in Family History research may not yet be fully aware of Genetic Genealogy. If you volunteer an hour to help your fellow genealogists understand this new tool, and help more people become knowledgeable, all of us will benefit as we seek participants for our testing. Look for social histories of the ethnic group you're researching.

When you're working with DNA, you can look for historical medical records. Only a few are open to the public. You might try the microfilmed collections at The Family History Library in Salt Lake City, UT, or rent one of the microfilms from any of the worldwide family history centers, usually found in a genealogical library in various cities around the world.

Look at records from the Eugenics Record Office (ERO) that operated from 1910 to 1944. The purpose of that office was to study human genetics in order to reduce inheritable genetic disorders. You can look over the 520 rolls of microfilm. Visit the Family History Library Catalog and look under United States—Medical Records—Eugenics.

Look up the state you want, and look under Medical Records. You might want to read up on the controversial Eugenics movement. Think about how DNA testing today differs in that the test results are used today either to find relatives, ancestors, or tailor individual therapies for individual genetic makeups—phenomics. How times have changed. Or have they? What do you see in your own DNA and family history research?

The Family History Library in Salt Lake City also has some historical hospital records. One example is the Northwestern Memorial Hospital record, Chicago in the Family History Library, dated 1896–1933. Perhaps one of your relatives is in those files. That's one other way of finding a maiden name from the days when many people were never given a birth certificate because they were born at home and never registered. That's what happened with my mom, born in 1904 at home in NY state.

Understanding your Results: Ethnic Origin

Whether you are just starting a Surname Project, or ordered a test for yourself to learn about DNA testing for genealogy, everyone experiences the situation of receiving the first test result, and what now? You have one test result, and what do you do with a string of 12 or 25 numbers? Can they tell you anything? Where can you find the information you need?

In the situation of the one or first test result, most likely you will not find others to whom you are related. The odds of a random match to some one to whom you are related when you are the first of your surname to test is slim to none. Instead, you might find some clues to your ethnic origin.

To find clues about your ethnic origin, Log into FamilyTreeDna.com, and at your Personal Page click on Recent Ethnic Origins to search this data base. The results show others whom you match, or who are a near match, and their ancestor's ethnic origin.

The information on their ethnic origin is provided by each person tested (testee). The information provided for ethnic origin is only as accurate as the knowledge held by the testee regarding their ancestors. Testees are instructed to answer unknown for ethnic origin when their ancestor's origin is not known, or not certain. Sometimes the origin the testees provided is incorrect.

Incorrect origins provided by testees may lead to search results that do not seem logical. For example: Assume your ancestors are from England, but your search results show the ethnic origin of your matches as England, France, AND one match shows an origin of Native American. Does that mean that your ancestor's relatives may have lived in England and France? Yes. Does it mean that your ancestor was also a Native American? No. It means that a settler in America had a child with a Native American woman, the child was brought up as a Native American, and that, over time, the family has "forgotten" the European ancestor, and believes their ancestry to be Native American.

During the span of generations people tend to move, as do borders, so nationality or ethnicity becomes subjective. For example, testees may enter Germany for ethnic origin, because the land of their ancestors is in Germany today, but the land had been held by Denmark for many centuries.

Your search should return at least one match, namely yourself. If your results show 3 matches from Ireland and 1 from Scotland, and you have reported to us that your ancestors came from Scotland, then you are the Scotland result. The other 3 matches are either from the Family Tree DNA database or from the databases Family Tree DNA have been supplied by the University of Arizona.

To see how your ethnic origin is recorded in our database, click on the link titled Update Contact Information. You can also update your paternal and maternal ethnic origin on this Update Contact Information page.

Exact matches show people who are the closest to you genetically. The Ethnic origin shows where they have been reported to have lived. Since many persons migrated since the beginning of time, you will typically see matches in more than one country.

For information purposes, the Recent Ethnic Origin search also displays results for those who are not exact matches, but are 'near matches'. A near match is either one step or two steps from your result. An exact match is 12/12 or 25/25. A one step match is 11/12 or 24/25. A two step match is 10/12 or 23/25. The value of the near matches is to see where those who may be related migrated over time.

Other databases available that you can search are:

European: http://ystr.charite.de/index_gr.html

US: http://www.ystr.org/usa/

In some cases you will not find any results. This is because only a very small percentage of the world population has been tested and is in the databases. The Ystr databases, plus the FamilyTreeDNA Recent Ethnic Origin database together hold about 21,000 test results. Every day more results are added, and it is only a matter of time before you will have some matches. Your test with Family Tree DNA includes access to our databases for matching.

If you do not find any results in the two YSTR databases shown above, try entering your result, and then eliminating a marker, and do this until you have a smaller set of markers that results in some matches. This might provide some clues regarding where your markers have occurred geographically.

The value of DNA testing comes from comparing your results to others. If you have started a Surname Project, you will most likely have results from others soon. If you only tested yourself, you may want to consider either using DNA testing to solve one of your Family History questions, or starting a Surname Project.

Haplotypes: Convergence

A Haplotype is the 12 Marker result from testing the Y chromosome. Some Haplotypes are common, with a high frequency of occurrence and some Haplotypes are rare, with a low frequency of occurrence.

Many people have common Haplotypes, which means that they would expect to find matches to those who do not have their surname. This occurs because we were all at one point related. As the different branches of the Adam +Eve tree evolved throughout time, mutations occurred, forming different Haplotypes. Thousands of years later, you have many different Haplotypes. Due to these mutations, you could have two branches that mutate to an identical Haplotype. This is called convergence.

If your Haplotype matches an individual with a different surname, and your genealogy research shows no evidence of an extra-marital event or adoption, your match may be the result of Convergence.

The example below shows convergence between the ABC surname and the XYZ surname, using just 3 markers to keep the example simple. Notice how the mutations over time bring two different Family Lines to the point that they match.

Time	ABC	XYZ
1000 A.D.	12 24 15	14 25 13
1200	13 24 15	14 25 13
1400	13 24 15	14 25 14
1600	13 24 15	14 24 14
1800	13 24 15	13 24 14
2000	13 24 14	13 24 14

Convergence explains why a haplotype will match others with a different surname. DNA testing for genealogy is not a substitute for genealogy research, but is instead a companion. Results that match must be considered in light of the genealogy research. If you match someone with a different surname, most likely there wasn't an adoption or extra marital event, and your match may be the result of convergence.

Case Studies in Genetic Genealogy

In each issue of the Newsletter, Family Tree DNA looks at what Genetic Genealogy will do for your Family History research. This article is a continuation of the topic, with situations, called *"Case Studies"*, followed by a recommendation. The objective of the case studies is to present different situations you may encounter in your family history research, and how DNA testing can be applied.

Case Study

From November 21, 2002 Volume 1, Issue 5, *Family Tree DNA Newsletter*, "I have participated in a Surname Project, and had quite surprising results. All the other Lines of my surname are related, except my Line. We have all traced our ancestors to England. Not only is my line not related, but also my ethnic origin is Eastern European. What do I do now?"

Recommendation

From November 21, 2002 Volume 1, Issue 5, *Family Tree DNA Newsletter*, " I am sure you were quite surprised, and perhaps disappointed. The first step is to validate the result for your Line or family tree. Since only one person was tested for your Line, we recommend testing additional males from each branch on your tree, to see if they all match each other. If they end up matching, your result is probably due to an extra marital event, an adoption, or a name or spelling change.

"In reviewing the surnames of Eastern Europe, your surname is pronounced as the surname in England, only the spelling is different. A review of your Family History shows that the research and documentation for the time period 1800–1850 is quite sparse. Many more records are available in England for this time period, including parish registers and wills. I would suggest that more family history research might shed some light on the situation."

***

39

Managing a Genetic Genealogy Project: Participants with Poor Documentation

Occasionally you might run across a willing participant for your Surname Project who has a poorly documented family tree, perhaps even built entirely out of the International Genealogical Index (IGI) by matching surnames. Your dilemma is that the prospective participant appears to be from a Line you haven't tested yet, but without better research you can't be sure. What comes first, the testing or the research?

This is a complex issue. If you turn away the participant and suggest that they do more research, they may become discouraged, and never return. If the participant tests, and gets unexpected results, they may become an unhappy participant.

One solution is to fill in the gaps of their research. You may not have the time to take this step. A better solution may be to communicate the situation to the participant, and let them make the decision to test now with the possibility of unexpected results, and also encourage them to do further research.

Perhaps from your research experience, you may be able to suggest to the participant specific sources for them to investigate. Most likely, they want to do more research, and just need some guidance and direction.

It will be a win-win for both the Surname Project and the participant if you are able to achieve both additional research on their part, and their participation.

Spot Light: Witt—Whitt Surname Project

Objective: Prove or disprove the genealogy research of the Witt/ Whitt Line from Old Virginia

There are three identified Lines or families of the Witt/Whitt surname in the US. One family Line that today spells their surname as both Witt and Whitt begins with German immigrants in both South Carolina and Pennsylvania in the early 1700s. A second family Line that today spells their surname as Witt and DeWitt, began in New England around 1640 with an English immigrant by the name of John Witt.

The third family Line that today spells their surname as Witt and Whitt began with an individual named John Witt or Whitt, who first appears in early records in 1670 in colonial Virginia. The records relating to John show the spelling of his name both as Witt and as Whitt. It was from John Witt-Whitt that the Witt-Whitt Family of Old Virginia began.

Early Virginia records, John Witt-Whitt was the father of at least four sons, John Witt II, William Witt, Edward Whitt, and Richard Whitt. The participants in the Witt-Whitt Surname Project are all documented descendants of the 4 identified sons of John Witt-Whitt: John Witt II, William Witt, Edward Whitt, and Richard Whitt. For each of these sons, at least two documented male descendants participated in the Project. All participants took the 25 marker test.

The results for this Surname Project are that the majority of participants matched 25/25, and a few matched 24/25. Therefore, the Project has confirmed the genealogy research, and shows that the participants are related and have a common ancestor.

When combined with surviving colonial Virginia records for the surnames Witt and Whitt, the Witt-Whitt DNA study determined John Witt II, William Witt, Edward Whitt, and Richard Whitt were brothers and their father was the immigrant John Witt-Whitt of Charles City County, Virginia. The Witt-Whitt DNA Surname Project also identified the common ancestor of these four men was from England, or possibly Scotland.

A DNA baseline for the Witt-Whitt family of Old Virginia has now been established. Other descendants who have incomplete records, or where records no longer exist and preclude the determination of a family's origin, may take the 25 marker DNA test to determine if they are related to the Witt-Whitt Line from old Virginia.

If other descendants find that they match, they can contact one of the participants in the baseline study to share the Witt-Whitt family of Old Virginia ancestral history for their family line. The next phase of the Witt-Whitt surname project is to identify the county in England or Scotland from which John Witt-Whitt originated.

If you're interested in receiving Facts & Genes newsletter, feel free to contact the editor at Family Tree DNA with your comments, feedback, questions to be addressed, as well as suggestions for future articles. If you would like your Surname Project featured in their *Spotlight* column in a future issue, please send an email telling them about your project. If you are a Project Manager and can help others with tips or suggestions, please contact the editor: editor@familytreedna.com

Reprint Policy:

Family Tree DNA encourages the circulation of "Facts & Genes" by newsletters and lists providing that you credit the author, include their copyright information (Copyright 2002, Family Tree DNA), and cite "Facts & Genes" (http://www.familytreeDNA.com/facts_genes.asp) as the source.

November 21, 2002 Volume 1, Issue 5

Reprinted with permission, from November 21, 2002 Volume 1, Issue 5, Family Tree DNA Newsletter.

***

40

Haplogroups and Markers

What's a Recessive Gene?

Sometimes a recessive gene is referred to as a form of a gene called a recessive allele. The recessive allele will not express itself if combined with a dominant allele. The recessive allele is expressed by a lower-case letter. Some traits may be caused by having two recessive alleles.

Markers

How many genetic markers can tell us something for various pairings between groups of people? Markers often are great for telling multiple groups apart. For example, one marker in particular can tell Africans apart from all the other groups, but that marker can't tell Europeans from East Asians. Scientists also will look at male Y chromosomes to study various markers.

Markers are different for various ethnic groups. However, there also is some overlap as peoples become mixed. A geneticist can tell the percentages of various races by looking at the markers, even though we have come to accept there really is no such thing as a particular race because of the diversity between peoples in any one race. However, you can still look at genetic markers to see various ethnic group traits for people who have been separated for thousands of years.

African-European	54 markers
African-East Asian	50 markers
African-Native American	50 markers
European-East Asian	45 markers
European-Native American	41 markers
East Asian-Native American	24 markers

What is a Haplogroup? How is it different from a Haplotype?

Your matrilineal or female ancestors inherit the same mtDNA sequences which form a haplogroup. Look at female lineages starting with your mtDNA haplogroup today. It will be the same haplogroup letter as your common ancestor with the same haplogroup letter who lived 21,000 years ago. You are looking at a connection from a single female ancestor to all your direct female line ancestors today.

The sequences within the haplogroup may be slightly different because of the slow mutation rate of the mtDNA, but the haplogroup will be the same. And in some cases, the sequences will be similar to your ancient ancestors. While in other cases, the mutation rate may have changed your mtDNA just a little over that long span of time.

What's a Haplotype?

Let's look at the female lineages, the mitochondrial DNA clues called mtDNA for short.

Individual mitochondrial DNA called for short, mtDNA sequences, is grouped into haplotypes. A haplotype defines a series of special mutations. The mutations when lumped together are called haplogroups. Each haplogroup contains a set of haplotypes descended from the same one common ancestor. How many haplogroups of mtDNA are there? According to Bryan Syke's book, *The Seven Daughters of Eve*, at least 35 mtDNA haplogroups represented by a letter of the alphabet are listed in one of the illustrated tables. Matrilineal (female ancestral) lineages contain the mitochondria.

You can look at ancient ancestry by tracing the mtDNA lines. Some mtDNA letters belong mostly to Africa, while others belong to East Asia. Some are specific to South West Asia (India) and others are found in central and west Eurasia, which includes Europe and the Middle East. Five different mtDNA haplogroups are found in the New World—the Americas, such as ABCD and X, but the differences between the European and Middle Eastern X and the X among some Native American peoples show that they have been separated for thousands of years.

For example, in the book *Mapping Human History*, by Steve Olson, a rare and unusual haplogroup, X showed up among the Algonquian-speaking Native Americans living around the Great Lakes. It also is present in small amounts in the Lakota and Sioux. Previously mtDNA haplogroup X had been found in Finland, and in Italian, Greek, and Druze (Israel and Lebanon) peoples. Haplogroup X so far has not been found in East Asians. How did it get to the New World?

The Native American X mtDNA differed very much from the European haplogroup X to be separated by only one or two thousand years. It had to have come to the New World tens of thousands of years ago. Scientist Douglas Wallace of the Center for Molecular Medicine at Emory University in Atlanta is one of the world's leading experts on mitochondrial genetics. So when he studied two skeletons that lived in the 1300s in Illinois among the Native Americans of that time, he found the skeletons contained traces of haplogroup X.

How did he know it wasn't from mixture with a European? It was the time divergence between the European and the Native American X haplogroup that gave the answer. The haplogroup X in North America had been there for more than 10,000 years. It wasn't a "modern" European who lived in Illinois in the 14th century.

Again, you might ask, perhaps it was a Viking from Finland since X is found in Finland? The tests showed this type of X differs from the European X by mutations that reveal the X that lived in America really had been there more than 10,000 years. So it could have come more than 10,000 years ago from anywhere—central Asia, Siberia. No X haplogroups are in Siberia today as far as one can tell.

Then again, not everyone has been tested there. However, you have to draw the line somewhere, and the differences between the old world and new world X haplogroup were great as if they had been separated more than 10,000 years. It's easy to imagine someone in north central Asia could have joined up with a group of people such as hunters and traveled with them over the Bering Strait while it was still a land bridge more than 12,000 years ago.

What's a Haplogroup?

A group of related haplotypes make up a haplogroup. Haplogroups are studied especially when referring to mitochondrial DNA and Y-chromosomes. If a set of haplotypes are placed into a tree determined by the minimum number of mutations that separate them, the main branches of that tree are haplogroups. Each haplogroup in theory contains haplotypes that are all descended from a single founding individual.

Haplotypes from other regions of the genome are not studied as much because they may not always group together. Recombination makes ancestor-descendant relationships not as specific to see. You have to look for connections when you study haplotypes.

Examples: The vast majority of Native Americans belong to one of four mtDNA haplogroups: A, B, C, and D, but a few Native Americans also belong to haplogroup X. Haplogroup X is found at a low percentage in Europe, but the differences between the European haplogroup X and the Native American haplogroup X show that they separated more than 10,000 years ago.

It's more likely that someone with haplogroup X mtDNA from Southern Siberia, the Caucasus, or central Asia joined a group of hunters headed north and east more than 12,000 years ago when there was a land bridge over the Bering Strait, and settled in what is now called the North American continent.

* What's an allele?

For lots of definitions of these terms, also see the Web site: library.thinkquest.org/18258/noframes/def-allele.htm. Or See: *www.apnet.com/inscight/08271998/allele1.htm*

An allele is a form of a gene. Alleles are located at the same position (locus) on homologous chromosomes and are separated from each other during meiosis. An allele is what is actually within a region of the chromosome, and is found within a gene. *An allele is any of two or more alternative forms of a gene occupying the same chromosomal locus; such as that which determines flower petal color in peas.*

* What's a Haplogroup?

Definition: A bunch of haplotypes make up a haplogroup. The term is used usually when referring to female lineages and mitochondrial DNA or mtDNA. You might call a form of a gene an allele. An allele is an alternative form of a genetic locus. A single allele for each locus is inherited from each parent (e.g., at a locus for eye color the allele might result in blue or brown eyes).

So when a group of alleles on a single chromosome are linked together and usually inherited as a unit, these genes make up a haplogroup. Haplotypes are particularly stable in mitochondrial DNA and on the Y-chromosome, because they are not subject to recombination.

Analyses of mtDNA and Y-chromosome variation usually focus on the haplotype or haplogroup level, rather than comparing exact base pair sequences. In this case, haplotypes are defined on the basis of particular mutations shared by various individual DNA lineages.

Examples: One study of Finnish Y-chromosome variation found that 40% belonged to one of two different haplotypes, which in turn each belonged to different haplogroups and were probably introduced by different founding populations.

* **Genome.** A person's genome is one set of his (or her) genes. The human genes, which control a cell's structure, operation, and division, are located in the cell's nucleus. The full human genome (estimated at 50,000 to 100,000 genes) is present in every cell-nucleus. Many genes are inactive in cells that have some specialized functions. Many cells are differentiated to perform certain functions only.

* **Genes and Chromosomes.** Genes are composed of segments of DNA. In normal cell-nuclei, the DNA is distributed among 46 chromosomes (23 inherited at conception from a person's dad and 23 from mom). Each chromosome consists of one very long strand of DNA and numerous proteins.

The proteins are needed to manage the long DNA molecule. The longest chromosomes each support thousands of genes. Every time a cell divides, the cell must duplicate the 46 chromosomes. Every cell must distribute one copy of each chromosome to the two new cells. When cells stop dividing, that's the end of them and the organism.

* **The DNA Code.** The DNA of each chromosome is composed of units—"nucleotides" of four different types (A, T, G, C). These nucleotides are linked to each other in linear fashion. The necessary sequence of the four types of nucleotides produces the "code" which first determines the function of each particular gene. Then the sequence identifies the gene's start-point and stop-point along the DNA strand. Finally, the sequence allows specified regulatory functions. The code of the human genome consists of more than a billion nucleotides.

The Mitochondrial DNA (mtDNA). Mitochondria are needed for energy in the cell. The mitochondria are inherited from the mother. When tracing ancient and modern ancestry, geneticists look at female lineages or mtDNA. Your mtDNA is passed from mother to daughter over tens of thousands of years with few changes.

MtDNA mutates slowly during thousands of years of migrations of people across the globe. Men inherit their mtDNA from their mothers, but pass on their Y chromosomes to their sons. Women pass their mtDNA to their daughters. On very rare occasions, a few women may inherit some mtDNA from their fathers, but almost all women inherit their mtDNA from their mothers.

What mtDNA Does Not Do: It is not junk DNA. Often DNA produces more copies than it needs to function. Sometimes this is called junk DNA. MtDNA is necessary for providing energy to the cell. Outside the nucleus, human cells also have some "foreign" DNA located in structures called the mitochondria. This small and separate set of DNA does not participate in the 46 human chromosomes.

The mitochondrial DNA (mtDNA) really is not part of "the genomic DNA." According to the "out of Africa" theory that's widely held in acceptance by most scientists, all the mtDNA in the world today came from a single woman called Mitochondrial Eve who had two daughters who survived to create a line of females that expanded all over the world. Similar histories are noted for the male line using the Y chromosome. According to the book, *Mapping Human History*, by Steve Olson, (page 56) "All non-Africans descend from Africans who left the continent within the past 100,000 years."

According to a number of the latest videos on whether people took the northern or southern route out of Africa, the southern route is most favored. According to the book, *Archaeogenetics*, the flow of people varied between then and now. Today, most scientists theorize that since the north route out of Africa most likely was blocked by an Ice Age that created a dry desert in the Middle East, those leaving Africa headed toward Yemen and then along a southern route to India, Malaysia, and finally Australia.

Only when the climate changed and the Fertile Crescent of the Middle East opened up, did people expand back from India toward where the rivers met, the Middle East, such as what today are Iraq and Iran, and the Levant, reaching the coast, and from there north to what today is Europe. About 21,000 years ago, a new Ice Age began, and people who moved up from the Middle East into Europe found refuge in only a few places such as Southwest France, Northern Spain facing the Mediterranean and the Pyrenees, the Balkans and Ukraine, until the last Ice Age ended about 12,500 years ago. Then populations expanded across Europe from Spain to the Urals.

By that time, the Far East had been populated for a long time, and Central Asia was the newest land to be seen. Then by 9,000 years ago, farmers from the Levant and Anatolia moved into Europe and introduced the idea of farming so that about 80 percent of Europeans today consist of the old Paleolithic hunters and about 20 to 26 percent from the more recent arrivals from the Middle East, the cereal belt grain farmers of the Neolithic era that started about 10,000 years ago in the Levant and Fertile Crescent of the Middle East.

Genealogy, history, folklore, oral history, memoirs writing, diary journaling, demography, anthropology, and archaeology are in the midst of a molecular revolution. Has archaeology become archaeogenetics? Actually, molecular genetics biotechnology is one more ***tool*** in the hands of the genealogist, historian, archaeologist, folklorist, prosopographer, onomasticist, demographer, videographer, anthropologist, or family historian. And that tool, molecular genetics, is used to untangle distantly ancestral as well as recent family roots.

Now you have computer technology and Web databases to research family ties. You have molecular genetics biotechnology—DNA testing, bioinformatics, and beyond.

From ancestry by DNA to racial percentages by markers and phenomics, experts can customize medicine or therapy to an individual's genes.

You can take a paternity test. Or find out whether you're related to a distant cousin you never met. Or you can study DNA as legal evidence. Your genes are used for matching bone marrow donors to recipients. The molecular revolution is enhancing research. From community colleges where students earn one or two-year certificates in biotechnology to perform DNA testing and bioinformatics processing on computers to the PhDs who work in research labs and universities, the molecular revolution has now joined science to history. How much do you want to know about your genome?

***

41

Cancer Genetics Network (CGN)

Reprinted with permission from the Cancer Genetics Network (CGN).

Have a Personal or Family History of Cancer? Consider Joining the Cancer Genetics Network

The Cancer Genetics Network (CGN) seeks individuals with a personal or family history of cancer who may be interested in participating in studies about inherited susceptibility to cancer. Nearly 8,500 individuals have enrolled in this unique program.

The Network is becoming an important vehicle to conduct studies that will provide much-needed clinical information to help individuals who may be at increased risk for cancer because of a personal or family history of the disease.

Eight U.S. centers, funded by the National Cancer Institute (NCI), joined forces 3 years ago to establish a national resource to support investigations into the genetic basis of cancer susceptibility. Together, the centers are working to make possible research that a single institution may not be able to accomplish because of insufficient numbers of participants, or the time needed to recruit them.

"The idea is to have a pool of interested individuals readily available so that important research questions can be answered, and studies can progress without unnecessary delay," said Deborah Winn, PhD, acting associate director of NCI's Epidemiology and Genetics Research Program (EGRP), Division of Cancer Control and Population Sciences (DCCPS). Participants may be invited to be part of specific studies, depending on the research requirements, and may choose to participate on a study-by-study basis.

Questions in Search of Answers

The Network's emphasis is on supporting research that brings the tremendous knowledge about genetics gained from laboratory research to bear on improving prevention, screening, diagnosis, and treatment of cancer in humans. "A wealth of new information on genetics has emerged over the past decade, and the challenge now is to find out how to make these findings meaningful in clinical practice and for public health programs," said Dr. Winn.

Some of the pressing questions that the Network aims to address are:

- How common are the genetic changes (alterations) that cause cancer in different groups?
- What determines whether someone with a genetic change gets cancer?
- What environmental exposures interact with genetic susceptibility to cause cancer?
- How can genetic discoveries be translated into better ways to prevent and treat cancer?
- What ethical, psychological, social, and family issues affect healthy individuals and their families who carry cancer susceptibility gene alterations?

"The Cancer Genetics Network is uniquely suited to support research centered on the study of key interactions between external environmental exposures and inherited susceptibility factors for cancer," said Joellen Schildkraut, Ph.D., of Duke University Medical Center, Durham, NC, and chair of the Network's Steering Committee. "This research can lead to the design of timely interventions, such as behavior modifications and chemoprevention strategies that prevent cancer or halt its progression."

Being Part of the Network

The Network offers individuals an opportunity to keep up to date on cancer genetics and potentially to participate in studies. All Network centers are enrolling eligible participants, and are especially interested in recruiting minorities, among whom membership lags. "We want all groups to be able to take advantage of this opportunity and to benefit from the studies," said Dr. Winn.

Participants provide information about their personal and family medical histories, which is entered into a central database that is operated by an informatics group. Presently, information on more than 134,000 family members is in the database. All information is kept private and is protected by the latest communications technology safeguards.

Network researchers and their centers have longstanding experience in working with individuals and families at increased risk for cancer, and will confidentially consult with individuals who are interested in joining. "The ultimate aim is to prevent cancer, and our best hope for developing effective cancer prevention programs lies in the early identification of high-risk populations and individuals at high risk," said Dr. Schildkraut.

Pilot Studies Under Way

Although still a young program, the Network is conducting a variety of pilot studies. It also has begun to work with other research groups, and welcomes new opportunities to cooperate on important research. Some of the pilot studies under way, or slated to begin soon, are to:

- Test the value of screening for ovarian cancer among women at high risk for the disease using a blood test for CA–125 (a chemical found in the blood) and transvaginal ultrasound;

- Search for novel regions on genes associated with susceptibility to colon cancer among siblings who have a history of the disease;

- Obtain and characterize biological specimens from families who have a history of onset of prostate cancer at an early age;

- Study genetic and environmental factors that may modify risk for developing breast or ovarian cancer among women who are carriers of BRCA1 and BRCA2 gene alterations; and

- Compare statistical models for estimating the likelihood that a woman has a BRCA1 or BRCA2 gene alteration based on her family history.

How to Contact the Network

Individuals may contact one of the Network centers to discuss enrollment. It is not necessary to live near a center in order to join. Some centers have hospital affiliates through which one can enroll, and much of the contact can be by telephone, mail, or e-mail. More information about the Network is available on NCI's Web site: http://epi.grants.cancer.gov/CGN on the Internet.

Carolina-Georgia Cancer Genetics Network Center

Institutions:	Duke University Medical Center, Durham, NC, in collaboration with the University of North Carolina at Chapel Hill, NC, and Emory University, Atlanta, GA
Principal investigator:	Joellen Schildkraut, Ph.D., Duke University Medical Center
CGN Web site:	http://cancer.duke.edu/CGN
Institution:	Duke University Medical Center
Contact:	Sydnee Steadman
Telephone:	888-681-4762 (toll free)
E-mail:	stead006@mc.duke.edu
Institution:	University of North Carolina at Chapel Hill
Contact:	Cindy Smith
Telephone:	877-692-6960 (toll free)
E-mail:	cesmith@med.unc.edu
Institution:	Emory University
Contact:	Lisa Susswein
Telephone:	800-366-1502 (toll free)
E-mail:	lrs@rw.ped.emory.edu

Georgetown University Medical Center's Cancer Genetics Network

Institution:	Georgetown University Lombardi Cancer Center, Washington, DC
Principal investigator:	Claudine Isaacs, M.D.
CGN Web site:	http://lombardi.georgetown.edu/research/areas/cancercontrol/cgn
Contact:	Camille Corio

Telephone:	202-687-8070
E-mail:	corioc@georgetown.edu

Mid-Atlantic Cancer Genetics Network Center

Institutions:	Johns Hopkins University, Baltimore, MD, in collaboration with the Greater Baltimore Medical Center
Principal investigator:	Constance Griffin, M.D., Johns Hopkins University
CGN Web site:	http://www.macgn.org
Contact:	CGN Staff
Telephone:	877-880-6188 (toll free)

Northwest Cancer Genetics Network

Institutions:	Fred Hutchinson Cancer Research Center, Seattle, WA, in collaboration with the University of Washington School of Medicine, Seattle
Principal investigator:	John D. Potter, M.D., Ph.D., Fred Hutchinson Cancer Research Center
CGN Web site:	http://www.fhcrc.org/science/phs/cgn
Contact:	CGN Staff
Telephone:	800-616-8347 (toll free)

Rocky Mountain Cancer Genetics Coalition

Institutions:	University of Utah, Salt Lake City, UT, in collaboration with the University of Colorado, Aurora, CO, and University of New Mexico, Albuquerque, NM
Principal investigator:	Geraldine Mineau, Ph.D., University of Utah, Salt Lake City
CGN Web site:	http://www.hci.utah.edu/cgn
Institution:	University of Utah
Contact:	Debra Dutson
Telephone:	877-585-0473 (toll free)
E-mail:	ddutson@hci.utah.edu

Institution:	University of New Mexico
Contact:	Lloryn Swan
Telephone:	505-272-5659
E-mail:	swan@nmtr.unm.edu
Institution:	University of Colorado
Contact:	Theresa Mickiewicz
Telephone:	877-700-0697 (toll free)
E-mail:	theresa.mickiewicz@uchsc.edu

Texas Cancer Genetics Consortium

Institutions:	University of Texas M.D. Anderson Cancer Center, Houston, TX, in collaboration with the University of Texas Health Science Center at San Antonio, University of Texas Southwestern Medical Center at Dallas, and Baylor College of Medicine, Houston
Principal investigator:	Louise C. Strong, M.D., M.D. Anderson Cancer Center
CGN Web site:	http://texas.cgnweb.org
Telephone:	877-900-8894 (toll free)
Institution:	Baylor College of Medicine
Contact:	Sharon Plon, M.D.
Telephone:	713-770-4251
E-mail:	splon@bcm.tmc.edu
Institution:	University of Texas Southwestern Medical Center
Contact:	Gail Tomlinson, M.D.
Telephone:	214-648-4907
E-mail:	tomlinson@simmons.swmed.edu
Institution:	University of Texas Health Sciences Center
Contact:	Susan Naylor, M.D.
Telephone:	210-567-3842
E-mail:	naylor@uthscsa.edu

Institution:	University of Texas M.D. Anderson Cancer Center
Contact:	Louise C. Strong, M.D.
Telephone:	713-792-7555
E-mail:	lstrong@mdanderson.org

UCI-UCSD Cancer Genetics Network Center

Institutions:	University of California at Irvine and University of California at San Diego
Principal investigator:	Hoda Anton-Culver, Ph.D., UC Irvine
Contact:	CGN Staff
Telephone:	949-824-7401 (collect calls accepted)

University of Pennsylvania Cancer Genetics Network

Institution:	University of Pennsylvania Cancer Center, Philadelphia, PA
Principal investigator:	Barbara Weber, M.D.
Contact:	Rhonda Kitlas
Telephone:	888-666-6002 (toll free)
E-mail:	kitlasr@mail.med.upenn.edu

Informatics Infrastructure

The CGN also has an Informatics and Information Technology Group to meet its information exchange and data management and statistical needs. The participating institutions and principal investigators are the University of California at Irvine, with Hoda Anton-Culver, Ph.D.; Massachusetts General Hospital, Boston, MA, with Dianne M. Finkelstein, Ph.D.; and Yale University, New Haven, CT, with Prakash M. Nadkarni, Ph.D.

SOURCES OF NATIONAL CANCER INSTITUTE INFORMATION

Cancer Information Service

Toll-free: 1–800–4–CANCER (1–800–422–6237)
TTY (for deaf and hard of hearing callers): 1–800–332–8615

NCI Online

Internet
Use http://cancer.gov to reach NCI's Web site.
LiveHelp
Cancer Information Specialists offer online assistance through the *LiveHelp* link on the NCI's Web site.

***

42

How to Open Your Own DNA Test Results or Molecular Genealogy Reporting Company

Did you ever wonder what the next money-making step for entrepreneurs in genealogy is—searching records for family history and ancestry? It's about opening a genealogy-driven DNA testing service. Take your pick: tracking ancestry by DNA for pets or people. You don't need any science courses or degrees to start or operate this small business. It can be done online, at home, or in an office. What should you charge per test? About $200 is affordable. You'll have to pay a laboratory to do the testing. Work out your budget with the laboratory.

Laboratories that do the testing can take up to fifty percent of what you make on each test unless they have research grants to test a particular ethnic group and need donors to give DNA for testing. Each lab is different. Shop around for an affordable, reputable laboratory. Your first step would be to ask the genetics and/or molecular anthropology departments of universities who's applying for a grant to do DNA testing. Also check out the oral history libraries which usually are based at universities and ethnic museums. You're bringing together two different groups—genealogists and geneticists.

You'd work with the laboratories that do the testing. Customers want to see online message boards to discuss their DNA test results and find people whose DNA sequences match their own.

So you'd need a Web site with databases of the customers, message boards, and any type of interactive communication system that allows privacy and communication. DNA database material would not show real names or identify the people. So you'd use numbers. Those who want to contact others could use regular email addresses. People want ethnic privacy, but at the same time love to find DNA matches. At this point you might want to work only with dogs, horses, or other pets or farm animals providing a DNA testing service for ancestry or nutrition.

Take your choice as an entrepreneur: sending the DNA of people to laboratories to be tested for ancestry or having the DNA of dogs, horses or other pets and animals sent out to be tested for ancestry and supplying reports to owners regarding ancestry or for information on how to tailor food to the genetic signatures of people or animals. For animals, you'd contact breeders.

For people, your next step is to contact genealogists and genealogy online and print publications. You'd focus on specific ethnic groups as a niche market. The major groups interested in ancestry using DNA testing include Northern European, Ashkenazi, Italian, Greek, Armenian, Eastern European, African, Asian, Latin American, and Middle Eastern.

Many successful entrepreneurs in the DNA testing for ancestry businesses started with a hobby of looking up family history records—genealogy. So if you're a history buff, or if your hobby is family history research, oral history, archaeology, or genealogy, you now can turn to DNA testing.

What you actually sell to customers are DNA test kits and DNA test reports. To promote your business, offer free access to your Web site database with all your clients listed by important DNA sequences. Keep names private and use only assigned numbers or letters to protect the privacy of your clients. Never give private and confidential genetic test information to insurance companies or employers. Clients who want to have their DNA tested for ancestry do not want their names and DNA stored to fall into the "wrong hands." So honor privacy requests. Some people will actually ask you to store DNA for future generations.

If you want to include this service, offer a time capsule. For your clients, you would create a time capsule, which is like a secure scrap book on acid-free paper and on technology that can be transferred in the future when technology changes. Don't store anything on materials that can't be transferred from one technology to another. For example, have reports on acid-free paper.

You can include a CD or DVD also, but make sure that in the future when the CD players aren't around any longer, the well-preserved report, perhaps laminated or on vellum or other acid-free materials that don't crumble with age can be put into the time capsule. You can include a scrap book with family photos and video on a CD if you wish, or simply offer the DNA test report and comments explaining to the customer what the DNA shows.

Use plain language and no technical terms unless you define them on the same page. Your goal is to help people find other people who match DNA sequences and to use this knowledge to send your customers reports. If no matches can be found, then supply your clients with a thorough report. Keep out any confusing jargon. Show with illustrations how your customer's DNA was tested. In plain language tell them what was done.

Your report will show the results, and tell simply what the results mean. You can offer clients a list of how many people in what countries have their same DNA sequences. Include the present day city or town and the geographic location using longitude and latitude. For example, when I had my mtDNA (maternal lineages) tested, the report included my DNA matches by geographic coordinates. The geographic center is 48.30N 4.65E, Bar sur Aube, France with a deviation of 669.62 miles as done by "Roots for Real," a London company that tests DNA for ancestry. The exact sequences are in the Roots for Real Database (and other mtDNA databases) for my markers.

You're going to ask, with no science background yourself, how will you know what to put in the report? That's the second step. You contact a university laboratory that does DNA testing for outside companies. They will generate all the reports for you. What you do with the report is to promote it by making it look visually appealing. Define any words you think the customer won't understand with simpler words that fully explain what the DNA sequences mean and what the various letters and numbers mean. Any dictionary of genetic terms will give you the meaning in one sentence using plain language. Use short sentences in your reports and plain language.

Your new service targets genealogists who help their own customers find lost relatives. Your secondary market is the general public. Most people taking a DNA test for ancestry want information on where their DNA roamed 20,000 years ago and in the last 10,000 years. DNA testing shows people only where their ancient ancestors camped. However, when sequences with other people match exactly, it could point the way to an ancient common ancestor whose descendants went in a straight line from someone with those sequences who lived 10,000 years ago to a common ancestor who lived only a few generations ago.

Those people may or may not actually be related, but they share the same sequences. The relationship could be back in a straight line 20,000 years or more or only a few centuries. Ancient DNA sequences are spread over a huge area, like mine—from Iceland to Bashkortostan in the Urals. DNA sequences that sprung up only a few generations ago generally are limited to a more narrow geographic area, except for those who lived in isolation in one area for thousands of years, such as the Basques.

You would purchase wholesale DNA kits from laboratory suppliers and send the kits to your customer. The customer takes a painless cheek scraping with a felt or cotton type swab or uses mouthwash put into a small container to obtain DNA that can help accurately determine a relationship with either a 99.9% probability of YES or a 100% certainly that no near term relationship existed.

The DNA sample is sealed and mailed to a laboratory address where it is tested. The laboratory then disposes of the DNA after a report is generated. Then you package the report like a gift card portfolio, a time capsule, or other fancy packaging to look like a gift. You add your promotional material and a thorough explanation of what to expect from the DNA test—the results.

The best way to learn this business is to check out on the Web all the businesses that are doing this successfully. Have your own DNA tested and look at the printout or report of the results. Is it thorough? Does it eliminate jargon? Include in the report materials the client would like to see. Make it look like a press kit. For example, you take a folder such as a report folder. On the outside cover print the name of your company printed and a logo or photograph of something related to DNA that won't frighten away the consumer. Simple graphic art such as a map or globe of the world, a prehistoric statue, for example the Willendorf Venus, or some other symbol is appropriate.

Inside, you'd have maps, charts, and locations for the client to look at. Keep the material visual. Include a CD with the DNA sequences if you can. The explanation would show the customer the steps taken to test the DNA.

Keep that visual with charts and graphs. Don't use small print fonts or scientific terminology to any extent so your customer won't feel your report is over his or her head. Instead use illustrations, geographic maps. Put colorful circles on the cities or geographic locations where that person's DNA is found.

Put a bright color or arrow on the possible geographic area of origin for those DNA sequences. Nobody can pinpoint an exact town for certain, but scientists know where certain DNA sequences are found and where they might have sat out the last Ice Age 20,000 years ago, and survived to pass those same DNA sequences on to their direct descendants, that customer of yours who has those sequences.

In the last decade, businesses have opened offering personality profilers. This decade, since the human genome code was cracked and scientists know a lot more about DNA testing for the courtroom, DNA testing businesses have opened to test DNA for information other than who committed a crime or to prove who's innocent. Applications of DNA testing now are used for finding ancient and not-so-ancient ancestry. DNA testing is not only used for paternity and maternity testing, but for tailoring what you eat to your genetic signature. The new field of pharmacogenetics also tests DNA for markers that allow a client to customize medicine to his or her genetic expression.

You may be an entrepreneur with no science background. That's okay as long as your laboratory contacts are scientists. Your most important contact and contract would be with a DNA testing laboratory. Find out who your competitors contract with as far as testing laboratories. For example, Family Tree DNA at the Web site: http://www.familytreedna.com/faq.html#q1 sends its DNA samples to be tested by the DNA testing laboratories at the University of Arizona.

Bennett Greenspan, President and CEO of Family Tree DNA founded Family Tree in 1999. Greenspan is an entrepreneur and life-long genealogy enthusiast. He successfully turned his family history and ancestry hobby into a full-time vocation running a DNA testing-for-ancestry company. Together with Max Blankfeld, they founded in 1997 GoCollege.com a website for college-bound students which survived the .COM implosion. Max Blankfeld is Greenspan's Vice President of Operations/Marketing. Before entering the business world, Blankfeld was a journalist. After that, he started and managed several successful ventures in the area of public relations as well as consumer goods both in Brazil and the US. Today, the highly successful Family Tree DNA is America's first genealogy-driven DNA testing service.

At the University of Arizona, top DNA research scientists such as geneticist, Mike Hammer, PhD, population geneticist Bruce Walsh, PhD, geneticist Max F. Rothschild, molecular anthropologist, Theodore G. Schurr, and lab manager, Matthew Kaplan along with the rest of the DNA testing team do the testing and analysis. So it's important if you want to open your own DNA for ancestry testing company to contract with a reputable laboratory to do the testing. Find out whether the lab you're going to be dealing with will answer a client's questions in case of problems with a test that might require re-testing. Clients will come to you to answer questions rather than go to the busy laboratory. Most laboratories are either part of a university, a medical school, or are independent DNA testing laboratories run by scientists and their technicians and technologists.

Your business will have a very different focus if you're only dealing with genealogy buffs testing their DNA for ancestry than would a business testing DNA for genetic risk markers in order to tailor a special diet or foods to someone's genetic risk markers. For that more specialized business, you'd have to partner with a nutritionist, scientist, or physician trained in customizing diets to genetic signatures. Many independent laboratories do test genes for the purpose of tailoring diets to genes. The new field is called nutrigenomics. Check out the various Web sites devoted to nutrigenomics if you're interested in this type of DNA testing business. For example, there is Alpha-Genetics at http://www.Alpha-Genics.com.

According to Dr. Fredric D. Abramson, PhD, S.M., President and CEO of AlphaGenics, Inc., "The key to using diet to manage genes and health lies in managing gene expression (which we call the Expressitype). Knowing your genotype merely tells you a starting point. Genotype is like knowing where the entrance ramps to an interstate can be found. They are important to know, but tell you absolutely nothing about what direction to travel or how the journey will go. That is why Expressitype must be the focus." You can contact AlphaGenics, Inc. at: http://www.Alpha-Genics.com or write to: Maryland Technology Incubator, 9700 Great Seneca Highway, Rockville, MD 20850.

Why open any kind of a DNA testing business? It's because the entrepreneur is at the forefront of a revolution in our concept of ancestry, diet, and medicines. Genes are tested to reveal how your body metabolizes medicine as well as food, and genes are tested for ancient ancestry or recent relationships such as paternity. Genes are tested for courtroom evidence.

So you have the choice of opening a DNA testing service focusing on diet, ancestry, skin care product matches, or medicine. You can have scientists contract with you to test genes for risk or relationships. Some companies claim to test DNA in order to determine whether the skin care products are right for your genetic signature. It goes beyond the old allergy tests of the eighties.

"Each of us is a unique organism, and for the first time in human history, genetic research is confirming that one diet is not optimum for everyone," says Abramson. Because your genes differ from someone else's, you process food and supplements in a unique way. Your ancestry is unique also.

Do you want to open a business that tunes nutrition to meet the optimum health needs of each person? If so, you need to contract with scientists to do the testing. If you have no science background, it would be an easier first step to open a business that tests DNA only for ancestry and contract with university laboratories who know about genes and ancestry.

Your client would receive a report on only the ancestry. This means the maternal and/or paternal sequences. For a woman it's the mtDNA that's tested. You're testing the maternal lineages. It's ancient and goes back thousands of years. For the man, you can have a lab test the Y-chromosome, the paternal lineages and the mtDNA, the maternal lineages.

What you supply your clients with is a printout report and explanation of the individual's sequences and mtDNA group called the haplogroup and/or the Y-chromosome ancestral genetic markers. For a male, you can test the Y-chromosome and provide those markers, usually 25 markers and the mtDNA. For a woman, you can only test the mtDNA, the maternal line for haplogroup letter and what is called the HVS-1 and HVS-2 sequences. These sequences show the maternal lineages back thousands of years. To get started, look at the Web sites and databases of all the companies that test for ancestry using DNA.

What most of the DNA testing entrepreneurs have in common is that they can do business online. People order the DNA testing kit online. The companies send out a DNA testing kit. The client sends back DNA to a lab to be tested. The process does not involve any blood drawing to test for ancestry. Then the company sends a report directly to the customer about what the DNA test revealed solely in regard to ancient ancestry—maternal or paternal lines.

Reports include the possible geographic location where the DNA sequences originated. Customers usually want to see the name of an actual town, even though towns didn't exist 10,000 years ago when the sequences might have arisen. The whole genome is not tested, only the few ancestral markers, usually 500 base pairs of genes. Testing DNA for ancestry does not have anything to do with testing genes for health risks because only certain genes are tested—genes related to ancestry. And all the testing is done at a laboratory, not at your online business.

If you're interested in a career in genetics counseling and wish to pursue a graduate degree in genetics counseling, that's another career route. For information, contact The American Board of Genetic Counseling. Sometimes social workers with some coursework in biology take a graduate degree in genetic counseling since it combines counseling skills with training in genetics and in interpreting genetics tests for your clients.

> The American Board of Genetic Counseling.
> 9650 Rockville Pike
> Bethesda, MD 20814-3998
> Phone: (301) 571-1825
> FAX: (301) 571-1895
> http://www.abgc.net/

Below is a list of several DNA-testing companies. Some of these companies test DNA only for ancestry. Other companies listed below test genes for personalized medicine and nutrigenomics, and some companies test for nutrigenomics, pharmacogenetics, and ancestry.

You'll also find several companies listed that only test the DNA of animals. So you have a choice of testing DNA for a variety of purposes, for testing human

DNA only, or for testing animal DNA. And the applications for testing genetic signatures are growing, since this science is still in its infancy in regard to applications of genetic and genomic testing.

Roots for Real
http://www.rootsforreal.com
address: PO Box 43708
London W14 8WG UK

Family Tree DNA—Genealogy by Genetics, Ltd.
World Headquarters
1919 North Loop West, Suite 110 Houston, Texas 77008, USA
Phone: (713) 868-1438 | Fax: (713) 868-4584
info@FamilyTreeDNA.com
http://www.familytreedna.com/

Oxford Ancestors
Oxford Ancestors, London,
http://www.oxfordancestors.com/

AncestrybyDNA, DNAPrint genomics, Inc.
900 Cocoanut Ave, Sarasota, FL 34236. USA
Tel: 941-366-3400 Fax: 941-952-9770 Web site: http://www.ancestrybydna.com/

GeneTree DNA Testing Center
2495 South West Temple
Salt Lake City, UT 84115
Toll Free: (888) 404-GENE
Phone: (801) 461-9757
Fax: (801) 461-9761, http://www.genetree.com/

Trace Genetics LLC
P.O. Box 2010
Davis, California 95617
info@tracegenetics.com
http://www.tracegenetics.com/aboutus.html

Predictive Genomics for Personalized Medicine including Nutrigenomics

AlphaGenics Inc.
9700 Great Seneca Highway
Rockville, Maryland 20850
Email: info@alpha-genics.com

http://www.alpha-genics.com/index.php

Genovations ™
Great Smokies Diagnostic Laboratory/Genovations™
63 Zillicoa Street
Asheville, NC 28801 USA
http://www.genovations.com/

Centre for Human Nutrigenomics
http://www.nutrigenomics.nl/
According to its Web site, "The Centre for Human NutriGenomics aims at establishing an international centre of expertise combining excellent pre-competitive research and high quality (post)graduate training on the interface of genomics, nutrition and human health."

Nutrigenomics Links: http://nutrigene.4t.com/nutrigen.htm

Veterinary DNA Testing

Veterinary Genetics Laboratory
University of California, Davis
One Shields Avenue
Davis, CA 95616-8744
http://www.vgl.ucdavis.edu/

According to their Web site: "The Veterinary Genetics Laboratory is internationally recognized for its expertise in parentage verification and genetic diagnostics for animals. VGL has provided services to breed registries, practitioners, individual owners and breeders since 1955." The Veterinary Genetics Laboratory performs contracted DNA testing.

Alpaca/Llama
Beefalo
Cat
Cattle
Dog
Elk
Goat
Horse
Sheep

DNA Testing of Dogs and Horses:
VetGen, 3728 Plaza Drive, Suite 1, Ann Arbor, Michigan, 48108 USA http://www.vetgen.com/

◆ ◆ ◆

Ethnic Genealogy Web Sites:

Acadian/Cajun: & French Canadian: http://www.acadian.org/tidbits.html
African-American: http://www.cyndislist.com/african.htm
African Royalty Genealogy: http://www.uq.net.au/~zzhsoszy/
Albanian Research List: http://feefhs.org/al/alrl.html
Armenian Genealogical Society: http://feefhs.org/am/frg-amgs.html
Asia and the Pacific: http://www.cyndislist.com/asia.htm
Austria-Hungary Empire: http://feefhs.org/ah/indexah.html
Baltic-Russian Information Center: http://feefhs.org/blitz/frgblitz.html
Belarusian—Association of the Belarusian Nobility: http://feefhs.org/by/frg-zbs.html
Bukovina Genealogy: http://feefhs.org/bukovina/bukovina.html
Carpatho-Rusyn Knowledge Base: http://feefhs.org/rusyn/frg-crkb.html
Chinese Genealogy: http://www.chineseroots.com.
Croatia Genealogy Cross Index: http://feefhs.org/cro/indexcro.html
Czechoslovak Genealogical Society Int'l, Inc.: http://feefhs.org/czs/cgsi/frg-cgsi.html
Eastern Europe: http://www.cyndislist.com/easteuro.htm
Eastern European Genealogical Society, Inc.: http://feefhs.org/ca/frg-eegs.html
Eastern Europe Ethnic, Religious, and National Index with Home Pages includes the FEEFHS Resource Guide that lists organizations associated with FEEFHS

from 14 Countries. It also includes Finnish and Armenian genealogy resources: http://feefhs.org/ethnic.html
Ethnic, Religious, and National Index 14 countries: http://feefhs.org/ethnic.html
Finnish Genealogy Group: http://feefhs.org/misc/frgfinmn.html
Galicia Jewish SIG: http://feefhs.org/jsig/frg-gsig.html
German Genealogical Digest: http://feefhs.org/pub/frg-ggdp.html
Greek Genealogy Sources on the Internet: http://www-personal.umich.edu/~cgaunt/greece.html
Genealogy Societies Online List: http://www.daddezio.com/catalog/grkndx04.html
German Research Association: http://feefhs.org/gra/frg-gra.html
Greek Genealogy (Hellenes-Diaspora Greek Genealogy): http://www.geocities.com/SouthBeach/Cove/4537/
Greek Genealogy Home Page: http://www.daddezio.com/grekgen.html
Greek Genealogy Articles: http://www.daddezio.com/catalog/grkndx01.html
India Genealogy: http://genforum.genealogy.com/india/
India Family Histories: http://www.mycinnamontoast.com/perl/results.cgi?region=79&sort=n
India-Anglo-Indian/Europeans in India genealogy: http://members.ozemail.com.au/~clday/
Irish Travellers: http://www.pitt.edu/~alkst3/Traveller.html
Japanese Genealogy: http://www.rootsweb.com/~jpnwgw/
Jewish Genealogy: http://www.jewishgen.org/infofiles/
Latvian Jewish Genealogy Page: http://feefhs.org/jsig/frg-lsig.html
Lebanese Genealogy: http://www.rootsweb.com/~lbnwgw/
Lithuanian American Genealogy Society: http://feefhs.org/frg-lags.html
Melungeon: http://www.geocities.com/Paris/5121/melungeon.htm
Mennonite Heritage Center: http://feefhs.org/men/frg-mhc.html
Middle East Genealogy: http://www.rootsweb.com/~mdeastgw/index.html
Middle East Genealogy by country: http://www.rootsweb.com/~mdeastgw/index.html#country
Native American: http://www.cyndislist.com/native.htm
Polish Genealogical Society of America: http://feefhs.org/pol/frg-pgsa.html
Quebec and Francophone: http://www.francogene.com/quebec/amerin.html
Romanian American Heritage Center: http://feefhs.org/ro/frg-rahc.html
Slovak World: http://feefhs.org/slovak/frg-sw.html
Slavs, South: Cultural Society: http://feefhs.org/frg-csss.html
Syrian and Lebanese Genealogy: http://www.genealogytoday.com/family/syrian/

Syria Genealogy: http://www.rootsweb.com/~syrwgw/
Tibetan Genealogy: http://www.distantcousin.com/Links/Ethnic/China/Tibetan.html
Turkish Genealogy Discussion Group: http://www.turkey.com/forums/forumdisplay.php3?forumid=18
Ukrainian Genealogical and Historical Society of Canada: http://feefhs.org/ca/frgughsc.html
Unique Peoples: http://www.cyndislist.com/peoples.htm Note: The Unique People's list includes: Black Dutch, Doukhobors, Gypsy, Romani, Romany & Travellers, Melungeons, Metis, Miscellaneous, and Wends/Sorbs

***

Appendix A

Name Frequency in the US

(Reprinted with permission from the U.S. Census Bureau, Population Division*)*

How Frequently Do Names Appear?

NOTE: No specific individual information is given

See the US Census Bureau Web site at: http://www.census.gov/genealogy/names/

For example:
US Census Bureau, 1990

(1). A "Name"

(2). Frequency in percent

(3). Cumulative Frequency in percent

(4). Rank

In the file (dist.all.last) one entry appears as:

MOORE   0.312   5.312   9

In our Search Area sample, MOORE ranks 9th in terms of frequency. 5.312 percent of the sample population is covered by MOORE and the 8 names occurring more frequently than MOORE. The surname, MOORE, is possessed by 0.312 percent of our population sample.
Detailed Methodology

Variables in Names Files:
name
freq = Frequency in percent
cum.freq = Cumulative Frequency in percent
rank

First ten entries in dist.all.last

name	freq	cum.freq	rank
SMITH	1.006	1.006	1
JOHNSON	0.810	1.816	2
WILLIAMS	0.699	2.515	3
JONES	0.621	3.136	4
BROWN	0.621	3.757	5
DAVIS	0.480	4.237	6
MILLER	0.424	4.660	7
WILSON	0.339	5.000	8
MOORE	0.312	5.312	9
TAYLOR	0.311	5.623	10

First ten entries in dist.female.first

name	freq	cum.freq	rank
MARY	2.629	2.629	1
PATRICIA	1.073	3.702	2
LINDA	1.035	4.736	3
BARBARA	0.980	5.716	4

name	freq	cum.freq	rank
ELIZABETH	0.937	6.653	5
JENNIFER	0.932	7.586	6
MARIA	0.828	8.414	7
SUSAN	0.794	9.209	8
MARGARET	0.768	9.976	9
DOROTHY	0.727	10.703	10

First ten entries in dist.male.first

name	freq	cum.freq	rank
JAMES	3.318	3.318	1
JOHN	3.271	6.589	2
ROBERT	3.143	9.732	3
MICHAEL	2.629	12.361	4
WILLIAM	2.451	14.812	5
DAVID	2.363	17.176	6
RICHARD	1.703	18.878	7
CHARLES	1.523	20.401	8
JOSEPH	1.404	21.805	9
THOMAS	1.380	23.185	10

Source: U.S. Census Bureau, Population Division,
Population Analysis & Evaluation Staff
Maintained By: Laura K. Yax (Population Division)
Last Revised: December 20, 1999 at 11:33:07 AM

APPENDIX B

Ethnic Genealogy Web Sites:

Acadian/Cajun: & French Canadian: http://www.acadian.org/tidbits.html
African-American: http://www.cyndislist.com/african.htm
African Royalty Genealogy: http://www.uq.net.au/~zzhsoszy/
Albanian Research List: http://feefhs.org/al/alrl.html
Armenian Genealogical Society: http://feefhs.org/am/frg-amgs.html
Asia and the Pacific: http://www.cyndislist.com/asia.htm
Austria-Hungary Empire: http://feefhs.org/ah/indexah.html
Baltic-Russian Information Center: http://feefhs.org/blitz/frgblitz.html
Belarusian—Association of the Belarusian Nobility: http://feefhs.org/by/frg-zbs.html
Bukovina Genealogy: http://feefhs.org/bukovina/bukovina.html
Carpatho-Rusyn Knowledge Base: http://feefhs.org/rusyn/frg-crkb.html
Chinese Genealogy: http://www.chineseroots.com.
Croatia Genealogy Cross Index: http://feefhs.org/cro/indexcro.html
Czechoslovak Genealogical Society Int'l, Inc.: http://feefhs.org/czs/cgsi/frg-cgsi.html
Eastern Europe: http://www.cyndislist.com/easteuro.htm
Eastern European Genealogical Society, Inc.: http://feefhs.org/ca/frg-eegs.html
Eastern Europe Ethnic, Religious, and National Index with Home Pages includes the FEEFHS Resource Guide that lists organizations associated with FEEFHS from 14 Countries. It also includes Finnish and Armenian genealogy resources: http://feefhs.org/ethnic.html
Ethnic, Religious, and National Index 14 countries: http://feefhs.org/ethnic.html
Finnish Genealogy Group: http://feefhs.org/misc/frgfinmn.html
Galicia Jewish SIG: http://feefhs.org/jsig/frg-gsig.html
German Genealogical Digest: http://feefhs.org/pub/frg-ggdp.html
Greek Genealogy Sources on the Internet: http://www-personal.umich.edu/~cgaunt/greece.html
Genealogy Societies Online List: http://www.daddezio.com/catalog/grkndx04.html

German Research Association: http://feefhs.org/gra/frg-gra.html
Greek Genealogy (Hellenes-Diaspora Greek Genealogy): http://www.geocities.com/SouthBeach/Cove/4537/
Greek Genealogy Home Page: http://www.daddezio.com/grekgen.html
Greek Genealogy Articles: http://www.daddezio.com/catalog/grkndx01.html
India Genealogy: http://genforum.genealogy.com/india/
India Family Histories: http://www.mycinnamontoast.com/perl/results.cgi?region=79&sort=n
India-Anglo-Indian/Europeans in India genealogy: http://members.ozemail.com.au/~clday/
Irish Travellers: http://www.pitt.edu/~alkst3/Traveller.html
Japanese Genealogy: http://www.rootsweb.com/~jpnwgw/
Jewish Genealogy: http://www.jewishgen.org/infofiles/
Latvian Jewish Genealogy Page: http://feefhs.org/jsig/frg-lsig.html
Lebanese Genealogy: http://www.rootsweb.com/~lbnwgw/
Lithuanian American Genealogy Society: http://feefhs.org/frg-lags.html
Melungeon: http://www.geocities.com/Paris/5121/melungeon.htm
Mennonite Heritage Center: http://feefhs.org/men/frg-mhc.html
Middle East Genealogy: http://www.rootsweb.com/~mdeastgw/index.html
Middle East Genealogy by country: http://www.rootsweb.com/~mdeastgw/index.html#country
Native American: http://www.cyndislist.com/native.htm
Polish Genealogical Society of America: http://feefhs.org/pol/frg-pgsa.html
Quebec and Francophone: http://www.francogene.com/quebec/amerin.html
Romanian American Heritage Center: http://feefhs.org/ro/frg-rahc.html
Slovak World: http://feefhs.org/slovak/frg-sw.html
Slavs, South: Cultural Society: http://feefhs.org/frg-csss.html
Syrian and Lebanese Genealogy: http://www.genealogytoday.com/family/syrian/
Syria Genealogy: http://www.rootsweb.com/~syrwgw/
Tibetan Genealogy: http://www.distantcousin.com/Links/Ethnic/China/Tibetan.html
Turkish Genealogy Discussion Group: http://www.turkey.com/forums/forumdisplay.php3?forumid=18
Ukrainian Genealogical and Historical Society of Canada: http://feefhs.org/ca/frgughsc.html
Unique Peoples: http://www.cyndislist.com/peoples.htm Note: The Unique People's list includes: Black Dutch, Doukhobors,

Gypsy, Romani, Romany & Travellers, Melungeons, Metis, Miscellaneous, and Wends/Sorbs

***

APPENDIX C

Genealogy, (General):

Ancestry.com: http://www.ancestry.com/main.htm?lfl=m
Cyndi's List of Genealogy on the Internet: http://www.cyndislist.com/
Cyndi's List is a categorized & cross-referenced index to genealogical resources on the Internet with thousands of links.
DistantCousin.com (Uniting Cousins Worldwide) http://distantcousin.com/Links/surname.html
Ellis Island Online: http://www.ellisisland.org/
Family History Library: http://www.familysearch.org/Eng/default.asp
http://www.familysearch.org/Eng/Search/frameset_search.asp
(The Church of Jesus Christ of Latter Day Saints) International Genealogical Index
Female Ancestors: http://www.cyndislist.com/female.htm
Genealogist's Index to the Web:
http://www.genealogytoday.com/GIWWW/?
Genealogy Web http://www.genealogyweb.com/
Genealogy Authors and Speakers: http://feefhs.org/frg/frg-a&l.html
Genealogy Today: http://www.genealogytoday.com/
My Genealogy.com http://www.genealogy.com/cgi-bin/my_main.cgi
Scriver, Dr. Charles: The Canadian Medical Hall of Fame http://www.virtualmuseum.ca/Exhibitions/Medicentre/en/scri_print.htm
Surname Sites: http://www.cyndislist.com/surn-gen.htm
National Genealogical Society: http://www.ngsgenealogy.org/index.htm
United States List of Local by State Genealogical Societies: http://www.daddezio.com/society/hill/index.html
United States Vital Records List: http://www.daddezio.com/records/room/index.html or http://www.cyndislist.com/usvital.htm

***

APPENDIX D

Bibliography 1:
Genealogy: Ethnic, Oral History, Internet, Instructional Workbooks, Intergenerational, and Women's Genealogical Resources.

A Bintel Brief: Sixty Years of Letters From the Lower East Side to the Jewish Daily Forward. Metzker, Isaac, ed. Doubleday and Co. 1971. Garden City, NY

Climbing Your Family Tree: Online and Offline Genealogy for Kids IRA Wolfman, Tim Robinson (Illustrator), Alex Haley (Introduction)/ Paperback/ Workman Publishing Company, Inc./ October 2001

Complete Beginner's Guide to Genealogy, the Internet, and Your Genealogy Computer Program Karen Clifford/ Paperback/ Genealogical Publishing Company, Incorporated/ February 2001

Complete Idiot's Guide(R) to Online Geneology Rhonda McClure/ Paperback/ Pearson Education/ January 2002

Creating Your Family Heritage Scrapbook : From Ancestors to Grandchildren, Your Complete Resource & Idea Book for Creating a Treasured Heirloom. Nerius, Maria Given, Bill Gardner ISBN: 0761530142 Published by Prima Publishing, Aug 2001

Cyndi's List: A Comprehensive List of 70,000 Genealogy Sites on the Internet (Vol. 1 & 2) Cyndi Howells/ Paperback/ Genealogical Publishing Company, Incorporated/ June 2001.

Discovering Your Female Ancestors: Special strategies for uncovering your hard-to-find information about your female lineage. Carmack, Sharon

DeBartolo. Conference Lecture on Audio Tape: Carmack, Sharon DeBartolo.

Folklife and Fieldwork: A Layman's Introduction to Field Techniques. Bartis, Peter. Washington, DC: Library of Congress, 1990.

Genealogy Online for Dummies Matthew L. Helm, April Leigh Helm, April Leigh Helm, Matthew L. Helm/ Paperback/ Wiley, John & Sons, Incorporated/ February 2001

Genealogy Online Elizabeth Powell Crowe/ Paperback/ McGraw-Hill Companies, November 2001

History From Below: How to Uncover and Tell the Story of Your Community, Association, or Union. Brecher, Jeremy. New Haven: Advocate Press/Commonwork Pamphlets, 1988.

My Family Tree Workbook: Genealogy for Beginners Rosemary A. Chorzempa/ Paperback/ Dover Publications, Incorporated/

National Genealogical Society Quarterly 79, no. 3 (September 19991): 183–93

"**Numbering Your Genealogy: Sound and Simple Systems**." Curran, Joan Ferris.

Oral History and the Law. Neuenschwander, John. Pamphlet Series #1. Albuquerque: Oral History Association, 1993.

Oral History for the Local Historical Society. Baum, Willa K. Nashville: American Association for State and Local History, 1987.

Scrapbook Storytelling: Save Family Stories & Memories with Photos, Journaling & Your Own Creativity Slan, Joanna Campbell, Published by EFG, Incorporated, ISBN: 0963022288 May 1999

"The Silent Woman: Bringing a Name to Life." NE-59. Boston, MA: New England Historic Genealogical Society Sesquicentennial Conference, 1995.

The Source: A Guidebook of American Genealogy Alice Eichholz, Loretto Dennis Szucs (Editor), Sandra Hargreaves Luebking (Editor), Sandra Hargreaves Luebking (Editor)/ Hardcover/ MyFamily.com, Incorporated/ February 1997

To Our Children's Children: Journal of Family Members, Bob Greene, D. G. Fulford 240pp. ISBN: 038549064X Publisher: Doubleday & Company, Incorporated: October 1998.

Transcribing and Editing Oral History. Nashville: American Association for State and Local History, 1991.

Using Oral History in Community History Projects. Buckendorf, Madeline, and Laurie Mercier. Pamphlet Series #4. Albuquerque: Oral History Association, 1992.

Unpuzzling Your Past: The Best-Selling Basic Guide to Genealogy (Expanded, Updated and Revised) Emily Anne Croom, Emily Croom/ Paperback/ F & W Publications, Incorporated/ August 2001

Writing a Woman's Life. Heilbrun, Carolyn G. New York: W.W. Norton, 1988

Your Guide to the Family History Library: How to Access the World's Largest Genealogy Resource Paula Stuart Warren, James W. Warren/ Paperback/ F & W Publications, Incorporated/ August 2001

Your Story: A Guided Interview Through Your Personal and Family History, 2nd ed., 64pp. ISBN: 0966604105 Publisher: Stack Resources, LLC

***

Appendix E

Bibliography 2: DNA Testing and Genetics.

A Biologist's Guide to Analysis of DNA Microarray Data Steen Knudsen/ Hardcover/ Wiley, John & Sons, Incorporated/ April 2002

Advances and Opportunities in DNA Testing and Gene Probes Business Communications Company Incorporated (Editor)/ Hardcover/ Business Communications/ September 1996

African Exodus, The Origins of Modern Humanity Stringer, Christopher and Robin McKie. Henry Holt And Company 1997

An A to Z of DNA Science: What Scientists Mean when They Talk about Genes and Genomes Jeffre L. Witherly, Galen P. Perry, Darryl L. Leja/ Paperback/ Cold Spring Harbor Laboratory Press/ September 2002

An Introduction to Forensic DNA Analysis Norah Rudin, Keith Inman/ Hardcover/ CRC Press/ December 2001

Archaeogenetics: DNA and the population prehistory of Europe, Ed. Colin Renfrew & Katie Boyle. McDonald Institute Monographs. Cambridge, UK, Distributed by Oxbow Books UK. In USA: The David Brown Book Company, Oakville, CT. 2000

Cartoon Guide to Genetics Gonick, Larry, With Mark Wheelis: Paperback/ HarperInformation/ July 1991

DNA Detectives, The—Working Against Time, novel, Hart, Anne. Mystery and Suspense Press, iuniverse.com paperback 248 pages at http://www.iuniverse.com or 1-877-823-9235.

DNA for Family Historians (ISBN 0-9539171-0-X). Savin, Alan of Maidenhead, England, is author of the 32-page book. See the Web site: http://www.savin.org/dna/dna-book.html

DNA Microarrays and Gene Expression Pierre Baldi, G. Wesley Hatfield, G. Wesley Hatfield/ Hardcover/ Cambridge University Press/ August 2002

Microarrays for an Integrative Genomics Isaac S. Kohane, Alvin Kho, Atul J. Butte/ Hardcover/ MIT Press/ August 2002

Does It Run in the Family?: A Consumers Guide to DNA Testing for Genetic Disorders Doris Teichler Zallen, Doris Teichler-Zallen, Doris Teichler Zallen/ Hardcover/ Rutgers University Press/ May 1997

Double Helix, The: A Personal Account of the Discovery of the Structure of DNA James D. Watson/ Paperback/ Simon & Schuster Trade Paperbacks/ June 2001

Genes, Peoples, and Languages Luigi Luca Cavalli-Sforza, Mark Seielstad (Translator).

Genetic Witness: Forensic Uses of DNA Tests DIANE Publishing Company (Editor)/ Paperback/ DIANE Publishing Company/ April 1993

History and Geography of Human Genes, The [ABRIDGED] L. Luca Cavalli-Sforza, Paolo Menozzi (Contributor), Alberto Piazza (Contributor).

How to DNA Test Our Family Relationships Terry Carmichael, Alexander Ivanof Kuklin, Ed Grotjan/ Paperback/ Acen Press/ November 2000

Introduction to Genetic Analysis Anthony J. Griffiths, Suzuki, Lewontin, Gelbart, David T. Suzuki, Richard C. Lewontin, Willi Gelbart, Miller, Jeffrey H. Miller/ Hardcover/ W. H. Freeman Company/ February 2000

Jefferson's Children: The Story of One American Family Shannon Lanier, Jane Feldman, Lucian K. Truscott (Introduction)/ Hardcover/ Random House Books for Young Readers/ September 2000

Medical Genetics Lynn B. B. Jorde, Michael J. Bamshad, Raymond L. White, Michael J. Bamshad, John C. Carey, John C. Carey, Raymond L. White, John C. Carey/ Paperback/ Mosby-Year Book, Inc./ July 2000

Molecule Hunt, The: Archaeology and the Search for Ancient DNA Martin Jones/ Hardcover/ Arcade/ April 2002

More Chemistry and Crime: From Marsh Arsenic Test to DNA Profile Richard Saferstein, Samuel M. Gerber (Editor)/ Hardcover/ American Chemical Society/ August 1998

1996, Quest For Perfection—The Drive to Breed Better Human Beings, Maranto, Gina. Scribner, 1996

Our Molecular Future: How Nanotechnology, Robotics, Genetics, and Artificial Intelligence Will Transform Our World Mulhall, Douglas./ Hardcover/ Prometheus Books/ March 2002

Paternity—Disputed, Typing, PCR and DNA Tests: Index of New Information Dexter Z. Franklin/ Hardcover/ Abbe Pub Assn of Washington Dc/ January 1998

Paternity in Primates: Tests and Theories R. D. Martin (Editor), A. F. Dickson (Editor), E. J. Wickings (Editor)/ Hardcover/ Karger, S Publishers/ December 1991

Queen Victoria's Gene: Hemophilia and the Royal Family (Pbk) D. M. Potts, W. T. Potts/ Paperback/ Sutton Publishing, Limited/ June 1999

Redesigning Humans: Our Inevitable Genetic Future Stock, Gregory./ Hardcover/ Houghton Mifflin Company/ April 2002

Rosalind Franklin: The Dark Lady of DNA, Brenda Maddox/ Hardcover/ HarperCollins Publishers/ October 2002

Schaum's Outline Of Genetics Susan Elrod, William D. Stansfield/ Paperback/ McGraw-Hill Companies, The/ December 2001

Seven Daughters of Eve, The: The Science That Reveals Our Genetic Ancestry. Sykes, Bryan. **ISBN:** 0393323145 **Publisher:** Norton, W. W. & Company, Inc. May 2002

Stedman's OB-GYN & Genetics Words Ellen Atwood (Editor), Stedmans/ Paperback/ Lippincott Williams & Wilkins/ December 2000

♦ ♦ ♦

Bibliography 3: Articles, Periodicals, and Books on DNA and Ancestry/Ethnicity

• Santachiara Benerecette AS, et al.	The common, Near-Eastern Origin of Ashkenazi and Sephardi Jews Supported by Y-Chromosome Similarity," Santachiara Benerecette AS, Semino O, Passarino G, Torroni A, Brdicka R, Fellous M., Modiano G, Diparrtimento de Biologia Cellulare, University della Calabria, Cosenza, Italy. Annals of Human Genetics Jan; 57 (Pt 1): 55–64, 1993
• Richards, Martin, Vincent Macaulay, et al.	**Tracing European Founder Lineages in the Near Eastern mtDNA Pool. American Journal of Human Genetics, 67: 1251–1276, 2000**
• Mourant, et al.	**The Genetics of the Jews. Clarendon Press, 1978**

Also see a list of more DNA-related articles to read at Vincent Macaulay's Web site: http://www.stats.ox.ac.uk/~macaulay/

Periodical: **Avotaynu**, A Journal of Jewish Genealogy http://www.avotaynu.com/

Articles: Genetic Research and Origins:

- Mishmar, D., Ruiz–Pesini, E., Golik, P., Macaulay, V., Clark, A. G., Hosseini, S., Brandon, M., Easley, K., Chen, E., Brown, M. D., Sukernik, R. I., Olckers, A. and Wallace, D. C. (2003). Natural selection shaped regional mtDNA variation in humans. *Proceedings of the National Academy of Sciences USA*, **100**, 171–176.

- Richards, M., Macaulay, V. and Bandelt, H.–J. (2003). Analyzing genetic data in a model-based framework: inferences about European prehistory. In *Examining the farming/language dispersal hypothesis* (eds. P. Bellwood and C. Renfrew), 459–466. McDonald Institute for Archaeological Research, Cambridge.

- Bandelt, H.-J., Macaulay, V. and Richards, M. (2003). What molecules can't tell us about the spread of languages and the Neolithic. In *Examining the farming/language dispersal hypothesis* (eds. P. Bellwood and C. Renfrew), 99–112. McDonald Institute for Archaeological Research, Cambridge.

- Richards, M., Macaulay, V., Torroni, A. and Bandelt, H.-J. (2002). In search of geographical patterns in European mtDNA. *American Journal of Human Genetics*, **71**, 1168–1174.

- Macaulay, V. A. (2002). Review: *Genetics and the search for modern human origins*, John H. Relethford. *Heredity*, **89**, 160.

- Bandelt, H.-J., Lahermo, P., Richards, M. and Macaulay, V. (2001). Detecting errors in mtDNA data by phylogenetic analysis. *International Journal of Legal Medicine*, **115**, 64–69.

- Torroni, A., Bandelt, H.-J., Macaulay, V., Richards, M., Cruciani, F., Rengo, C., Martinez–Cabrera, V., Villems, R., Kivisild, T., Metspalu, E., Parik, J., Tolk, H.-V., Tambets, K., Forster, P., Karger, B., Francalacci, P., Rudan, P., Janicijevic, B., Rickards, O., Savontaus, M.-L., Huoponen, K., Laitinen, V., Koivumäki, S., Sykes, B., Hickey, E., Novelletto, A., Moral, P., Sellitto, D., Coppa, A., Al–Zaheri, N., Santachiara–Benerecetti, A. S., Semino, O. and Scozzari, R. (2001). A signal, from human mtDNA, of postglacial recolonization in Europe. *American Journal of Human Genetics*, **69**, 844–852.

- Scozzari, R., Cruciani, F., Pangrazio, A., Santolamazza, P., Vona, G., Moral, P., Latini, V., Varesi, L., Memmi, M. M., Romano, V., De Leo, G., Gennarelli, M., Jaruzelska, J., Villems, R., Parik, J., Macaulay, V. and Torroni, A. (2001). Human Y-chromosome variation in the western Mediterranean area: implications for the peopling of the region. *Human Immunology*, **62**, 871–884.

- Cooper, A., Rambaut, A., Macaulay, V., Willerslev, E., Hansen, A. J. and Stringer, C. (2001). Human origins and ancient human DNA. *Science*, **292**, 1655–1656.

- Richards, M. and Macaulay, V. (2001). The mitochondrial gene tree comes of ages. *American Journal of Human Genetics*, **68**, 1315–1320.

- Richards, M. and Macaulay, V. (2000). Genetic data and the colonization of Europe: genealogies and founders. In *Archaeogenetics: DNA and the Population Prehistory of Europe* (eds. C. Renfrew and K. Boyle), pp. 139–151. McDonald Institute for Archaeological Research, Cambridge.

- Richards, M., Macaulay, V., Hickey, E., Vega, E., Sykes, B., Guida, V., Rengo, C., Sellitto, D., Cruciani, F., Kivisild, T., Villems, R., Thomas, M., Rychkov, S., Rychkov, O., Rychkov, Y., Gölge, M., Dimitrov, D., Hill, E., Bradley, D., Romano, V., Calì, F., Vona, G., Demaine,A., Papiha, S., Triantaphyllidis, C., Stefanescu, G., Hatina, J., Belledi, M., Di Rienzo, A., Novelletto, A., Oppenheim, A., Nørby, S., Santachiara–Benerecetti, S., Scozzari, R., Torroni, A., Bandelt, H.–J. (2000). Tracing European founder lineages in the Near Eastern mtDNA pool. *American Journal of Human Genetics*, **67**, 1251–1276.

- Torroni, A., Richards, M., Macaulay, V., Forster, P., Villems, R., Nørby,S., Savontaus, M.–L., Huoponen, K., Scozzari, R. and Bandelt, H.–J. (2000). mtDNA haplogroups and frequency patterns in Europe. *American Journal of Human Genetics*, **66**, 1173–1177.

- Macaulay, V., Richards, M. and Sykes, B. (1999). Mitochondrial DNA recombination—no need to panic. *Proceedings of the Royal Society of London* B, **266**, 2037–2039.

- Macaulay, V., Richards, M., Hickey, E., Vega, E., Cruciani, F., Guida,V., Scozzari, R., Bonné–Tamir, B., Sykes, B. and Torroni, A. (1999). The emerging tree of west Eurasians mtDNAs: a synthesis of control-region sequences and RFLPs. *American Journal of Human Genetics*, **64**, 232–249.

- Richards, M. B., Macaulay, V. A., Bandelt, H.–J. and Sykes, B. C. (1998). Phylogeography of mitochondrial DNA in Western Europe. *Annals of Human Genetics*, **62**, 241–260.

- Wexler, Paul, Ph.D, Two-tiered relexification in Yiddish: Jews, Sorbs, Khazars and the Kiev-Polessian dialect, Berlin-NY 2002: Mouton de Gruyter Press, 2002.

***

APPENDIX F

Permissions:

Dear Anne,

You have permission to use the primer text and the photo of the DNA molecule for your book on DNA testing for genealogists. Please prominently credit the U.S. Department of Energy Human Genome Program as the source for both and also include our website for more information on the Human Genome Project and its applications: www.ornl.gov/hgmis. Yes, you can use the *Dictionary of Genetic Terms* at the end of the online version of the Primer. Please provide source citations when you use each part of the document. We would appreciate having a copy of your book when it is completed.

Sincerely,

Denise Casey

Denise K. Casey
Science Writer/Editor
Human Genome News
Human Genome Management Information System
Oak Ridge National Laboratory
1060 Commerce Park, MS 6480
Oak Ridge, TN 37830
865/574-0597; Fax: 865/574-9888; Email: caseydk@ornl.gov
HGMIS World Wide Web URL: http://www.ornl.gov/hgmis
Sponsor: U.S. Department of Energy

Harry Ostrer, M.D.
Professor of Pediatrics, Pathology, and Medicine
Director, Human Genetics Program
New York University School of Medicine

550 First Avenue, MSB 136
New York, NY 10016
tel 212 263-7596
fax 212 263-3477
email harry.ostrer@med.nyu.edu

The text of National Cancer Institute (NCI) material is in the public domain when the content was written by a government employee. Such content is not subject to copyright restrictions. One does not need special permission to reproduce or translate written text created by NCI staff. However, we would appreciate a credit line and a copy of any translated material.

Likewise, permission is not needed to link to NCI Web sites. If you wish to use material on NCI Web sites, we strongly suggest linking directly to that information to be sure that you have the most up-to-date version.

***

APPENDIX G

List of Published Paperback Books in Print Written by Anne Hart

1. Title: How to Interpret Family History and Ancestry DNA Test Results for Beginners: The Geography and History of Your Relatives
   ISBN: 0-595-31684-0

2. Title: Cover Letters, Follow-Ups, and Book Proposals: Samples with Templates
   ISBN: 0-595-31663-8

3. Title: Writer's Guide to Book Proposals: Templates, Query Letters, & Free Media Publicity
   ISBN: 0-595-31673-5

4. Title: Search Your Middle Eastern & European Genealogy: In the Former Ottoman Empire's Records and Online
   ISBN:0-595-31811-8

5. Title: Is Radical Liberalism or Extreme Conservatism a Character Disorder, Mental Disease, or Publicity Campaign?—A Novel of Intrigue—
   ISBN: 0-595-31751-0

6. How to Write Plays, Monologues, and Skits from Life Stories, Social Issues, and Current Events—for all Ages.
   ISBN: 0-595-31866-5

7. Title: How to Make Money Organizing Information
   ISBN: 0-595-23695-2

8. Title: How To Stop Elderly Abuse: A Prevention Guidebook
   ISBN: 0-595-23550-6

9. Title: How to Make Money Teaching Online With Your Camcorder and PC: 25 Practical and Creative How-To Start-Ups To Teach Online
    ISBN: 0-595-22123-8

10. Title: A Private Eye Called Mama Africa: What's an Egyptian Jewish Female Psycho-Sleuth Doing Fighting Hate Crimes in California?
    ISBN: 0-595-18940-7

11. Title: The Freelance Writer's E-Publishing Guidebook: 25+ E-Publishing Home-based Online Writing Businesses to Start for Freelancers
    ISBN: 0-595-18952-0

12. Title: The Courage to Be Jewish and the Wife of an Arab Sheik: What's a Jewish Girl from Brooklyn Doing Living as a Bedouin?
    ISBN: 0-595-18790-0

13. Title: The Year My Whole Country Turned Jewish: A Time-Travel Adventure Novel in Medieval Khazaria
    ISBN: 0-75967-251-2

14. Stage Play Book: ISBN:

15. Title: The Day My Whole Country Turned Jewish: The Silk Road Kids
    ISBN: 0-7596-6380-7

16. Title: Four Astronauts and a Kitten: A Mother and Daughter Astronaut Team, the Teen Twin Sons, and Patches, the Kitten: The Intergalactic Friendship Club
    ISBN: 0-595-19202-5

17. Title: The Writer's Bible: Digital and Print Media: Skills, Promotion, and Marketing for Novelists, Playwrights, and Script Writers. Writing Entertainment Content for the New and Print Media.
    ISBN: 0-595-19305-6

18. Title: New Afghanistan's TV Anchorwoman: A novel of mystery set in the New Afghanistan
    ISBN: 0-595-21557-2

19. Title: Tools for Mystery Writers: Writing Suspense Using Hidden Personality Traits
    ISBN: 0-595-21747-8

20. Title: The Khazars Will Rise Again!: Mystery Tales of the Khazars
    ISBN: 0-595-21830-X

21. Title: Murder in the Women's Studies Department: A Professor Sleuth Novel of Mystery
    ISBN: 0-595-21859-8

22. Title: Make Money With Your Camcorder and PC: 25+ Businesses: Make Money With Your Camcorder and Your Personal Computer by Linking Them.
    ISBN: 0-595-21864-4

23. Title: Writing What People Buy: 101+ Projects That Get Results
    ISBN: 0-595-21936-5

24. Title: Anne Joan Levine, Private Eye: Internal adventure through first-person mystery writer's diary novels
    ISBN: 0-595-21860-1

25. Title: Verbal Intercourse: A Darkly Humorous Novel of Interpersonal Couples and Family Communication
    ISBN: 0-595-21946-2

26. Title: The Date Who Unleashed Hell: If You Love Me, Why Do You Humiliate Me?
    "The Date" Mystery Fiction Series
    ISBN: 0-595-21982-9

27. Title: Cleopatra's Daughter: Global Intercourse
    ISBN: 0-595-22021-5

28. Title: Cyber Snoop Nation: The Adventures Of Littanie Webster, Sixteen-Year-Old Genius Private Eye On Internet Radio
    ISBN: 0-595-22033-9

29. Title: Counseling Anarchists: We All Marry Our Mirrors—Someone Who Reflects How We Feel About Ourselves. Folding Inside Ourselves: A Novel of Mystery
ISBN: 0-595-22054-1

30. Title: Sacramento Latina: When the One Universal We Have In Common Divides Us
ISBN: 0-595-22061-4

31. Title: Astronauts and Their Cats: At night, the space station is cat-shadow dark
ISBN: 0-595-22330-3

32. Title: How Two Yellow Labs Saved the Space Program: When Smart Dogs Shape Shift in Space
ISBN: 0-595-23181-0

33. Title: The DNA Detectives: Working Against Time
ISBN: 0-595-25339-3

34. Title: How to Interpret Your DNA Test Results For Family History & Ancestry: Scientists Speak Out on Genealogy Joining Genetics
ISBN: 0-595-26334-8

35. Title: Roman Justice: SPQR: Too Roman To Handle
ISBN: 0-595-27282-7

36. Title: How to Make Money Selling Facts: to Non-Traditional Markets
ISBN: 0-595-27842-6

37. Title: Tracing Your Jewish DNA For Family History & Ancestry: Merging a Mosaic of Communities
ISBN: 0-595-28127-3

38. Title: The Beginner's Guide to Interpreting Ethnic DNA Origins for Family History: How Ashkenazi, Sephardi, Mizrahi & Europeans Are Related to Everyone Else
ISBN: 0-595-28306-3

39. Title: Nutritional Genomics—A Consumer's Guide to How Your Genes and Ancestry Respond to Food: Tailoring What You Eat to Your DNA
ISBN: 0-595-29067-1

40. Title: How to Safely Tailor Your Food, Medicines, & Cosmetics to Your Genes: A Consumer's Guide to Genetic Testing Kits from Ancestry to Nourishment
ISBN: 0-595-29403-0

41. Title: One Day Some Schlemiel Will Marry Me, Pay the Bills, and Hug Me.: Parents & Children Kvetch on Arab & Jewish Intermarriage
ISBN: 0-595-29826-5

42. Title: Find Your Personal Adam And Eve: Make DNA-Driven Genealogy Time Capsules
ISBN: 0-595-30633-0

43. Title: Creative Genealogy Projects: Writing Salable Life Stories
ISBN: 0-595-31305-1

44. Title: Power Dating Games: What's Important to Know About the Person You'll Marry
ISBN: 0-595-19186-X

APPENDIX H

Reprinted with Permission Human Genome Project Information Web site
Check out the updates on their site.

Need a Glossary of Genetic Terms Defined?

Dictionary of Genetic Terms

Genomics and Its Impact on Medicine and Society: A 2001 Primer

A

Acquired genetic mutation
See: somatic cell genetic mutation

Additive genetic effects
When the combined effects of alleles at different loci are equal to the sum of their individual effects.
See also: anticipation, complex trait

Adenine (A)
A nitrogenous base, one member of the base pair AT (adenine-thymine).
See also: base pair, nucleotide

Affected relative pair
Individuals related by blood, each of whom is affected with the same trait. Examples are affected sibling, cousin, and avuncular pairs.
See also: avuncular relationship

Aggregation technique
A technique used in model organism studies in which embryos at the 8-cell stage of development are pushed together to yield a single embryo (used as an alternative to microinjection).
See also: model organisms

Allele
Alternative form of a genetic locus; a single allele for each locus is inherited from each parent (e.g., at a locus for eye color the allele might result in blue or brown eyes).
See also: locus, gene expression

Allogeneic
Variation in alleles among members of the same species.

Alternative splicing
Different ways of combining a gene's exons to make variants of the complete protein

Amino acid
Any of a class of 20 molecules that are combined to form proteins in living things. The sequence of amino acids in a protein and hence protein function are determined by the genetic code.

Amplification
An increase in the number of copies of a specific DNA fragment; can be in vivo or in vitro.
See also: cloning, polymerase chain reaction

Animal model
See: model organisms

Annotation
Adding pertinent information such as gene coded for, amino acid sequence, or other commentary to the database entry of raw sequence of DNA bases.
See also: bioinformatics

Anticipation
Each generation of offspring has increased severity of a genetic disorder; e.g., a grandchild may have earlier onset and more severe symptoms than the parent, who had earlier onset than the grandparent.
See also: additive genetic effects, complex trait

Antisense
Nucleic acid that has a sequence exactly opposite to an mRNA molecule made by the body; binds to the mRNA molecule to prevent a protein from being made.
See also: transcription

Apoptosis
Programmed cell death, the body's normal method of disposing of damaged, unwanted, or unneeded cells.
See also: cell

Arrayed library
Individual primary recombinant clones (hosted in phage, cosmid, YAC, or other vector) that are placed in two-dimensional arrays in microtiter dishes. Each primary clone can be identified by the identity of the plate and the clone location (row and column) on that plate. Arrayed libraries of clones can be used for many applications, including screening for a specific gene or genomic region of interest.
See also: library, genomic library, gene chip technology

Assembly
Putting sequenced fragments of DNA into their correct chromosomal positions.

Autoradiography
A technique that uses X-ray film to visualize radioactively labeled molecules or fragments of molecules; used in analyzing length and number of DNA fragments after they are separated by gel electrophoresis.

Autosomal dominant
A gene on one of the non-sex chromosomes that is always expressed, even if only one copy is present. The chance of passing the gene to offspring is 50% for each pregnancy.
See also: autosome, dominant, gene

Autosome
A chromosome not involved in sex determination. The diploid human genome consists of a total of 46 chromosomes: 22 pairs of autosomes, and 1 pair of sex chromosomes (the X and Y chromosomes).
See also: sex chromosome

Avuncular relationship
The genetic relationship between nieces and nephews and their aunts and uncles.

B

Backcross
A cross between an animal that is heterozygous for alleles obtained from two parental strains and a second animal from one of those parental strains. Also used to describe the breeding protocol of an outcross followed by a backcross.
See also: model organisms

Bacterial artificial chromosome (BAC)
A vector used to clone DNA fragments (100-to 300-kb insert size; average, 150 kb) in *Escherichia coli* cells. Based on naturally occurring F-factor plasmid found in the bacterium *E. coli*.
See also: cloning vector

Bacteriophage
See: phage

Base
One of the molecules that form DNA and RNA molecules.
See also: nucleotide, base pair, base sequence

Base pair (bp)
Two nitrogenous bases (adenine and thymine or guanine and cytosine) held together by weak bonds. Two strands of DNA are held together in the shape of a double helix by the bonds between base pairs.

Base sequence
The order of nucleotide bases in a DNA molecule; determines structure of proteins encoded by that DNA.

Base sequence analysis
A method, sometimes automated, for determining the base sequence.

Behavioral genetics
The study of genes that may influence behavior.

Bioinformatics
The science of managing and analyzing biological data using advanced computing techniques. Especially important in analyzing genomic research data.
See also: informatics

Bioremediation
The use of biological organisms such as plants or microbes to aid in removing hazardous substances from an area.

Biotechnology
A set of biological techniques developed through basic research and now applied to research and product development. In particular, biotechnology refers to the use by industry of recombinant DNA, cell fusion, and new bioprocessing techniques.

Birth defect
Any harmful trait, physical or biochemical, present at birth, whether a result of a genetic mutation or some other nongenetic factor.
See also: congenital, gene, mutation, syndrome

BLAST
A computer program that identifies homologous (similar) genes in different organisms, such as human, fruit fly, or nematode.

C

Cancer
Diseases in which abnormal cells divide and grow unchecked. Cancer can spread from its original site to other parts of the body and can be fatal.
See also: hereditary cancer, sporadic cancer

Candidate gene

A gene located in a chromosome region suspected of being involved in a disease.

See also: positional cloning, protein

Capillary array

Gel-filled silica capillaries used to separate fragments for DNA sequencing. The small diameter of the capillaries permit the application of higher electric fields, providing high speed, high throughput separations that are significantly faster than traditional slab gels.

Carcinogen

Something which causes cancer to occur by causing changes in a cell's DNA.

See also: mutagene

Carrier

An individual who possesses an unexpressed, recessive trait.

cDNA library

A collection of DNA sequences that code for genes. The sequences are generated in the laboratory from mRNA sequences.

See also: messenger RNA

Cell

The basic unit of any living organism that carries on the biochemical processes of life.

See also: genome, nucleus

Centimorgan (cM)

A unit of measure of recombination frequency. One centimorgan is equal to a 1% chance that a marker at one genetic locus will be separated from a marker at a second locus due to crossing over in a single generation. In human beings, one centimorgan is equivalent, on average, to one million base pairs.

See also: megabase

Centromere

A specialized chromosome region to which spindle fibers attach during cell division.

Chimera (pl. chimaera)
An organism that contains cells or tissues with a different genotype. These can be mutated cells of the host organism or cells from a different organism or species.

Chimeraplasty
An experimental targeted repair process in which a desirable sequence of DNA is combined with RNA to form a chimeraplast. These molecules bind selectively to the target DNA. Once bound, the chimeraplast activates a naturally occurring gene-correcting mechanism. Does not use viral or other conventional gene-delivery vectors.
See also: gene therapy, cloning vector

Chloroplast chromosome
Circular DNA found in the photosynthesizing organelle (chloroplast) of plants instead of the cell nucleus where most genetic material is located.

Chromomere
One of the serially aligned beads or granules of a eukaryotic chromosome, resulting from local coiling of a continuous DNA thread.

Chromosomal deletion
The loss of part of a chromosome's DNA.

Chromosomal inversion
Chromosome segments that have been turned 180 degrees. The gene sequence for the segment is reversed with respect to the rest of the chromosome.

Chromosome
The self-replicating genetic structure of cells containing the cellular DNA that bears in its nucleotide sequence the linear array of genes. In prokaryotes, chromosomal DNA is circular, and the entire genome is carried on one chromosome. Eukaryotic genomes consist of a number of chromosomes whose DNA is associated with different kinds of proteins.

Chromosome painting
Attachment of certain fluorescent dyes to targeted parts of the chromosome. Used as a diagnositic for particular diseases, e.g. types of leukemia.

Chromosome region p
A designation for the short arm of a chromosome.

Chromosome region q
A designation for the long arm of a chromosome.

Clone
An exact copy made of biological material such as a DNA segment (e.g., a gene or other region), a whole cell, or a complete organism.

Clone bank
See: genomic library

Cloning
Using specialized DNA technology to produce multiple, exact copies of a single gene or other segment of DNA to obtain enough material for further study. This process, used by researchers in the Human Genome Project, is referred to as cloning DNA. The resulting cloned (copied) collections of DNA molecules are called clone libraries. A second type of cloning exploits the natural process of cell division to make many copies of an entire cell. The genetic makeup of these cloned cells, called a cell line, is identical to the original cell. A third type of cloning produces complete, genetically identical animals such as the famous Scottish sheep, Dolly.
See also: cloning vector

Cloning vector
DNA molecule originating from a virus, a plasmid, or the cell of a higher organism into which another DNA fragment of appropriate size can be integrated without loss of the vector's capacity for self-replication; vectors introduce foreign DNA into host cells, where the DNA can be reproduced in large quantities. Examples are plasmids, cosmids, and yeast artificial chromosomes; vectors are often recombinant molecules containing DNA sequences from several sources.

Code
See: genetic code

Codominance
   Situation in which two different alleles for a genetic trait are both expressed.
   See also: autosomal dominant, recessive gene

Codon
   See: genetic code

Coisogenic or congenic
   Nearly identical strains of an organism; they vary at only a single locus.

Comparative genomics
   The study of human genetics by comparisons with model organisms such as mice, the fruit fly, and the bacterium *E. coli*.

Complementary DNA (cDNA)
   DNA that is synthesized in the laboratory from a messenger RNA template.

Complementary sequence
   Nucleic acid base sequence that can form a double-stranded structure with another DNA fragment by following base-pairing rules (A pairs with T and C with G). The complementary sequence to GTAC for example, is CATG.

Complex trait
   Trait that has a genetic component that does not follow strict Mendelian inheritance. May involve the interaction of two or more genes or gene-environment interactions.
   See also: Mendelian inheritance, additive genetic effects

Computational biology
   See: bioinformatics

Confidentiality
   In genetics, the expectation that genetic material and the information gained from testing that material will not be available without the donor's consent.

Congenital
>Any trait present at birth, whether the result of a genetic or nongenetic factor.
>*See also:* birth defect

Conserved sequence
>A base sequence in a DNA molecule (or an amino acid sequence in a protein) that has remained essentially unchanged throughout evolution.

Constitutive ablation
>Gene expression that results in cell death.

Contig
>Group of cloned (copied) pieces of DNA representing overlapping regions of a particular chromosome.

Contig map
>A map depicting the relative order of a linked library of overlapping clones representing a complete chromosomal segment.

Cosmid
>Artificially constructed cloning vector containing the cos gene of phage lambda. Cosmids can be packaged in lambda phage particles for infection into *E. coli*; this permits cloning of larger DNA fragments (up to 45kb) than can be introduced into bacterial hosts in plasmid vectors.

Crossing over
>The breaking during meiosis of one maternal and one paternal chromosome, the exchange of corresponding sections of DNA, and the rejoining of the chromosomes. This process can result in an exchange of alleles between chromosomes.
>*See also:* recombination

Cytogenetics
>The study of the physical appearance of chromosomes.
>*See also:* karyotype

Cytological band
>An area of the chromosome that stains differently from areas around it.
>*See also:* cytological map

Cytological map
A type of chromosome map whereby genes are located on the basis of cytological findings obtained with the aid of chromosome mutations.

Cytoplasmic (uniparental) inheritance
See: cytoplasmic trait

Cytoplasmic trait
A genetic characteristic in which the genes are found outside the nucleus, in chloroplasts or mitochondria. Results in offspring inheriting genetic material from only one parent.

Cytosine (C)
A nitrogenous base, one member of the base pair GC (guanine and cytosine) in DNA.
See also: base pair, nucleotide

D

Data warehouse
A collection of databases, data tables, and mechanisms to access the data on a single subject.

Deletion
A loss of part of the DNA from a chromosome; can lead to a disease or abnormality.
See also: chromosome, mutation

Deletion map
A description of a specific chromosome that uses defined mutations—specific deleted areas in the genome—as 'biochemical signposts,' or markers for specific areas.

Deoxyribonucleotide
See: nucleotide

Deoxyribose
A type of sugar that is one component of DNA (deoxyribonucleic acid).

Diploid
A full set of genetic material consisting of paired chromosomes, one from

each parental set. Most animal cells except the gametes have a diploid set of chromosomes. The diploid human genome has 46 chromosomes.
See also: haploid

Directed evolution
A laboratory process used on isolated molecules or microbes to cause mutations and identify subsequent adaptations to novel environments.

Directed mutagenesis
Alteration of DNA at a specific site and its reinsertion into an organism to study any effects of the change.

Directed sequencing
Successively sequencing DNA from adjacent stretches of chromosome.

Disease-associated genes
Alleles carrying particular DNA sequences associated with the presence of disease.

DNA (deoxyribonucleic acid)
The molecule that encodes genetic information. DNA is a double-stranded molecule held together by weak bonds between base pairs of nucleotides. The four nucleotides in DNA contain the bases adenine (A), guanine (G), cytosine (C), and thymine (T). In nature, base pairs form only between A and T and between G and C; thus the base sequence of each single strand can be deduced from that of its partner.

DNA bank
A service that stores DNA extracted from blood samples or other human tissue.

DNA probe
See: probe

DNA repair genes
Genes encoding proteins that correct errors in DNA sequencing.

DNA replication
The use of existing DNA as a template for the synthesis of new DNA

strands. In humans and other eukaryotes, replication occurs in the cell nucleus.

DNA sequence
The relative order of base pairs, whether in a DNA fragment, gene, chromosome, or an entire genome.
See also: base sequence analysis

Domain
A discrete portion of a protein with its own function. The combination of domains in a single protein determines its overall function.

Dominant
An allele that is almost always expressed, even if only one copy is present.
See also: gene, genome

Double helix
The twisted-ladder shape that two linear strands of DNA assume when complementary nucleotides on opposing strands bond together.

Draft sequence
The sequence generated by the HGP as of June 2000 that, while incomplete, offers a virtual road map to an estimated 95% of all human genes. Draft sequence data are mostly in the form of 10,000 base pair-sized fragments whose approximate chromosomal locations are known.
See also: sequencing, finished DNA sequence, working draft DNA sequence.

E

Electrophoresis
A method of separating large molecules (such as DNA fragments or proteins) from a mixture of similar molecules. An electric current is passed through a medium containing the mixture, and each kind of molecule travels through the medium at a different rate, depending on its electrical charge and size. Agarose and acrylamide gels are the media commonly used for electrophoresis of proteins and nucleic acids.

Electroporation
A process using high-voltage current to make cell membranes permeable to

allow the introduction of new DNA; commonly used in recombinant DNA technology.
See also: transfection

Embryonic stem (ES) cells
An embryonic cell that can replicate indefinitely, transform into other types of cells, and serve as a continuous source of new cells.

Endonuclease
See: restriction enzyme

Enzyme
A protein that acts as a catalyst, speeding the rate at which a biochemical reaction proceeds but not altering the direction or nature of the reaction.

Epistasis
One gene interfers with or prevents the expression of another gene located at a different locus.

Escherichia coli
Common bacterium that has been studied intensively by geneticists because of its small genome size, normal lack of pathogenicity, and ease of growth in the laboratory.

Eugenics
The study of improving a species by artificial selection; usually refers to the selective breeding of humans.

Eukaryote
Cell or organism with membrane-bound, structurally discrete nucleus and other well-developed subcellular compartments. Eukaryotes include all organisms except viruses, bacteria, and bluegreen algae.
See also: prokaryote, chromosome.

Evolutionarily conserved
See: conserved sequence

Exogenous DNA
DNA originating outside an organism that has been introduced into the organism.

Exon
    The protein-coding DNA sequence of a gene.
    See also: intron

Exonuclease
    An enzyme that cleaves nucleotides sequentially from free ends of a linear nucleic acid substrate.

Expressed gene
    See: gene expression

Expressed sequence tag (EST)
    A short strand of DNA that is a part of a cDNA molecule and can act as identifier of a gene. Used in locating and mapping genes.
    See also: cDNA, sequence tagged site

F

Filial generation (F1, F2)
    Each generation of offspring in a breeding program, designated F1, F2, etc.

Fingerprinting
    In genetics, the identification of multiple specific alleles on a person's DNA to produce a unique identifier for that person.
    See also: forensics

Finished DNA Sequence
    High-quality, low error, gap-free DNA sequence of the human genome. Achieving this ultimate 2003 HGP goal requires additional sequencing to close gaps, reduce ambiguities, and allow for only a single error every 10,000 bases, the agreed-upon standard for HGP finished sequence.
    See also: sequencing, draft sequence

Flow cytometry
    Analysis of biological material by detection of the light-absorbing or fluorescing properties of cells or subcellular fractions (i.e., chromosomes) passing in a narrow stream through a laser beam. An absorbance or fluorescence profile of the sample is produced. Automated sorting devices, used to fractionate samples, sort successive droplets of the analyzed stream into different fractions depending on the fluorescence emitted by each droplet.

Flow karyotyping
Use of flow cytometry to analyze and separate chromosomes according to their DNA content.

Fluorescence in situ hybridization (FISH)
A physical mapping approach that uses fluorescein tags to detect hybridization of probes with metaphase chromosomes and with the less-condensed somatic interphase chromatin.

Forensics
The use of DNA for identification. Some examples of DNA use are to establish paternity in child support cases; establish the presence of a suspect at a crime scene, and identify accident victims.

Fraternal twin
Siblings born at the same time as the result of fertilization of two ova by two sperm. They share the same genetic relationship to each other as any other siblings.
See also: identical twin

Full gene sequence
The complete order of bases in a gene. This order determines which protein a gene will produce.

Functional genomics
The study of genes, their resulting proteins, and the role played by the proteins the body's biochemical processes.

G

Gamete
Mature male or female reproductive cell (sperm or ovum) with a haploid set of chromosomes (23 for humans).

GC-rich area
Many DNA sequences carry long stretches of repeated G and C which often indicate a gene-rich region.

Gel electrophoresis
See: electrophoresis

Gene
The fundamental physical and functional unit of heredity. A gene is an ordered sequence of nucleotides located in a particular position on a particular chromosome that encodes a specific functional product (i.e., a protein or RNA molecule).
See also: gene expression

Gene amplification
Repeated copying of a piece of DNA; a characteristic of tumor cells.
See also: gene, oncogene

Gene chip technology
Development of cDNA microarrays from a large number of genes. Used to monitor and measure changes in gene expression for each gene represented on the chip.

Gene expression
The process by which a gene's coded information is converted into the structures present and operating in the cell. Expressed genes include those that are transcribed into mRNA and then translated into protein and those that are transcribed into RNA but not translated into protein (e.g., transfer and ribosomal RNAs).

Gene family
Group of closely related genes that make similar products.

Gene library
See: genomic library

Gene mapping
Determination of the relative positions of genes on a DNA molecule (chromosome or plasmid) and of the distance, in linkage units or physical units, between them.

Gene pool
All the variations of genes in a species.
See also: allele, gene, polymorphism

Gene prediction
> Predictions of possible genes made by a computer program based on how well a stretch of DNA sequence matches known gene sequences

Gene product
> The biochemical material, either RNA or protein, resulting from expression of a gene. The amount of gene product is used to measure how active a gene is; abnormal amounts can be correlated with disease-causing alleles.

Gene testing
> *See:* genetic testing, genetic screening

Gene therapy
> An experimental procedure aimed at replacing, manipulating, or supplementing nonfunctional or misfunctioning genes with healthy genes.
> *See also:* gene, inherit, somatic cell gene therapy, germ line gene therapy

Gene transfer
> Incorporation of new DNA into and organism's cells, usually by a vector such as a modified virus. Used in gene therapy.
> *See also:* mutation, gene therapy, vector

Genetic code
> The sequence of nucleotides, coded in triplets (codons) along the mRNA, that determines the sequence of amino acids in protein synthesis. A gene's DNA sequence can be used to predict the mRNA sequence, and the genetic code can in turn be used to predict the amino acid sequence.

Genetic counseling
> Provides patients and their families with education and information about genetic-related conditions and helps them make informed decisions.

Genetic discrimination
> Prejudice against those who have or are likely to develop an inherited disorder.

Genetic engineering
> Altering the genetic material of cells or organisms to enable them to make new substances or perform new functions.

Genetic engineering technology
See: recombinant DNA technology

Genetic illness
Sickness, physical disability, or other disorder resulting from the inheritance of one or more deleterious alleles.

Genetic informatics
See: bioinformatics

Genetic map
See: linkage map

Genetic marker
A gene or other identifiable portion of DNA whose inheritance can be followed.
See also: chromosome, DNA, gene, inherit

Genetic material
See: genome

Genetic mosaic
An organism in which different cells contain different genetic sequence. This can be the result of a mutation during development or fusion of embryos at an early developmental stage.

Genetic polymorphism
Difference in DNA sequence among individuals, groups, or populations (e.g., genes for blue eyes versus brown eyes).

Genetic predisposition
Susceptibility to a genetic disease. May or may not result in actual development of the disease.

Genetic screening
Testing a group of people to identify individuals at high risk of having or passing on a specific genetic disorder.

Genetic testing
Analyzing an individual's genetic material to determine predisposition to a particular health condition or to confirm a diagnosis of genetic disease.

Genetics
> The study of inheritance patterns of specific traits.

Genome
> All the genetic material in the chromosomes of a particular organism; its size is generally given as its total number of base pairs.

Genome project
> Research and technology-development effort aimed at mapping and sequencing the genome of human beings and certain model organisms.
> *See also:* Human Genome Initiative

Genomic library
> A collection of clones made from a set of randomly generated overlapping DNA fragments that represent the entire genome of an organism.
> *See also:* library, arrayed library

Genomic sequence
> *See:* DNA

Genomics
> The study of genes and their function.

Genotype
> The genetic constitution of an organism, as distinguished from its physical appearance (its phenotype).

Germ cell
> Sperm and egg cells and their precursors. Germ cells are haploid and have only one set of chromosomes (23 in all), while all other cells have two copies (46 in all).

Germ line
> The continuation of a set of genetic information from one generation to the next.
> *See also:* inherit

Germ line gene therapy
An experimental process of inserting genes into germ cells or fertilized eggs to cause a genetic change that can be passed on to offspring. May be used to alleviate effects associated with a genetic disease.
See also: genomics, somatic cell gene therapy.

Germ line genetic mutation
See: mutation

Guanine (G)
A nitrogenous base, one member of the base pair GC (guanine and cytosine) in DNA.
See also: base pair, nucleotide

Gyandromorph
Organisms that have both male and female cells and therefore express both male and female characteristics.

H

Haploid
A single set of chromosomes (half the full set of genetic material) present in the egg and sperm cells of animals and in the egg and pollen cells of plants. Human beings have 23 chromosomes in their reproductive cells.
See also: diploid

Haplotype
A way of denoting the collective genotype of a number of closely linked loci on a chromosome.

Hemizygous
Having only one copy of a particular gene. For example, in humans, males are hemizygous for genes found on the Y chromosome.

Hereditary cancer
Cancer that occurs due to the inheritance of an altered gene within a family.
See also: sporadic cancer

Heterozygosity
The presence of different alleles at one or more loci on homologous chromosomes.

Heterozygote
See: heterozygosity

Highly conserved sequence
DNA sequence that is very similar across several different types of organisms.
See also: gene, mutation

High-throughput sequencing
A fast method of determining the order of bases in DNA.
See also: sequencing

Homeobox
A short stretch of nucleotides whose base sequence is virtually identical in all the genes that contain it. Homeoboxes have been found in many organisms from fruit flies to human beings. In the fruit fly, a homeobox appears to determine when particular groups of genes are expressed during development.

Homolog
A member of a chromosome pair in diploid organisms or a gene that has the same origin and functions in two or more species.

Homologous chromosome
Chromosome containing the same linear gene sequences as another, each derived from one parent.

Homologous recombination
Swapping of DNA fragments between paired chromosomes.

Homology
Similarity in DNA or protein sequences between individuals of the same species or among different species.

Homozygote
An organism that has two identical alleles of a gene.
See also: heterozygote

Homozygous
See: homozygote

Human artificial chromosome (HAC)
A vector used to hold large DNA fragments.
See also: chromosome, DNA

Human gene therapy
See: gene therapy

Human Genome Initiative
Collective name for several projects begun in 1986 by DOE to create an ordered set of DNA segments from known chromosomal locations, develop new computational methods for analyzing genetic map and DNA sequence data, and develop new techniques and instruments for detecting and analyzing DNA. This DOE initiative is now known as the Human Genome Program. The joint national effort, led by DOE and NIH, is known as the Human Genome Project.

Human Genome Project (HGP)
Formerly titled Human Genome Initiative.
See also: Human Genome Initiative

Hybrid
The offspring of genetically different parents.
See also: heterozygote

Hybridization
The process of joining two complementary strands of DNA or one each of DNA and RNA to form a double-stranded molecule.

I

Identical twin
Twins produced by the division of a single zygote; both have identical genotypes.
See also: fraternal twin

Immunotherapy
Using the immune system to treat disease, for example, in the development of vaccines. May also refer to the therapy of diseases caused by the immune system.
See also: cancer

Imprinting
A phenomenon in which the disease phenotype depends on which parent passed on the disease gene. For instance, both Prader-Willi and Angelman syndromes are inherited when the same part of chromosome 15 is missing. When the father's complement of 15 is missing, the child has Prader-Willi, but when the mother's complement of 15 is missing, the child has Angelman syndrome.

In situ hybridization
Use of a DNA or RNA probe to detect the presence of the complementary DNA sequence in cloned bacterial or cultured eukaryotic cells.

In vitro
Studies performed outside a living organism such as in a laboratory.

In vivo
Studies carried out in living organisms.

Independent assortment
During meiosis each of the two copies of a gene is distributed to the germ cells independently of the distribution of other genes.
See also: linkage

Informatics
See: bioinformatics

Informed consent
An individual willingly agrees to participate in an activity after first being advised of the risks and benefits.
See also: privacy

Inherit
In genetics, to receive genetic material from parents through biological processes.

Inherited
See: inherit

Insertion
A chromosome abnormality in which a piece of DNA is incorporated into a gene and thereby disrupts the gene's normal function.
See also: chromosome, DNA, gene, mutation

Insertional mutation
See: insertion

Intellectual property rights
Patents, copyrights, and trademarks.
See also: patent

Interference
One crossover event inhibits the chances of another crossover event. Also known as positive interference. Negative interference increases the chance of a second crossover.
See also: crossing over

Interphase
The period in the cell cycle when DNA is replicated in the nucleus; followed by mitosis.

Intron
DNA sequence that interrupts the protein-coding sequence of a gene; an intron is transcribed into RNA but is cut out of the message before it is translated into protein.
See also: exon

Isoenzyme
An enzyme performing the same function as another enzyme but having a different set of amino acids. The two enzymes may function at different speeds.

J

Junk DNA
Stretches of DNA that do not code for genes; most of the genome consists of so-called junk DNA which may have regulatory and other functions. Also called non-coding DNA.

K

Karyotype
A photomicrograph of an individual's chromosomes arranged in a standard format showing the number, size, and shape of each chromosome type; used in low-resolution physical mapping to correlate gross chromosomal abnormalities with the characteristics of specific diseases.

Kilobase (kb)
Unit of length for DNA fragments equal to 1000 nucleotides.

Knockout
Deactivation of specific genes; used in laboratory organisms to study gene function.
See also: gene, locus, model organisms

L

Library
An unordered collection of clones (i.e., cloned DNA from a particular organism) whose relationship to each other can be established by physical mapping.
See also: genomic library, arrayed library

Linkage
The proximity of two or more markers (e.g., genes, RFLP markers) on a chromosome; the closer the markers, the lower the probability that they will be separated during DNA repair or replication processes (binary fission

in prokaryotes, mitosis or meiosis in eukaryotes), and hence the greater the probability that they will be inherited together.

Linkage disequilibrium

Where alleles occur together more often than can be accounted for by chance. Indicates that the two alleles are physically close on the DNA strand.

See also: Mendelian inheritance

Linkage map

A map of the relative positions of genetic loci on a chromosome, determined on the basis of how often the loci are inherited together. Distance is measured in centimorgans (cM).

Localize

Determination of the original position (locus) of a gene or other marker on a chromosome.

Locus (pl. loci)

The position on a chromosome of a gene or other chromosome marker; also, the DNA at that position. The use of locus is sometimes restricted to mean expressed DNA regions.

See also: gene expression

Long-Range Restriction Mapping

Restriction enzymes are proteins that cut DNA at precise locations. Restriction maps depict the chromosomal positions of restriction-enzyme cutting sites. These are used as biochemical "signposts," or markers of specific areas along the chromosomes. The map will detail the positions where the DNA molecule is cut by particular restriction enzymes.

M

Macrorestriction map

Map depicting the order of and distance between sites at which restriction enzymes cleave chromosomes.

Mapping

See: gene mapping, linkage map, physical map

Mapping population
The group of related organisms used in constructing a genetic map.

Marker
See: genetic marker

Mass spectrometry
An instrument used to identify chemicals in a substance by their mass and charge.

Megabase (Mb)
Unit of length for DNA fragments equal to 1 million nucleotides and roughly equal to 1 cM.
See also: centimorgan

Meiosis
The process of two consecutive cell divisions in the diploid progenitors of sex cells. Meiosis results in four rather than two daughter cells, each with a haploid set of chromosomes.
See also: mitosis

Mendelian inheritance
One method in which genetic traits are passed from parents to offspring. Named for Gregor Mendel, who first studied and recognized the existence of genes and this method of inheritance.
See also: autosomal dominant, recessive gene, sex-linked

Messenger RNA (mRNA)
RNA that serves as a template for protein synthesis.
See also: genetic code

Metaphase
A stage in mitosis or meiosis during which the chromosomes are aligned along the equatorial plane of the cell.

Microarray
Sets of miniaturized chemical reaction areas that may also be used to test DNA fragments, antibodies, or proteins.

Microbial genetics
> The study of genes and gene function in bacteria, archaea, and other microorganisms. Often used in research in the fields of bioremediation, alternative energy, and disease prevention.
> *See also:* model organisms, biotechnology, bioremediation

Microinjection
> A technique for introducing a solution of DNA into a cell using a fine microcapillary pipet.

Micronuclei
> Chromosome fragments that are not incorporated into the nucleus at cell division.

Mitochondrial DNA
> The genetic material found in mitochondria, the organelles that generate energy for the cell. Not inherited in the same fashion as nucleic DNA.
> *See also:* cell, DNA, genome, nucleus

Mitosis
> The process of nuclear division in cells that produces daughter cells that are genetically identical to each other and to the parent cell.
> *See also:* meiosis

Model organisms
> A laboratory animal or other organism useful for research.

Modeling
> The use of statistical analysis, computer analysis, or model organisms to predict outcomes of research.

Molecular biology
> The study of the structure, function, and makeup of biologically important molecules.

Molecular farming
> The development of transgenic animals to produce human proteins for medical use.

Molecular genetics
The study of macromolecules important in biological inheritance.

Molecular medicine
The treatment of injury or disease at the molecular level. Examples include the use of DNA-based diagnostic tests or medicine derived from DNA sequence information.

Monogenic disorder
A disorder caused by mutation of a single gene.
See also: mutation, polygenic disorder

Monogenic inheritance
See: monogenic disorder

Monosomy
Possessing only one copy of a particular chromosome instead of the normal two copies.
See also: cell, chromosome, gene expression, trisomy

Morbid map
A diagram showing the chromosomal location of genes associated with disease.

Mouse model
See: model organisms

Multifactorial or multigenic disorder
See: polygenic disorder

Multiplexing
A laboratory approach that performs multiple sets of reactions in parallel (simultaneously); greatly increasing speed and throughput.

Murine
Organism in the genus Mus. A rat or mouse.

Mutagen
An agent that causes a permanent genetic change in a cell. Does not include changes occurring during normal genetic recombination.

Mutagenicity
   The capacity of a chemical or physical agent to cause permanent genetic alterations.
   See also: somatic cell genetic mutation

Mutation
   Any heritable change in DNA sequence.
   See also: polymorphism

N

Nitrogenous base
   A nitrogen-containing molecule having the chemical properties of a base. DNA contains the nitrogenous bases adenine (A), guanine (G), cytosine (C), and thymine (T).
   See also: DNA

Northern blot
   A gel-based laboratory procedure that locates mRNA sequences on a gel that are complementary to a piece of DNA used as a probe.
   See also: DNA, library

Nuclear transfer
   A laboratory procedure in which a cell's nucleus is removed and placed into an oocyte with its own nucleus removed so the genetic information from the donor nucleus controls the resulting cell. Such cells can be induced to form embryos. This process was used to create the cloned sheep "Dolly".
   See also: cloning

Nucleic acid
   A large molecule composed of nucleotide subunits.
   See also: DNA

Nucleolar organizing region
   A part of the chromosome containing rRNA genes.

Nucleotide
   A subunit of DNA or RNA consisting of a nitrogenous base (adenine, guanine, thymine, or cytosine in DNA; adenine, guanine, uracil, or cytosine in RNA), a phosphate molecule, and a sugar molecule (deoxyribose in DNA

and ribose in RNA). Thousands of nucleotides are linked to form a DNA or RNA molecule.
See also: DNA, base pair, RNA

Nucleus
The cellular organelle in eukaryotes that contains most of the genetic material.

O

Oligo
See: oligonucleotide

Oligogenic
A phenotypic trait produced by two or more genes working together.
See also: polygenic disorder

Oligonucleotide
A molecule usually composed of 25 or fewer nucleotides; used as a DNA synthesis primer.
See also: nucleotide

Oncogene
A gene, one or more forms of which is associated with cancer. Many oncogenes are involved, directly or indirectly, in controlling the rate of cell growth.

Open reading frame (ORF)
The sequence of DNA or RNA located between the start-code sequence (initiation codon) and the stop-code sequence (termination codon).

Operon
A set of genes transcribed under the control of an operator gene.

Overlapping clones
See: genomic library

P

P1-derived artificial chromosome (PAC)
One type of vector used to clone DNA fragments (100-to 300-kb insert

size; average, 150 kb) in *Escherichia coli* cells. Based on bacteriophage (a virus) P1 genome.
See also: cloning vector

Patent
In genetics, conferring the right or title to genes, gene variations, or identifiable portions of sequenced genetic material to an individual or organization.
See also: gene

Pedigree
A family tree diagram that shows how a particular genetic trait or disease has been inherited.
See also: inherit

Penetrance
The probability of a gene or genetic trait being expressed. "Complete" penetrance means the gene or genes for a trait are expressed in all the population who have the genes. "Incomplete" penetrance means the genetic trait is expressed in only part of the population. The percent penetrance also may change with the age range of the population.

Peptide
Two or more amino acids joined by a bond called a "peptide bond."
See also: polypeptide

Phage
A virus for which the natural host is a bacterial cell.

Pharmacogenomics
The study of the interaction of an individual's genetic makeup and response to a drug.

Phenocopy
A trait not caused by inheritance of a gene but appears to be identical to a genetic trait.

Phenotype
> The physical characteristics of an organism or the presence of a disease that may or may not be genetic.
> *See also:* genotype

Physical map
> A map of the locations of identifiable landmarks on DNA (e.g., restriction-enzyme cutting sites, genes), regardless of inheritance. Distance is measured in base pairs. For the human genome, the lowest-resolution physical map is the banding patterns on the 24 different chromosomes; the highest-resolution map is the complete nucleotide sequence of the chromosomes.

Plasmid
> Autonomously replicating extra-chromosomal circular DNA molecules, distinct from the normal bacterial genome and nonessential for cell survival under nonselective conditions. Some plasmids are capable of integrating into the host genome. A number of artificially constructed plasmids are used as cloning vectors.

Pleiotropy
> One gene that causes many different physical traits such as multiple disease symptoms.

Pluripotency
> The potential of a cell to develop into more than one type of mature cell, depending on environment.

Polygenic disorder
> Genetic disorder resulting from the combined action of alleles of more than one gene (e.g., heart disease, diabetes, and some cancers). Although such disorders are inherited, they depend on the simultaneous presence of several alleles; thus the hereditary patterns usually are more complex than those of single-gene disorders.
> *See also:* single-gene disorder

Polymerase chain reaction (PCR)
> A method for amplifying a DNA base sequence using a heat-stable polymerase and two 20-base primers, one complementary to the (+) strand at one end of the sequence to be amplified and one complementary to the (-) strand at the other end. Because the newly synthesized DNA strands can

subsequently serve as additional templates for the same primer sequences, successive rounds of primer annealing, strand elongation, and dissociation produce rapid and highly specific amplification of the desired sequence. PCR also can be used to detect the existence of the defined sequence in a DNA sample.

Polymerase, DNA or RNA
Enzyme that catalyzes the synthesis of nucleic acids on preexisting nucleic acid templates, assembling RNA from ribonucleotides or DNA from deoxyribonucleotides.

Polymorphism
Difference in DNA sequence among individuals that may underlie differences in health. Genetic variations occurring in more than 1% of a population would be considered useful polymorphisms for genetic linkage analysis.
See also: mutation

Polypeptide
A protein or part of a protein made of a chain of amino acids joined by a peptide bond.

Population genetics
The study of variation in genes among a group of individuals.

Positional cloning
A technique used to identify genes, usually those that are associated with diseases, based on their location on a chromosome.

Premature chromosome condensation (PCC)
A method of studying chromosomes in the interphase stage of the cell cycle.

Primer
Short preexisting polynucleotide chain to which new deoxyribonucleotides can be added by DNA polymerase.

Privacy
In genetics, the right of people to restrict access to their genetic information.

Probe
> Single-stranded DNA or RNA molecules of specific base sequence, labeled either radioactively or immunologically, that are used to detect the complementary base sequence by hybridization.

Prokaryote
> Cell or organism lacking a membrane-bound, structurally discrete nucleus and other subcellular compartments. Bacteria are examples of prokaryotes.
> *See also:* chromosome, eukaryote

Promoter
> A DNA site to which RNA polymerase will bind and initiate transcription.

Pronucleus
> The nucleus of a sperm or egg prior to fertilization.
> *See also:* nucleus, transgenic

Protein
> A large molecule composed of one or more chains of amino acids in a specific order; the order is determined by the base sequence of nucleotides in the gene that codes for the protein. Proteins are required for the structure, function, and regulation of the body's cells, tissues, and organs; and each protein has unique functions. Examples are hormones, enzymes, and antibodies.

Proteome
> Proteins expressed by a cell or organ at a particular time and under specific conditions.

Proteomics
> The study of the full set of proteins encoded by a genome.

Pseudogene
> A sequence of DNA similar to a gene but nonfunctional; probably the remnant of a once-functional gene that accumulated mutations.

Purine
> A nitrogen-containing, double-ring, basic compound that occurs in nucleic acids. The purines in DNA and RNA are adenine and guanine.
> *See also:* base pair

Pyrimidine

A nitrogen-containing, single-ring, basic compound that occurs in nucleic acids. The pyrimidines in DNA are cytosine and thymine; in RNA, cytosine and uracil.

See also: base pair

R

Radiation hybrid

A hybrid cell containing small fragments of irradiated human chromosomes. Maps of irradiation sites on chromosomes for the human, rat, mouse, and other genomes provide important markers, allowing the construction of very precise STS maps indispensable to studying multifactorial diseases.

See also: sequence tagged site

Rare-cutter enzyme

See: restriction-enzyme cutting site

Recessive gene

A gene which will be expressed only if there are 2 identical copies or, for a male, if one copy is present on the X chromosome.

Reciprocal translocation

When a pair of chromosomes exchange exactly the same length and area of DNA. Results in a shuffling of genes.

Recombinant clone

Clone containing recombinant DNA molecules.
See also: recombinant DNA technology

Recombinant DNA molecules

A combination of DNA molecules of different origin that are joined using recombinant DNA technologies.

Recombinant DNA technology

Procedure used to join together DNA segments in a cell-free system (an environment outside a cell or organism). Under appropriate conditions, a recombinant DNA molecule can enter a cell and replicate there, either autonomously or after it has become integrated into a cellular chromosome.

Recombination
The process by which progeny derive a combination of genes different from that of either parent. In higher organisms, this can occur by crossing over.
See also: crossing over, mutation

Regulatory region or sequence
A DNA base sequence that controls gene expression.

Repetitive DNA
Sequences of varying lengths that occur in multiple copies in the genome; it represents much of the human genome.

Reporter gene
See: marker

Resolution
Degree of molecular detail on a physical map of DNA, ranging from low to high.

Restriction enzyme, endonuclease
A protein that recognizes specific, short nucleotide sequences and cuts DNA at those sites. Bacteria contain over 400 such enzymes that recognize and cut more than 100 different DNA sequences.
See also: restriction enzyme cutting site

Restriction fragment length polymorphism (RFLP)
Variation between individuals in DNA fragment sizes cut by specific restriction enzymes; polymorphic sequences that result in RFLPs are used as markers on both physical maps and genetic linkage maps. RFLPs usually are caused by mutation at a cutting site.
See also: marker, polymorphism

Restriction-enzyme cutting site
A specific nucleotide sequence of DNA at which a particular restriction enzyme cuts the DNA. Some sites occur frequently in DNA (e.g., every several hundred base pairs); others much less frequently (rare-cutter; e.g., every 10,000 base pairs).

Retroviral infection
　　The presence of retroviral vectors, such as some viruses, which use their recombinant DNA to insert their genetic material into the chromosomes of the host's cells. The virus is then propogated by the host cell.

Reverse transcriptase
　　An enzyme used by retroviruses to form a complementary DNA sequence (cDNA) from their RNA. The resulting DNA is then inserted into the chromosome of the host cell.

Ribonucleotide
　　See: nucleotide

Ribose
　　The five-carbon sugar that serves as a component of RNA.
　　See also: ribonucleic acid, deoxyribose

Ribosomal RNA (rRNA)
　　A class of RNA found in the ribosomes of cells.

Ribosomes
　　Small cellular components composed of specialized ribosomal RNA and protein; site of protein synthesis.
　　See also: RNA

Risk communication
　　In genetics, a process in which a genetic counselor or other medical professional interprets genetic test results and advises patients of the consequences for them and their offspring.

RNA (Ribonucleic acid)
　　A chemical found in the nucleus and cytoplasm of cells; it plays an important role in protein synthesis and other chemical activities of the cell. The structure of RNA is similar to that of DNA. There are several classes of RNA molecules, including messenger RNA, transfer RNA, ribosomal RNA, and other small RNAs, each serving a different purpose.

S

Sanger sequencing
A widely used method of determining the order of bases in DNA. *See also:* sequencing, shotgun sequencing

Satellite
A chromosomal segment that branches off from the rest of the chromosome but is still connected by a thin filament or stalk.

Scaffold
In genomic mapping, a series of contigs that are in the right order but not necessarily connected in one continuous stretch of sequence.

Segregation
The normal biological process whereby the two pieces of a chromosome pair are separated during meiosis and randomly distributed to the germ cells.

Sequence
See: base sequence

Sequence assembly
A process whereby the order of multiple sequenced DNA fragments is determined.

Sequence tagged site (STS)
Short (200 to 500 base pairs) DNA sequence that has a single occurrence in the human genome and whose location and base sequence are known. Detectable by polymerase chain reaction, STSs are useful for localizing and orienting the mapping and sequence data reported from many different laboratories and serve as landmarks on the developing physical map of the human genome. Expressed sequence tags (ESTs) are STSs derived from cDNAs.

Sequencing
Determination of the order of nucleotides (base sequences) in a DNA or RNA molecule or the order of amino acids in a protein.

Sequencing technology
>The instrumentation and procedures used to determine the order of nucleotides in DNA.

Sex chromosome
>The X or Y chromosome in human beings that determines the sex of an individual. Females have two X chromosomes in diploid cells; males have an X and a Y chromosome. The sex chromosomes comprise the 23rd chromosome pair in a karyotype.
>*See also:* autosome

Sex-linked
>Traits or diseases associated with the X or Y chromosome; generally seen in males.
>*See also:* gene, mutation, sex chromosome

Shotgun method
>Sequencing method that involves randomly sequenced cloned pieces of the genome, with no foreknowledge of where the piece originally came from. This can be contrasted with "directed" strategies, in which pieces of DNA from known chromosomal locations are sequenced. Because there are advantages to both strategies, researchers use both random (or shotgun) and directed strategies in combination to sequence the human genome.
>*See also:* library, genomic library

Single nucleotide polymorphism (SNP)
>DNA sequence variations that occur when a single nucleotide (A, T, C, or G) in the genome sequence is altered.
>*See also:* mutation, polymorphism, single-gene disorder

Single-gene disorder
>Hereditary disorder caused by a mutant allele of a single gene (e.g., Duchenne muscular dystrophy, retinoblastoma, sickle cell disease).
>*See also:* polygenic disorders

Somatic cell
>Any cell in the body except gametes and their precursors.
>*See also:* gamete

Somatic cell gene therapy
Incorporating new genetic material into cells for therapeutic purposes. The new genetic material cannot be passed to offspring.
See also: gene therapy

Somatic cell genetic mutation
A change in the genetic structure that is neither inherited nor passed to offspring. Also called acquired mutations.
See also: germ line genetic mutation

Southern blotting
Transfer by absorption of DNA fragments separated in electrophoretic gels to membrane filters for detection of specific base sequences by radio-labeled complementary probes.

Spectral karyotype (SKY)
A graphic of all an organism's chromosomes, each labeled with a different color. Useful for identifying chromosomal abnormalities.
See also: chromosome

Splice site
Location in the DNA sequence where RNA removes the noncoding areas to form a continuous gene transcript for translation into a protein.

Sporadic cancer
Cancer that occurs randomly and is not inherited from parents. Caused by DNA changes in one cell that grows and divides, spreading throughout the body.
See also: hereditary cancer

Stem cell
Undifferentiated, primitive cells in the bone marrow that have the ability both to multiply and to differentiate into specific blood cells.

Structural genomics
The effort to determine the 3D structures of large numbers of proteins using both experimental techniques and computer simulation

Substitution
In genetics, a type of mutation due to replacement of one nucleotide in a

DNA sequence by another nucleotide or replacement of one amino acid in a protein by another amino acid.
See also: mutation

Suppressor gene
A gene that can suppress the action of another gene.

Syndrome
The group or recognizable pattern of symptoms or abnormalities that indicate a particular trait or disease.

Syngeneic
Genetically identical members of the same species.

Synteny
Genes occurring in the same order on chromosomes of different species.
See also: linkage, conserved sequence

T

Tandem repeat sequences
Multiple copies of the same base sequence on a chromosome; used as markers in physical mapping.
See also: physical map

Targeted mutagenesis
Deliberate change in the genetic structure directed at a specific site on the chromosome. Used in research to determine the targeted region's function.
See also: mutation, polymorphism

Technology transfer
The process of transferring scientific findings from research laboratories to the commercial sector.

Telomerase
The enzyme that directs the replication of telomeres.

Telomere
The end of a chromosome. This specialized structure is involved in the replication and stability of linear DNA molecules.
See also: DNA replication

Teratogenic
Substances such as chemicals or radiation that cause abnormal development of a embryo.
See also: mutatgen

Thymine (T)
A nitrogenous base, one member of the base pair AT (adenine-thymine).
See also: base pair, nucleotide

Toxicogenomics
The study of how genomes respond to environmental stressors or toxicants. Combines genome-wide mRNA expression profiling with protein expression patterns using bioinformatics to understand the role of gene-environment interactions in disease and dysfunction.

Transcription
The synthesis of an RNA copy from a sequence of DNA (a gene); the first step in gene expression.
See also: translation

Transcription factor
A protein that binds to regulatory regions and helps control gene expression.

Transcriptome
The full complement of activated genes, mRNAs, or transcripts in a particular tissue at a particular time

Transfection
The introduction of foreign DNA into a host cell.
See also: cloning vector, gene therapy

Transfer RNA (tRNA)
A class of RNA having structures with triplet nucleotide sequences that are complementary to the triplet nucleotide coding sequences of mRNA. The role of tRNAs in protein synthesis is to bond with amino acids and transfer them to the ribosomes, where proteins are assembled according to the genetic code carried by mRNA.

Transformation
A process by which the genetic material carried by an individual cell is altered by incorporation of exogenous DNA into its genome.

Transgenic
An experimentally produced organism in which DNA has been artificially introduced and incorporated into the organism's germ line.
See also: cell, DNA, gene, nucleus, germ line

Translation
The process in which the genetic code carried by mRNA directs the synthesis of proteins from amino acids.
See also: transcription

Translocation
A mutation in which a large segment of one chromosome breaks off and attaches to another chromosome.
See also: mutation

Transposable element
A class of DNA sequences that can move from one chromosomal site to another.

Trisomy
Possessing three copies of a particular chromosome instead of the normal two copies.
See also: cell, gene, gene expression, chromosome

U

Uracil
A nitrogenous base normally found in RNA but not DNA; uracil is capable of forming a base pair with adenine.
See also: base pair, nucleotide

V

Vector
See: cloning vector

Virus
A noncellular biological entity that can reproduce only within a host cell. Viruses consist of nucleic acid covered by protein; some animal viruses are also surrounded by membrane. Inside the infected cell, the virus uses the synthetic capability of the host to produce progeny virus.
See also: cloning vector

W

Western blot
A technique used to identify and locate proteins based on their ability to bind to specific antibodies.
See also: DNA, Northern blot, protein, RNA, Southern blotting

Wild type
The form of an organism that occurs most frequently in nature.

Working Draft DNA Sequence
See: Draft DNA Sequence

X

X chromosome
One of the two sex chromosomes, X and Y.
See also: Y chromosome, sex chromosome

Xenograft
Tissue or organs from an individual of one species transplanted into or grafted onto an organism of another species, genus, or family. A common example is the use of pig heart valves in humans.

Y

Y chromosome
One of the two sex chromosomes, X and Y.
See also: X chromosome, sex chromosome

Yeast artificial chromosome (YAC)
Constructed from yeast DNA, it is a vector used to clone large DNA fragments.
See also: cloning vector, cosmid

Z

Zinc-finger protein
A secondary feature of some proteins containing a zinc atom; a DNA-binding protein.

Updated 12-Oct-02

The online presentation of this publication is a special feature of the Human Genome Project Information Web site.

This document may be cited in the following style:
Human Genome Program, U.S. Department of Energy, *Genomics and Its Impact on Medicine and Society: A 2001 Primer*, 2001.

For printed copies, please contact Laura Yust at Oak Ridge National Laboratory. Send questions or comments to the author, Denise K. Casey. Site designed by Marissa Mills.

Reprinted with permission....from Human Genome Program, U.S. Department of Energy, *Genomics and Its Impact on Medicine and Society: A 2001 Primer*, 2001, for my book *How to Interpret Your DNA Test Results for Family History & Ancestry*, Dec. 31, 2002.

◆ ◆ ◆

Those interested in molecular genealogy/DNA for ancestry and similar subjects will find this glossary/dictionary helpful for definitions. It also is of interest to teachers of genealogy, biology, genetics, biological anthropology, family history and DNA, bioscience communications, oral history, and several other related areas of humanities and sciences.

The Dictionary of Genetics Terms from 2001 was updated in 2003. So check out the original source Web site for updates from the *Human Genome Project Information* Web site. Check for current updates as *The Dictionary of Genetics Terms* of the Human Genome Project Information Web site is a special feature of the Human Genome Project Information Web site.

The site has many wonderful articles on genetics and pharmacogenomics as well as related topics in genomics and all types of informational articles and publications about the Human Genome Project. One the Human GenomeProject Web site, check for updates at: **http://www.ornl.gov/TechResources/Human_Genome/publicat/primer2001/glossary.html**

Also, I highly recommend the publication at: http://www.ornl.gov/TechResources/Human_Genome/publicat/primer2001/5.html titled: **Genome Sequences Launch a New Level of Scientific Challenges.** Check out all these wonderful government publications available for your education about genomics and the Human Genome Project. Isn't reading about DNA exhilarating?

If you're a student thinking about a possible career in genetic counseling, think about what new discoveries the future holds.

Charting human variation is one thing. Designing a menu of foods and supplements or a time capsule of DNA reports for future generations is quite a project for the bioscience communicators, let alone the scientists and physicians. And then there are those pharmacogenomics projects.

Tailoring what you eat, your medicines, supplements, and even your skin care products to your DNA presents quite a story and challenge for the media to research....especially when several genes can act in numerous ways to achieve the same result. Whichever angle you look at the results of your DNA tests, for ancestry or for nutrition or medical information, it's an exciting look into the future. The fields of science journalism, video, and communications focused on genetics are growing also as are the science careers and the DNA-driven genealogy research. Go for it. Know your prehistoric past and genomics future.

***

Index

A

Allele 7, 74, 122, 135, 278, 291, 422, 450, 451, 537, 540, 594, 605, 609, 633
Ancestry.com 16, 413, 445, 573
AncestryByDNA 2, 5, 23, 24, 373, 450, 451, 463, 464, 525, 526, 559
Appendix A
  Name Frequency in the US 565
Appendix B
  Ethnic Genealogy Web Sites 569
Appendix C
  Genealogy, General 573
Appendix D
  Bibliography 1, Genealogy 575
Appendix E
  Bibliography 2, DNA Testing and Genetics 579
Appendix F
  Permissions 585
Appendix G
  Books by Anne Hart 587
Archaeogenetics 41, 91, 113, 114, 197, 198, 205, 206, 207, 211, 246, 251, 269, 270, 330, 331, 339, 340, 345, 379, 401, 430, 454, 466, 542, 543, 579, 583
Aristocracy 69
Armenia 31, 32, 48, 53, 56, 64, 82, 375, 381, 383, 389, 392, 432, 455
Ashkenazi 27, 28, 29, 30, 31, 32, 33, 34, 36, 37, 49, 51, 52, 53, 55, 57, 58, 59, 61, 63, 64, 66, 67, 70, 71, 72, 73, 74, 75, 76, 77, 78, 79, 80, 81, 84, 87, 380, 386, 387, 407, 408, 418, 419, 433, 434, 440, 553, 582, 590
Austin, John 471

B

Babylonia 66
Barbujani, Guido 430
Bialystock 50, 54, 66, 67, 82, 380
BioGeographical Ancestry 463, 464
Bioinformatics 196, 204, 212, 217, 222, 329, 338, 346, 351, 356, 429, 481, 543, 594, 597, 601, 611, 616, 636
Bukhara 410, 419

C

Cambridge Reference Sequence (CRS) 32, 49, 60, 371, 379, 386, 453, 461
Cancer 88, 89, 103, 105, 107, 111, 116, 119, 120, 151, 152, 153, 157, 164, 168, 169, 170, 172, 213, 220, 248, 249, 263, 265, 266, 267, 272, 275, 276, 306, 307, 308, 312, 347, 354, 414, 415, 479, 480, 487, 490, 491, 544, 545, 546, 547, 548, 549, 550, 551, 586, 597, 598, 613, 615, 624, 634
Cells 53, 113, 116, 117, 118, 158, 159, 174, 217, 218, 224, 239, 269, 272, 273, 313, 351, 357, 423, 424, 425, 436, 449, 474, 475, 476, 541, 542, 595, 596, 597, 599, 600, 604, 606, 607, 609, 610, 611, 612, 613, 616, 620, 621, 623, 624, 628, 631, 632, 633, 634
City Directories 9, 438, 439, 444, 450, 467, 512
Cohanim 27, 28, 66, 73, 433, 435, 443, 446
Cohen 29, 52, 54, 56, 60, 65, 78, 79, 81, 84, 399, 400, 432, 434
Cohen Modal Haplotype 29, 54, 56, 60, 65, 78, 79, 81, 84, 399, 400, 432, 434

Communities 25, 26, 28, 29, 31, 32, 33, 34, 35, 37, 38, 49, 50, 51, 52, 53, 54, 55, 56, 57, 58, 59, 60, 61, 62, 63, 64, 65, 66, 67, 69, 70, 71, 72, 73, 74, 75, 76, 77, 78, 79, 80, 82, 83, 84, 86, 87, 244, 387, 391, 398, 408, 409, 410, 418, 424, 427, 431, 433, 441, 445, 495, 590

Computational biology 481, 601

Court Jew 67, 68

Craycraft Surname Project 469

D

Demography 159, 412, 428, 441, 543

Diamond, Stanley M. 412, 413

Dictionary of Genetic Terms 554, 585, 593

Diploid 474, 596, 603, 604, 613, 614, 620, 633

Division of Cancer Control and Population Sciences (DCCPS) 544

DNA 1, 2, 3, 4, 5, 6, 7, 8, 9, 13, 16, 18, 21, 23, 24, 25, 26, 29, 30, 32, 34, 35, 36, 37, 38, 39, 40, 41, 42, 47, 50, 54, 56, 57, 58, 59, 60, 62, 63, 67, 68, 69, 74, 75, 76, 77, 80, 81, 83, 84, 86, 88, 89, 90, 91, 92, 93, 94, 95, 96, 97, 98, 99, 100, 101, 102, 103, 104, 105, 106, 107, 108, 109, 111, 113, 114, 115, 116, 119, 120, 123, 128, 135, 137, 142, 143, 144, 146, 154, 155, 156, 157, 158, 159, 160, 161, 163, 164, 173, 174, 181, 182, 183, 184, 185, 186, 187, 188, 189, 192, 193, 194, 195, 196, 197, 198, 199, 200, 206, 207, 209, 210, 211, 212, 214, 215, 216, 217, 218, 220, 221, 222, 223, 225, 226, 228, 230, 232, 233, 234, 235, 236, 238, 239, 242, 243, 244, 245, 246, 247, 248, 249, 250, 251, 252, 253, 254, 255, 256, 257, 258, 259, 260, 261, 262, 263, 264, 265, 266, 267, 269, 270, 271, 272, 274, 275, 276, 279, 283, 291, 293, 298, 299, 300, 302, 309, 310, 311, 312, 313, 314, 315, 316, 317, 318, 319, 320, 321, 322, 323, 325, 326, 327, 328, 329, 330, 331, 332, 333, 334, 340, 341, 343, 344, 345, 346, 349, 350, 351, 352, 353, 354, 355, 356, 357, 359, 360, 362, 363, 365, 367, 368, 369, 370, 371, 373, 375, 379, 385, 386, 387, 388, 398, 400, 401, 403, 404, 405, 406, 407, 408, 409, 410, 411, 412, 413, 414, 419, 420, 421, 422, 423, 424, 426, 428, 429, 430, 431, 432, 433, 435, 436, 438, 439, 440, 441, 442, 443, 444, 445, 446, 447, 449, 450, 451, 452, 453, 454, 455, 456, 458, 461, 462, 463, 464, 465, 466, 468, 469, 470, 471, 472, 474, 475, 476, 479, 480, 481, 483, 484, 485, 486, 487, 488, 489, 490, 493, 499, 501, 512, 513, 517, 518, 524, 525, 526, 528, 529, 530, 531, 532, 533, 535, 536, 538, 539, 540, 541, 542, 543, 552, 553, 554, 555, 556, 557, 558, 559, 560, 561, 579, 580, 581, 582, 583, 584, 585, 590, 591, 594, 595, 596, 597, 598, 599, 600, 601, 602, 603, 604, 605, 606, 607, 608, 609, 610, 611, 612, 613, 614, 615, 616, 617, 618, 619, 620, 621, 622, 623, 624, 625, 626, 627, 628, 629, 630, 631, 632, 633, 634, 635, 636, 637, 638, 639, 640

DNA for Family Historians 443, 580

DNA testing 580, 585

E

Epidemiology and Genetics Research Program (EGRP) 544

Ethnic Genealogy Web Sites 561, 569

F

Family history 1, 2, 13, 15, 17, 18, 20, 22, 25, 26, 69, 75, 88, 91, 92, 95, 97, 107, 109, 111, 143, 155, 164, 176, 181, 186, 188, 204, 218, 245, 248, 251, 252, 255, 256, 266, 269, 299, 310, 315, 319, 322, 338, 352, 387, 388, 389, 398, 401, 403, 406, 407, 411, 415, 419, 421, 428, 429, 430, 431, 435, 436, 438, 439, 440, 441, 442, 443, 447, 448, 475, 490, 498, 499, 513, 520, 523, 528, 529, 531, 532, 533, 544, 546, 552, 553, 556, 573, 577, 590, 639

FamilyTreeDNA 23, 371, 469, 472, 525, 526, 530, 531, 536, 556, 559

Forensic DNA testing 580
Founders 25, 31, 32, 33, 34, 39, 57, 58, 59, 61, 74, 386, 387, 388, 424, 427, 464, 583
Frudakis, Tony, PhD 463

G

Genealogy 1, 2, 3, 5, 8, 9, 10, 12, 13, 16, 18, 19, 20, 21, 23, 24, 25, 26, 29, 36, 40, 41, 42, 51, 67, 68, 69, 75, 82, 85, 87, 88, 89, 91, 95, 98, 99, 109, 113, 143, 151, 156, 159, 161, 185, 187, 188, 192, 197, 198, 216, 218, 234, 244, 246, 248, 250, 251, 252, 255, 258, 259, 269, 299, 306, 311, 318, 320, 322, 326, 330, 331, 345, 352, 367, 386, 388, 389, 392, 393, 394, 395, 396, 397, 400, 401, 404, 405, 406, 407, 413, 416, 417, 418, 419, 420, 421, 422, 423, 424, 426, 428, 429, 430, 439, 440, 441, 442, 443, 444, 445, 446, 447, 449, 454, 458, 461, 465, 466, 467, 468, 469, 470, 472, 494, 503, 512, 513, 516, 517, 520, 521, 523, 524, 525, 527, 528, 529, 530, 532, 534, 535, 543, 552, 553, 556, 559, 561, 562, 563, 565, 569, 570, 573, 575, 576, 577, 582, 587, 590, 591, 639, 640

Genes 24, 26, 29, 33, 35, 36, 37, 49, 51, 52, 53, 54, 55, 56, 57, 61, 62, 66, 70, 74, 75, 77, 83, 85, 86, 87, 88, 89, 90, 91, 92, 93, 94, 96, 97, 98, 99, 101, 104, 105, 106, 107, 108, 114, 115, 116, 117, 119, 120, 124, 130, 131, 135, 136, 142, 143, 145, 154, 155, 156, 157, 158, 159, 160, 161, 162, 163, 164, 165, 166, 167, 168, 169, 170, 171, 172, 173, 174, 175, 177, 178, 179, 181, 182, 183, 185, 186, 188, 189, 190, 191, 193, 194, 197, 198, 201, 203, 204, 205, 206, 208, 210, 212, 214, 215, 216, 217, 218, 219, 220, 221, 222, 223, 224, 225, 226, 227, 228, 229, 230, 231, 232, 233, 234, 235, 236, 237, 238, 239, 241, 242, 243, 244, 245, 247, 248, 249, 250, 251, 252, 253, 254, 255, 256, 257, 258, 261, 264, 265, 266, 267, 268, 270, 271, 272, 273, 275, 280, 286, 291, 292, 298, 299, 301, 309, 310, 311, 312, 314, 315, 316, 318, 320, 321, 322, 323, 324, 325, 326, 327, 328, 331, 335, 336, 337, 338, 339, 342, 344, 346, 348, 349, 350, 351, 352, 353, 354, 355, 356, 357, 358, 359, 360, 361, 362, 363, 365, 366, 367, 368, 369, 370, 381, 387, 399, 400, 401, 403, 406, 408, 409, 410, 411, 412, 417, 420, 423, 424, 425, 426, 428, 430, 431, 436, 438, 445, 446, 447, 450, 451, 456, 466, 474, 475, 476, 478, 479, 480, 481, 483, 484, 485, 487, 488, 489, 490, 491, 493, 525, 536, 540, 541, 543, 546, 556, 557, 558, 579, 580, 590, 591, 597, 598, 599, 601, 603, 604, 605, 607, 608, 609, 610, 611, 612, 613, 614, 616, 617, 618, 620, 622, 623, 624, 625, 626, 627, 629, 630, 635, 636, 640

GeneTree DNA Testing Center 468, 469, 559

Genome 6, 42, 43, 51, 64, 75, 81, 84, 90, 93, 97, 116, 119, 120, 135, 156, 158, 159, 162, 163, 169, 170, 172, 173, 175, 181, 182, 191, 192, 194, 203, 208, 209, 210, 212, 214, 215, 216, 218, 219, 220, 221, 222, 223, 224, 225, 226, 230, 232, 233, 234, 235, 236, 238, 239, 241, 242, 248, 250, 253, 256, 272, 275, 291, 311, 315, 316, 324, 326, 328, 336, 342, 343, 344, 346, 348, 349, 350, 352, 353, 354, 355, 356, 357, 358, 359, 360, 364, 366, 367, 368, 370, 417, 428, 429, 430, 431, 436, 438, 439, 445, 451, 452, 462, 463, 464, 465, 474, 475, 476, 478, 479, 480, 481, 482, 483, 484, 485, 486, 487, 488, 492, 526, 539, 541, 543, 555, 558, 585, 593, 596, 598, 599, 600, 603, 604, 605, 606, 607, 611, 612, 615, 617, 621, 624, 626, 628, 630, 632, 633, 636, 637, 639, 640

Genomics 93, 101, 117, 142, 143, 154, 155, 156, 158, 159, 160, 161, 163, 164, 166, 168, 169, 175, 176, 180, 181, 182, 183, 185, 186, 187, 188, 189, 191, 192, 194, 195, 196, 197, 198, 199, 200, 201, 203, 204, 205, 206, 207, 208, 209, 210, 211, 212, 214, 215, 216, 217, 218, 219, 220, 221, 223, 225, 226, 228, 229, 230, 231,

232, 233, 234, 235, 236, 237, 238, 239, 241, 242, 243, 253, 260, 273, 298, 299, 309, 310, 311, 313, 314, 315, 316, 318, 319, 320, 321, 322, 323, 325, 328, 329, 330, 331, 332, 333, 334, 336, 337, 338, 339, 340, 342, 343, 344, 345, 346, 348, 349, 350, 352, 353, 354, 357, 358, 359, 360, 361, 363, 364, 365, 366, 367, 368, 369, 370, 450, 454, 463, 475, 478, 481, 482, 483, 488, 489, 492, 525, 526, 559, 560, 580, 590, 601, 608, 612, 613, 634, 639, 640

Germany 29, 31, 32, 33, 42, 47, 48, 50, 52, 53, 54, 56, 61, 62, 66, 67, 69, 70, 71, 72, 73, 77, 82, 84, 85, 86, 372, 373, 380, 381, 382, 383, 384, 409, 410, 433, 434, 446, 447, 451, 452, 454, 530

Goldstein 37, 49, 50, 59, 398, 399

Greece 31, 47, 51, 55, 56, 80, 376, 384, 389, 394, 410, 433, 460, 562, 569

H

Hammer 27, 28, 65, 79, 80, 386, 398, 399, 461, 556

Haplogroup 3, 23, 32, 33, 38, 39, 41, 42, 48, 49, 53, 55, 56, 57, 58, 59, 64, 65, 74, 79, 84, 94, 245, 246, 247, 255, 371, 372, 375, 376, 377, 378, 379, 380, 381, 382, 383, 385, 386, 401, 408, 418, 419, 427, 446, 447, 453, 454, 455, 456, 458, 459, 460, 461, 538, 539, 540, 557, 558

Haplotype 6, 7, 8, 9, 29, 30, 38, 54, 56, 57, 59, 60, 65, 73, 78, 79, 81, 84, 205, 245, 339, 377, 386, 399, 400, 421, 422, 423, 426, 432, 433, 434, 529, 531, 532, 538, 540, 613

High resolution 5, 42, 73, 371, 385, 452, 455, 461, 462

Human Genome Management Information System 475, 585

Human Genome News 585

Human Genome Program 475, 478, 482, 483, 585, 615, 639

Human Genome Project Goals 480

Human Genome Sequence 478, 483

Hungary 33, 53, 69, 389, 409, 432, 433, 454, 561, 569

I

Informatics and Information Technology Group, (CGN) 550

Institute of Molecular and Cell Biology, Tartu, Estonia 445

Iran 30, 58, 64, 433, 542

Iraq 30, 33, 34, 54, 56, 62, 63, 75, 377, 383, 384, 385, 389, 391, 410, 460, 542

Italy 31, 33, 47, 48, 50, 51, 52, 54, 56, 64, 70, 72, 80, 375, 380, 381, 383, 384, 398, 401, 402, 407, 410, 427, 430, 432, 447, 451, 452, 455, 582

K

Khait, Barbara 412, 413

L

Levite 29, 52, 62, 66, 73, 77, 81, 82, 84, 86, 433

Lineages 3, 6, 7, 25, 30, 37, 52, 53, 58, 61, 64, 77, 86, 91, 244, 248, 251, 388, 408, 410, 418, 421, 424, 426, 438, 439, 440, 444, 445, 449, 450, 459, 474, 538, 540, 541, 554, 557, 558, 582, 584

Low resolution 4, 5, 10, 23, 24, 42, 371, 452, 466

M

Macaulay, Vincent 386, 410, 455, 582

MadSci.ORG 474

Marks, Jonathan 432

Maternal 1, 2, 3, 6, 10, 25, 38, 39, 40, 52, 57, 59, 66, 76, 89, 226, 245, 246, 248, 250, 360, 385, 387, 388, 408, 418, 420, 421, 424, 426, 438, 444, 450, 462, 531, 554, 557, 558, 602

Matrilineal 5, 23, 25, 35, 38, 47, 52, 58, 59, 60, 61, 77, 86, 94, 99, 244, 245, 248, 255, 259, 372, 374, 375, 379, 398, 418,

429, 439, 440, 444, 452, 453, 459, 462, 538
Medical history 214, 235, 349, 369, 429
Medieval 27, 28, 29, 31, 33, 47, 50, 52, 53, 54, 55, 56, 58, 59, 61, 66, 67, 69, 70, 71, 72, 73, 75, 76, 80, 82, 83, 85, 86, 87, 376, 378, 384, 387, 404, 409, 432, 433, 434, 464, 588
Melton, Terry 444
Mitochondria 30, 436, 449, 474, 538, 541, 542, 603, 621
Mitotyping Technologies, LLC. 444
Mizrahi 67, 392, 427, 590
Molecular genetics 24, 51, 75, 95, 106, 107, 109, 110, 111, 114, 115, 156, 200, 255, 265, 267, 268, 270, 271, 311, 334, 419, 428, 430, 441, 484, 543, 621
Morocco 32, 34, 62, 63, 75, 376, 377
mtDNA 2, 3, 4, 5, 6, 7, 9, 23, 24, 25, 28, 29, 30, 31, 32, 33, 34, 35, 38, 39, 40, 41, 42, 47, 48, 49, 50, 51, 52, 53, 54, 55, 56, 57, 58, 59, 60, 61, 62, 63, 64, 65, 68, 70, 71, 74, 76, 77, 80, 83, 84, 86, 87, 94, 99, 163, 174, 223, 226, 241, 244, 245, 246, 248, 255, 259, 356, 360, 371, 372, 374, 375, 376, 377, 378, 379, 380, 381, 382, 383, 384, 385, 386, 388, 398, 400, 401, 408, 410, 418, 419, 420, 421, 424, 426, 427, 436, 438, 439, 442, 444, 445, 446, 447, 449, 450, 451, 452, 453, 454, 455, 456, 458, 459, 460, 461, 462, 464, 466, 468, 469, 474, 538, 539, 540, 541, 542, 554, 557, 558, 582, 583, 584
Mutations 4, 8, 28, 35, 40, 42, 48, 50, 58, 74, 75, 80, 95, 105, 106, 111, 117, 133, 154, 157, 191, 198, 216, 222, 241, 246, 248, 255, 265, 266, 273, 289, 309, 312, 325, 331, 356, 371, 372, 379, 380, 382, 385, 386, 401, 412, 418, 419, 421, 423, 427, 429, 446, 453, 454, 455, 461, 462, 486, 489, 490, 532, 538, 539, 541, 603, 604, 628, 634

N

National Cancer Institute (NCI) 544

National Center for Health Statistics in Hyattsville 467
Nature Reviews/Genetics 440
Neolithic 36, 55, 247, 382, 458, 459, 542, 583
Network 92, 131, 132, 149, 152, 159, 182, 200, 217, 224, 252, 287, 288, 304, 307, 315, 333, 350, 358, 510, 544, 545, 546, 547, 548, 550

O

Oak Ridge National Laboratory 475, 482, 585, 639
Olson, Steve 406, 538, 542
Onomastics 155, 159, 310, 428
Oral histories 429, 440, 441, 500, 501, 502, 504, 515, 520
Oxford Ancestors 3, 5, 6, 23, 374, 406, 447, 453, 455, 456, 457, 461, 462, 463, 559

P

Paternal 1, 2, 6, 8, 10, 12, 14, 25, 28, 42, 52, 59, 73, 78, 89, 226, 250, 360, 407, 408, 409, 413, 420, 421, 422, 432, 435, 438, 453, 531, 557, 558, 602
Patrilineal 25, 35, 37, 47, 56, 73, 99, 244, 259, 429, 433, 439, 440, 462
Perego, Ugo 424
Phenomics 154, 159, 204, 209, 221, 222, 232, 309, 338, 343, 354, 356, 366, 417, 428, 449, 463, 464, 465, 523, 529, 543
Pilot studies 546
Poland 29, 32, 33, 36, 42, 50, 52, 53, 54, 56, 61, 62, 66, 67, 68, 69, 70, 71, 72, 73, 77, 81, 82, 84, 85, 86, 372, 373, 380, 381, 384, 409, 414, 415, 416, 434, 451
Polin 53
Population genetics 91, 98, 99, 119, 154, 188, 197, 208, 220, 248, 251, 252, 258, 259, 274, 309, 322, 330, 342, 353, 400, 408, 410, 417, 431, 466, 627
Prosopography 155, 159, 310, 411, 412, 428

R

Reed, Peter 424, 447
Renfrew, Colin 454, 579
Richards, Martin 582
Roman Jews 386, 410
Roper Surname Project 471
Rosenstein, Dr. Neil 67, 68
Russia 4, 47, 52, 53, 67, 69, 70, 71, 82

S

Savin, Alan 580
Schildkraut, Joellen 545, 547
Science 597, 640
Science 33, 41, 43, 52, 59, 60, 64, 75, 89, 90, 91, 92, 93, 94, 95, 97, 98, 102, 103, 104, 107, 111, 113, 114, 115, 116, 119, 129, 131, 142, 143, 149, 155, 156, 158, 159, 160, 161, 162, 164, 167, 182, 183, 188, 191, 192, 193, 194, 196, 198, 199, 200, 203, 204, 205, 206, 208, 209, 211, 212, 214, 217, 223, 224, 225, 228, 229, 230, 232, 233, 234, 236, 237, 238, 239, 246, 250, 251, 252, 253, 254, 255, 256, 257, 258, 261, 263, 264, 267, 269, 270, 271, 272, 274, 287, 298, 299, 304, 310, 311, 316, 322, 324, 325, 326, 327, 328, 330, 332, 333, 334, 337, 338, 339, 340, 342, 343, 345, 346, 348, 350, 351, 357, 358, 359, 362, 363, 364, 366, 367, 368, 370, 398, 400, 401, 406, 407, 428, 429, 440, 442, 443, 444, 468, 474, 475, 481, 482, 483, 484, 502, 543, 548, 549, 552, 554, 556, 557, 559, 579, 581, 583, 585
Scriver, Dr. Charles 573
Sephardim 25, 26, 27, 68, 70, 79, 244, 392, 393, 396, 407, 408, 409
Sequence variation 4, 480, 486, 633
Sequences 1, 2, 3, 4, 5, 6, 24, 26, 39, 40, 57, 61, 64, 71, 80, 81, 246, 248, 371, 372, 374, 375, 376, 377, 380, 381, 382, 383, 384, 385, 386, 401, 418, 419, 429, 430, 438, 440, 441, 442, 447, 451, 452, 453, 454, 455, 456, 457, 462, 464, 466, 476, 480, 481, 485, 486, 487, 489, 493, 538, 540, 552, 553, 554, 555, 557, 558, 584, 598, 600, 604, 608, 610, 614, 623, 626, 630, 632, 634, 635, 636, 637, 640
Sequencing technology 480, 633
Sethi, Ricky J. 474
Single Nucleotide Polymorphisms (SNPs) 100, 135, 220, 260, 291, 354
Slavs 27, 32, 50, 56, 61, 71, 72, 74, 82, 83, 85, 86, 397, 434, 562, 570
SNP Consortium 479
Sorb 28, 29, 32, 55, 61, 66, 72, 74, 83, 84, 85, 86, 409, 434
Sykes, Bryan 581

T

Torroni 407, 447, 582, 583, 584
Transitions 374, 378, 386, 429, 453, 454, 462

U

U.S. Department of Energy 475, 478, 482, 585, 639
Underhill, Peter 444
US Census 16, 439, 466, 527, 528, 565

V

Villems, Richard 445, 446

W

Winn, Deborah, PhD 544

Y

Y-Chromosome 2, 5, 6, 7, 8, 9, 10, 13, 16, 23, 25, 28, 29, 30, 32, 33, 35, 36, 38, 39, 40, 41, 51, 55, 59, 77, 83, 86, 87, 99, 174, 226, 241, 244, 245, 246, 259, 360, 386, 388, 398, 400, 407, 408, 419, 420, 421, 422, 423, 424, 427, 431, 432, 433, 434, 435, 438, 450, 452, 458, 459, 461,

468, 469, 539, 540, 541, 557, 558, 582, 583

Yiddish 29, 50, 53, 54, 55, 57, 71, 73, 77, 81, 82, 83, 84, 85, 380, 388, 398, 434, 584

0-595-31684-0